F THE ELEMENTS

IB	IIB	IIIA	IVA	VA	VIA	VIIA	VIIIA
							2 4.00260 He — 54.4 $1s^2$
		5 10.81 B — 8.3 $2s^22p$	6 12.011 C — 11.3 $2s^22p^2$	7 14.0067 N — 14.5 $2s^22p^3$	8 15.9994 O — 13.6 $2s^22p^4$	9 18.998403 F — 17.4 $2s^22p^5$	10 20.179 Ne — 21.6 $2s^22p^6$
		13 26.98154 Al 1.42 6.0 $3s^23p$	14 28.0855 Si — 8.15 $3s^23p^2$	15 30.97376 P — 10.5 $3s^23p^3$	16 32.06 S — 10.4 $3s^23p^4$	17 35.453 Cl — 13.0 $3s^23p^5$	18 39.948 Ar — 15.8 $3s^23p^6$
9 63.546 Cu 8 3 $3d^{10}4s$	30 65.38 Zn 1.37 9.39 $3d^{10}4s^2$	31 69.72 Ga 1.35 6.0 $4s^24p$	32 72.59 Ge 1.39 7.9 $4s^34p^2$	33 74.9216 As — 9.81 $4s^24p^3$	34 78.95 Se — 9.75 $4s^24p^4$	35 79.904 Br — 11.8 $4s^24p^5$	36 83.80 Kr — 14.0 $4s^24p^6$
7 107.868 Ag 44 58 $4d^{10}5s$	48 112.41 Cd 1.52 8.99 $4d^{10}5s^2$	49 114.82 In 1.67 5.79 $5s^24p$	50 118.69 Sn 1.58 7.34 $5s^25p^2$	51 121.75 Sb 1.61 8.64 $5s^25p^3$	52 124.60 Te — 9.01 $5s^25p^4$	53 126.9045 I — 10.5 $5s^25p^5$	54 131.30 Xe — 12.1 $5s^25p^6$
9 196.9665 Au 44 23 $5d^{10}6s$	80 200.59 Hg 1.55 10.4 $5d^{10}6s^2$	81 204.37 Tl 1.71 6.11 $6s^26p$	82 207.2 Pb 1.74 7.42 $6s^26p^2$	83 208.9804 Bi 1.82 7.29 $6s^26p^3$	84 (209) Po — 8.42 $6s^26p^4$	85 (210) At — — $6s^26p^5$	86 (222) Rn — 10.7 $6s^26p^6$

5 158.9254 Tb 77 36 f^85d6s^2	66 162.50 Dy 1.75 7.94 $4f^{10}6s^2$	67 164.9304 Ho 1.76 6.01 $4f^{11}6s^2$	68 167.26 Er 1.73 6.10 $4f^{12}6s^2$	69 168.9342 Tm 1.74 6.18 $4f^{13}6s^2$	70 173.04 Yb 1.93 6.25 $4f^{14}6s^2$	71 174.967 Lu 1.73 5.43 $4f^{14}5d6s^2$
7 (247) Bk 23	98 (251) Cf 6.30	99 (252) Es 6.42	100 (257) Fm 6.50	101 (258) Md 6.58	102 (259) No 6.65	103 (260) Lr

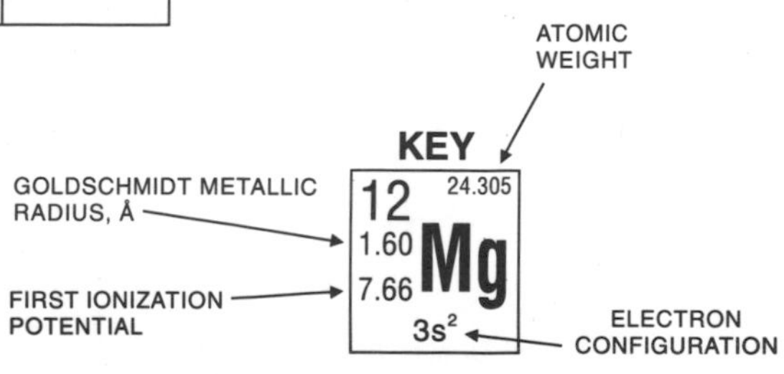

The Physical Chemistry *of* Solids

This book is for Iris and Margaret.

The Physical Chemistry *of* Solids

Richard J. Borg

Lawrence Livermore National Laboratory
Livermore, California

G. J. Dienes

Brookhaven National Laboratory
Upton, New York

ACADEMIC PRESS, INC.
Harcourt Brace Jovanovich, Publishers
Boston San Diego New York
London Sydney Tokyo Toronto

This book is printed on acid-free paper ∞

ACADEMIC PRESS, INC.
1250 Sixth Avenue, San Diego, CA 92101

United Kingdom Edition published by
ACADEMIC PRESS LIMITED
24–28 Oval Road, London NW1 7DX

Library of Congress Cataloging-in-Publication Data

Borg, R. J. (Richard John), date.
The physical chemistry of solids / Richard J. Borg, G. J. Dienes
p. cm.
Includes bibliographical references and index.
ISBN 0-12-118420-X
1. Solid state chemistry. I. Dienes, G. J. (George Julian), 1918- . II. Title.
QD478.B67 1991
541.2'8—dc20 90-23735
CIP

Printed in the United States of America
91 92 93 94 9 8 7 6 5 4 3 2 1

Contents

Preface

This book was written to encourage the teaching of solid state chemistry in university chemistry and materials science departments. Although a large number, perhaps even a preponderance, of physical chemists find ultimate employment in the realm of solid state science, few have been exposed to the subject during their student years. Perhaps the most widely read book in all of chemistry[†] is subtitled "The Structure of Molecules and Crystals" and devotes considerable space to solids. However, the teaching of solid state chemistry, particularly at the undergraduate level, has never quite caught on. As new technologies continue to emerge, materials science promises to be an expanding field, and physical chemists will have a major stake in the action. An undergraduate or an early post-graduate introduction to the subject will certainly assist in their participation.

The only prerequisite for understanding this text is a reasonable grounding in chemical and elementary statistical thermodynamics plus upper division standing in science or engineering. We have tried to write a self-contained treatise, but suspect the student will find it helpful to keep a modern, advanced physical chemistry text within convenient reach. We have omitted discussion of electron transport properties, magnetism, and optical properties, which seem more properly assigned to solid state physics. Also, mechanical properties and dislocation theory are excluded. We have tried, whenever possible, to use a chemical approach with explanations couched in terms of the chemical bond and thermodynamics.

Empirical or semiempirical methods are presented in detail when they have a useful predictive value. Unfortunately, much, if not most, of modern solid state theory is

[†] *The Nature of the Chemical Bond*, Linus Pauling.

beyond the level of this text. Our task would have been much easier if watered-down versions of such advanced treatments could be presented in a way that would prove enlightening to the student and provide a better tool for making numerical estimates; such is seldom the situation. Rightly or wrongly, we have consciously avoided abbreviated presentations of complex theories, thinking that this leads more frequently to confusion rather than understanding.

Finally, we suggest that a student planning a career in solid state chemistry will profit from additional courses in solid state physics and physical metallurgy. Such course work will furnish a review of some of the material that is in this book, as well as an introduction to much important material that is not.

The authors wish to acknowledge the friendly criticism of several friends and colleagues. In particular we want to thank Professor Lawrence M. Slifkin, who has advised us on every chapter. Selected chapters were critiqued by Drs. Richard van Konynenburg, Arthur C. Damask, Malcolm H. MacGregor, Laurence Passel, Donald Hildenbrand, and James C. Selser. In addition, we give our heartfelt thanks to the careful efforts of Mrs. Diane L. Governor and Ms. Sue E. Garber, whose contributions far exceed mere typing of the manuscript. Most of the illustrations were prepared by Steven M. Wright, whose artistic and graphics capabilities enhance the text. The cover was designed by Mrs. Ellen L. Baldwin. We also wish to thank our technical editor Ms. B. Jane Ellis of Academic Press for her help and, above all, her patience.

Chapter 1

Crystallography and Structure of the Elements

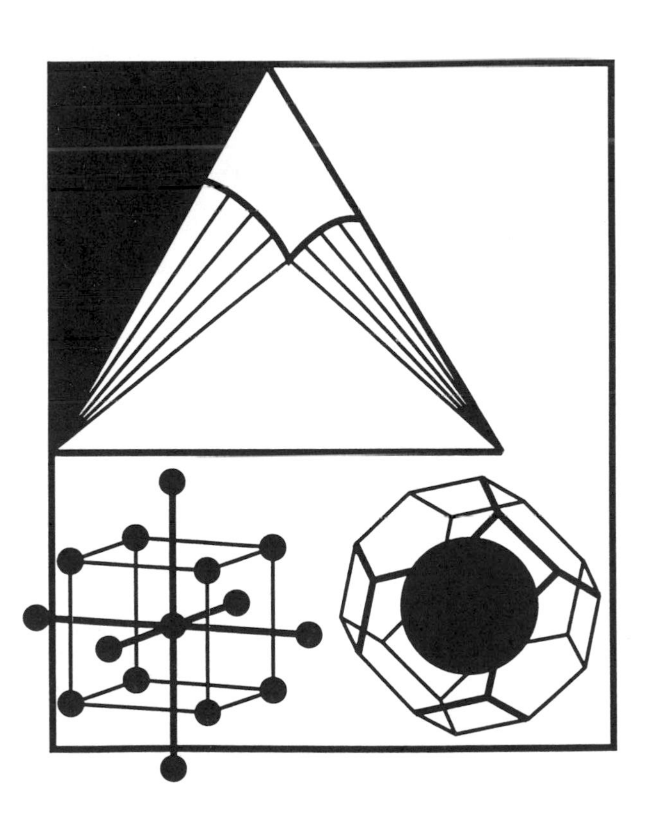

Chapter 1

Contents

Don't think there are no crocodiles just because the water is calm.
—Sam Wyche, Coach, Cincinnati Bengals

Crystallography and Structure of the Elements

The most distinguishing feature of crystals is the very long range periodic order of the atoms, leading to a high degree of macroscopic symmetry, which, in turn, reflects the symmetry of the crystal's physical properties. Of course, not all inorganic solids are crystalline, glass being a prominent exception, and most liquids exhibit at least some short-range molecular order; in fact, that class of substances known as liquid crystals provides a spectacular example of rather long-range order in the fluid state. Nevertheless, it was the obvious macroscopic symmetry that first attracted man's attention to crystals and continues to do so for esthetic, as well as scientific, reasons. Many of the common substances that we encounter daily are crystalline without appearing to be. The soil under our feet is composed mostly of tiny crystals of various minerals. Solid metals, regardless of the shapes into which they have been fabricated, are made up of crystallites. Because the realm of inorganic solids consists mainly of crystalline, rather than amorphous, substances we can adduce that maximum thermodynamic stability is achieved by the creation of the spatial periodicity of the atoms.

The external symmetry of a crystal directly or indirectly reflects the nature of the bonding between the constituent atoms, ions, or molecules. Indeed, there are fundamental qualitative differences between ionic, covalent, and metallic chemical bonds that dictate the symmetry of the resultant crystal structure. In addition to crystals that are held together by strong bonds, such as the aforementioned, there are molecular crystals in which the fundamental building blocks are molecules bonded through weaker dipolar forces. Crystals of solidified rare gases are bonded via very weak induced dipolar forces, whereas crystals such as graphite and MoS_2

have both covalent and dipolar bonds. In all cases the symmetry, on the atomic scale, is manifest in the overall symmetry of the crystal structure, though it need not be the same.

Before delving into the physical and chemical behavior of solids, we need the vocabulary and concepts necessary to describe the geometry of the atomic arrangements in crystals. It will not be necessary to review in detail a great deal of classical crystallography in order to understand solid state chemistry, but it will be helpful to have in mind some of the more important ideas and terminology.

1. Symmetry

The geometric relationships between the planar faces determined the macroscopic symmetry by which crystals were first grouped and cataloged. Niels Stensen, professor of anatomy at Copenhagen, noted in 1669 that the angles between corresponding planes were always the same in a collection of quartz crystals. His observation has been duly generalized to include all crystals under the fitting title of "Steno's Law," sometimes called the first law of crystallography. The degree and type of symmetry possessed by any object is determined by certain symmetry operations, e.g., rotation, translation, and reflection. A symmetry operation is defined by the particular manipulation that brings the object into congruence with itself.

By way of illustrating this procedure, let us first consider a sphere, the most symmetrical of all shapes. Rotation of a perfect sphere about any of the infinite number of axes passing through its center leaves the object unchanged with reference to any arbitrary set of external coordinates. There is, in fact, no symmetry operation about the center of the symmetry of a sphere that does not leave it congruent with its previous position. Let us now examine the two-dimensional constructs shown in Fig. 1.1. One should imagine a single axis emerging from the center of each figure, which is perpendicular to the plane of the page. The symmetry operation will be

Figure 1.1.

Planar objects with varying degrees of symmetry categorized by the number of positions of congruence attained during a single complete rotation.

simple rotation about this imaginary axis, as indicated by an arrow, and the number of positions of congruence is listed beneath each figure. The central three figures possess, respectively from left to right, fourfold, threefold, and twofold axes of rotation. The last figure on the right, although geometrically quite regular, has no elements of rotational symmetry but the trivial identity element.

The symmetry operations of reflection, translation, and inversion are illustrated in Fig. 1.2.

A mirror-of-reflection plane has an identical set of points equidistant from it on both sides. Translation is the reproduction of the initial configuration at another place in space by simple linear motion. The rotation-inversion operation, Fig. 1.2, is equivalent to having a mirror plane, *m*, between the initial and final positions. Crystals can have two-, three-, four-, and sixfold rotation and rotation-inversion axes, but no five- or sevenfold. A pseudo-fivefold symmetry exists for substances called quasicrystals, but it would be inappropriate to take up this phenomenon so early in our treatise. The nonexistence of five- and sevenfold axes stems from the fact that objects with this symmetry are inherently unable to fill all space, as is illustrated in Fig. 1.3. The difficulty results from the requirement that the rotational and translational symmetries be commensurate, which they clearly are not. That is to say, one cannot perform both the translational repetition of the pentagon and the rotational operation in a way that maintains the postulated fivefold symmetry and yet leaves no voids.

We can recognize, simply by inspection, the maximum number of planes of symmetry and axes of rotation possessed by a simple cube, as shown in Fig. 1.4.

One should not conclude from this that all cubic structures are described by a simple cube possessing all possible symmetry elements. There are innumerable complex cubic structures, such as the spinels (e.g., Fe_3O_4, $MgFeO_4$, $HgK_2(CN)_4$ or

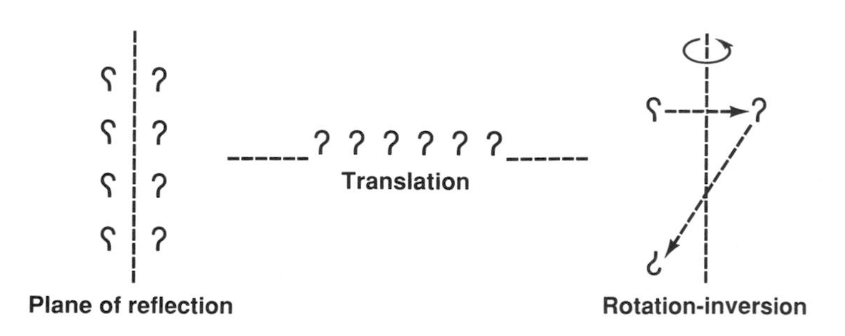

Figure 1.2.

Symmetry operations, reflection, translation, and rotation-inversion.

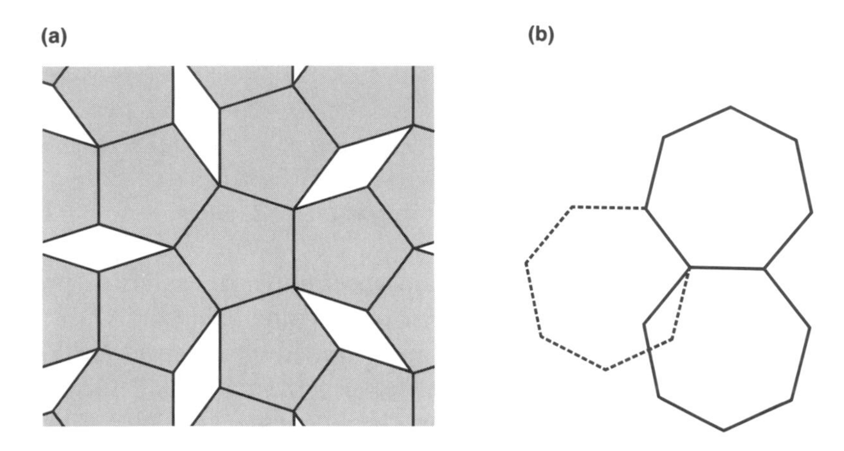

Figure 1.3

(a) A fivefold axis of symmetry does not exist in a lattice because it is impossible to fill all space with a connected array of pentagons. (b) Kepler's original demonstration that no lattice can possess a sevenfold axis of symmetry. (After C. Kittel, *Introduction to Solid State Physics*, Wiley, New York, p. 2, 1967.)

intermetallic compounds, including examples such as $CuCd_5$, that are complex cubic and contain only a minimum number of the symmetry elements shown in Fig. 1.4.

There are two commonly used sets of symbols that stand for the symmetry elements, viz. the Schoenflies and the Hermann-Mauguin; the latter are listed in Table 1.1, and we shall not elaborate the former because it is not commonly used. Although this system of symbols provides a convenient and efficient way of cataloging crystal types according to their symmetry elements, they are infrequently applied in the study of solid state chemistry where attention is focused on the atomic structure of the unit cell. This brings us to the subject of crystal lattices.

Table 1.1. The international symbols for the symmetry elements of crystals.

Identity element (no symmetry)	1	Fourfold rotation axis (rotor)	4
Mirror plane of symmetry	$m\ (=\bar{2})$	Fourfold rotatory inverter	$\bar{4}$
Twofold rotation axis (rotor)	2	Sixfold rotation axis (rotor)	6
Threefold rotation axis (rotor)	3	Sixfold rotatory inverter	$\bar{6}$
Threefold rotatory inverter	$\bar{3}$	Center of symmetry (inverter)	$\bar{1}$

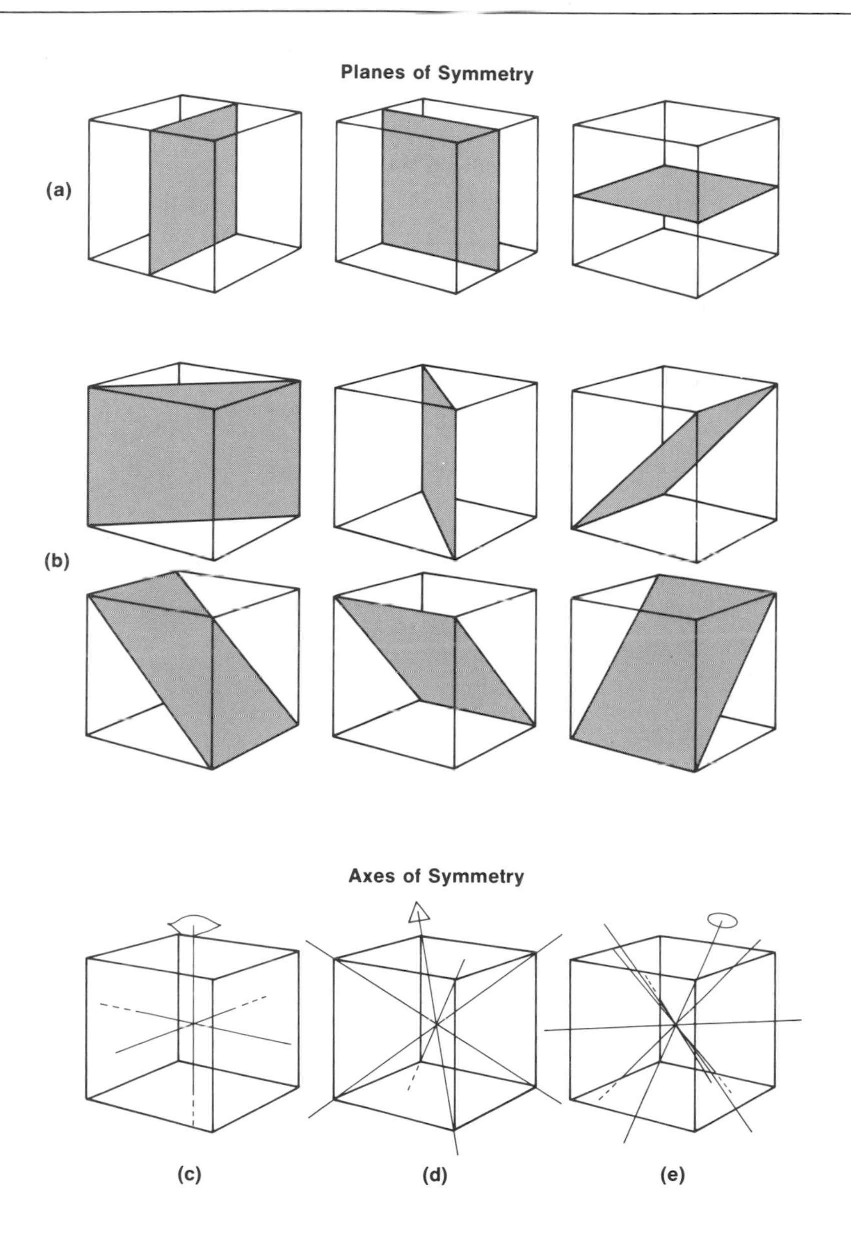

Figure 1.4.

Major planes and axes in the cubic system: (a) the three symmetry planes parallel to the cube faces, (b) the six diagonal mirror planes, (c) the three fourfold rotation axes ◇, (d) the four threefold axes △, and (e) the six twofold axes of rotation ○.

2. Lattices

Although all crystal structures can be described in terms of their respective symmetry elements, they are more easily visualized as the actual three-dimensional arrays of ions or atoms. Such periodic structures, in turn, are most conveniently depicted by their appropriate three-dimensional lattices. Strictly speaking, a lattice is a theoretical symmetrical array of structureless points, a purely mathematical concept not to be interchanged with the term "crystal structure." This rigorous differentiation is largely ignored in common practice, and we will conform to popular usage.

Lattices are constructed by combining the symmetry operations described in Fig. 1.1. The simplest lattice is a row of equally spaced dots separated by the translational vector **t** such as

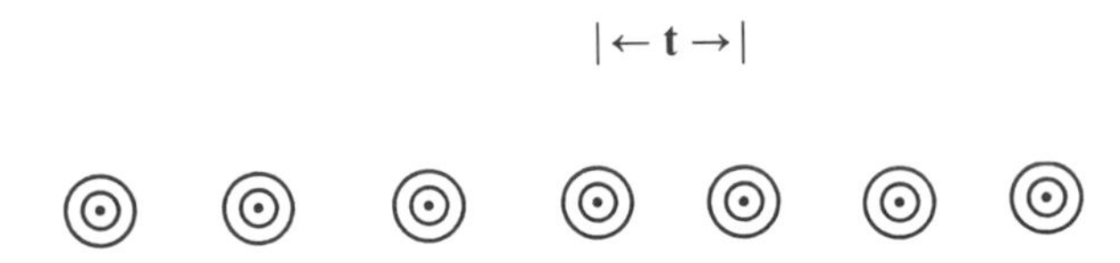

An example of a two-dimensional lattice is the planar array shown in Fig. 1.5. All points in this figure can be reached by the straightforward addition of the conjugate translations $\mathbf{t}_1$ and $\mathbf{t}_2$. Together these vectors can be combined to form a variety of unit cells, three examples of which are shown in Fig. 1.5. The symbol P stands for

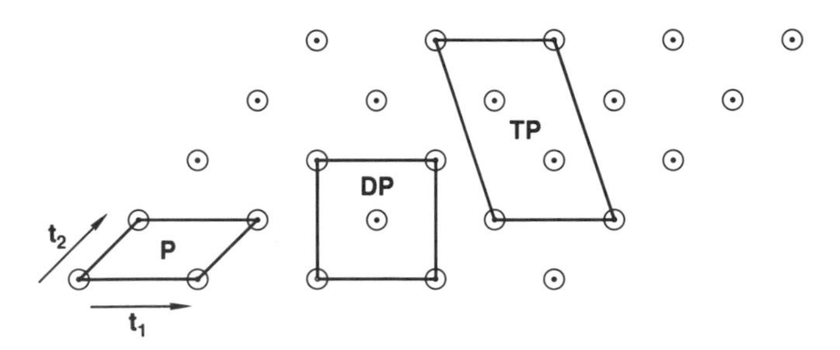

Figure 1.5.

All points are connected by a linear combination of the primitive translation vectors $\mathbf{t}_1$ and $\mathbf{t}_2$. P, DP, and TP designate the outlined primitive, doubly primitive, and triply primitive two-dimensional unit cells.

primitive, and the primitive two-dimensional unit cell by definition contains but a single atom resulting from the summation of the fractional atomic volumes at each of the four corners. Consistent with this convention, DP stands for doubly primitive, TP for triply primitive, containing two and three atoms, respectively. Unit cells that describe actual crystal structures are selected on the basis of convenience, the only requirement being that such a cell embody all the symmetry elements necessary to reproduce the entire crystal structure. It is usual to select the smallest unit cell satisfying this requirement.

The lattice concept is expanded to include three dimensions by the addition of a third conjugate vector. Now, any rational lattice point can be related to the origin of the coordinates, 0, 0, 0, by the vector sum

$$\mathbf{T} = u\mathbf{t}_1 + v\mathbf{t}_2 + w\mathbf{t}_3, \tag{1.1}$$

where u, v, and w are just the numerical coefficients of the unitary translations. Naively, one might assume that an infinitude of lattices of different symmetry is possible, whereas in actuality a limited number suffices to include all geometrical possibilities.

3. Bravais Lattices

There are but 14 ways of arranging structureless points in three-dimensional space such that all points bear the same symmetrical relation to the remainder of the lattice. That is to say, each point bears the same geometrical relation to all the others, and consequently, the points are indistinguishable. These 14 space lattices, called Bravais lattices, are illustrated in Fig. 1.6. It is, perhaps, obvious that only monoelemental structures can be described by a simple Bravais lattice. As we shall see later, the three cubic or isometric lattices combined with the tetragonal and hexagonal ones do, in fact, account for almost all the elemental solids. However, it is easy to see how the symmetry of the Bravais lattice is reduced by replacing the lattice points by ions, such as Na^+ and Cl^-, and reduced even further by the introduction of foreign ions, such as CO_3^{2-}, SO_4^{2-}, or NO_3^-, and their associated cations.

4. Crystal Systems

The 14 space lattices are collected into six crystal systems, and crystal structures are commonly assigned to these general categories. They are listed in Table 1.2, where a, b, c are the unitary translations, and α, β, and γ are the opposing angles as defined in Fig. 1.7. Alternative definitions of the six crystal systems are furnished by the

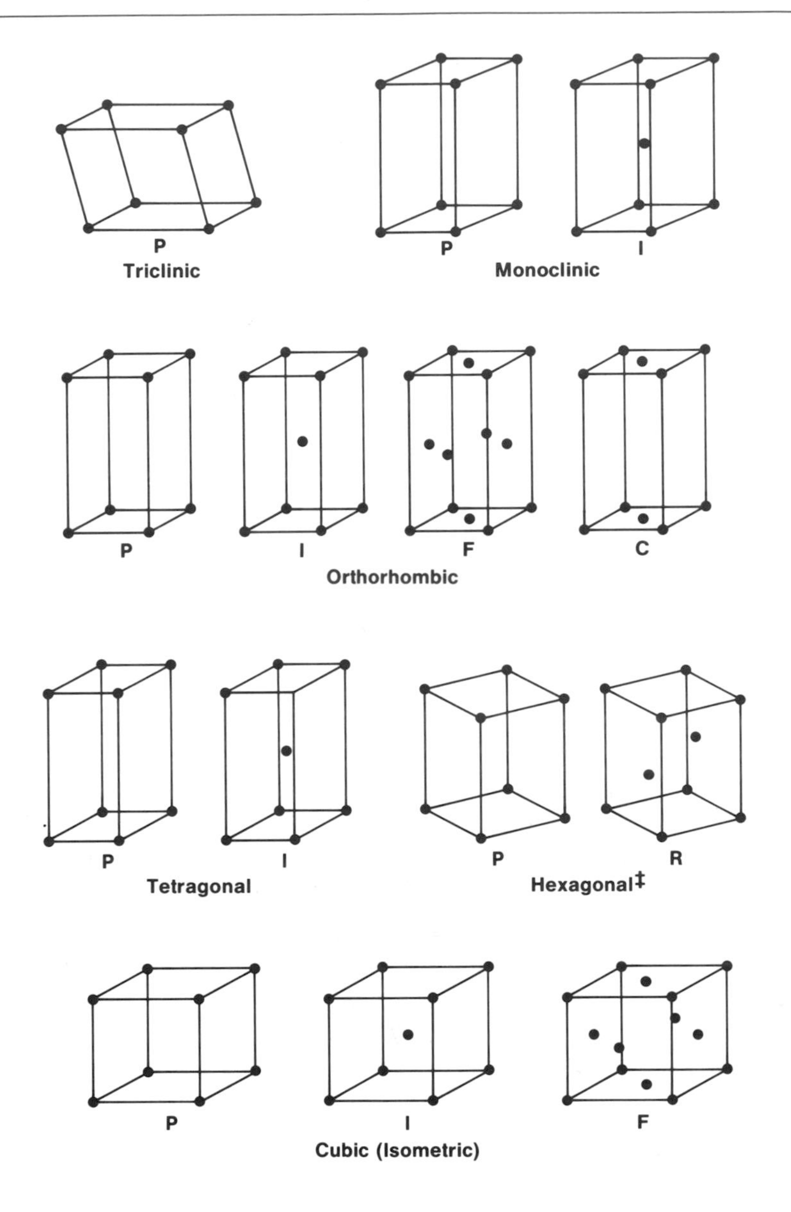

Figure 1.6.

The 14 Bravais lattices with the letters designating P (primitive), I (body-centered), F (face centered), C (end-centered), and R (rhombohedral). ‡Note that a rhomb, which is 1/6 the hexagonal unit cell (see Fig. 1.20), contains all the symmetry elements of this structure.

Table 1.2. The six crystal systems.

System	Lattices	Coordinates	Examples
Cubic (isometric)	3	$a = b = c$ $\alpha = \beta = \gamma = 90^\circ$	α-Fe NaCl
Hexagonal			
Hexagonal subsystem	1	$a = b \neq c$ $\alpha = \beta = 90^\circ$	Graphite MoS_2
Rhombohedral subsystem (trigonal)			$\gamma = 120^\circ$
Tetragonal	2	$a = b \neq c$ $\alpha = \beta = \gamma = 90^\circ$	White Sn PtS
Rhombohedral (trigonal) subsystem[a]	1	$a = b = c$ $\alpha = \beta = \gamma \neq 90^\circ < 120^\circ$	Calcite $K_2Cr_2O_4$
Orthorhombic	4	$a \neq b \neq c$ $\alpha = \beta = \gamma = 90^\circ$	Rhombic S α-Np
Monoclinic	2	$a \neq b \neq c$ $\alpha = \gamma = 90^\circ \neq \beta$	Monoclinic S $KICl_2$
Triclinic	1	$a \neq b \neq c$ $\alpha \neq \beta \neq \gamma$	$K_2Cr_2O_7$ $Cu(ClO_4)_2 \cdot 6H_2O$

[a] The rhombohedral subsystem is sometimes categorized as an independent system; however, American crystallographers treat it as a hexagonal subgroup, since the rhombohedral cell can also be described with hexagonal coordinates.

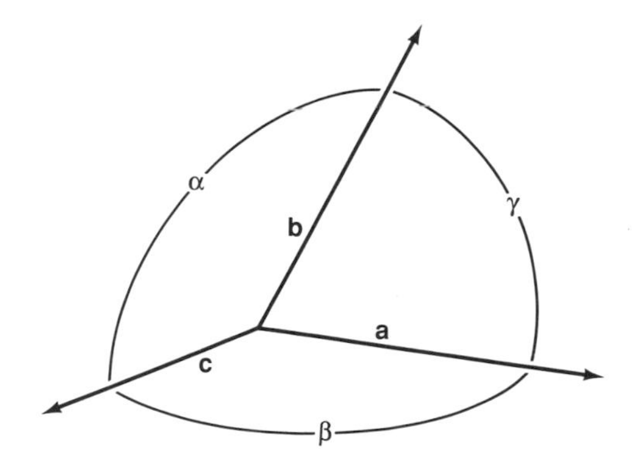

Figure 1.7.

The unit vectors a, b, c and the angles α, β, and γ define the coordinates of the six crystal systems.

minimum number of symmetry elements required by each of the different unit cells. They are:

Cubic: four axes of threefold symmetry arranged parallel to the body diagonals of a cube, which necessarily requires either three twofold axes or three fourfold axes that are mutually perpendicular and parallel to the cube edges

Tetragonal: single axis of fourfold symmetry

Trigonal and hexagonal: single axis of threefold or sixfold symmetry

Orthorhombic: three mutually perpendicular twofold axes or two perpendicular planes of symmetry

Monoclinic: one twofold axis and/or a plane of symmetry perpendicular to it

Triclinic: no axes of symmetry, maximum symmetry $\bar{1}$

5. Space Groups

There are 32 distinguishable ways of arranging points about a central point so that application of all symmetry operations to the group, save translation, leaves at least one point invariant. The 32 groups are called *point groups*. Configuring the point groups about the points of the Bravais lattices gives rise to a large number of *space groups*. Additional space groups are generated when translation is included, so that there are 230 space groups in all. All crystals structures, perforce, must belong to one or more of these groups.

6. Miller Indices

A system of notation called Miller indices provides a concise, unambiguous numerical label for all rational crystal planes. Rational crystallographic planes are defined by a set of rational lattice points, which in turn are specified by the primitive lattice vectors. To derive the basis for Miller indices in two dimensions, we begin with the intercept equation of the line *pq* shown in Fig. 1.8.

The equation of *pq* is

$$x/A + y/B = 1. \tag{1.1}$$

Equation (1.1) can be verified by noting that $y = B$ when $x = 0$ and, conversely, $x = A$ when $y = 0$. This procedure can be extended to three dimensions as shown in Fig. 1.9, and Eq. (1.2) is then the equation of a plane.

$$x/A + y/B + z/C = 1. \tag{1.2}$$

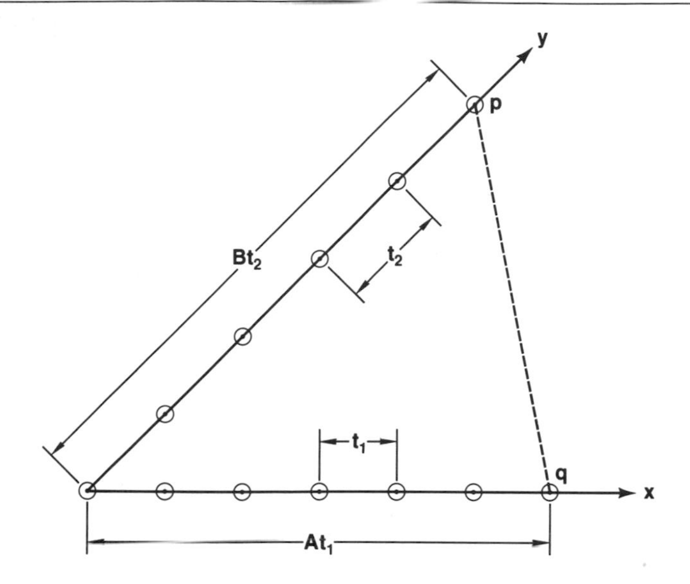

Figure 1.8.

The conjugate translational vectors $\mathbf{t}_1$ and $\mathbf{t}_2$ multiplied by their respective coefficients A and B specify the line pq.

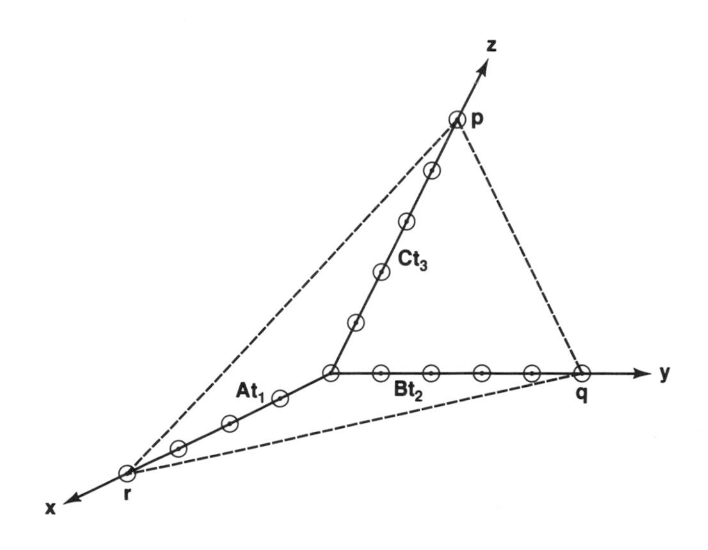

Figure 1.9.

The rational lattice plane pqr is delineated by the intercepts $A\mathbf{t}_1$, $B\mathbf{t}_2$, and $C\mathbf{t}_3$.

We can again test this equation in the limits by setting x, y, and z, pairwise, equal to zero. Although Eq. (1.2) provides an exact mathematical description of the plane pqr, it is inconvenient because the coefficients of x, y, and z are fractions and it is, furthermore, not a particularly concise notation. Consequently, we can multiply Eq. (1.2) by ABC, giving

$$BCx + ACy + ABz = ABC. \tag{1.3}$$

In the generalized Miller index notation, $BC = h$, $AC = k$, and $AB = l$, where the letters hkl are the conventional symbols for any unspecified plane. Consequently,

$$hx + ky + lz = ABC \tag{1.4}$$

denotes a plane of the same form as

$$hx + ky + lz = 1, \tag{1.5}$$

except that it is ABC times farther removed from the origin. This is again demonstrated by allowing x, y, and z to go to zero, pairwise, and noting that the remaining intercept is ABC units removed from the origin. The lowest values of h, k, and l that preserve their proper ratios for a given plane are called Miller indices. For example, the plane ABC shown in Fig. 1.9 corresponds to $A = 4$, $B = 5$, $C = 5$. Therefore

$$\frac{100x}{4} + \frac{100y}{5} + \frac{100z}{5} = 25x + 20y + 20z = 100. \tag{1.6}$$

Reducing the numerical coefficients to their lowest values by dividing by the largest common denominator yields $hkl = 544$ for the Miller indices.

A plane with intercepts on the x, y, and z axes at A, B, and C, respectively, has ABC times as many identical planes lying between itself and the origin unless two or more of the intercepts have the same numerical values. (We made use of this fact to justify the writing of Eq. (1.5), which was stated without explanation.) If two or all of the intercepts have the same value, we shall see that the total number of planes is reduced. This concept is difficult to perceive in three dimensions (and even more difficult to draw), so let us pursue this principle in two dimensions. We amplify Fig. 1.9 by including the square planar lattice underlying it in Fig. 1.10.

Looking at Fig. 1.10, the reader is asked to imagine a line with intercepts $x = y = 6$ and, hence, a Miller index (66), which is the altitude of a right triangle. By connecting equivalent points between this intercept line and the origin, we find that there are not 36 identical lines, but only 6 with the origin included. From this we can infer the formula AB/q = total number of equivalent planes between AB and the origin, where q is the highest common denominator (in this case 6).

In three dimensions ABC/qrs = total number of equivalent planes between ABC

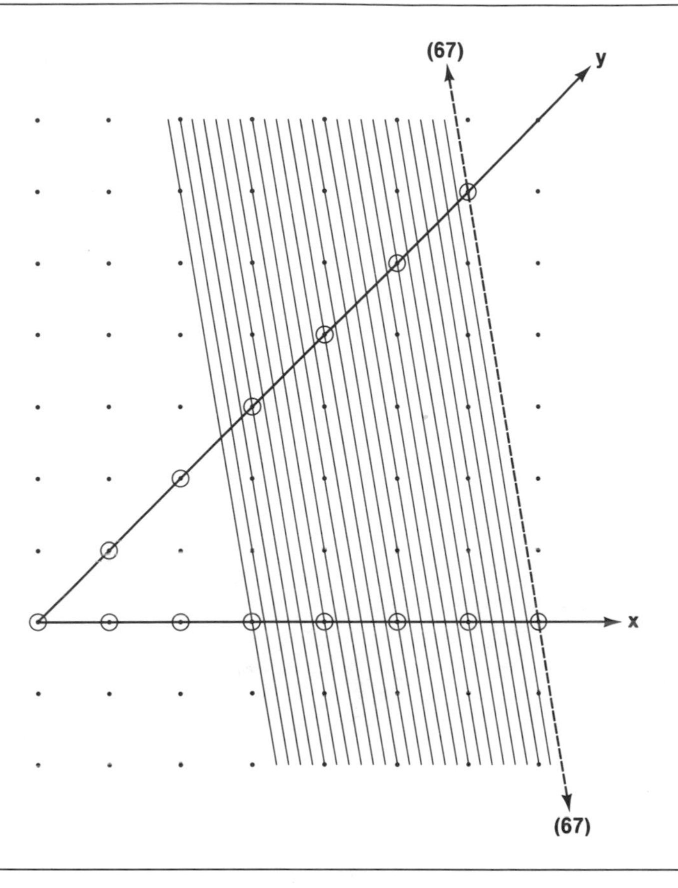

Figure 1.10.

Illustrating the nature of the 42 lines [$AB = 42$, see Eq. (1.4)] with Miller indices (67) lying between the intercept $x = 7$, $y = 6$, and the origin.

and the origin, and q, r, and s are the highest common denominators between the pairs AB, BC, and CA, respectively. For example, the number of equivalent planes between (422) and the origin is $(4 \times 2 \times 2)/(2 \times 2 \times 2) = 2$.

We have emphasized the concept of the extended lattice, which is reproduced by virtue of the translational symmetry operation, because so much of our thinking will be restricted to the unit cell that one is apt to lose sight of its importance in such things as diffraction phenomena, which provide the basic structural determinations.

Miller indices are always enclosed in some sort of brackets in order to signify that this is not simply a three-digit number. Furthermore, one always speaks of the indices as "five-four-four" and never as "five hundred and forty-four." When enclosed by parentheses, e.g., (100), the Miller indices refer to a single specific plane, whereas when enclosed by braces, e.g., {100}, they denote all planes of the form (100), viz. (100), (010), (001, ($\bar{1}$00), (0$\bar{1}$0), (00$\bar{1}$), where the bar over the numeral indicates a negative intercept. Directions are indicated by brackets, i.e., [*uvw*]. Thus, in the cubic system, [100] indicates the direction normal to (100), and [111] the direction normal to (111), and so on. The complete set of equivalent directions is written ⟨*uvw*⟩. Hence, ⟨100⟩ represents all six directions normal to the {100} planes in the cubic system. Square brackets also designate zone axes, i.e., a direction common to a group of faces. An index of zero is registered by a plane that is parallel to a coordinate axis and, hence, does not intercept it, i.e., ∞^{-1}. Examples of various planes in the cubic system labeled with their appropriate indices are shown in Fig. 1.11.

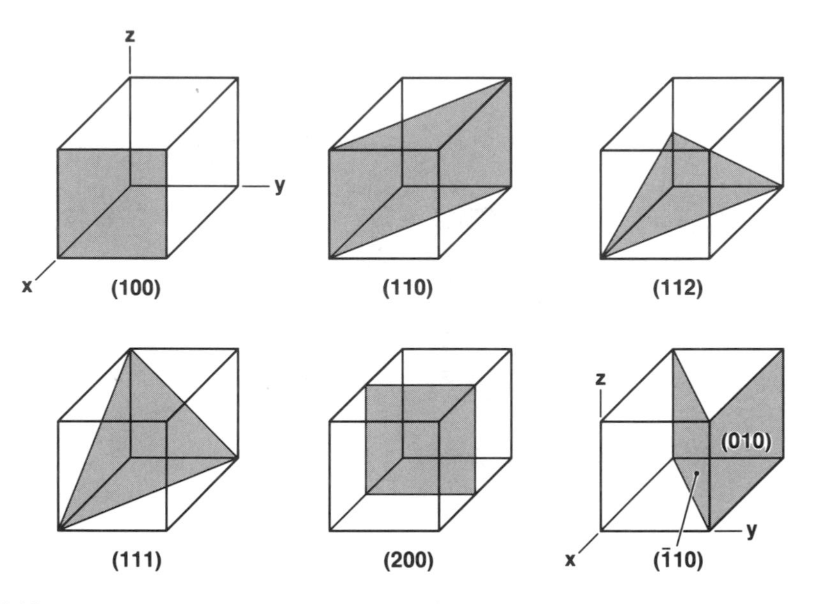

Figure 1.11.

Prominent planes of the cubic system.

7. The Unit Cell

The smallest three-dimensional unit of a particular crystal structure that contains all the symmetry elements necessary to reproduce the extended structure is generally chosen to be its unit cell. Again, when several unit cells can be defined, convention is to take the smallest. It is, so to speak, the fundamental building block from which the entire crystal can be constructed. One cannot infer the atomic structure that defines the unit cell from the morphology of a macroscopic crystal. The former is deduced from diffraction measurements most frequently employing x rays, or by neutron diffraction. We will present representative examples of a variety of unit cells in the following sections, commencing with an examination of the structure of the elements.

8. Crystal Structures of the Metallic Elements

At room temperature and 1 atm of pressure, the metallic state is the equilibrium form for more than 70 of the naturally occurring elements. Several others, including prominent examples such as phosphorus and iodine, become metallic under the application of modest pressures, whereas practically all elements assume the metallic state at sufficiently high pressures. In addition to the naturally occurring elements, the man-made transuranic elements, insofar as can be determined, are all metals under standard conditions. The metallic properties, which as a matter of course define the metallic state, principally derive from the delocalized electrons found in the conduction bands. It is they that provide the high degree of electrical and heat conductivity as well as the silvery sheen common to most metals. It is also the virtually nondirectional bonding via this sea of delocalized electrons that dictates the high degree of symmetry found in elemental metal crystals and their solid solutions. However, nature seems always to defy the assignment of rigidly exclusive classifications, and elemental metals frequently combine to form intermetallic compounds that are nonmetallic, and not infrequently react with such nonmetallic elements as oxygen and sulfur to create metallic oxides and sulfides.

All but a few elemental metals crystallize in one of three highly symmetric structures, viz., body-centered cubic, face-centered cubic, or hexagonal close-packed. Several of the elements possess stable allotropes of at least two of these three structures, depending upon the temperature and pressure regime. These three simple crystal structures are characteristic of ions and atoms that have little or no tendency toward covalent bonding and, hence, have no preferred bonding directions. Instead, they exert a centrosymmetric attractive potential, and stability is achieved through maximization of the coordination number.

A. Body-Centered Cubic (bcc)

The bcc lattice is the easiest of all configurations to visualize, with the exception of the primitive cube (only the element polonium is reported to have the latter structure

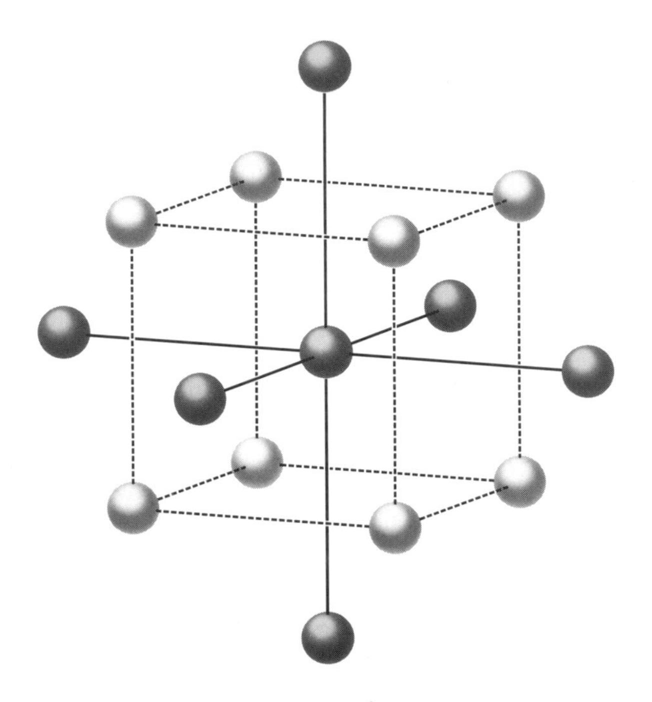

Figure 1.12.

The body-centered cubic structure showing the eight nearest neighbors and six next-nearest neighboring atoms.

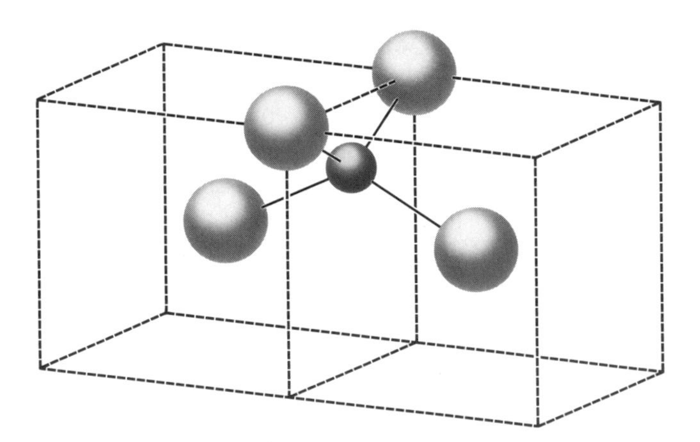

Figure 1.13.

The tetrahedrally coordinated interstitial atom in the bcc structure.

under normal conditions). Most of the refractory transition elements with very high melting temperatures have the bcc structure, and several having other structures at lower temperatures transform to the bcc as they approach their melting points. This is especially true of the rare earth elements.

The extended bcc lattice is constructed of two interpenetrating primitive cubic lattices. The distance of closest approach between adjacent atoms is $a_0\sqrt{3}/2$ and lies along the cube diagonals, where a_0 is the length of the unit cell (Fig. 1.11). Consequently, the geometric radius of an atom in this structure is $a_0\sqrt{3}/4$. Each atom has eight nearest neighbors at this distance and another six next-nearest neighbors at a distance a_0, the length of the cube edge. Because these intervals differ by only 13.4%, the coordination number is sometimes written as 8, 6 rather than 8 alone. This consideration is displayed in Fig. 1.12.

Small atoms, such as atomic hydrogen, nitrogen, and oxygen, as well as carbon, can take up interstitial positions in the bcc crystal structure. This has important chemical implications because interstitial atoms generally diffuse much more rapidly than substitutional ones, causing certain solid state reactions to proceed more rapidly than otherwise. For humans, no doubt the single most important consequence of interstitial atoms is steel, which is an interstitial solid solution of carbon in iron. The bcc lattice has two interstitial sites. The interstitial atom in a tetrahedral site, as shown in Fig. 1.13, has a coordination number of 4. Connecting these four atoms by imaginary lines outlines a regular tetrahedron. With respect to the coordinates of a single unit cell, the tetrahedral sites lying in the *xz* plane have coordinates 1/2, 0, 3/4 or 1/2, 0, 1/4. Each of the six interfaces between adjacent unit cells contains two such positions.

The other interstitial site contained in the bcc lattice has a coordination number of 8, and these eight ligands mark the corners of a regular octahedron. This octahedral interstitial position is shown in Fig. 1.14. An easy way to remember the location of the octahedral site is to notice that it lies midway between each pair of rational lattice points separated by the distance a_0 (i.e., all pairs along the cube edges). There are 18 such sites shared between adjacent unit cells. The coordinates of the interstitial atom shown in Fig. 1.14 are 1/2, 1/2, 1.

Table 1.3 lists most of the bcc elements along with their unit cell dimensions, a_0, and T_m, their respective melting temperatures.

B. Close-Packed Structures

There are two basically different ways of stacking equidiameter spheres so that maximum density is achieved. The spheres are arranged in close-packed two-dimensional layers, which are then stacked one upon the other. The stacking

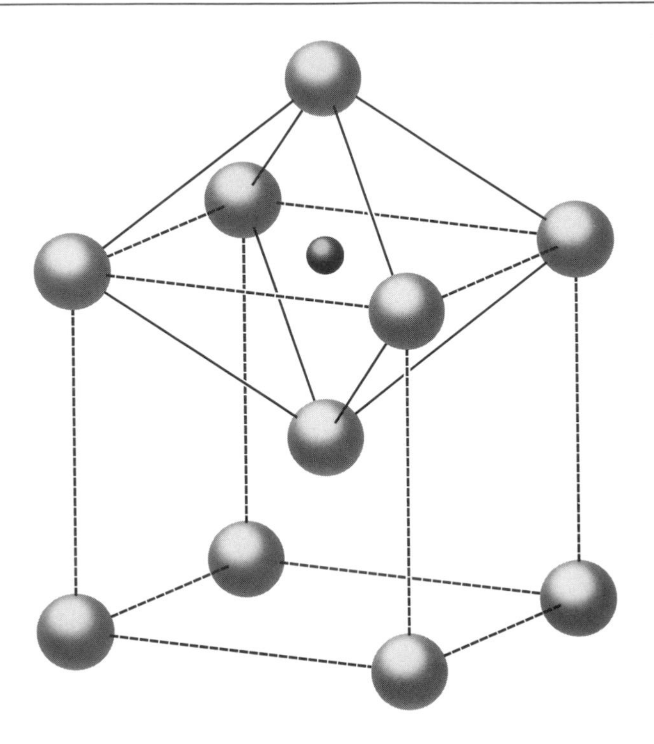

Figure 1.14.

The octahedral interstitial site in the bcc crystal structure.

sequence, that is, the coordinates of the atoms in a given layer plane, can be repeated every third or fourth layer, producing the hexagonal close-packed (hcp) or face-centered (fcc) cubic structure, respectively. Let us examine the two-dimensional analog shown in Fig. 1.15. Let the dashed-line circles represent the first-layer plane, and the solid-line circles the second. If the third plane is placed over the gray-colored interstices, it will have the same coordinates as the first one; and if the fourth layer is placed in the black interstices, it reproduces the positions in the second layer. Arbitrarily labeling the bottommost layer A and the second one B, the stacking sequence that occupies the gray and black positions is described by ABAB. The infinite reproduction of the AB sequence produces the hcp configuration.

The fcc structure makes use of the third set of unique spatial coordinates, represented by the black interstices. Placing the third layer on the black positions results in the ABC stacking sequence, which generates the fcc lattice.

Because of the geometrical similarity of the two close-packed structures, it is generally easy to interrupt the perfect periodicity of either by mechanical deformation. This leads to *stacking faults*, which are a common feature of close-packed structures.

Table 1.3. Bcc elements.

Element	Temperature Range (K)[a]	a_0 (Å)[b]	T_m (K)	Element	Temperature Range (K)	a_0 (Å)	T_m (K)
Ba		5.025	1002	β-Sm	1190–T_m	—	1345
β-Ca	720–T_m	5.38 (773)	1112	γ-Sr	830–T_m	4.85 (887)	1041
δ-Ce	999–T_m	—	1071	Ta		3.3058	3269
Cr		2.8839	2130	β-Tb	150–T_m	—	1630
Cs		6.067 (78)	301.5	β-Th	1636–T_m	4.11 (1723)	2028
β-Dy	1657–T_m	—	1682	β-Ti	1155–T_m	3.3065 (1173)	1943
Eu		4.606	1090	β-Tl	507–T_m	3.882	577
α-Fe	<1183	2.8665		γ-U	1048–T_m	3.474	1405
δ-Fe	1667–T_m	2.94 (1698)	1809	V		3.0240	2175
β-Gd	1533–T_m	—	1595	W		3.16469	3653
β-Hf	2013–T_m	—	2227	β-Yb	1033–T_m	—	1097
β-Ho	1701–T_m	—	1843	β-Zr	1036–T_m	3.62 (1123)	2125
K		5.247 (78)	336.4				
γ-La	1134–T_m	—	1193				
β-Li	80–T_m	3.5093	454				
δ-Mn	1410–T_m	3.0806 (1407)	1517				
Mo		3.1473	2890				
β-Na	40–T_m	4.2906	371				
Nb		3.3004	2740				
β-Nd	1128–T_m	4.13 (1156)	1283				
β-Pr	1068–T_m	4.13 (1094)	1204				
ε-Pu	753–T_m	3.638 (773)	913				
Rb		5.605 (78)	312				
β-Sc	1608–T_m	—	1812				

[a] Where no temperature range is given, the element has only one crystal structure.
[b] All lattice parameters are given for 298 K unless otherwise noted in parentheses.

An example of a stacking fault might be described by —ABABABCBCBABAB—, displaying the insertion of the BC sequence into the otherwise perfect hcp array.

Another variation on the close-packed structure is delineated by —ABACABAC—; which effectively doubles the length of the c-axis. Neodynium is an example of this *double close-packed* hexagonal structure. It is perhaps worth emphasizing that because a crystal has the hcp or fcc structure does not ensure that it is close-packed; in fact, it very seldom is. Only when all the atoms or ions are of equal size can each one have a point of tangency with each of its 12 nearest neighbors, which is required for closest packing. However, this alone does not guarantee close packing, and all the elements deviate somewhat from the ideal configuration. The close-packed hexagonal structure is shown in terms of close-packed spheres in Fig. 1.16.

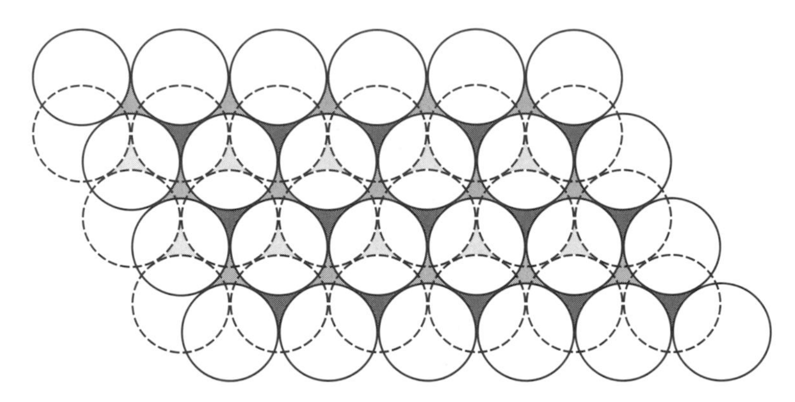

Figure 1.15.

Placing the next layer of circles over the gray interstices leads to the hcp structure; placing them over the black produced the fcc arrangement.

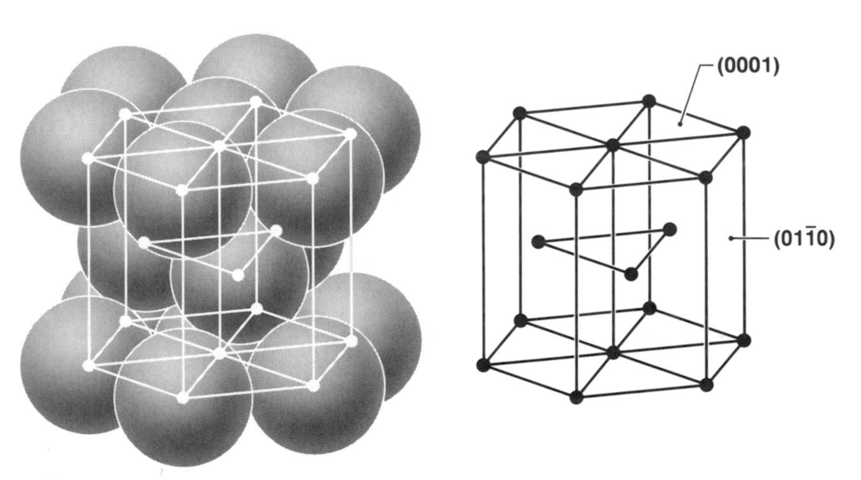

Figure 1.16.

The hcp unit cell as constructed from equidimensional spheres and the lattice upon which this structure is based.

C. The Face-Centered Cubic (fcc) Structure

The fcc lattice is a Bravais lattice as depicted in Fig. 1.6. As the name implies, each face of the primitive cube contains an atom at its center and each is separated from its 12 nearest neighbors by a distance of $a_0\sqrt{2}/2$. Each atom also has six next-nearest neighbors at a distance a_0, which is about 31% greater than the nearest-neighbor interval. It is somewhat more difficult to see the fcc unit cell in the hard sphere representation than the hcp one. Two prominent planes are depicted and outlined in Fig. 1.17: the (100) containing the face center atom, and the triangular segment of the (111) plane.

In addition to the rational lattice points, there are two different types of interstitial cavities that are capable of holding solute atoms of relatively small diameter, or of the host atoms in special circumstances such as might result from radiation damage. The largest of these is located at the center of the cube or the geometrically equivalent sites along the cube edges. This site has six nearest neighboring atoms that occupy the centers of the six cube faces and that define the corners of a regular octahedron, as shown in Fig. 1.18. The site shown by this illustration has coordinates 1/2, 1/2, 1/2. The octahedral interstice will accommodate a sphere whose maximum radius is $0.41r$, r being the radius of the constituent close-packed spheres. Surprisingly, this is larger than the spheres that can fit into the octahedrally coordinated interstice of the more open bcc lattice.

There are eight smaller tetrahedrally coordinated interstices in each unit. They reside in the center of each octant, as shown in Fig. 1.19, and the one nearest the origin then has coordinates 1/4, 1/4, 1/4. Table 1.4 lists most of the fcc elements along with the unit cell dimension and melting temperature.

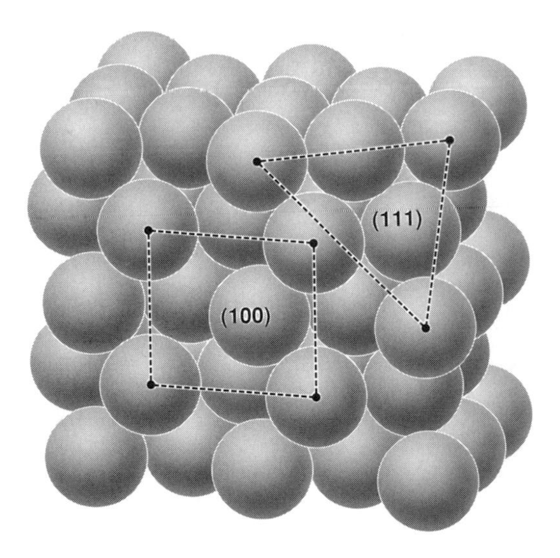

Figure 1.17.

The face-centered (100) plane and a section of the close-packed (111) plane.

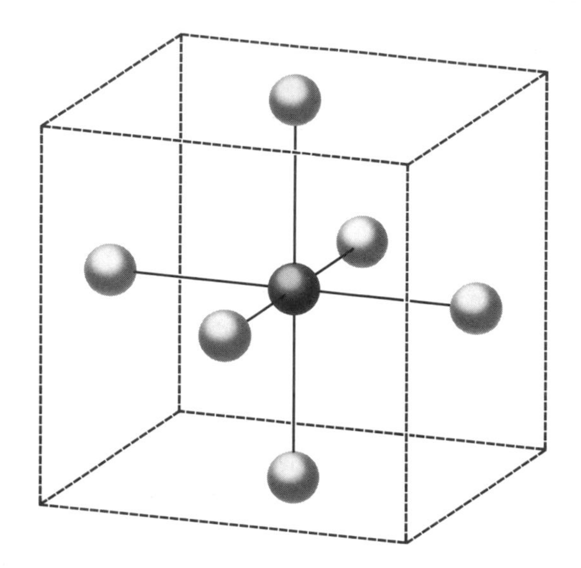

Figure 1.18.

The solid circle is the octahedrally coordinated interstice at the center of a face-centered cubic cell.

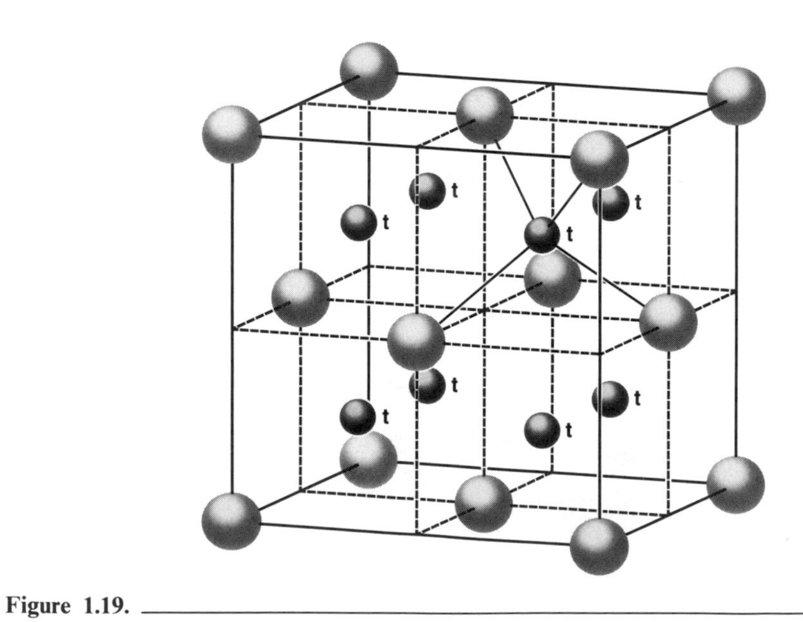

Figure 1.19.

Illustrating the eight tetrahedral interstitial positions in the fcc unit cell.

Table 1.4. Fcc elements.

Element	Temperature Range (K)	a_0 (Å)[a]	T_m(K)
Ac		5.311	1323
Ag		4.0862	1234
Al		4.04958	933
Au		4.07825	1337
α-Ca	<720	5.576	
α-Ce	<125	4.85 (77)	
γ-Ce	160–999	5.1601	
β-Co	700–T_m	3.548	1768
Cu		3.61496	1356
γ-Fe	1184–1665	3.5910 (320)	
Ir		3.8394 (321)	2716
β-La	550–1134	5.296	
γ-Mn	1360–1410	3.52	
Ni		3.52387	1726
Pb		4.9505	601
Pd		3.8896	1825
Pt		3.9231	2042
δ-Pu	480–588	4.6370	
Rh		3.8031	2233
α-Sr	<830	6.0847	
α-Th	<1636	5.0843	
α-Yb	<1033	5.4862	

[a] All lattice parameters are given for 298 K unless otherwise noted in parentheses.

D. The Hexagonal Structure

The close packing of identical inelastic spheres in the hexagonal structure must have an axial ratio, c/a, equal to 1.633. Most elements differ significantly from this ideal value, as can be seen in Table 1.5. Magnesium is the element in closest agreement, its ratio being 1.623, and as a result it is taken as the prototypic hcp element. Cadmium, with $c/a = 1.886$, deviates the most from the ideal ratio. The ideal ratio is a consequence of all the atoms being strictly spherically symmetrical, as might be expected from simple ions or attractive induced dipole interactions. A departure from the ideal ratio reflects the influence of nonspherically symmetrical atomic wave functions, where the directionality arises from the hybridization with the *p*, *d*, or *f* electrons.

The ideal hexagonal unit cell, like the body- and face-centered cubic ones, is a Bravais lattice. Figure 1.16 shows the unit cell in expanded form and defines the two

Table 1.5 Hcp elements.

Element	Temperature Range (K)	a_0 (Å)[a]	c_0 (Å)	T_m (K)
Be		2.2866	3.5833	1527
Cd		2.97887	5.61765	594
β-Ce	125–350 (dhcp)[b]	3.65	5.96	
α-Co	<700	2.5071	4.0686 (293)	
α-Dy	<1657	3.5903	5.6475 (293)	
Er		3.5588	5.5874	1795
α-Gd	<1533	3.6360	5.7026 (293)	
α-Hf	<2013	3.1967	5.578 (299)	
α-Ho	<1701	3.5773	5.6158 (293)	
α-La	<550 (dhcp)	3.75	6.07	
α-Li	<80	3.111	5.093 (78)	
Lu		3.5031	5.5509	
Mg		3.20927	5.21033	
α-Na	<40	3.767	6.154 (5)	
α-Nd	<1128 (dhcp)	3.657	5.902	
Os		2.7352	4.3190	3300
α-Pr	<1068 (dhcp	3.669	5.920	
Re		2.7608	4.4582	3453
Ru		2.70389	4.28168	2523
α-Sc	<1608	3.3090	5.2733 (293)	
α-Tb	<1560	3.6010	5.6930 (293)	
Tc		2.735	4.388	2473
α-Ti	<1155	2.950	4.686 (298)	
α-Tl	<507	3.456	5.525	
Tm		3.5375	5.5546 (293)	1818
α-Y	<1752	3.6474	5.7306 (293)	

[a] All lattice parameters are given for 298 K unless otherwise noted in parentheses.
[b] Double hexagonal close-packed.

coordinate axes. In the close-packed structure, the distance between atoms in the basal plane is a and the separation of each atom from the three nearest neighbors above and below is given by

$$d = \left(\frac{a^2}{3} + \frac{c^2}{4}\right)^{1/2}. \tag{1.7}$$

When $c/a = 1.633$, all 12 nearest neighbors are at a distance a from the reference atom, i.e., the distance corresponding to closest packing.

The indices for the hexagonal lattice are based upon four axes: c, a_1, a_2, and a_3, as defined by Fig. 1.20. Under this labeling system the general expression for the

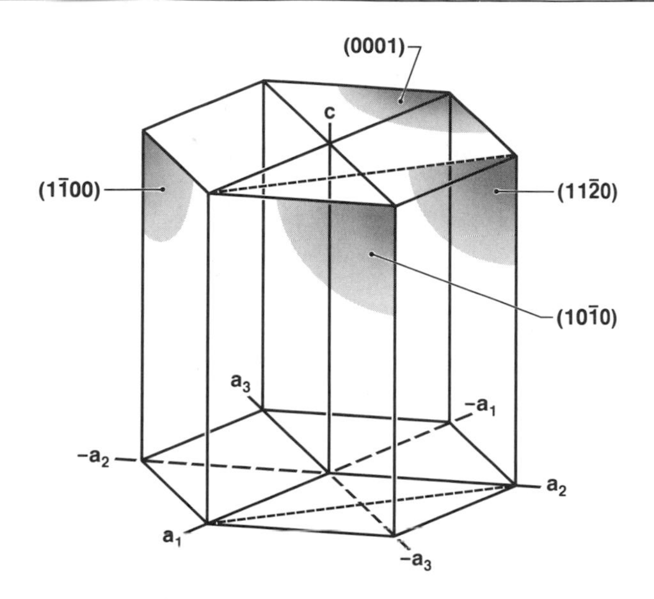

Figure 1.20.

The axes and indices of some of the prominent planes in the hexagonal system.

indices is $(hkil)$, corresponding to the reduced intercepts on the four axes in the order listed above. The index along a_3 is always related to the other two by

$$i = -(h + k).$$

9. Covalently Bonded Elements

The term "metallic bond" should not be taken to imply a total absence of covalent character, nor vice versa. It is, in fact, somewhat arbitrary at what point we cease to call a substance metallic and term it covalent instead. Bearing this in mind, we now examine a variety of elements that traverse the entire gamut of properties from modestly metallic to being nearly ideal insulators, yet all manifesting a recognizable degree of covalence.

A. The Diamond Cubic Structure

The group IVA elements are prone to the diamond cubic structure, which directly reflects the tetrahedral bond symmetry produced by the sp^3 hybridization. The diamond cubic unit cell, as well as a sketch emphasizing the tetrahedral coordination

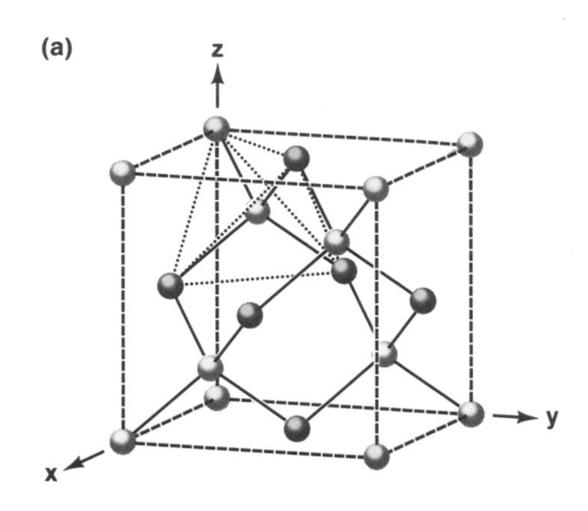

Figure 1.21a.

The diamond cubic unit cell presents a face-centered cubic exterior.

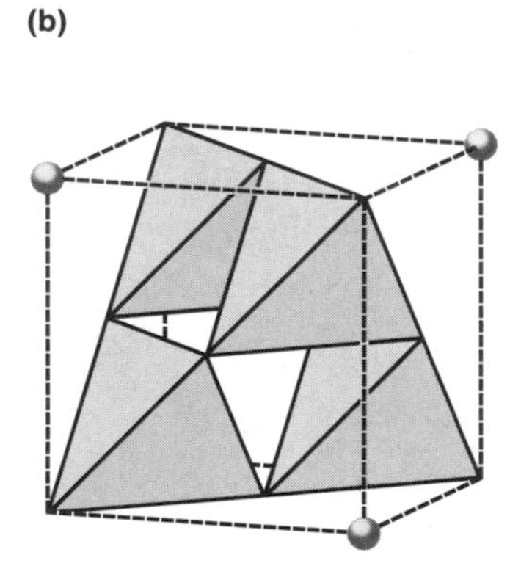

Figure 1.21b.

The sp^3 hybridization leads to tetrahedral bonding with a bond angle of 109°28′.

of all the carbon atoms, is shown in Figs. 1.21a,b. Diamond is itself the prototype that gives its name to this structure.

The four outer electrons of each carbon atom are each paired in one of the four covalent bonds. The empty conduction band lies, energetically, considerably above the filled valence band; hence, diamond is an extremely poor conductor of electricity. To the contrary, Si and Ge have relatively low-lying conduction bands that can be populated by electrons as a result of thermal excitation. Accordingly, these diamond cubic elements are semiconductors.

The diamond cubic structure is not confined solely to elemental crystals, but finds expression in several compounds, particularly in the wurtzite structure, of which ZnS,

AlN, and CuCl are representative. These structures have the same tetrahedral bond angles as diamond and, hence, the same overall arrangement of atoms. However, because wurtzite has two types of atoms, it lacks some of the symmetry possessed by monoelemental diamond.

As already noted, diamond is an insulator, and the next two group IVA elements, Si and Ge, are semiconductors. The next element is Sn. In its low-temperature diamond cubic modification, gray Sn is an insulator, but above 291°K it transforms to the metallic tetragonal white Sn. Perhaps diamond cubic Sn would also become an intrinsic semiconductor at higher temperatures were it not for the intervention of the transformation.

The last element in this column of the periodic table is lead. Lead is a true fcc metal, albeit for a metal one of the poorest conductors of electricity.

Thus, a clear trend is apparent from the least to the most metallic element as the atomic number increases within this group. The same trend is maintained by groups V, VI, and VII.

The group IVA elements, Pb excepted, illustrate the 8-N rule, which defines the effective covalency of the Nth group. This can apply only to those elements that form covalent bonds requiring that $N >$ IV. By way of example, diamond forms four covalent bonds corresponding to 8-IV. This is nothing more than the elementary observation that elements and ions are stabilized by having a full complement of electrons in the outer shell. Groups IV through VII require eight outermost electrons to satisfy this requirement.

B. Graphite

Graphite is the stable allotrope of carbon, except at very high pressure. Under standard conditions of 25°C and 1 atm of pressure, the free energy of 1 g atom of diamond is 693 cal greater than that of graphite. The hexagonal structure, as shown in Fig. 1.22, far from being close-packed, has an interplanar spacing of 3.35 Å, and the carbon–carbon nearest-neighbor distance within the basal plane is 1.42 Å. Thus, $c/a = 4.72$ and is approximately four times greater than the ideal close-packed value.

The basal plane of graphite is made up of benzene-like hexagons. Three of the outer electrons of carbon hybridize to form localized sp^2 bonds (or, in terms of molecular orbital theory, σ bonds). The remaining electrons unite to make delocalized π bonds. The basal planes are bonded to one another by weak van der Waals forces. As a consequence of these two inherently different types of binding, the physical and mechanical properties of graphite are extremely anisotropic. As expected, it cleaves readily along the basal plane, and, in fact, small perfect crystals of graphite are

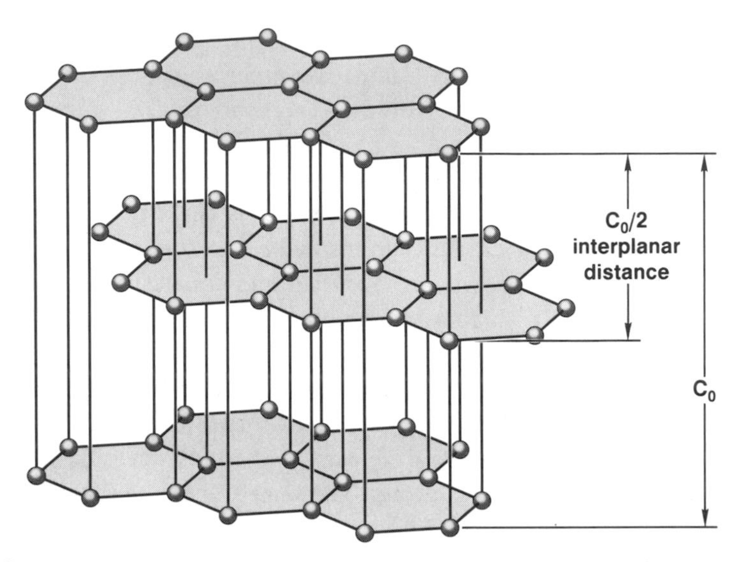

Figure 1.22.

The hexagonal crystal structure of graphite.

generally handled by flotation to avoid inadvertent damage. The electrical and thermal conductivity in the *a* directions can exceed that of pure copper in a carefully grown crystal of pyrolytic graphite, whereas it is an excellent insulator parallel to the *c*-axis. The graphitic structure is not confined to carbon, but is even found in compounds such as MoS_2, which is the naturally occurring mineral molybdenite, as well as in WS_2 and $MoSe_2$.

C. Elements of Group VA

The elements of Group V have five outer electrons and, hence, a covalent valency of 3. Nitrogen, the first element of this group, has the structure :N:::N:, where the dots, as usual, represent electrons in the outermost shell—in this case, $3s^2 3p^3$ in the atomic state. The N_2 molecule has great stability because of the strong triple bond, and, thus, nitrogen gas is mainly inert at ordinary temperatures.

At the triple point, 63.14 K, β-nitrogen crystallizes with a hexagonal structure having two N_2 molecules per unit cell. At yet a lower temperature of 35.6°K, the hexagonal structure transforms to the cubic modification, having four N_2 molecules per unit cell. These crystalline substances belong to the class of materials generally referred to as molecular crystals. The bonds between molecules are of the weak van der Waals type resulting from the interaction of permanent or induced dipoles or a mixture of both. Continuing from left to right from group I to group VII, molecular crystals occur with increasing frequency. We will not deal with molecular crystals in much detail in this text, but it is important to remember that essentially all organic crystals fall into this category.

Phosphorus crystallizes in at least five polymorphic structures. Possibly, P_4 is the simplest and among the most stable of these molecules. In this case, the atoms are located at the corners of a regular tetrahedron, each being directly bonded to the remaining three atoms, as shown in Fig. 1.23. There are apparently two forms of white phosphorus, both of which are molecular crystals with P_4 as the elementary molecule unit.

The bonding in P_4 cannot be explained by the straightforward pairing of the three half-filled p orbitals, as the p_x, p_y, and p_z atomic wave functions are normal to one another, whereas the tetrahedral bonds subtend angles of 60°. It has been suggested that these bonds may result from sp^2 hybridization, which produces a bond angle of 66° and, hence, much closer to the observed value.

The next two elements in subgroup V, arsenic and antimony, are frequently referred to as metalloids, implying a partially metallic character. Bismuth, the last element in

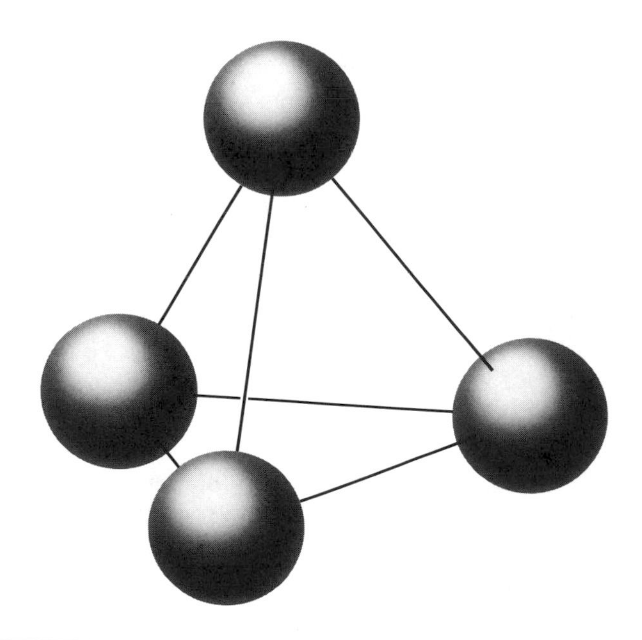

Figure 1.23.

The structure of P_4 is a regular tetrahedron.

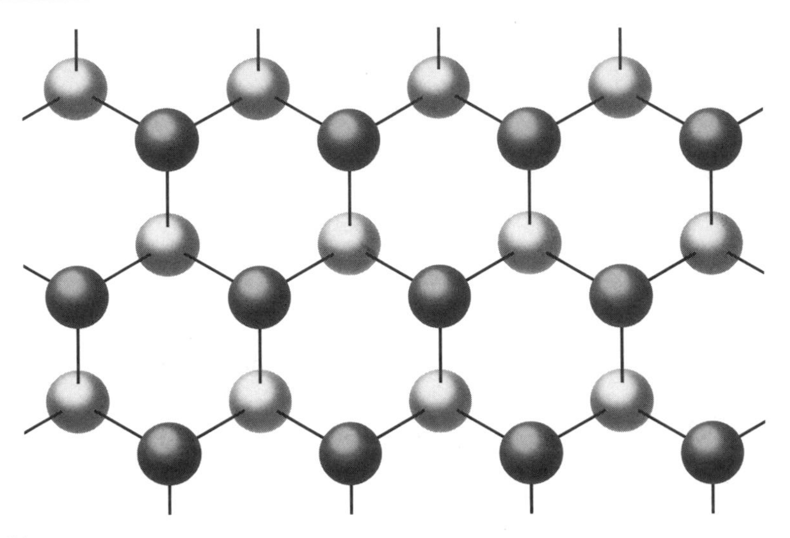

Figure 1.24.

As, Sb, and Bi form rhombohedral crystals from puckered sheets of hexagons. The shaded atoms are above the plane of the page.

this group, more properly fits the description of a true metal. All three of these elements possess the rhombohedral structure. Each atom is bonded to three others, creating semi-infinite sheets of puckered hexagons. The puckering is expressed by the elevation (or depression) of every other atom out of the plane in which it resides; in other words, all atoms are covalently bonded to three atoms above or below the median plane. This is illustrated in Fig. 1.24. The bond angle is almost exactly 90°, strongly suggesting straightforward bonding via the *p* orbitals. The sheets are themselves held together by van der Waals forces. Each atom is covalently bonded to three nearest neighbors and has an additional three next-nearest neighbors in one of its adjacent sheets. The difference between these two separations, d_1 and d_2, is greatest for As and least for Bi, as listed in Table 1.6.

Table 1.6. Distance between nearest and next-nearest neighbors.

	d_1 (Å)	d_2 (Å)	$d_2 - d_1$ (Å)
As	2.51	3.15	0.64
Sb	2.87	3.37	0.50
Bi	3.10	3.47	0.37

D. Elements of Group VIA

The elements of this group are all divalent, which is most clearly exemplified by oxygen, which either forms two covalent bonds or becomes the doubly charged oxide ion. Solid oxygen has at least three solid modifications, but the atomic arrangement of none is known with certainty.[1]

Sulfur forms stable S_2 molecules at elevated temperatures, but below 390 K it exists as S_8 molecules, which then condense with an orthorhombic structure (see Fig. 1.25). These S_8 units are bonded via van der Waals forces in the solid. The unit cell of rhombic sulfur is extremely large. It contains 128 atoms, and its dimensions are $a_0 = 10.467$ Å, $b_0 = 12.870$ Å, and $c_0 = 24.493$ Å at 298 K (see Fig. 1.26).

The crystal structures of selenium are also extremely complicated. They form two modifications of the eight-membered puckered rings analogous to that of sulfur. There are two monoclinic isomorphs with different and individually varying bond lengths, as presented in Fig. 1.27. Both the α and β forms have 32 atoms per unit cell, but with significantly different cell dimensions:

$$\alpha\text{-Se} \quad a_0 = 9.05\ \text{Å},\ b_0 = 9.07\ \text{Å},\ c_0 = 11.61\ \text{Å},\ \beta = 90°46';$$

$$\beta\text{-Se} \quad a_0 = 12.85\ \text{Å},\ b_0 = 8.07\ \text{Å},\ c_0 = 9.31\ \text{Å},\ \beta = 93°8'.$$

The eight-membered rings condense to form long spiral chains with van der Waals forces, as usual, connecting the individual molecules, as shown in Figs. 1.28 and 1.29.

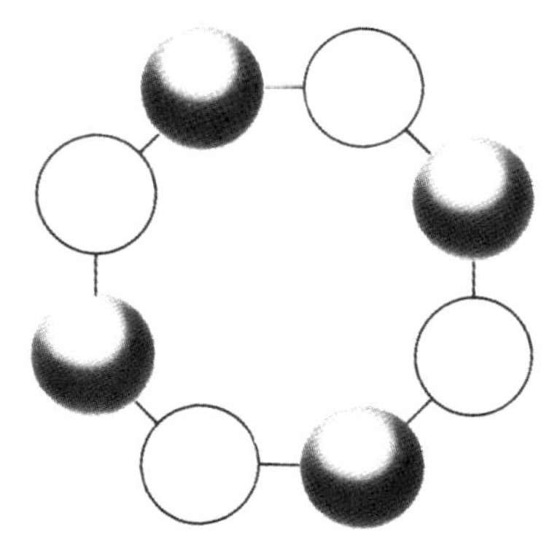

Figure 1.25.

The atoms in the S_8 molecule alternate above and below the plane of the page, forming a puckered ring.

[1] W. G. Wyckoff, *Crystal Structures*, p. 32, Wiley, New York, 1971.

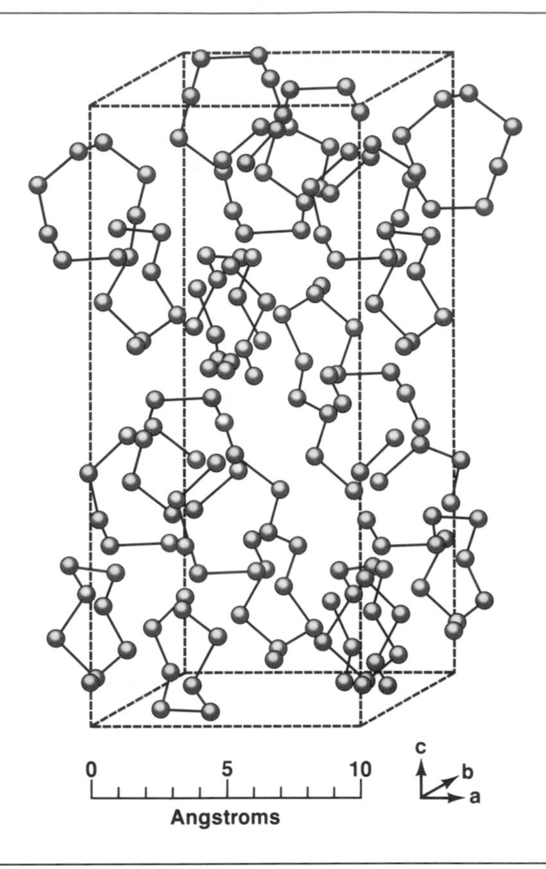

Figure 1.26.

The unit cell of rhombic sulfur.

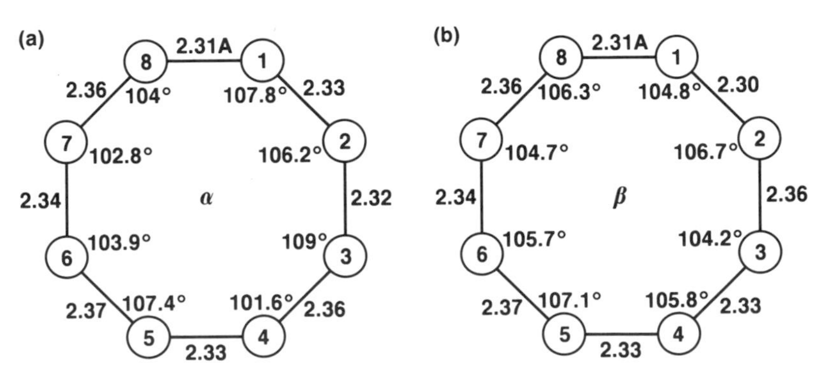

Figure 1.27.

The two different puckered rings that go to form the α and β monoclinic modifications of Se.

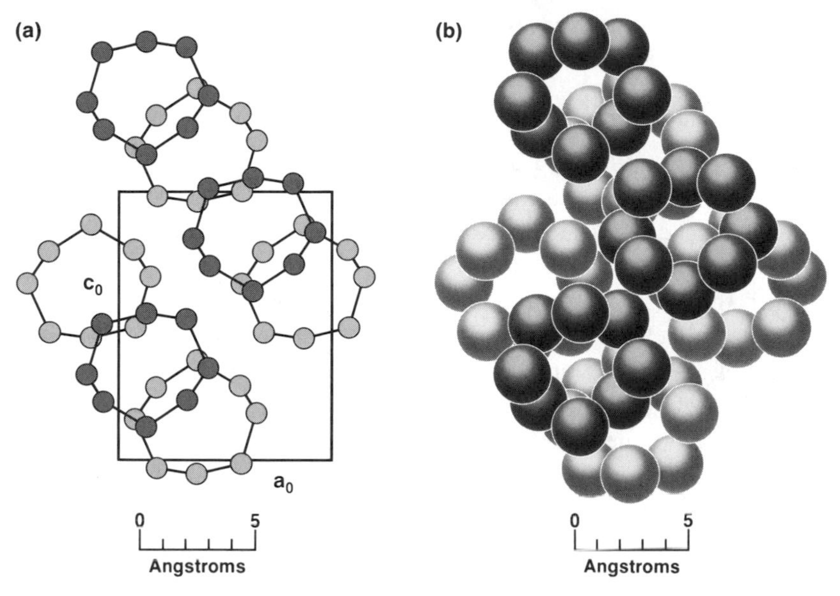

Figure 1.28.

The crystal structure of α-Se projected on the (100) plane; (b) further illustration of the puckered ring structure. (After R. W. G. Wyckoff, *Crystal Structures*, Interscience, New York, pp. 38, 39, 1963.)

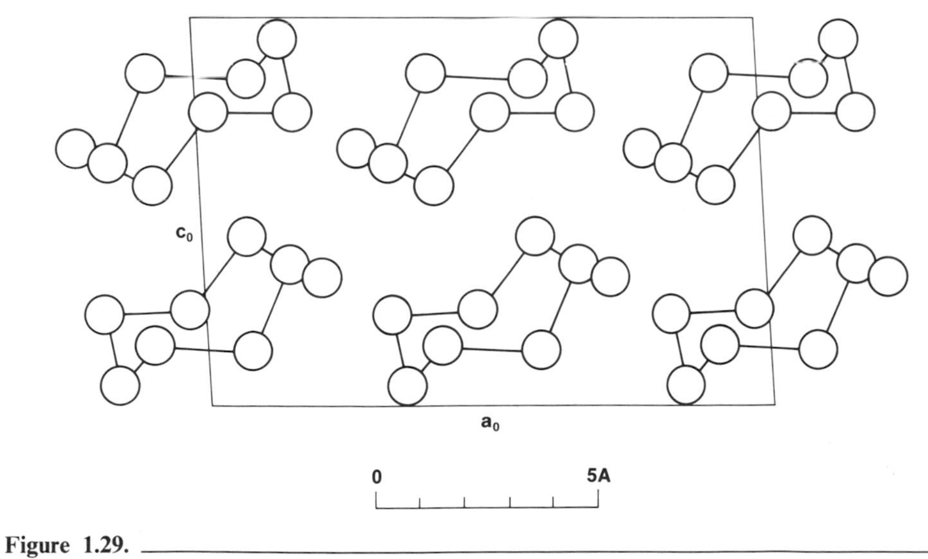

Figure 1.29.

A projection of monoclinic β-Se along the b_0 plane.

These chains are close-packed and, in the words of one author, "like rods in a bundle."[2]

E. Elements of Groups VII and VIII

The halogens are all univalent, existing only as diatomic molecules at normal temperatures. Only I_2 is a solid at room temperature, and it has an orthorhombic unit cell with the molecular arrangement shown in Fig. 1.30.[3]

We will not deal at length with the solid form of the rare gases, group VIII, although they are of considerable theoretical interest. However, rare gas crystals do serve to emphasize, once again, that purely centrosymmetric bonding forces, as also exemplified by s-electron metals and simple monatomic ions, lead to hexagonal or face-centered cubic structures. The melting of the rare gas solids, all of which are

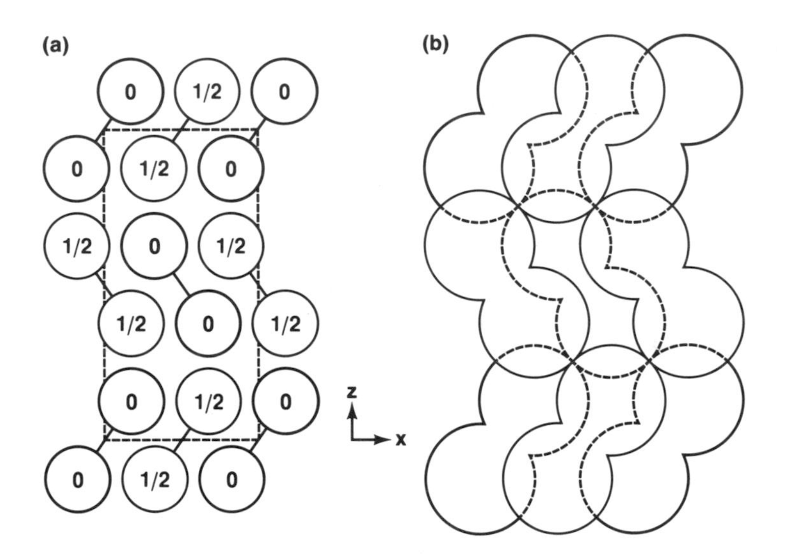

Figure 1.30.

(a) The orthorhombic unit cell of I_2 projected on the xy plane, the numbers denoting relative distance in and above plane of the page. (b) The atoms drawn with correct van der Waals radii to illustrate the close packing. (After Ref. 3.)

---

[2] R. C. Evans, *An Introduction to Crystal Chemistry*, p. 125, Cambridge University Press, 1964.

[3] Ibid., p. 122.

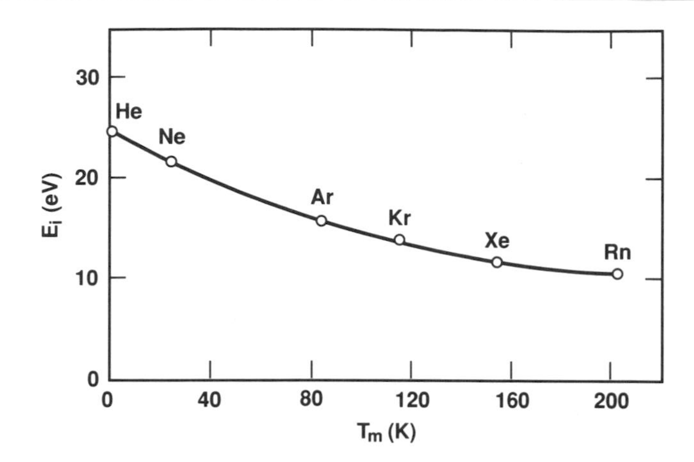

Figure 1.31.

Illustrating the decreasing ionization energy needed to remove the first electron plotted versus the melting points of the rare gas crystals.

close-packed fcc structures, progresses in a regular manner from lowest to highest with increasing atomic number, as shown in Fig. 1.31.

Also, Fig. 1.31 shows the decrease in ionization potential with increasing atomic number resulting from the increasing screening of the nuclear charge. However, if the spherical charge distribution around each atom were invariant, there would be no attractive forces exerted by the atoms upon one another. However, because of the uncertainty principle, there are weak electrostatic forces that arise from oscillations in the electronic charge distributions. These weak electrostatic forces that arise from fluctuating atomic or molecular dipoles are called van der Waals or dispersion forces. In an atom with no permanent moment, the rapidly varying dipole moments average to zero over a large number of configurations and, hence, are centrosymmetric. However, the instantaneous electric field associated with the temporary dipoles will induce a dipole moment on neighboring atoms. As a result, there is a net time-independent attractive interaction arising from the average interaction of the induced and original moments. These interactions are additive for all the pairs of molecules or atoms in the gas and, hence, account for cohesion. As noted above, these forces are also known as dispersion forces because the characteristic molecular frequency pertinent to the interaction energy is the same one that determines the dispersion of light by the molecule (see Appendix A).

Although the condensed state of the rare gases is of considerable theoretical interest, and this is especially true for He, we will not deal with them in this treatise. Suffice it to repeat that all form close-packed cubic solids at appropriately low temperatures.

Chapter 1 Exercises

1. Calculate the maximum and minimum dimensions of the irregular octahedral interstitial site in a bcc structure. Answer: $0.633r$, $0.154r$ (Hint: $r = \sqrt{3}a/4$).

2. Calculate the size of the largest sphere that can occupy the octahedral tetrahedrally coordinated interstice in fcc and hcp structures composed of atoms of radius r.

3. What are the atomic densities of the $\{100\}$, $\{110\}$, and $\{111\}$ planes of the bcc and fcc structures. Areas can be expressed in terms of a_0.

4. Derive Eq. (1.7) and use it to prove that the value of the ideal c/a ratio is $(8/3)^{1/2}$.

5. For the three cubic lattices (primitive, fcc, and bcc) calculate, in terms of the lattice parameter a_0, the following:
   a. volume of unit cell
   b. atoms per unit cell
   c. nearest-neighbor distance and number of nearest neighbors
   d. second-nearest-neighbor distance and number of second-nearest-neighbors.

6. Show that the perpendicular distances in a cubic lattice between two neighboring planes of classes $\{100\}$, $\{110\}$, and $\{111\}$ are in the ratios

$$d\{100\}:d\{110\}:d\{111\} = 1:\frac{1}{2^{1/2}}:\frac{1}{3^{1/2}},$$

   and derive the general formula

$$d(hkl) = \frac{a_0}{(h^2 + k^2 + l^2)^{1/2}}.$$

7. Iron in the bcc structure at 298 K has a density of 7.86 g/cm^3 and a lattice spacing $a_0 = 2.867$ Å. Calculate Avogadro's number from these data.

8. The unit cell dimension of diamond is 0.891 Å. Calculate the C—C bond length.

9. Prove that in the cubic system the angle θ between directions $[h_1, k_1, l_1]$ and $[h_2, k_2, l_2]$ is given by

$$\cos\theta = \frac{h_1h_2 + k_1k_2 + l_1l_2}{[(h_1^2 + k_1^2 + l_1^2)(h_2^2 + k_2^2 + l_2^2)]^{1/2}}.$$

10. Calculate the fractional volume of space filled by equidimensional spheres that have a point of tangency with all of their nearest neighbors and are in the following configurations:
a. primitive cubic
b. body-centered cubic
c. face-centered cubic
d. diamond cubic
e. hexagonal close-packed.

Additional Reading

G. Burns, *Solid State Physics*, Academic Press (1985). The first three chapters of this text present a thorough introduction to classical crystallography with an exhaustive treatment of symmetry and the attendant symbols.

R. C. Evans, *An Introduction to Crystal Chemistry*, Cambridge University Press (1964). As the title suggests, this volume is concerned with the relations between chemical bonding and crystal structure. Chapter 8 reviews the structure of the elements, but Chapters 1 through 4 provide a lucid, readable introduction to the relations between atomic structure, chemical bonding, and crystal structure.

W. Hume-Rothery and G. V. Raynor, *The Structure of Metals and Alloys*, The Institute of Metals (1956). Chapters 1 through 3 in some ways closely resemble those of the preceding reference and the same statements apply; however, this book is restricted to dealing only with metals.

R. W. G. Wyckoff, *Crystal Structures*, Wiley (1971). An encyclopedic work of several volumes. It is more than a compendium of structure data and can be referred to for explanatory material.

F. Donald Bloss, *Crystallography and Crystal Chemistry*. Holt, Rinehart and Winston (1971). Chapters 1 to 7 present the material of classical crystallography in a clear readable style. For students with strong interests in crystal structures, this book is highly recommended.

Chapter 2

Thermodynamic Equations of State

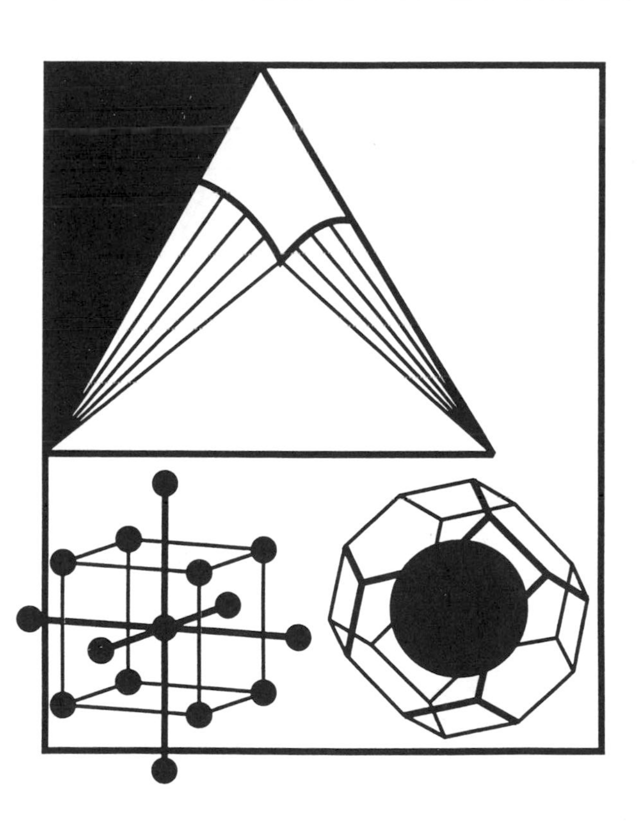

Chapter 2

Contents

It ain't restful but you gotta do it anyway in case you miss something.

—Satchel Paige

Thermodynamic Equations of State

An equation of state connects all the physical variables that control the free energy of a system such as temperature, pressure, volume, and magnetic and electric fields. An equation of state describing a specific substance is not derivable as a strict thermodynamic entity but must be extracted from laboratory measurements or derived from models such as, for example, "the perfect gas." Nevertheless, there are equations of general applicability, provided their parameters are individually adjusted to conform with the experimental results. Consequently, these relationships are intrinsically empirical in nature, although the existence of an equation of state for each and every substance is a thermodynamic necessity. However, equations of state (EOS) provide more information about a substance than just a simple relation of the pertinent variables. They furnish the basis for the potential functions needed for theoretical modeling and, from thence, to more accurate quantitative description of the chemical bonding. In fact, our atomistic interpretation of the compressibility leads directly to the calculation of the repulsive forces between atoms or molecules. The following treatment starts here with a brief review of the needed thermodynamic functions. In the following chapters, we discuss model-dependent vibrational effects, followed by potential functions, and, finally, chemical bonding.

1. Thermodynamic Functions

The equations of state, which are developed in the next section, are expressed solely in terms of pressure (P), volume (V), and temperature (T). These equations are then used to describe the P-V-T dependence of the state functions F, G, E, H, and S, which stand for the Helmholtz and Gibbs free energies, the internal energy, the enthalpy, and entropy, respectively. In order to review the underlying thermodynamic principles, we will here present detailed derivations of the fundamental relations.

We commence by writing the most general expression for the first law of thermodynamics:

$$dE = dw + dq, \tag{2.1}$$

which states that the change in the internal energy of a system is the sum of the work done on the system and the heat absorbed by the system. The work and heat are not true state functions, and dw and dq cannot be integrated. Hence, they are not derivatives in the usual sense. One can visualize an essentially infinite number of paths combining the work and heat consisting of arbitrary compressions, expansions, heatings, and coolings that, when summed, could yield the same change in the internal energy. The internal energy, however, is a state function and is the sum of all the kinetic and potential energies contained in the rotational, vibrational, translational, and electronic degrees of freedom by an equilibrium ensemble of atoms or molecules. If we now change the internal energy by keeping our system in quasi-equilibrium, the internal energy is partitioned among all the aforementioned modes in an equilibrium manner at all points along this path of change. If one of the three variables, P, V, or T, is held constant, then there is only one such path and it is called the reversible path or process. Each point along this pathway obeys an equation of state, from which we can correctly infer that an equation of state is applicable only to the equilibrium state of the system. This is intuitively appealing for, in principle, there is an infinity of nonequilibrium states that are possible for a given system, and it is difficult to perceive how a single equation, based upon equilibrium conditions, could be extended to include them.

Returning now to (2.1), let us apply this equation to the special case wherein all work is of the $P\,dV$ sort; i.e., we exclude magnetic, electrical, gravitational, and other types of work. We further specify that the change in internal energy proceeds along a reversible pathway or some combination of reversible pathways. Now, remembering that the entropy is defined by

$$dq_{rev}/T \equiv dS,$$

we substitute into (2.1) and obtain

$$dE = -P\,dV + T\,dS, \tag{2.2}$$

which proves to be a more useful formulation of the first law. The minus sign precedes the $P\,dV$ term because work done on the system raises the internal energy and, hence, is positive, although this compression perforce results in a negative ΔV. Looking at

(2.2), we see that the partial derivative of E, with respect to V at constant temperature, is

$$\left(\frac{\partial E}{\partial V}\right)_T = -P + T\left(\frac{\partial S}{\partial V}\right)_T. \tag{2.3}$$

Because our equations of state will be expressed only in terms of P, V, and T, we next need to evaluate $(\partial S/\partial V)_T$ in terms of these quantities. We commence with the definition of the Helmholtz free energy, which is

$$F \equiv E - TS. \tag{2.4}$$

Taking the differential of (2.4) and substituting for dE from (2.2) leads to

$$dF = -P\,dV - S\,dT. \tag{2.5}$$

Consequently,

$$\left(\frac{\partial F}{\partial V}\right)_T = -P \tag{2.6}$$

and

$$\frac{\partial^2 F}{\partial V\,\partial T} = -\left(\frac{\partial P}{\partial T}\right)_V. \tag{2.7}$$

However, F is itself a state function having been defined solely in terms of state functions. Hence, it does not matter whether we proceed, as just noted, by first reversibly changing the free energy with respect to volume at constant temperature (i.e., a reversible isothermal compression) and then heating the system, holding this resultant volume constant, or whether we reverse the order of these steps. If we do the latter, we obtain, from (2.5),

$$\left(\frac{\partial F}{\partial T}\right)_V = -S \tag{2.8}$$

and

$$\frac{\partial^2 F}{\partial T\,\partial V} = -\left(\frac{\partial S}{\partial V}\right)_T. \tag{2.9}$$

Thus, because

$$\frac{\partial^2 F}{\partial V\,\partial T} = \frac{\partial^2 F}{\partial T\,\partial V},$$

we finally obtain the desired result:

$$\left(\frac{\partial S}{\partial V}\right)_T = \left(\frac{\partial P}{\partial T}\right)_V. \tag{2.10}$$

Substitution of (2.10) into (2.3) gives us the P-V-T dependence of E in terms of the desired physical parameters:

$$\left(\frac{\partial E}{\partial V}\right)_T = -P + T\left(\frac{\partial P}{\partial T}\right)_V. \tag{2.11}$$

We will now derive an analogous expression for the pressure dependence of the enthalpy. We recall that the enthalpy is defined by

$$H \equiv E + PV. \tag{2.12}$$

Hence,

$$dH = dE + P\,dV + V\,dP. \tag{2.13}$$

Substituting for dE from (2.2) yields

$$dH = T\,dS + V\,dP, \tag{2.14}$$

and the partial derivative of H with respect to P is

$$\left(\frac{\partial H}{\partial P}\right)_T = T\left(\frac{\partial S}{\partial P}\right)_T + V. \tag{2.15}$$

Now, the Gibbs free energy is defined as

$$G \equiv H - TS. \tag{2.16}$$

Hence,

$$dG = dH - T\,dS - S\,dT. \tag{2.17}$$

Now, substituting (2.13) for dH and then (2.2) for dE, we arrive at

$$dG = V\,dP - S\,dT. \tag{2.18}$$

In the same manner as used in (2.5)–(2.10), we have

$$\left(\frac{\partial G}{\partial P}\right)_T = V, \tag{2.19}$$

$$\frac{\partial^2 G}{\partial P\,\partial T} = \left(\frac{\partial V}{\partial T}\right)_P, \tag{2.20}$$

$$\left(\frac{\partial G}{\partial T}\right)_P = -S,$$

$$\frac{\partial^2 G}{\partial T\,\partial P} = -\left(\frac{\partial S}{\partial P}\right)_T. \tag{2.21}$$

Thus, we finally obtain

$$\left(\frac{\partial S}{\partial P}\right)_T = -\left(\frac{\partial V}{\partial T}\right)_P. \tag{2.22}$$

Substituting into (2.15) gives us the pressure dependence of the enthalpy:

$$\left(\frac{\partial H}{\partial P}\right)_T = V - T\left(\frac{\partial V}{\partial T}\right)_P. \tag{2.23}$$

We will now develop the empirical equations of state that will be used to express (2.6), (2.10), (2.11), and (2.22) over a wide range of temperature and pressure.

2. Equations of State

In the following discussion we limit our attention to the three physical variables P, V, and T. Because an equation of state exists, any two of these three serve to define the energy state of the system.

We next develop two polynomial EOS, one with volume as the dependent variable, the other with pressure, following a procedure put forth by Slater.[1] Experimental data or, in some cases, theoretically calculated values are then substituted into either of these equations, which then can be used to calculate the P-V-T dependence of such thermodynamic quantities as F, G, E, H, S, C_P, and C_V.

Taking, first, V as the dependent variable, let us write

$$V = V_0[1 + a_0(T) - a_1(T)P + a_2(T)P^2 \ldots]. \tag{2.24}$$

The numerical coefficients a_i, which are functions of the temperature, are to be determined experimentally, and vary with every solid. V_0 is defined as the volume of 1 mol of substance at 0 K. By taking the derivative of (2.24), with respect to temperature at zero pressure, we obtain

$$\frac{1}{V_0}\left(\frac{\partial V}{\partial T}\right)_P = \left(\frac{\partial a_0}{\partial T}\right)_P$$

or

$$\frac{1}{V}\left(\frac{\partial V}{\partial T}\right)_P = \frac{1}{1 + a_0}\left(\frac{\partial a_0}{\partial T}\right)_P \simeq \left(\frac{\partial a_0}{\partial T}\right)_P \equiv \alpha. \tag{2.25}$$

[1] J. C. Slater, "Introduction to Chemical Physics", Chapter 13, McGraw-Hill Book Co. Inc. (1939).

As $a_0 \ll 1$, we can identify α, the volume thermal expansivity, with the derivative of a_0 with respect to temperature. In a similar manner one finds the isothermal compressibility to first order in P to be given by

$$\chi \equiv -\frac{1}{V}\left(\frac{\partial V}{\partial P}\right)_T = \frac{a_1}{1 + a_0} \simeq a_1. \tag{2.26}$$

If we now choose temperature and volume to be the independent variables, we can write

$$P = P_0(T) + P_1(T)\left[\frac{V_0 - V}{V_0}\right] + P_2(T)\left[\frac{V_0 - V}{V_0}\right]^2 + \cdots. \tag{2.27}$$

The numerical coefficients P_i, like the a_i, are unique for each substance and must be experimentally determined. The initial term, P_0, is the hydrostatic pressure one must apply at elevated temperatures to reduce the solid to the volume it would have at 0 K at zero pressure. At ordinary temperatures, P_0 is not negligible, yet it is significantly less than P_1 and P_2. Rearranging (2.24) now gives us

$$\frac{V_0 - V}{V_0} = -a_0(T) + a_1(T)P - a_2(T)P^2. \tag{2.28}$$

Substituting (2.28) into (2.27) and collecting terms gives

$$P = P_0 + P_1(-a_0 + a_1 P - a_2 P^2 \cdots) + P_2(-2a_0 a_1 P + a_0 a_2 P^2 + a_1^2 P^2 \cdots). \tag{2.29}$$

Assuming that we can neglect the quadratic and higher powers of a_0 and P_0, we equate coefficients of P, arriving at

$$\begin{aligned} P_0 - P_1 a_0 &= 0, \\ P_1 a_1 - 2P_2 a_0 a_1 &= 1, \\ -P_1 a_2 + 2P_2 a_0 a_2 + P_2 a_1^2 &= 0. \end{aligned} \tag{2.30}$$

Solving for the a_i gives

$$\begin{aligned} a_0 &= \frac{P_0}{P_1}, \\ a_1 &= \frac{1}{P_1 - 2P_2 a_0} = \frac{1}{P}\left(1 + \frac{2P_0 P_2}{P_1^2}\right), \\ a_2 &= \left(\frac{P_2 a_1^2}{P_1 - 2P_2 a_0}\right) = \frac{P_2}{P_1^3}\left(1 + \frac{6P_0 P_2}{P_1^2}\right). \end{aligned} \tag{2.31}$$

In a similar manner, solving for the P_i in terms of a_i yields

$$\begin{aligned} P_0 &= a_0/a_1, \\ P_1 &= (1/a_1)(1 + 2a_0 a_2/a_1^2), \\ P_2 &= a_2/a_1^3. \end{aligned} \tag{2.32}$$

Further discussions of specific equations of state and crystals energies will make use of the foregoing expressions.

The general form of the volume change of the alkali metal elements with increasing pressure is shown in Fig. 2.1.

It is clear from Fig. 2.1 that the alkali metals become more compressible with increasing atomic number. Thus, for this series of elements, the volume decrease

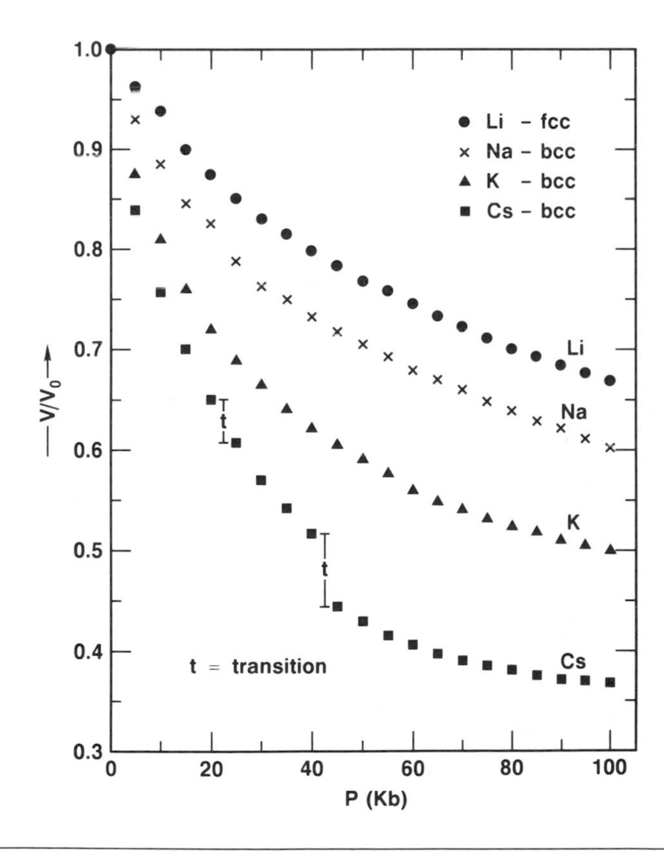

Figure 2.1.

The pressure dependence of the specific volume for certain alkali metals. (Data from *American Institute of Physics Handbook*, McGraw-Hill, New York, pp. 4.38–4.45, 1972.)

produced by a given value of the applied pressure varies directly with the size of the atom. However, at high pressures the actual compressibilities (i.e., the slopes in Fig. 2.1) appear quite similar for all of them.

It is possible to achieve very high pressures, exceeding even 10^6 atm, with the extremely simple device shown in Fig. 2.2. The pressure is applied by manually rotating the hexagonal nut protruding above the Belleville spring. The major disadvantage of this technique is the extremely small sample that is required, generally no more than about 150 μm in diameter. It is also frequently difficult to minimize the pressure gradients that develop across the sample. Nevertheless, this small, inexpensive device has extended the static pressure range by nearly an order of magnitude beyond that available before.

The cognizant reader might well be struck by the similarity between the temperature dependence of the thermal expansivity and the heat capacity at low temperature. Compare, for example, Fig. 2.3 with Fig. 3.7. Since both α and C_P vary with the thermal excitation of the phonon spectrum, this is perhaps not too surprising. The coefficients of thermal expansion shown in Fig. 2.3 run from cryogenic to room temperature. A glance at the high-temperature coefficients that are included in Fig.

(a)

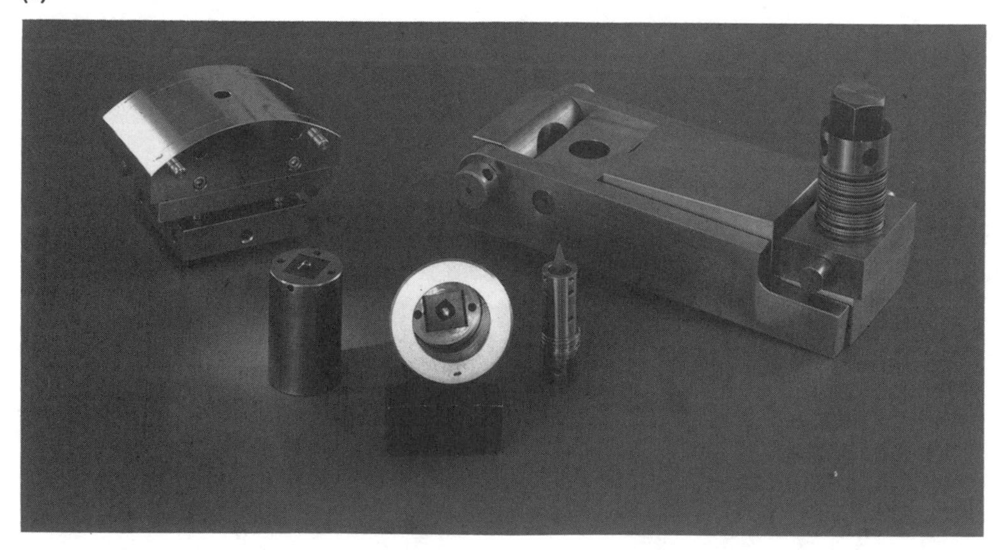

Figure 2.2.

Diamond-anvil pressure cell for studies of crystal structure at high pressures: (a) disassembled to show diamonds and x-ray collimator; (b) cutaway view showing how the various components fit together. (Courtesy of Vagannadham Akela, ECTR, Feb. 1983, p. 13.)

2.4 reveals another interesting feature, viz., that the dependence of α upon temperature shows a significant increase in $d\alpha/dT$ at elevated temperatures. This increasing expansivity with increasing temperature is attributed primarily to the increase in the equilibrium concentration of vacant lattice sites as well as to an increase in the anharmonic contribution.

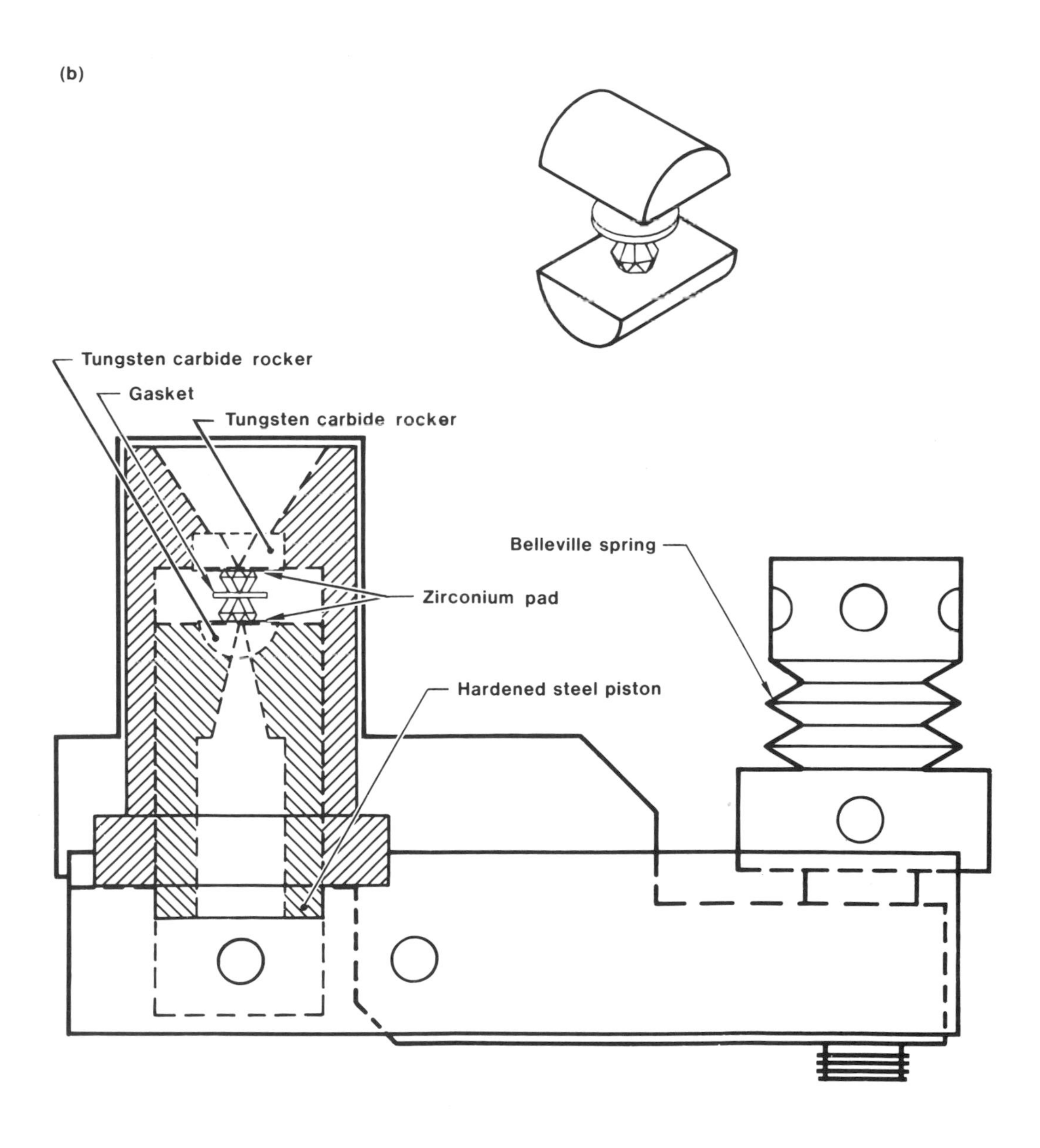

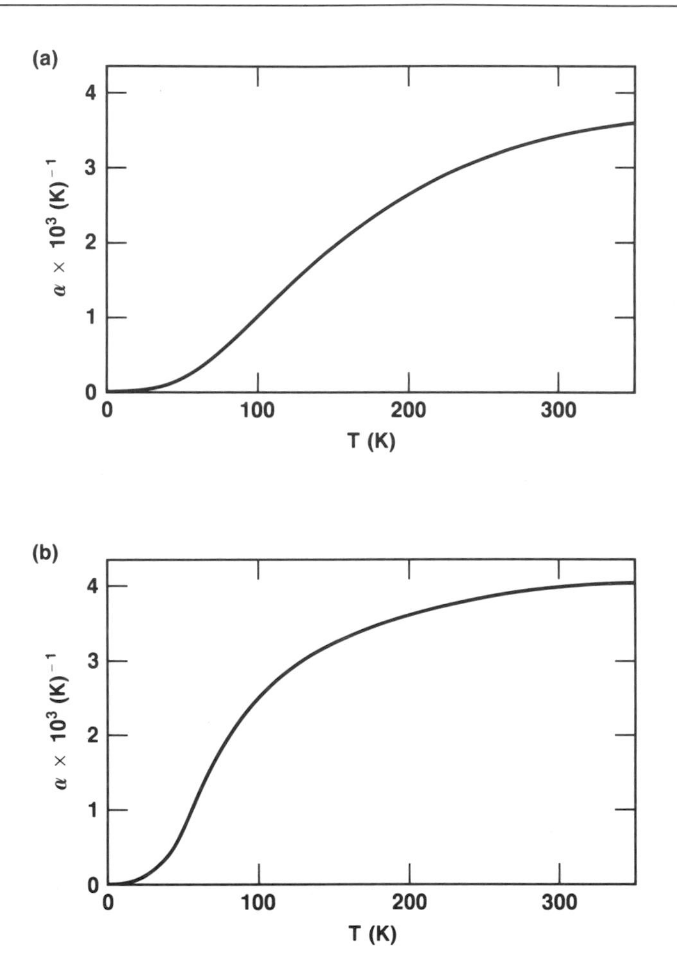

Figure 2.3.

The linear coefficients of thermal expansion for (a) LiF and (b) NaCl at temperatures well below their respective melting points of 1118 K and 1074 K.

3. The Effect of Pressure on Heat Capacity

Using the equations of state derived in the preceding section, we will now examine the effect of pressure upon the heat capacity. The principal equation is

$$\left(\frac{\partial C_P}{\partial P}\right)_T = -T\left(\frac{\partial^2 V}{\partial T^2}\right)_P, \tag{2.33}$$

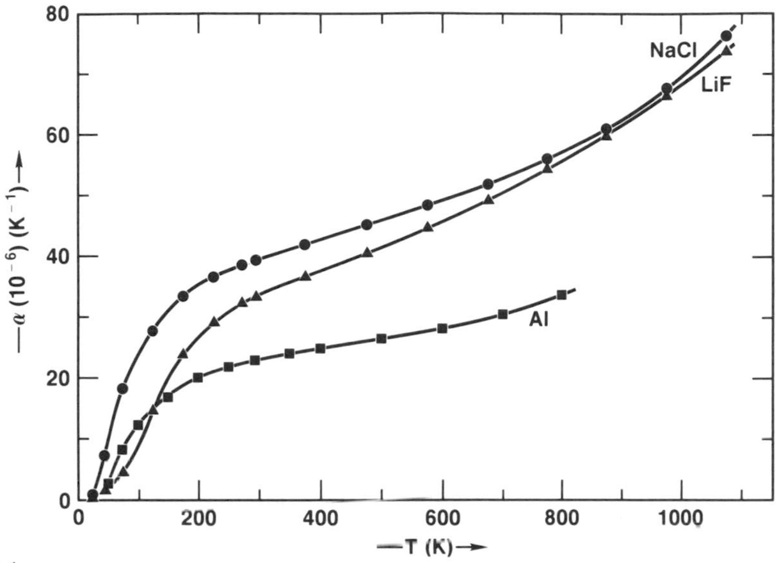

Figure 2.4

Coefficients of thermal expansion as a function of temperature for NaCl, LiF, and Al. Note the upward concavity of the curves at higher temperatures. (Data from *American Institute of Physics Handbook*, McGraw-Hill, New York, pp. 4.120–4.139, 1972.)

which is derived as follows:

$$C_P = \left(\frac{\partial H}{\partial T}\right)_P,$$

and

$$\left(\frac{\partial C_P}{\partial P}\right)_T = \frac{\partial}{\partial P}\left[\left(\frac{\partial H}{\partial T}\right)_P\right]_T = \frac{\partial}{\partial T}\left[\left(\frac{\partial H}{\partial P}\right)_T\right]_P.$$

Substituting for $(\partial H/\partial P)_T$ using (2.23) and then differentiating with respect to T gives us (2.33).

Substituting for V from (2.28) and integrating yields

$$C_P = C_P^\circ - V_0 T\left(\frac{d^2 a_0}{dT^2} P - \frac{1}{2}\frac{d^2 a_1}{dT^2} P^2 + \frac{1}{3}\frac{d^2 a_2}{dT^2} P^3 \ldots\right). \tag{2.34}$$

The term C_P° is the heat capacity at zero pressure, and if a_0, a_1, and a_2 can be approximated by a linear function, the second derivatives vanish. $d^2 a_0/dT^2$ is essentially the temperature coefficient of the thermal expansivity. As vibrations other than the zero-point vibrations cease at 0 K, the thermal expansivity must also go to zero. As the temperature is increased, it rises first slowly, becoming more rapid with increasing temperature, and becoming nearly linear in T in the highest temperature range (see Fig. 2.5). Consequently, as d^2a_0/dT^2 is the dominant term in (2.34), we expect C_P to decrease with increasing applied pressure.

We will now derive the effect of pressure upon C_V, the heat capacity at constant volume. Because E, the internal energy, is the appropriate state function for constant-volume processes, we write the first law of thermodynamics as

$$dE = T\,dS - P\,dV, \tag{2.2}$$

and so

$$\left(\frac{\partial E}{\partial V}\right)_T = T\left(\frac{\partial S}{\partial V}\right)_T - P = T\left(\frac{\partial P}{\partial T}\right)_V - P. \tag{2.11}$$

Thus,

$$\frac{\partial^2 E}{\partial V\,\partial T} = \left(\frac{\partial C_V}{\partial V}\right)_T = T\,\frac{\partial^2 P}{\partial T^2}. \tag{2.35}$$

since C_V is defined as

$$\left(\frac{\partial E}{\partial T}\right)_V.$$

Now, substituting for P in (2.35) from (2.27) and integrating gives

$$\int_{C_V^\circ}^{C_V} dC_V = -T\int\left[\frac{d^2P_0}{dT^2} + \frac{d^2P_1}{dT^2}\left(\frac{V_0 - V}{V_0}\right) + \frac{d^2P_2}{dT^2}\left(\frac{V_0 - V}{V_0}\right)^2 + \cdots\right] d\left(\frac{V_0 - V}{V_0}\right),$$

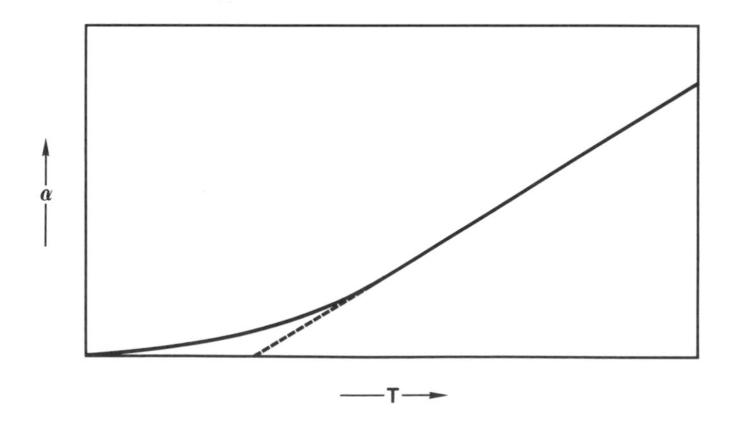

Figure 2.5.

The general form of the temperature dependence of the thermal expansion for a typical solid at very low temperatures.

and thus,

$$C_V = C_V^\circ - V_0 T\left[\frac{d^2P_0}{dT^2}\left(\frac{V_0 - V}{V_0}\right) + \frac{1}{2}\frac{d^2P_1}{dT^2}\left(\frac{V_0 - V}{V_0}\right)^2 + \frac{1}{3}\frac{d^2P_2}{dT^2}\left(\frac{V_0 - V}{V_0}\right)^3 + \cdots\right]. \tag{2.37}$$

It is easier to perform statistical thermodynamic calculations at constant volume, whereas experimental measurements are much more readily done at constant pressure. Consequently, it is extremely useful to be able to relate a measured value of C_P to the corresponding theoretical calculation of C_V. This can be done exactly by first writing the total differential for the entropy at constant volume and constant pressure and then multiplying each by the temperature yielding:

$$T\,dS = T\left(\frac{\partial S}{\partial T}\right)_V dT + T\left(\frac{\partial S}{\partial V}\right)_T = C_V\,dT + T\left(\frac{\partial P}{\partial T}\right)_V dV$$

and

$$T\,dS = T\left(\frac{\partial S}{\partial T}\right)_P dT + T\left(\frac{\partial S}{\partial P}\right)_T dP = C_P\,dT - T\left(\frac{\partial V}{\partial T}\right)_P dP,$$

and subtraction yields

$$(C_P - C_V)\,dT = T\left(\frac{\partial P}{\partial T}\right)_V dV + T\left(\frac{\partial V}{\partial T}\right)_P dP,$$

and at either constant temperature or at constant volume

$$C_P - C_V = T\left(\frac{\partial P}{\partial T}\right)_V\left(\frac{\partial V}{\partial T}\right)_P. \tag{2.38}$$

Using now our two equations of state we have, from (2.27),

$$\left(\frac{\partial P}{\partial T}\right)_V = \left(\frac{\partial P_0}{\partial T}\right)_V - P_1\left[\frac{\partial(V/V_0)}{\partial T}\right]_V + \left(\frac{V_0 - V}{V_0}\right)\left(\frac{\partial P_1}{\partial T}\right)_V + \cdots.$$

Setting $V = V_0$ and neglecting the last term gives

$$\left(\frac{\partial P}{\partial T}\right)_{V_0} \cong \left(\frac{\partial P_0}{\partial T}\right)_V = \frac{\partial(a_0/a_1)}{\partial T}. \tag{2.39}$$

Similarly, at constant pressure

$$\left(\frac{\partial V}{\partial T}\right)_P = V_0\left[\left(\frac{\partial a_0}{\partial T}\right)_P - \left(\frac{\partial a_1}{\partial T}\right)_P P + \left(\frac{\partial a_2}{\partial T}\right)_P P^2 + \cdots\right],$$

and at $P = 0$,

$$\left(\frac{\partial V}{\partial T}\right)_{P=0} = V_0\left(\frac{\partial a_0}{\partial T}\right)_{P=0} \cong \alpha V_0. \tag{2.40}$$

Substituting into (2.38), one arrives at a remarkably convenient formula:

$$C_P^\circ - C_V^\circ = \frac{TV_0\alpha^2}{a_1} \cong \frac{TV_0\alpha^2}{\chi}, \tag{2.41}$$

since, from (2.26),

$$-\frac{1}{V}\left(\frac{\partial V}{\partial P}\right)_T = \chi \simeq a_1.$$

Thus, if one knows either the heat capacity at constant volume or pressure, the unknown quantity may be calculated, providing the thermal expansivity, compressibility, and molar volume are available.

4. The Effect of Pressure Upon *S*, *E*, and *G*

The effect of pressure upon other thermodynamic state functions proceeds in a manner analogous to the preceding section. The total differential is first written and then use is made of thermodynamic identities expressed strictly in terms of the measurable quantities *T*, *P*, and *V*.

Commencing with the entropy and substituting from (2.37), we obtain

$$S = \int_0^T C_V \frac{dT}{T} = \int_0^T C_V^\circ \, d \ln T$$

$$- \int_0^T V_0\left[\frac{d^2P_0}{dT^2}\left(\frac{V_0 - V}{V_0}\right) + \frac{1}{2}\frac{d^2P_1}{dT^2}\left(\frac{V_0 - V}{V_0}\right)^2 + \frac{1}{3}\frac{d^2P_2}{dT^2}\left(\frac{V_0 - V}{V_0}\right)^3\right] dT$$

$$= C_V^\circ \ln T - V_0\left[\frac{dP_0}{dT}\left(\frac{V_0 - V}{V_0}\right) + \frac{1}{2}\frac{dP_1}{dT}\left(\frac{V_0 - V}{V_0}\right) + \frac{1}{3}\frac{dP_2}{dT}\left(\frac{V_0 - V}{V_0}\right)\right]. \tag{2.42}$$

Turning next to the internal energy, applying

$$\left(\frac{\partial E}{\partial V}\right)_T = T\left(\frac{\partial P}{\partial T}\right)_V - P, \tag{2.11}$$

substituting for P from (2.27), and integrating gives

$$E = E_{00} + \int_0^T C_V^\circ \, dT - V_0\left[\left(T\frac{dP_0}{dT} - P_0\right)\left(\frac{V_0 - V}{V_0}\right) + \frac{1}{2}\left(T\frac{dP_1}{dT} - P_1\right)\left(\frac{V_0 - V}{V_0}\right)^2 + \frac{1}{3}\left(T\frac{dP_2}{dT} - P_2\left(\frac{V_0 - V}{V_0}\right)^3\right]. \quad (2.43)$$

The Helmholtz free energy is defined by

$$F = E - TS,$$

and we can obtain F in terms of P, T, and V, using the previously derived functions for E and S. However, a more compact expression is obtained if we start with

$$dF = \left(\frac{\partial F}{\partial T}\right)_V dT + \left(\frac{\partial F}{\partial V}\right)_T dV = -S\, dT - P\, dV. \quad (2.5)$$

Equation (2.5) tells us to raise the temperature at constant volume, $V = V_0$, thus increasing the entropy, after which we hold the temperature constant at its new value while allowing the volume to expand to its equilibrium value. As

$$S = S_0 + \int_0^T C_{V_0} \, d \ln T$$

and $S_0 = 0$ for an ordered solid at $T = 0$ and $C_{V_0} = C_V^\circ$, we are allowed to write

$$F = E_{00} - \int_0^T \left[\int_0^T C_V^\circ \, d \ln T\right] dT - \int_{P_0}^P P\, dV, \quad (2.44)$$

where E_{00} is the internal energy at V_0 and 0 K, which is the binding energy. Substituting for P from (2.27) and writing

$$dV = -V_0\, d\left(\frac{V_0 - V}{V_0}\right)$$

leads to

$$-\int_{V_0}^V P\, dV = V_0\left[P_0\left(\frac{V_0 - V}{V_0}\right) + \frac{1}{2}P_1\left(\frac{V_0 - V}{V_0}\right)^2 + \frac{1}{3}P_2\left(\frac{V_0 - V}{V_0}\right)^3\right],$$

and (2.44) becomes

$$F = E_{00} - \int_0^T \left[\int_0^T C_V^\circ \, d \ln T\right] dT + V_0\left[P_0\left(\frac{V_0 - V}{V_0}\right) + \frac{1}{2}P_1\left(\frac{V_0 - V}{V_0}\right)^2 + \frac{1}{3}P_2\left(\frac{V_0 - V}{V_0}\right)^3\right]. \quad (2.45)$$

An experimental example is shown in Fig. 2.6 for the variation of F with V for sodium at several constant temperatures. These results were obtained from experimental data using (2.45). Note that basic thermodynamics, as given by (2.11) at constant temperature, tells us that the negative slopes of the curves in this figure are the pressures, and the minima correspond to the volumes where the pressure is zero. The minima move to higher volumes at the higher temperatures, as expected, from thermal expansion. The negative pressures to the right of the minima have no physical significance.

For our last example we derive the dependence of the Gibbs free energy, G, upon P, V, and T. Although G is defined in terms of quantities for which we have already

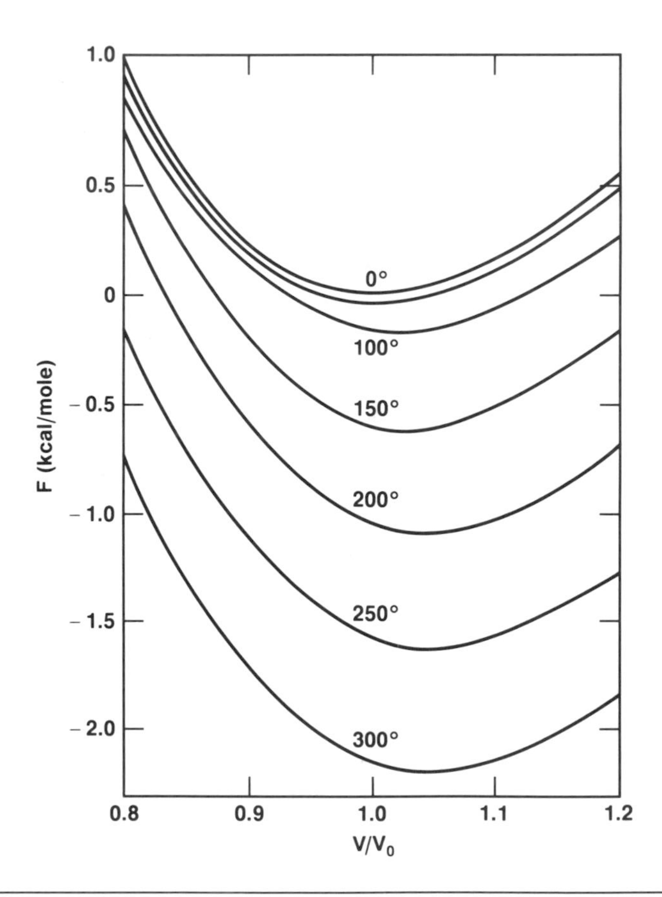

Figure 2.6.

Helmholtz free energy of solid sodium as a function of the volume at constant temperature. (After Slater, loc. cit.)

found expressions in terms of our equations of state, viz.,

$$G = F + PV,$$

it is more convenient again to begin with

$$dG = -S\,dT + V\,dP. \tag{2.18}$$

Remembering that $G = E_{00}$ at 0 K and $P = 0$, we substitute (2.24) for V and obtain

$$\begin{aligned} G &= E_{00} - \int_0^T \left[\int_0^T C_P^\circ \, d\ln T \right] dT + \int_0^P \left[V_0 \left(1 + a_0 - a_1 P + a_2 P^2\right) \right] dP \\ &= E_{00} - \int_0^T \left[\int_0^T C_P^\circ \, d\ln T \right] dT + PV_0 \left(1 + a_0 - \frac{a_1}{2} P + \frac{a_2}{3} P^2 \right). \end{aligned} \tag{2.46}$$

The sodium data treated this way are shown in Fig. 2.7. On such a plot, G is very nearly linear in P.

We have now completed our transformation of P and V, as supplied by our empirical equations of state, into useful thermodynamic functions. Our next task is to examine some of the more frequently employed semiempirical equations of state.

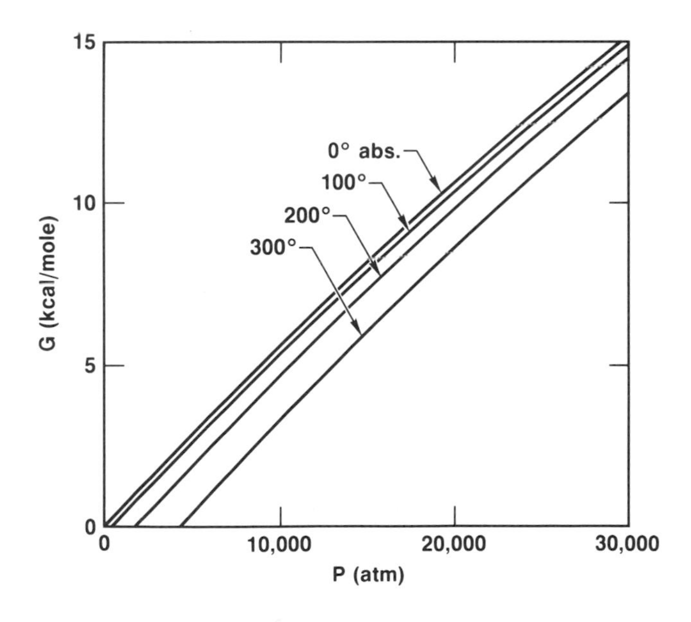

Figure 2.7.

Gibbs free energy of a solid (sodium) as a function of the pressure at constant temperature. (After Slater, loc. cit.)

5. Matter at Extremely High Temperatures and Pressures

Although the diamond-anvil cell, as previously described (Fig. 2.2) can attain pressures in the vicinity of 1 Mbar ($\sim 10^6$ atm), chemical explosives can produce much higher pressures, up to 13 Mbar. Such values significantly exceed that at the earth's center, which is about 3.5 Mbar, but fall far below the $\sim$100 Mbar believed to exist at the center of Jupiter. In this section we examine the methods by which equations of state for matter at extremely high temperatures and pressures are deduced from shock wave experiments. Under these extreme conditions, solids tend to behave more like fluids, and the noncommittal term "matter" is customarily employed without specification of its state.

The P-V relations obtained by shock methods are not directly comparable to those derived from the more conventional static methods. Because the shock process is essentially adiabatic, temperatures in excess of several thousands of degrees are reached, well above the melting temperatures of all solids. Consequently, equations of state are required to bring the results of these different techniques into the same P-V-T region.

A. The Hugoniot Equations

In deriving the fundamental equations that ultimately give us the pressure and volume of the shock-compressed state, let us refer first to Fig. 2.8, which will assist in the visualization of the physical parameters. Consider a solid (or liquid) initially at rest with a density ρ_0 and a pressure P_0, confined within a cylinder and contacted by a piston of cross section A. The solid vertical line in this figure defines the shock front.

If the piston is propelled toward the right, a shock wave is generated, which also travels to the right, and behind which the material is compressed at a constant velocity up to the piston velocity. In fact, a series of shock waves is generated. The first wave is generated in the initially uncompressed material, but, with continuing compression,

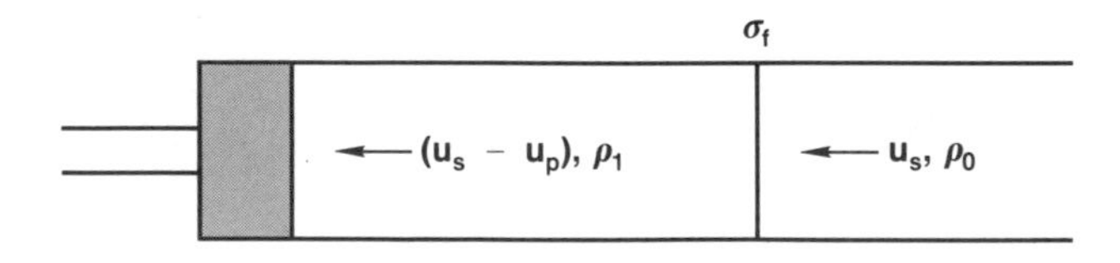

Figure 2.8.

Steady shock wave as viewed by an observer on the shock front. The fluid streams toward the shock at a velocity u_s and rushes to the left away at the velocity $u_s - u_P$.

subsequent shocks of higher velocity form, which overtake their predecessors but cannot overtake them into the, as yet, uncompressed solid. This is because the continuing compression creates material of increasing density with an accompanying increase in shock velocity. This is explained diagrammatically in Fig. 2.9.

This train of waves eventually coalesces into a single sharp wavefront, a few interatomic spacings in width, that proceeds with the speed of sound. If we center our coordinates on the shock front so that it remains stationary with respect to the laboratory frame of reference, then matter in the state ρ_0 enters the shock front from right to left with a velocity u_s and exits in the state ρ_1 with a velocity $u_s - u_P$ traveling in the same direction. Conservation of mass now requires

$$A\rho_0 u_s \delta t = A\rho_1(u_s - u_P)\delta t, \tag{2.47}$$

which, after substitution of $V = \rho^{-1}$, leads to

$$\frac{\rho_0}{\rho_1} = 1 - \frac{u_p}{u_s} = \frac{V}{V_0}. \tag{2.48}$$

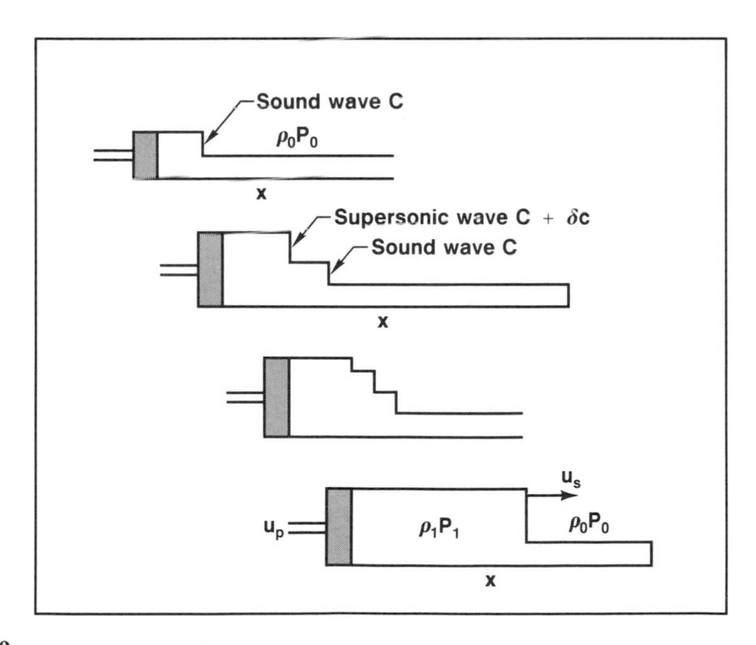

Figure 2.9.

Formation of a one-dimensional planar shock wave, shown by pressure–density–distance *P*-*x* plots at successive times. Supersonic refers to the shock velocity in the compressed region. (Taken from M. Ross, "Physics of Dense Fluids," in *High Pressure Chemistry and Biochemistry*, D. Reidel, 1987.)

The pressure differential across the shock front is now determined from the conservation of momentum. The momentum flow from left to right of the unshocked state ρ_0 through the shock front is $(A\rho_0 u_s \delta t)u_s$, and the momentum flow away from it is given by $A\rho_1(u_s - u_P)\delta t(u_s - u_P)$. Using (2.48), we obtain $A\rho_0 u_s \delta t(u_s - u_P)$. The change in momentum must equal the difference in force across the front, which, in terms of pressure, is $(P_1 - P_0)A\delta t$. Thus,

$$A\rho_0 u_s \delta t u_s - A\rho_0 u_s \delta t(u_s - u_P) = (P_1 - P_0)A\delta t$$

or

$$P_1 - P_0 = \rho_0 u_s u_P = u_s u_P / V_0. \tag{2.49}$$

By an analogous line of reasoning applied to the conservation of energy, we arrive at

$$\rho_1(u_s - u_p)\left[E_1 + \frac{P_1}{\rho_1} + \frac{(u_s - u_P)^2}{2}\right] = \rho_0 u_s\left[E_0 + \frac{P_0}{\rho_0} + \frac{u_s^2}{2}\right], \tag{2.50}$$

and collecting terms, rearranging, and again substituting $1/\rho_0 = V_0$ and $1/\rho_1 = V_1$ leads to

$$E_1 - E_0 = (1/2)(P_1 + P_0)(V_0 - V_1). \tag{2.51}$$

Equations (2.48), (2.49), and (2.51) are called the Hugoniot equations. Experimental measurements of pressure versus density (Hugoniots) are plotted in Fig. 2.10 for several elements. Remembering that $H = E + PV$, we can rewrite (2.51) in terms of the enthalpy:

$$H_1 - H_0 = (1/2)(P_1 - P_0)(V_0 - V_1). \tag{2.52}$$

The unavoidable irreversibility of shock compression is illustrated in Fig. 2.11.

The Hugoniot line is constructed from a series of discrete, irreversible shock compressions. Although the final states are at equilibrium (because of the high rate at which the energy partitions itself among the available degrees of freedom) the irreversible work expended to achieve this state is necessarily greater than is required by the path S. However, the difference in the PV work between the reversible and irreversible paths is small except at very high pressures. Equations of state such as (2.27) will give the same value of P for either static or dynamic compression up to and including the quadratic term and the energy values are the same up to and including the cubic term.

Taking the experimental material to be of unit mass, we can combine (2.48) and (2.49), obtaining for the shock velocity

$$u_s^2 = V_0^2\left(\frac{P_1 - P_0}{V_0 - V_1}\right) \tag{2.53}$$

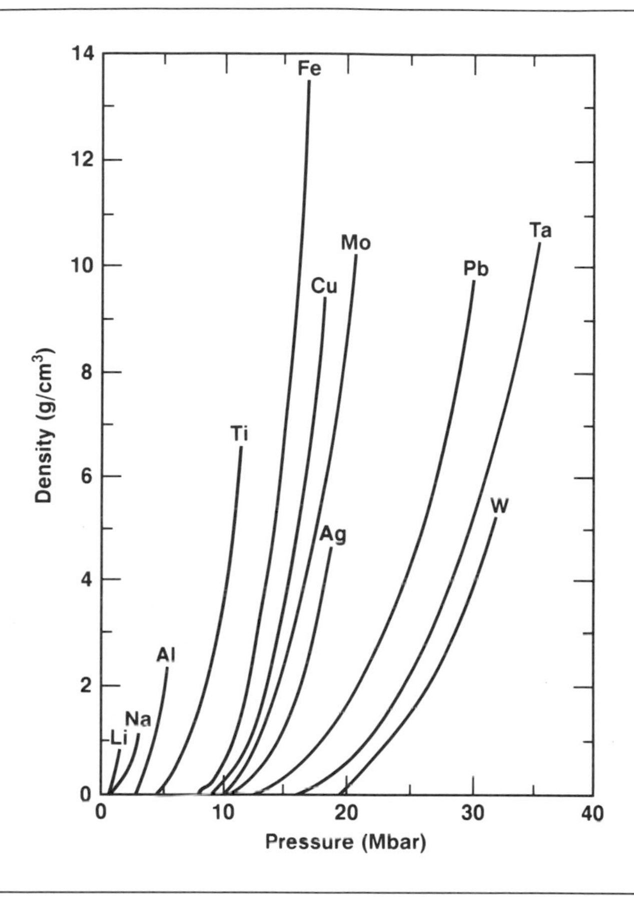

Figure 2.10.

The Hugoniots for various elements, which are the loci of all the states that can be reached by shock compression from a specified initial (P-V-T) state. (M. Ross, *High Pressure Chemistry and Biochemistry*, D. Reidel, pp. 9–49, 1987.)

and for the particle or piston velocity

$$u_p^2 = (V_0 - V_1)(P_1 - P_0). \tag{2.54}$$

However, (2.54) gives the total kinetic energy imparted to the compressed material, which in turn is equivalent to the irreversible energy given by the shaded area in Fig. 2.11. Thus, if $P_1 \gg P_0$, we may omit P_0 and write

$$E_1 - E_0 = (1/2)P_1(V_0 - V_1) = u_P^2/2. \tag{2.55}$$

However, the reader is once again reminded that the shock process is irreversible, and the change in E given by (2.55) is not the same as that obtained from a reversible path.

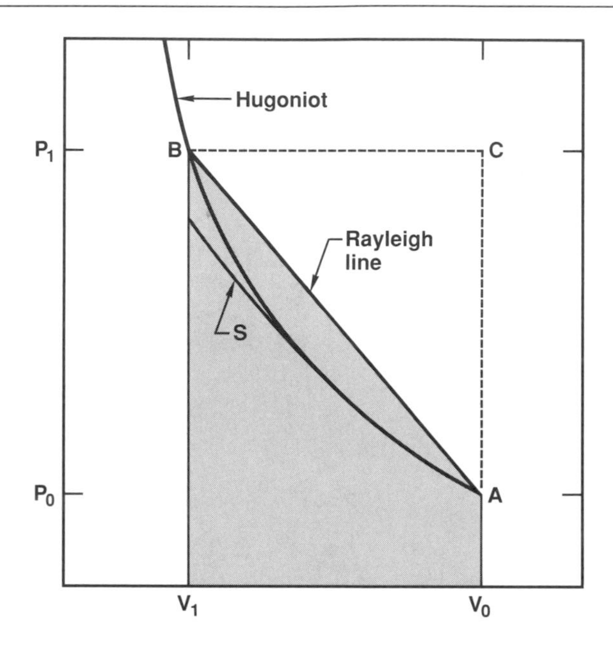

Figure 2.11.

The shaded area under the Rayleigh line corresponds to the change in internal energy in going from state A to state B. The area ABC above the Rayleigh line is the accompanying change in kinetic energy. The line S designates the reversible isentrope connecting V_0 with V_1.

B. Mie–Grüneisen Equation of State

A more complete treatment of semiempirical equations of state follows our discussions of vibrational properties and potential functions. However, for the sake of continuity, we present here the Mie–Grüneisen, without derivation, as it is the equation most frequently used to convert Hugoniots into valid thermodynamic data. Such calculations do not rest upon self-consistent theoretical foundations, but are tailored in a somewhat ad hoc manner to give the best fit to the experimental measurements. Generally, specific equations will apply to but a single class of materials, for example metals, ionic compounds, rare gases, molecular solids, etc., thus varying with the basic nature of the chemical bond. At sufficiently high temperatures the Mie–Grüneisen equation of state (2.56) offers a good approximation for both ionic and metallic substances, although this simple EOS takes into account only the vibrational contributions to the internal energy and omits electronic excitations.

At temperatures above the characteristic temperature, where the system behaves classically, E and P of a Mie–Grüneisen solid are given by

$$E - E_0 = E(V) - E(V_0) + 3k(T - T_0), \tag{2.56}$$

$$P - P_0 = P(V) - P(V_0) + 3k[T\gamma(V)/V - T_0\gamma(V_0)/V_0], \tag{2.57}$$

where the zero subscripts designate the initial values. In these equations the thermal pressure was separated out as the second term on the right-hand side of (2.57), where γ is the Grüneisen "constant," which in reality is generally a function of volume, but can often be approximated as a constant at high temperatures. A detailed discussion of the Grüneisen γ is given in the next chapter. The important point is that $E(V)$, $P(V)$, and $\gamma(V)$ are functions of the volume only. Combining (2.56) and (2.57) with the basic energy equation of the Hugoniot, (2.51), gives for the temperature

$$3k(T - T_0) = \frac{\frac{1}{2}ZV_0[P(V) + P(V_0)] - [E(V) - E(V_0)] + \frac{3}{2}kT_0Z[\gamma(V_0) + \gamma(V)/(1 - Z)]}{1 - \gamma(V)Z/2(1 - Z)} \tag{2.58}$$

where, for convenience, we put $Z = (V_0 - V)/V_0$. This equation gives the temperature reached in a shock wave experiment to the purely volume-dependent part of the equation of state and, hence, the static thermodynamic pressure. One can see right away another important point from (2.58), namely, that there is a limiting value to the compression given by

$$Z_C = \frac{1}{1 + \gamma(V_C)/2}, \tag{2.59}$$

since at this compression the denominator of (2.58) approaches zero and the temperature, therefore, approaches infinity. This "thermal catastrophe" results, of course, in an infinite thermal pressure. This thermal catastrophe is a consequence of the unbalance between the shock wave power input and that absorbed directly by the compression. The inability of the solid to respond rapidly enough to absorb all of the shock wave power leads to ever-increasing excess energy, resulting in an ever-increasing temperature and, hence, a thermal catastrophe. In the real world something happens, of course, before the temperature reaches infinity. Usually the solid undergoes a phase transition, most often that of melting.

There are a variety of Grüneisen constants that, although similar to one another, vary slightly depending on their derivation. The most direct way of calculating a γ is from

$$\gamma_{\text{th}} = \alpha V / x C_V,$$

as described in Chapter 3 and Appendix B. This so-called thermal gamma is a reasonably good approximation of the γ defined by (2.57), and its substitution into (2.58) provides a means for correcting the Hugoniot pressures.

More elaborate equations of state contain additional terms accounting for the electronic contributions. Nevertheless, (2.58), in principle, allows one to convert a Hugoniot pressure into the corresponding value of the true static thermodynamic pressure.

Chapter 2 Exercises

1. Analogous to the derivation of (2.10) and (2.22) derive the expressions

$$\left(\frac{\partial T}{\partial P}\right)_S = \left(\frac{\partial V}{\partial S}\right)_P, \qquad \left(\frac{\partial T}{\partial V}\right)_S = -\left(\frac{\partial P}{\partial S}\right)_V.$$

2. Starting with $G = H - TS$ and $(\partial H/\partial T)_P = C_P$, derive the following expressions:

$$\left[\frac{\partial(G/T)}{\partial T}\right]_P = -\frac{H}{T^2} \quad \text{and} \quad \left(\frac{\partial S}{\partial T}\right)_P = \frac{C_P}{T}.$$

3. Starting with (2.13) derive the expression

$$C_P = C_V = \left(\frac{\partial V}{\partial T}\right)_P \left[\left(\frac{\partial E}{\partial V}\right)_T + P\right]$$

and show that it is equivalent to (2.38).

4. Show that $(\partial H/\partial P)_V - (\partial H/\partial P)_T < 0$.

5. Explain in words and equations why an equation of state is necessary before one can establish an accurate temperature scale.

6. Explain why for any substance $C_P > C_V$.

7. Can polycrystalline substances be described by a single equation of state. Discuss in detail.

8. Show that at $T = 0$ K,

$$\left(\frac{\partial V}{\partial T}\right)_P = 0 \quad \text{and} \quad \left(\frac{\partial P}{\partial T}\right)_V = 0.$$

9. Estimate the change in the equilibrium melting point of Cu caused by increasing the pressure from 1.0 to 20.0 atm. The molar volume of molten Cu is 8.0 cm^3 and is 7.6 cm^3 for the solid at the melting temperature, which is 1085°C. What approximations are entailed in making this estimate?

10. Prove that $(\partial E/\partial P)_V = C_V\chi/\alpha$, where χ is the isothermal compressibility and α is the thermal expansivity.

11. The volume of a quartz crystal at 30°C is given by

$$V = V_0(1 - 2.658 \times 10^{-10}P + 24.4 \times 10^{-20}P^2),$$

where V_0 is the volume at $P = 0$, and P is given in kilograms per square meter.

a. How large a pressure would be required to reduce the volume by 1%?

b. Plot the compressibility at 30°C vs. P.

c. Calculate the work done on a 1-kg crystal in an isothermal reversible compression by 1%.

12. Calculate $C_P - C_V$ for Na if the compressibility is 12.3×10^{-12} cm^2/dyne and the linear coefficient of thermal expansion is 6.22×10^{-5}. Compare your answer with $C_P - C_V$ for a monoatomic gas. Also calculate the Grüneisen constant for Na.

Additional Reading

J. C. Slater, *Introduction to Chemical Physics*, Chaps. XIII, and XIV, McGraw-Hill, New York 1939. Probably still the best introduction to the subject of equations of state.

D. C. Wallace, *Thermodynamics of Crystals*, Chap. 1, Wiley, New York 1972.

Reviews on Shock Waves

M. H. Rice, R. G. McQueen, and J. M. Walsh, *Solid State Physics* (F. Seitz and D. Turnbull, eds.), 6, 1–63, Academic Press, New York 1958.

D. G. Doran and R. K. Linde, *Solid State Physics* (F. Seitz and D. Turnbull, eds.), Vol. 19, pp. 230–290, Academic Press, New York 1966.

M. Ross, *High Pressure Chemistry and Biochemistry*, D. Reidel, pp. 9–49 1987.

Chapter 3

The Vibrational Properties of Solids

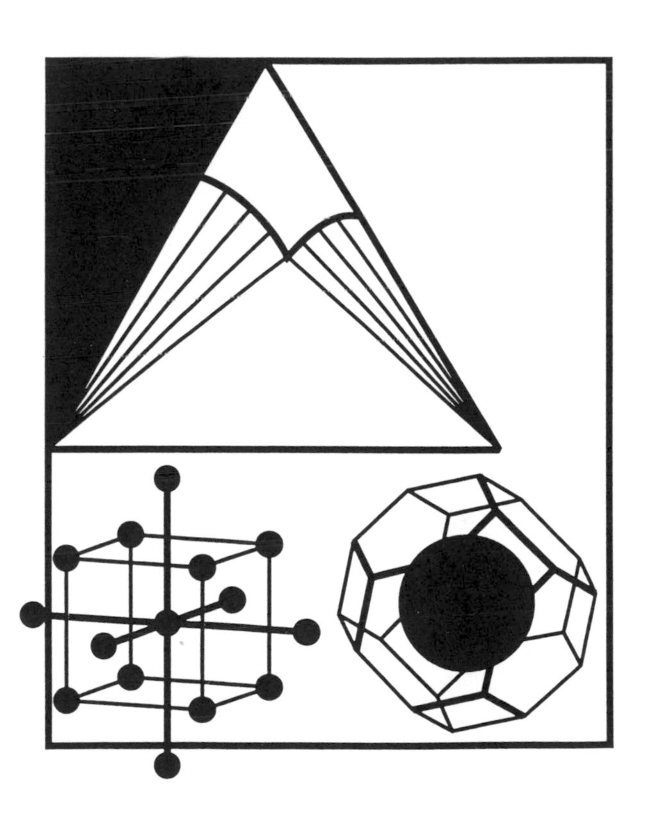

Chapter 3

Contents

Nothing puzzles me more than time and space, and yet nothing troubles me less as I never think about them.

—Charles Lamb (circa 1820)

The Vibrational Properties of Solids

The collective interplay of the attractive and repulsive forces between the atoms or molecules in a crystal leads to a state of equilibrium in which each occupies a definite position with no, time-averaged, net force acting upon it. At temperatures greater than absolute zero, the atoms will vibrate about the equilibrium positions, and thermal expansion will cause displacements of these positions themselves. These thermally induced vibrations in the main are responsible for the heat capacity, as well as the thermal expansivity, and hence are an important component of the equation of state.

1. The Grüneisen Constant

In the preceding chapter we introduced, without derivation, the Grüneisen equation of state to show how the temperature of a shock-compressed solid might be calculated. Here, we begin our discourse on vibrational properties by deriving the Grüneisen gamma.

Our thermodynamic system is a crystal consisting of N particles held together by elastic forces. This collection has $3N - 6$ vibrational degrees of freedom. However, as $N \sim 10^{20}$, we can safely discard the 6 and consider the system to have exactly $3N$ degrees of freedom. Our model assumes the vibrations to be caused by the simple harmonic oscillation of the atoms. Hence, the restoring forces are linear, and the frequency is independent of the amplitude. The vibrations are quantized and take energy values according to $(n_j + 1/2)h\nu_j$, where n_j is the quantum number associated with the jth oscillator and the 1/2 accounts for the zero-point energy. Thus, the eigenvalues for the frequency ν_j, and hence the vibrational energy are

$$(1/2)h\nu_j,\ (3/2)h\nu_j,\ (5/2)h\nu_j, \ldots.$$

Thus, for a single oscillator, summing over the $3N$ frequencies gives

$$\varepsilon_{n_1} + \varepsilon_{n_2} + \cdots + \varepsilon_{3N} = \sum_{j=1}^{3N}\left(n_j + \frac{1}{2}\right)h\nu_j + \varepsilon_0, \tag{3.1}$$

where ε_0 is the binding energy at absolute zero temperature. The binding energy is defined as the energy necessary to break all the bonds between the atoms, ions, or molecules in the solid and to remove them to infinite separation from one another.

Because all attractive interatomic forces vary as r^{-n} at large separations, where n varies with the type of interactive potential (i.e., ionic, dipolar, etc.), infinite separation defines a standard state of no interparticle forces. The ε_0 term can be combined with the zero-point vibrational[1] term to give a temperature-independent energy term ε_{00}:

$$\varepsilon_{00} \equiv \varepsilon_0 + \sum_{j=1}^{3N}\frac{h\nu_j}{2}. \tag{3.2}$$

The vibrational partition function (p.f.) is formed by summing the $3N$ frequencies over the $3N$ quantum numbers. This double sum is

$$(\text{p.f.}) = \exp\left(-\frac{F}{kT}\right) = e^{-\varepsilon_{00}/kT}\sum_{n_1}\cdots\sum_{n_{3N}}\exp\left(-\sum_j\frac{n_jh\nu_j}{kT}\right). \tag{3.3}$$

Because each sum in (3.3) can be regarded as the sum of a geometric series,

$$\begin{aligned}\sum_{j=1}^{3N}\exp\left(-\sum_j\frac{n_jh\nu_j}{kT}\right) &= (1 + e^{-h\nu_j/kT} + e^{-2h\nu_j/kT} + \cdots)\\ &= (1 + e^{-h\nu_j/kT} + (e^{-h\nu_j/kT})^2 + \cdots)\\ &= \frac{1}{1 - e^{-h\nu_j/kT}}.\end{aligned} \tag{3.4}$$

Thus, we obtain

$$(\text{p.f.}) = e^{-\varepsilon_{00}/kT}\prod_{j=1}^{3N}\frac{1}{1 - e^{-h\nu_j/kT}}, \tag{3.5}$$

---

[1] See Appendix C for the derivation of the zero-point vibrational energy of a harmonic oscillator.

and from (3.3) the Helmholtz free energy is given by

$$F = \varepsilon_{00} + \sum_{j=1}^{3N} kT \ln(1 - e^{-h\nu_j/kT}). \tag{3.6}$$

We now proceed with the calculation of the Grüneisen gamma by solving for the pressure. We reintroduce the first two 0 K terms from (2.27) and add the temperature dependence by differentiating (3.6) with respect to volume. Thus,

$$-\left(\frac{\partial F}{\partial V}\right) = P = P_1^\circ\left(\frac{V_0 - V}{V_0}\right) + P_2^\circ\left(\frac{V_0 - V}{V_0}\right)^2 + \frac{1}{V}\sum_{j=1}^{3N} \gamma_j \frac{h\nu_j}{e^{h\nu_j/kT} - 1}. \tag{3.7}$$

Here γ_j is implicitly defined by

$$\gamma_j = -\frac{V}{\nu_j}\left(\frac{\partial \nu_j}{\partial V}\right)_T = -\left(\frac{\partial \ln \nu_j}{\partial \ln V}\right)_T. \tag{3.8}$$

Consequently, the summation in (3.7) represents the thermal pressure, and differs from zero only because of the volume dependence of the vibrational frequencies. This is intuitively appealing in that if, contrary to fact, the frequencies were volume-independent, it would mean that the interatomic forces must also be independent of volume and hence would not vary with the interatomic spacing. Consequently, the compressibility would be a volume-independent constant, which clearly it is not.

At the high temperature limit

$$\lim_{T\to\infty}\left[\frac{h\nu_j}{e^{h\nu_j/kT} - 1}\right] = h\nu_j\left[\frac{1}{1 + h\nu_j/kT + \cdots - 1}\right] = kT. \tag{3.9}$$

Because there are $3N$ such terms in (3.7), we must multiply the last term by this factor, thus obtaining in the classical limit

$$P = P_1^\circ\left(\frac{V_0 - V}{V_0}\right) + P_2^\circ\left(\frac{V_0 - V}{V_0}\right)^2 + \frac{3N\gamma kT}{V}. \tag{3.10}$$

Equation (3.10) embodies Grüneisen's major approximation, viz. that γ is a single constant for each element or chemical compound. The last term in (3.10) has been previously used to correct the pressure derived from shock wave measurements for the temperature effect, viz. (2.57), and demonstrates a fair degree of accuracy in most cases. Another equivalent form of (3.10) is

$$P = -\left(\frac{\partial \varepsilon_{00}}{\partial V}\right) + \frac{3\gamma}{V} RT, \tag{3.11}$$

which also, generally, is called the Mie–Grüneisen equation of state.

With (3.6) as the starting point, we can obtain other thermodynamic state functions by performing the appropriate standard operations. The entropy is derived by taking the derivative with respect to temperature as

$$\left(\frac{\partial F}{\partial T}\right)_V = -S = k \sum_{j=1}^{3N} \left[\ln(1 - e^{-h\nu_j/kT}) - \frac{h\nu_j}{e^{h\nu_j/kT} - 1} \right], \tag{3.12}$$

or

$$S = k \sum_{j=1}^{3N} - \left[\ln(1 - e^{-h\nu_j/kT}) + \frac{h\nu_j}{e^{h\nu_j/kT} - 1} \right].$$

The internal energy is given by

$$E = F + TS,$$

and substituting from (3.6) and (3.12) for F and S, respectively, yields

$$E = \varepsilon_{00} + \sum_{j=1}^{3N} \frac{h\nu_j}{e^{h\nu_j/kT} - 1}. \tag{3.13}$$

An approximate expression for the thermal expansion can now be obtained, starting with (2.25) of Chapter 2, and using (2.31), as

$$\alpha = \left(\frac{\partial a_0}{\partial T}\right)_P \simeq \frac{d(P_0/P_1)}{dT} \simeq \chi^{-1} \frac{dP_0}{dT}. \tag{2.25}$$

We have seen that the compressibility $P_1^{-1} \simeq \chi$. Thus, with this additional approximation we substitute for P in (3.7); thus, since $V_0 = V$ when $P = P_0$, the terms containing $V_0 - V$ vanish, leaving us with

$$P_0 = \frac{1}{V_0} \sum_{j=1}^{3N} \gamma_j \frac{h\nu_j}{e^{h\nu_j/kT} - 1}. \tag{3.14}$$

Now, taking the derivative with respect to temperature gives

$$\alpha = \frac{\chi k}{V_0} \sum_{j=1}^{3N} \gamma_j \left(\frac{h\nu_j}{kT}\right)^2 e^{h\nu_j/kT} (e^{h\nu_j/kT} - 1)^{-2}. \tag{3.15}$$

The quantity under the summation, after multiplying by k and omitting γ_j, is C_V, as is easily demonstrated by taking the derivative of (3.13) at constant volume. Hence, making use of Grüneisen's approximation, we obtain

$$\alpha = \gamma \chi C_V / V_0, \tag{3.16}$$

a relation important in hydrodynamics.

The γ derived from this particular approximation is usually referred to as the *thermodynamic* γ. We now have a relation that allows us to determine γ from the measured values of the thermal expansivity, compressibility, and heat capacity. The experimental values of γ are generally in the range $1 \leq \gamma \leq 3$. There are other useful expressions for γ that, however, need not be numerically identical. These are discussed in Appendix B.

2. The Einstein Heat Capacity

The simplest statistical mechanical treatment of the vibrational heat capacity is furnished by the Einstein model. Unlike our model in the previous section, all frequencies are taken to be the same, and our single summation is over the quantum numbers only. As before, the energy levels are quantized according to

$$\varepsilon_i = (n_i + 1/2)h\nu,$$

and n_i accepts integer values running from zero to infinity. Consequently, the vibrational partition function is given by

$$\begin{aligned}
(\text{p.f.}) &= \sum_{n=0}^{\infty} e^{-(n+1/2)h\nu/kT} \\
&= e^{-h\nu/2kT}(1 + e^{-h\nu/kT} + e^{-2h\nu/kT} + \cdots) \\
&= {}^{-h\nu/2kT}(1 + e^{-h\nu/kT} + (e^{-h\nu/kT})^2 + \cdots) \\
&= \frac{e^{-h\nu/2kT}}{1 - e^{-h\nu/kT}}.
\end{aligned} \tag{3.17}$$

The vibrational internal energy for N localized oscillators is given by

$$E_v = \frac{3N \sum_0^{\infty} (n + 1/2)h\nu e^{-(n+1/2)h\nu/kT}}{\sum_0^{\infty} e^{-(n+1/2)h\nu/kT}}. \tag{3.18}$$

Dividing and multiplying (3.18) by $e^{h\nu/2kT}$ yields

$$\begin{aligned}
E_v &= \frac{3}{2h\nu} + \frac{3N \sum_0^{\infty} nh\nu e^{-nh\nu/kT}}{\sum_0^{\infty} e^{-nh\nu/kT}} \\
&= \frac{3}{2h\nu} + 3kT^2N \frac{\partial \ln}{\partial T} \sum_0^{\infty} e^{-nh\nu/kT} \\
&= \frac{3}{2Nh\nu} - 3NkT^2 \frac{\partial \ln}{\partial T} (1 - e^{-h\nu/kT}) \\
&= \frac{3}{2Nh\nu} + \frac{3Nh\nu}{e^{h\nu/kT} - 1},
\end{aligned} \tag{3.19}$$

where the temperature-independent zero-point energy is $(3/2)Nh\nu$. We now apply the identity

$$\left(\frac{\partial E}{\partial T}\right)_V = C_V$$

to (3.19) and obtain for the purely vibrational component of the heat capacity

$$C_V = 3Nk\left(\frac{h\nu}{kT}\right)^2 \frac{e^{h\nu/kT}}{(e^{h\nu/kT} - 1)^2}. \tag{3.20}$$

This expression is seen to be identical to the summation in (3.15), omitting γ_j and giving all ν_i a single fixed value.

The Einstein heat capacity works best for relatively high values of T/T_m, T_m being the melting temperature of the substance in question. By expanding $e^{h\nu/kT}$, we can show that, to first order, (3.19) passes to the classical limit of E_ν and C_V,

$$\begin{aligned} E_\nu(T) &= 3N \frac{h\nu}{(1 + h\nu/kT + \cdots) - 1} \\ &\simeq 3NkT = 3RT \end{aligned} \tag{3.21}$$

and again, by taking the derivative of (3.21) with respect to T at constant volume, we obtain $C_V = 3R$ as expected. It is clear that this approximation is most suitable at high temperatures, where the neglect of higher-order terms in the expansion of $e^{h\nu/kT}$ causes a smaller error. Thus, we see that at high temperatures the specific heat becomes independent of temperature, and the same is true for all solids. This is in agreement with the experimental law of Dulong and Petit, which states that many elemental crystals have a constant specific heat of about 6 cal deg^{-1} mol^{-1}, which is $\sim 3R$, as predicted by (3.21). However, the Einstein model takes no account of the electronic specific heat. Though generally small, it is nevertheless significant at temperatures sufficiently high to promote electrons into more energetic quantum states, and it dominates at low temperatures because of the rapid decline of the vibrational contribution.

Equation (3.20) is frequently written in terms of the so-called characteristic temperature, $\Theta_E = h\nu/k$, appearing then as

$$C_V = 3R\left(\frac{\Theta_E}{T}\right)^2 \frac{e^{\Theta_E/T}}{(e^{\Theta_E/T} - 1)^2}. \tag{3.22}$$

The quantity Θ_E can be obtained from experiment and, combined with (3.22), provides a useful method for extrapolating (or interpolating) into regions for which there are no experimental data. It must be remembered that this will almost always give erroneous results at low temperatures, but the accuracy of the theory, in view

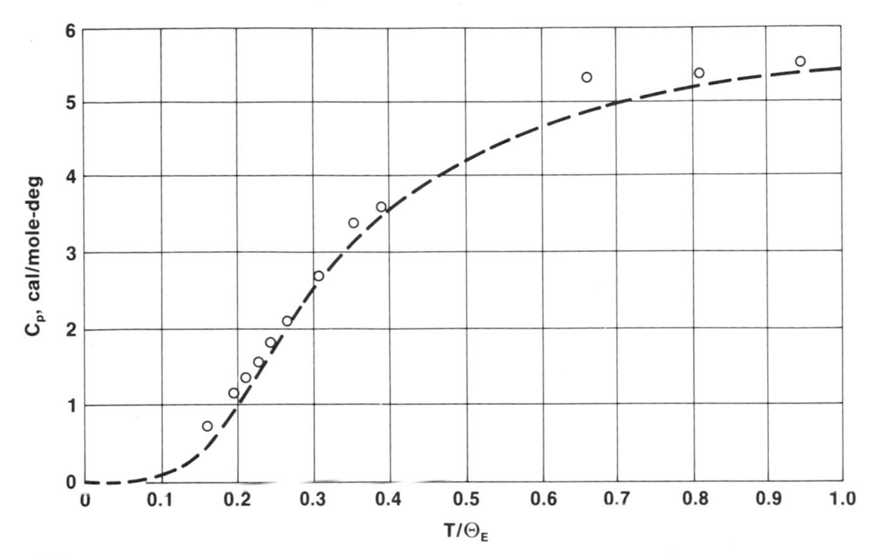

Figure 3.1.

Comparison of experimental values of the heat capacity of diamond with values calculated by the Einstein model, using the characteristic temperature $\Theta_E = h\omega/k_B = 1320$ K. [After A. Einstein, *Ann. Phys.*, **22**, 180 (1907).]

of its simplicity, is remarkable, as shown by the example presented by Einstein himself (see Fig. 3.1).

If we do not make all vibrations equal in frequency but, in a more realistic fashion, retain the spectrum and hence the summation, we then have

$$C_V = R \sum_j \left(\frac{\Theta_j}{T}\right)^2 \frac{e^{\Theta_j/T}}{(e^{\Theta_j/T} - 1)^2}, \tag{3.23}$$

and now θ_j may be thought of as the characteristic temperature $h\nu_j/k$ associated with each value of ν_j. We will expand on this concept in the next two sections, culminating with the Debye theory of the heat capacity.

3. Normal Vibrational Modes

The model developed by Debye for the heat capacity is more sophisticated than the Einstein one in that it maintains a finite distribution of frequencies rather than assigning the same value to all. This is illustrated graphically in Fig. 3.2.

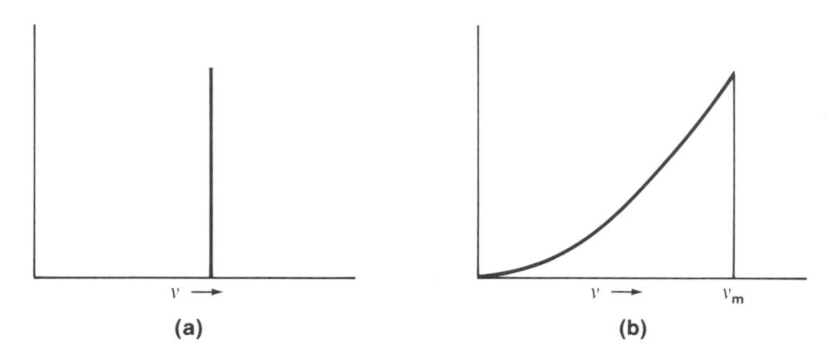

Figure 3.2.

The frequency distributions N_ν assumed for the Einstein heat capacity (a) and for the Debye heat capacity (b).

Before presenting the Debye theory, let us examine in some detail how vibrational waves are produced and how their frequencies are a function of the interatomic forces and crystal dimensions. The following discussion is simplified by considering only crystal structures of cubic symmetry. Treatment of crystals of lower symmetry adds nothing to our understanding of the physics of the model but greatly complicates the mathematics.

Let us commence with a crystal containing N atoms and, thus, having a total of $3N$ different values for the allowable frequencies. All of the $3N$ separate frequencies vibrate simultaneously so that the instantaneous displacement of each atom is the resultant of the vector sum of the displacements accruing from the superposition of $3N$ waves at a lattice point. Despite the collective action produced by the superposition, each fundamental harmonic can be taken to be a single, separately quantized oscillator. Since $E(\nu) = h\nu$, the oscillators of lower frequency contain less energy and will attain their respective limiting classical energy values, $3N(\nu)kT$, at temperatures lower than oscillators of higher frequency. Consequently, there is a sort of sliding scale with more and more frequencies becoming fully excited as the temperature is raised. The lowest harmonics are fully excited just above 0 K, and the highest ones become so a few hundred degrees above room temperature for most elemental solids. The $3N$ standing waves with properties as described above are called normal modes of vibration.

Thermally excited vibrations lie well within the elastic strain region and so obey Hooke's Law. The force F necessary to cause the displacement of an atom from its equilibrium position is related to the strain ε by a simple constant of proportionality C, which is termed the elastic stiffness, by

$$F = -C\varepsilon. \tag{3.24}$$

This is the harmonic approximation in which each atom vibrates within a parabolic potential energy well, as can be seen from the equation for the potential energy, E_{pot}:

$$E_{pot} = -\int F \, d\varepsilon = C \int \varepsilon \, d\varepsilon = \tfrac{1}{2} C\varepsilon^2 \tag{3.25}$$

The normal mode approximation considers the x, y, z displacements of a given atom to be independent of one another, and, for our analysis, we define the strain in terms of the local displacement

$$\varepsilon = \frac{\partial u}{\partial x}, \frac{\partial u}{\partial y}, \frac{\partial u}{\partial z},$$

where the x, y, z coordinates specify the rational equilibrium lattice points. Hence, ε is a dimensionless parameter that defines the strain for a specific atom from its equilibrium ($F = 0$) position.

Before taking up the equations of motion for atoms in a crystal, let us consider a simpler analogous example. We begin with a one-dimensional elastic solid of uniform density ρ (*the string model*), and write for the resultant force on the infinitesimal mass m between positions x and $x + \Delta x$

$$\begin{aligned} \lim_{\Delta x \to 0} [F_{x+\Delta x} - F_x] &= C \lim_{\Delta x \to 0} \frac{[(\partial u/\partial x)_{x+\Delta x} - (\partial u/\partial x)_x]}{2\,\Delta x} \\ &= C \frac{\partial^2 u}{\partial x^2} \Delta x. \end{aligned} \tag{3.26}$$

From a purely dimensional analysis, we now have

$$F = ma = \rho \, \Delta x \frac{\partial^2 u}{\partial t^2} = C \frac{\partial^2 u}{\partial x^2} \Delta x \tag{3.27}$$

or

$$\frac{\partial^2 u}{\partial x^2} = \frac{\rho}{C} \frac{\partial^2 u}{\partial t^2}, \tag{3.28}$$

where t has its usual meaning of time. This is the familiar vibrating string equation, which has a variety of sine and cosine solutions obtained by changing variables and integrating.

Our model for atomic vibrations in a solid is based, by analogy, on the string model. The allowable frequencies, which we will calculate later in this section, are standing waves made up of combinations of solutions to (3.28), which, individually, are for traveling waves. At first, we need only consider the longitudinal waves, whose displacements are collinear with the direction of the traveling wave. The transverse components, which are normal to the longitudinal, do not interfere, either constructively or destructively, with the latter. The allowable frequencies are determined by the size and shape of the solid. For a cube, the lowest frequency corresponds to a wavelength twice the length of the cube edge, and the highest to a wavelength corresponding to twice the interatomic spacing along a major axis. These individual frequencies, which are the product of the collective action of $3N$ atoms, are quantized, although the motion of the individual atoms about their respective equilibrium positions is not. Further, our model assumes that the forces acting upon each atom are entirely accounted for by only nearest-neighbor interactions, which is not as poor an approximation as it might first appear. Generally, 90% or more of the binding energy between atoms in a solid is accounted for by the nearest-neighbor potential. Also, the displacements produced by thermal vibration are small and are not significantly influenced by the potential function linking more distant atoms to one another.

The displacements produced by the standing waves are measured with respect to the rigid lattice coordinates. These local displacements, when projected on the x-, y-, z-axes of the laboratory reference frame, will then conform to a solution of the wave equation. It should be recalled that, as long as the displacements remain within the elastic region, the oscillations about the equilibrium lattice point will obey the equation of motion described by the well-known spring equation (see Appendix D).

Symbols that we will use in the following derivation are defined, in part, by Fig. 3.3, which illustrates the variation in Δu along the wavefront. The force is assumed to vary linearly with the local displacement where the proportionality constant, σ_x, is the spring constant. Thus,

$$F_x = -\frac{\sigma_x}{2}(u_j - u_{j-1}) - \frac{\sigma_x}{2}(u_j - u_{j+1}). \tag{3.29}$$

Because of the symmetry relations, similar equations hold for F_y and F_z.

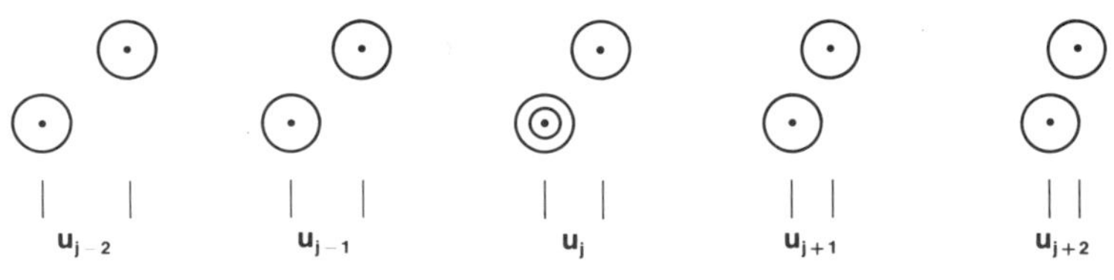

Figure 3.3.

A two-dimensional representation of the varying magnitudes of the relative displacements between neighboring atoms. Because we are considering only the longitudinal waves, the x coordinate provides a complete description for a cubic crystal structure.

Taking (3.29) to the limit of very small Δu leads to an equation analogous to (3.28), which is directly integrable. We are interested only in those solutions that represent standing waves, such as

$$u = A \sin 2\pi\nu t \sin(2\pi x/\lambda), \tag{3.30}$$

where A is a numerical constant and λ is the wavelength. Specifying the displacement by the notation introduced by Fig. 3.3 gives

$$u_j = A \sin 2\pi\nu t \sin\frac{2\pi jd}{\lambda}, \tag{3.31}$$

$$u_{j\pm 1} = A \sin 2\pi\nu t\left(\sin\frac{2\pi jd}{\lambda}\cos\frac{2\pi d}{\lambda} \pm \cos\frac{2\pi jd}{\lambda}\cos\frac{2\pi d}{\lambda}\right). \tag{3.32}$$

Then

$$u_{j-1} + u_{j+1} = A \sin 2\pi\nu t\left(2\sin\frac{2\pi jd}{\lambda}\cos\frac{2\pi d}{\lambda}\right). \tag{3.33}$$

Collecting terms in (3.29) yields

$$F_x = -\sigma_x u_j + (\sigma_x/2)(u_{j-1} + u_{j+1}).$$

Substituting (3.31) and (3.33) for the displacements and equating the restoring force F_x to ma, as in (3.27), leads to

$$F_x = m\frac{d^2u_j}{dt^2} = A\sigma_x \sin 2\pi\nu t \sin\frac{2\pi jd}{\lambda}\left(\cos\frac{2\pi d}{\lambda} - 1\right). \tag{3.34}$$

Substituting for u_j, according to (3.30), and taking the second derivative with respect to t allows us to reduce (3.34) to

$$\nu = \left[\frac{\sigma_x}{4\pi^2 m}\left(1 - \cos\frac{2\pi d}{\lambda}\right)\right]^{1/2} = \frac{1}{2\pi}\left(\frac{2\sigma_x}{m}\right)^{1/2}\sin\frac{\pi d}{\lambda}. \tag{3.35}$$

For large values of λ, $\sin(\pi d/\lambda) \simeq \pi d/\lambda$, and

$$v = \lambda\nu = \sqrt{\sigma d^2/2m}, \tag{3.36}$$

and the wave velocity is a frequency independent constant.

4. The Spectrum of the Normal Modes

The spectrum of allowed vibrational frequencies naturally depends upon the crystal structure. Clearly, the periodicity of the atoms must govern the periodicity of the standing waves of atomic displacements. It is convenient to recall that the solution to the wave equation in three dimensions can be obtained by the method of separation of variables. Representing the displacement as a function of time by u, with orthogonal components u_x, u_y, and u_z, we have

$$u(x, y, z, t) = u(x, y, z)\phi(t), \tag{3.37}$$

where $\phi(t)$ is some periodic function in t, such as $\phi(t) = \sin 2\pi\nu t$, which is independent of the spatial coordinates. Also in the normal mode approximation, the displacements along the x-, y-, z-directions are mutually independent. Upon separation of variables (see Appendix D) in the wave equation (3.28), the equation for the coupled displacements in three dimensions becomes

$$\nabla^2 u + k^2 u = 0, \tag{3.38}$$

which can be decomposed into

$$\begin{aligned} \frac{d^2u_x}{dx^2} + k_x^2 u_x &= 0, \\ \frac{d^2u_y}{dy^2} + k_y^2 u_y &= 0, \\ \frac{d^2u_z}{dz^2} + k_z^2 u_z &= 0, \end{aligned} \tag{3.39}$$

which are the same equations as for a freely oscillating spring. The solutions to Eqs. (3.39) are of the form

$$u_x = A_x \sin k_x x + B_x \cos k_x x. \tag{3.40}$$

Preserving only the real part, B_x being imaginary, leaves us with

$$u_x = A_x \sin k_x x, \tag{3.41}$$

and in like manner

$$u_y = A_y \sin k_y y, \qquad u_z = A_z \sin k_z z.$$

The superposition of these three sine functions results in a sinusoidal plane wave. In order to have standing waves in all three directions, the wave must have nodes at the extrema; that is, $u_x = u_y = u_z = 0$ at $x = 0, X$, $y = 0, Y$, and $z = 0, Z$. The first boundary condition is automatically satisfied, and the second is met if

$$k_x = \frac{n_x\lambda}{2X}, \qquad k_y = \frac{n_y\lambda}{2Y}, \qquad k_z = \frac{n_z\lambda}{2Z}, \tag{3.42}$$

where n_x, n_y, n_z are integers. However,

$$k_x x + k_y y + k_z z = k_x k_y k_z \tag{3.43}$$

is the equation of a plane $k_x k_y k_z$ units distant from the origin [refer to Eq. (1.3)], and k_x, k_y, and k_z are the components of the unit vectors along x, y, and z. Thus, the plane nearest the origin is described by

$$\left(\frac{\lambda}{2}\right)^2\left[\left(\frac{n_x}{X}\right)^2 + \left(\frac{n_y}{Y}\right)^2 + \left(\frac{n_z}{Z}\right)^2\right] = 1. \tag{3.44}$$

Rearranging (3.43) leads to

$$\lambda^{-1} = \left[\left(\frac{n_x}{2X}\right)^2 + \left(\frac{n_y}{2Y}\right)^2 + \left(\frac{n_z}{2Z}\right)^2\right]^{1/2}. \tag{3.45}$$

The maximum number of harmonics for a given direction is twice the maximum length, i.e., $2X$, $2Y$, or $2Z$ divided by the minimum wavelength, so that

$$n_x(\text{max}) = \frac{2X}{\lambda_{\text{min}}} = \frac{2N_x d}{2d} = N_x, \tag{3.46}$$

where N_x is the number of atoms between $x = 0$ and $x = X$. Similar equations exist for n_y and n_z for a simple (primitive) cubic crystal structure. Hence, the total number of allowed overtones is equal to the total number of atoms:

$$n_x n_y n_z(\text{max}) = N_x N_y N_z = N. \tag{3.47}$$

However, there are two transverse modes of vibration for each longitudinal mode; hence, the total number of modes is $3N$, as stated before in our discussion of the Einstein heat capacity.

The spectrum of wavelengths is generally derived in reciprocal space, which is a more natural representation in view of the fact that the coefficients in (3.43) are expressed in terms of reciprocal distance. The volume occupied by each overtone in the reciprocal lattice is given by

$$\left(\frac{1}{X}\right)\left(\frac{1}{Y}\right)\left(\frac{1}{Z}\right) = \frac{1}{V} = \left(\frac{1}{N_x d}\right)\left(\frac{1}{N_y d}\right)\left(\frac{1}{N_z d}\right), \tag{3.48}$$

which, after multiplying as indicated and rearranging, becomes

$$1/V = 1/Nd^3. \tag{3.49}$$

Hence, a single harmonic (i.e., $1/N$th of the total number) is contained in the volume d^3/V in reciprocal space. We now substitute $\lambda = v/\nu$, where v is the velocity, into (3.45) and rearrange to obtain

$$\nu = \frac{v}{2}\left[\left(\frac{n_x}{X}\right)^2 + \left(\frac{n_y}{Y}\right)^2 + \left(\frac{n_z}{Z}\right)^2\right]^{1/2} = \frac{rv}{2}, \tag{3.50}$$

where r can be considered as the radius vector in reciprocal space. Now the number of overtones whose frequencies lie in the range $d\nu$ form a spherical shell whose volume is

$$d\nu \propto 4\pi r^2\, dr = 32\pi\nu^2\, d\nu/v^3. \tag{3.51}$$

However, only the octant in which n_x, n_y and n_z all are positive represent vibrations in real space, and we must divide (3.50) by 8. The number of overtones in reciprocal space has been shown to equal V, and so the number of allowed harmonics in the range $d\nu$ in the longitudinal direction is given by

$$dN = \frac{4\pi\nu^2\, d\nu}{v^3}\, V. \tag{3.52}$$

We now add the two transverse modes to the longitudinal one to obtain the total number of overtones:

$$dN = 4\pi\nu^2 V\left(\frac{1}{v_l^3} + \frac{2}{v_t^3}\right) d\nu. \tag{3.53}$$

Equations (3.52) and (3.53) are valid only when n_x/X, n_y/X, and n_z/Z are less than d^{-1}, thereby lying within a cube of volume d^{-3}. If we plot $V^{-1}\, dN/d\nu$ versus ν, it commences with a parabolic rise given by

$$V^{-1}\, dN/d\nu = \alpha\nu^2, \tag{3.54}$$

where $\alpha = 4\pi(v_l^{-3} + v_t^{-3})$.

In one direction the spectrum appears as shown in Fig. 3.4. The number of allowed frequencies decreases sharply after the sphere intersects the cube wall at $\nu = v/2d$. Frequencies lying beyond $\lambda = 1/2d$ are outside the cube and do not exist. $dN/d\nu$ goes to zero when $\nu = (2d)^{-1}(3v)^{1/2}$, and the sphere reaches the farthest corners, completely filling the reciprocal lattice volume d^{-3}.

If the crystal structure is more complicated than simple cubic, the volumes in reciprocal space containing the $3N$ overtones are, perforce, also more complicated. The polyhedra are called Brillouin zones, and only the primitive cube has the same symmetry as its Brillouin zone. Also, in this regard, it must be emphasized that in actual crystals elastic waves are not propagated in all directions with the same velocity, because the elastic stiffness is anisotropic; nor is the velocity independent of wavelength. These are the approximations assumed for the construction of the foregoing model.

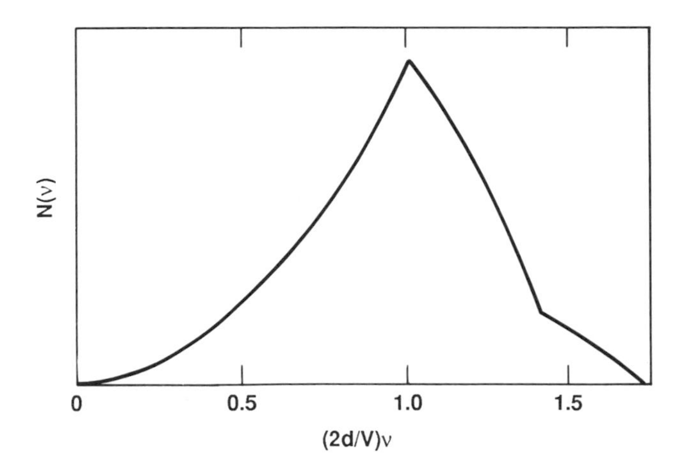

Figure 3.4.

The spectrum of vibrational frequencies in one direction for a primitive cubic lattice with interatomic spacing d and constant velocity of propagation v.

5. The Debye Model of the Heat Capacity

The Debye approximation treats the solid as an isotropic elastic medium. The heat energy is taken to be equivalent to the energy of $3N$ elastic waves, which vary in frequency from 0 to a maximum ν_{max}, characteristic of the solid. The frequency distribution, illustrated in Fig. 3.5, is quadratic up to the cutoff ν_{max}. This frequency, ν_{max}, is obtained by extrapolating ν according to (3.53) until the area under the curve equals that of the $3N$ normal modes. To determine ν_{max}, we set the integral of (3.53) equal to $3N$ and perform the integration:

$$\frac{3N}{V} = 4\pi\left(\frac{1}{v_l^3} + \frac{2}{v_t^3}\right)\int_0^{\nu_{max}} \nu^2\, d\nu$$

$$= \frac{4\pi}{3}\left(\frac{1}{v_l^3} + \frac{2}{v_t^3}\right)\nu_{max}^3. \tag{3.55}$$

Therefore,

$$\nu_{max} = \left[\frac{9N}{4\pi V}\,\frac{1}{1/v_l^3 + 2/v_t^3}\right]^{1/3}. \tag{3.56}$$

Now the heat capacity of an ensemble of coupled oscillators is given by

$$C_V = k\int_0^{\nu_{max}}\left(\frac{h\nu}{kT}\right)^2 \frac{g(\nu)e^{h\nu/kT}}{(e^{h\nu/kT} - 1)^2}\, d\nu. \tag{3.57}$$

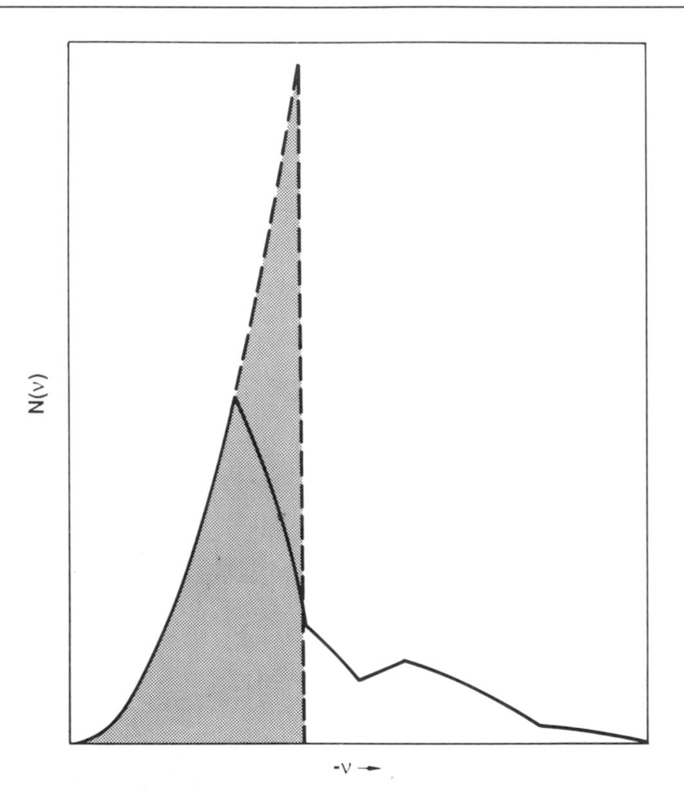

Figure 3.5.

Number of vibrations per unit frequency range in a simple cubic lattice with constant velocity of propagation. It is assumed that the velocity of the longitudinal wave is twice that of the transverse waves. (Dashed curve indicates Debye's assumption.)

The Einstein model sets the distribution function, $g(\nu) = 1$, but the Debye model substitutes (3.53), which results in

$$C_V = 4\pi V\left(\frac{1}{v_l^3} + \frac{2}{v_t^3}\right)\int_0^{\nu_{\max}}\left(\frac{h\nu}{kT}\right)^2 \frac{e^{h\nu/kT}\nu^2}{(e^{h\nu/kT} - 1)^2}\,d\nu. \tag{3.58}$$

Defining $\zeta \equiv h\nu/kT$ and $\zeta_0 = h\nu_{\max}/kT$, using (3.56), substituting into (3.58), and collecting terms gives

$$C_V = \frac{9Nk}{\zeta_0^3}\int_0^{\zeta_0} \frac{\zeta^4 e^\zeta}{(e^\zeta - 1)^2}\,d\zeta. \tag{3.59}$$

The so-called Debye temperature, Θ_D, is defined by

$$\Theta_D = h\nu_{\max}/k \tag{3.60}$$

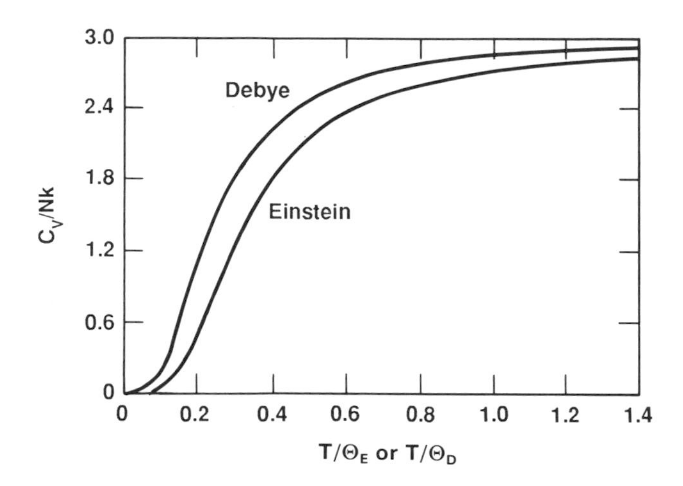

Figure 3.6.

Heat capacity of monatomic crystals according to the theories of Einstein, T/Θ_E, and Debye, T/Θ_D.

and is a measure of the cutoff frequency ν_{max}. Although (3.59) cannot be integrated, good approximations are possible at low and high temperatures, where it yields values comparable to the Einstein model, as seen in Fig. 3.6.

For temperatures where $T \ll \Theta_D$, $\zeta \gg 1$ and the integral in (3.59) becomes

$$\int_0^\infty \frac{\zeta^4 e^\zeta}{(e^\zeta - 1)^2}\, d\zeta = \frac{4}{15}\pi^4, \tag{3.61}$$

thus giving for the heat capacity

$$C_V = \frac{12}{5}\pi^4 Nk\left(\frac{T}{\Theta_D}\right)^3. \tag{3.62}$$

However, (3.59) can be evaluated by numerical methods, and Fig. 3.7 compares the heat capacities of several materials as a function of the reduced temperature.

This graphical illustration indicates that the theory is quite satisfactory since the data fall on a single universal curve, as required by Eq. (3.59), and the shape clearly fits the experimental data very well. This figure is somewhat misleading, however, appearing very accurate because of the scales used. A more stringent test of the theory is to inquire how closely Θ_D remains a temperature independent constant over a wide temperature range. By this criterion, the Debye theory is good to only 5–10%. It is excellent, for example, for Pb, Ag, Cu, W, Na, and K with Θ_D constant to within a

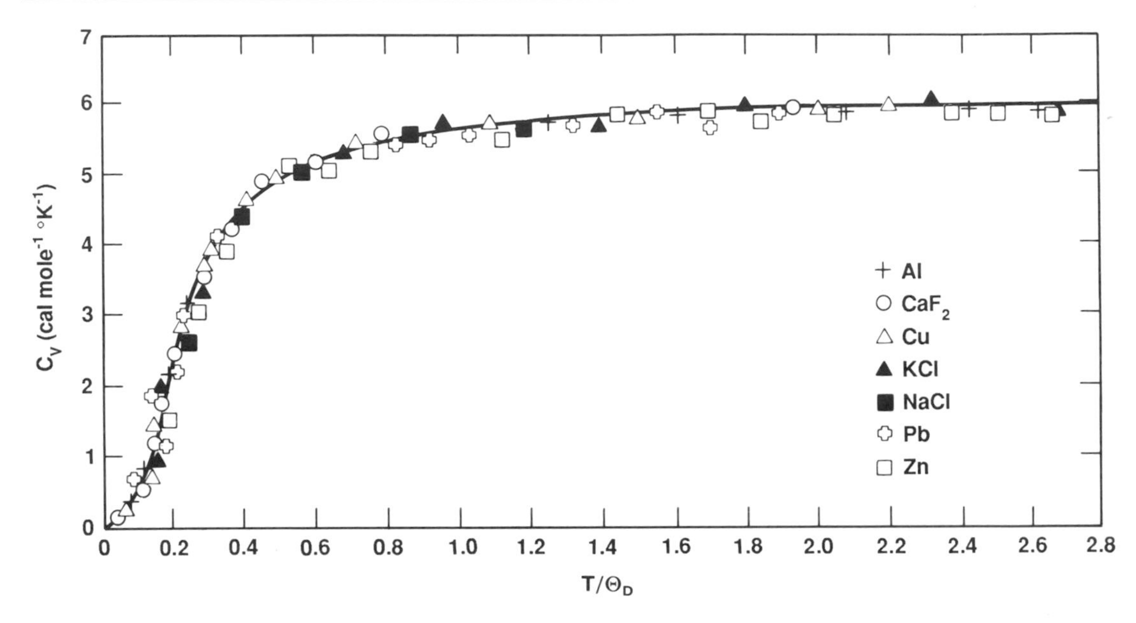

Figure 3.7.

Heat capacity vs. reduced temperature for a number of materials. (After G. Burns, *Solid State Physics*, Academic Press, New York, p. 369, 1985.)

few percent. It is not nearly as good, for example, for Li and Au, with Θ_D varying in the 10% range.

One should not minimize, however, the power and the usefulness of the Debye theory; it is certainly one of the triumphs of quantum theory. Also, the Debye temperature plays an important role in a number of additional solid state phenomena, such as electrical resistivity, thermal conductivity, and to describe diffraction line broadening.

Representative values of the Debye temperature for the elements are listed in Table 3.1.

6. The Electronic Specific Heat

The electronic specific heat is discussed here briefly, although the electron theory of metals is postponed until Chapter 8. It is an experimental fact, and a theoretical result of quantum statistics, that the electronic contribution to the total specific heat is very small at ordinary temperatures, but is important at low temperatures where

Table 3.1. Representative values of Θ_D for the elements.

Element	Θ_D (K)	Element	Θ_D (K)
A	85	Ir	283
Ag	215	K	99.5
Al	398	Li	328–430
Au	180	Mg	290
Be	1000	Mo	379
C (diamond)	1860	Na	159
Ca	230	Ne	63
Cd	160	Ni	370
Cr	485	Pb	88
Cu	315	Pt	225
Fe	420	Ta	245
Ge	290	W	310
		Zn	235

the vibrational contribution decreases. At low temperatures the total heat capacity of a metal can be represented by a sum of two terms as

$$C_V = \gamma T + BT^3, \tag{3.63}$$

where the first term arises from the conduction electrons and the BT^3 term from the lattice vibrations in accord with (3.62). At low temperatures the linear term dominates.

Experimentally, γ is found to be in the $10^{-4}R$ range, clearly a very small number compared to Dulong–Petit values of $C_V \cong 3R$ at ordinary temperatures. This large discrepancy was the original dilemma in the classical electron theory of metals, which regarded electrons as a classical ensemble of gaseous particles. The answer, as intimated above, is the inapplicability of classical statistics and the necessity of using quantum statistics in discussing the energy distribution of electrons. The electron gas must be treated by Fermi–Dirac statistics, according to which the fraction of electrons raised in energy in going from 0 K to a temperature T is about kT/E_F, where E_F is the Fermi energy. The increase in energy of these electrons is about kT, and the average thermal energy per mole, E_{th}, for $T < T_F$ is approximately

$$E_{\text{th}} \cong RT(kT/E_F), \tag{3.64}$$

and the electronic molar heat capacity, C_v^e, is given by

$$C_V^e = \frac{\partial E_{\text{th}}}{\partial T} \cong \frac{RT}{T_F}, \tag{3.65}$$

where $T_F = E_F/k$ is the Fermi temperature. This equation is clearly of the same form as the experimental one (i.e. linear in T). The Fermi temperature in metals is about 10^4 to 10^5 K, giving values of the required magnitude.[2]

7. Anharmonicity

Anharmonicity was already included in an implicit way when we treated thermal expansion. This is because a harmonic solid cannot produce thermal expansion. It is instructive to consider this point in some detail. The potential energy of a classical harmonic oscillator is, by definition, given by

$$E_{\text{pot}} = C\varepsilon^2/2, \tag{3.66}$$

which describes a purely parabolic potential energy well. The force acting on the system at any instant is

$$F = -C\varepsilon. \tag{3.67}$$

The equation of motion for a particle of mass m is

$$m\frac{d^2\varepsilon}{dt^2} = -C\varepsilon, \tag{3.68}$$

whose general solution is

$$\varepsilon = A\cos[(C/m)^{1/2}t + \delta], \tag{3.69}$$

where δ is the integration constant. The motion clearly repeats itself whenever $(C/m)^{1/2}t = 2\pi$ and the oscillation frequency is

$$\nu = (1/2\pi)(C/m)^{1/2}, \tag{3.70}$$

and, therefore,

$$\varepsilon = A\cos(2\pi\nu t + \delta). \tag{3.71}$$

The average value of the displacement, $\bar{\varepsilon}$, is given by

$$\bar{\varepsilon} = \frac{\int \varepsilon\, dt}{\int dt} = \frac{(a/2\pi\nu)[\sin(2\pi\nu t) + \delta]_{t_1}^{t_2}}{[t]_{t_1}^{t_2}}. \tag{3.72}$$

For the average over a complete vibration, the upper limit is taken as $t = 1/\nu$ and the lower as $t = 0$, giving $\bar{\varepsilon} = 0$. Thus, there is no change in $\bar{\varepsilon}$ for any value of ν, and hence there is no thermal expansion.

---

[2] For detailed calculations of γ see, for example, L. A. Girifalco, *Statistical Physics of Materials*, Chap. 3, Wiley, New York, 1973.

Anharmonicity arises from terms higher than quadratic in the potential energy. E_{pot} is modified by the next two terms of the expansion as (with $A = C/2$)

$$E(\varepsilon) = A\varepsilon^2 - B\varepsilon^3 - C\varepsilon^4, \tag{3.73}$$

where the cubic term represents the asymmetry of mutual repulsion, and the fourth-order one comes from large amplitude weakening of the vibrations. Consequently, the potential energy well is no longer parabolic, and the restoring force acting on the vibrating atom, ion, or molecule no longer obeys Hooke's law (3.24). It is the cubic term that gives rise to thermal expansion.

Let us calculate $\bar{\varepsilon}$ now by using the Boltzmann distribution function:

$$\bar{\varepsilon} = \int_{-\infty}^{\infty} \varepsilon e^{-E(\varepsilon)/kT}\, d\varepsilon \Big/ \int_{-\infty}^{\infty} e^{-E(\varepsilon)/kT}\, d\varepsilon. \tag{3.74}$$

The exponential term in the integrands may be expanded and evaluated as

$$\int_{-\infty}^{\infty} e^{-E(\varepsilon)/kT}\, d\varepsilon \cong \int_{-\infty}^{\infty} e^{-A\varepsilon^2/kT}\, d\varepsilon = \left(\frac{\pi kT}{A}\right)^{1/2}$$

$$\int_{-\infty}^{\infty} \varepsilon e^{-E(\varepsilon)RT}\, d\varepsilon \cong \int e^{-A\varepsilon^2/kT}\varepsilon\left[1 + \frac{B\varepsilon^3}{kT} + \cdots\right] d\varepsilon$$

$$= \left(\frac{b}{kT}\right)\left(\frac{kT}{A}\right)^{5/2}\left(\frac{3\pi^{1/2}}{4}\right), \tag{3.75}$$

with the use of standard tables of definite integrals. Thus,

$$\bar{\varepsilon} = 3kTB/4A^2, \tag{3.76}$$

and, therefore, the linear expansion coefficient $3kB/4A^2$ is a constant. In this relation kT is the classical mean energy of the harmonic oscillator.

One may obtain an approximate quantum mechanical result by replacing kT in (3.76) with the corresponding quantum mechanical expression [see Eq. (3.19)]

$$h\nu/(e^{h\nu/kT} - 1). \tag{3.77}$$

With this substitution it is easy to see that the thermal expansion coefficient goes to zero at $T \to 0$, as required by the third law.

8. Optical Vibrations

We now extend our discussion of wave motion to vibrations in a one-dimensional *diatomic* crystal in order to illustrate the development of optical vibrations. These vibrations arise when two different kinds of atoms in a diatomic chain vibrate out of phase with each other. They are called optical vibrations, or the optical branch, because they can be observed by optical absorptions in the infrared.

For simplicity of discussion, we shall limit ourselves to two atoms of different masses, m and M, located along a chain as sketched in Fig. 3.8.

The displacements are identified as u_j belonging to atom m and v_j belonging to atom M. We shall assume that a single force constant, β, describes the interaction and that it is sufficient to consider only nearest-neighbor interactions. The distance between identical atoms is d, as shown in Fig. 3.8. The essential physics can be illustrated by the behavior of this simple system. The equations of motion are

$$m\frac{d^2u_j}{dt^2} = -\beta(u_j - v_{j-1}) - \beta(u_j - v_j) = \beta(v_j + v_{j-1} - 2u_j),$$

$$M\frac{d^2v_j}{dt^2} = -\beta(v_j - u_{j+1}) - \beta(v_j - u_j) = \beta(u_j + u_{j+1} - 2v_j). \tag{3.78}$$

We look for solutions describing traveling waves—that is, solutions of the form

$$u_j = A\exp[i(\omega t + 2jkd)],$$

$$v_j = B\exp[i(\omega t + (2j+1)kd)], \tag{3.79}$$

where A and B are the amplitudes of the particle displacements and k is the wave number. Substitution into the equations of motion (3.78) gives

$$-\omega^2 mA = \beta B(e^{ikd} + e^{-ikd}) - 2\beta A,$$

$$-\omega^2 MB = \beta A(e^{ikd} + e^{-ikd}) - 2\beta B. \tag{3.80}$$

Solutions of these homogeneous equations exist if, and only if, the determinant of the coefficients of A and B vanishes. This condition is stated by

$$\begin{vmatrix} 2\beta - m\omega^2 & -2\beta\cos kd \\ -2\beta\cos kd & 2\beta - M\omega^2 \end{vmatrix} = 0. \tag{3.81}$$

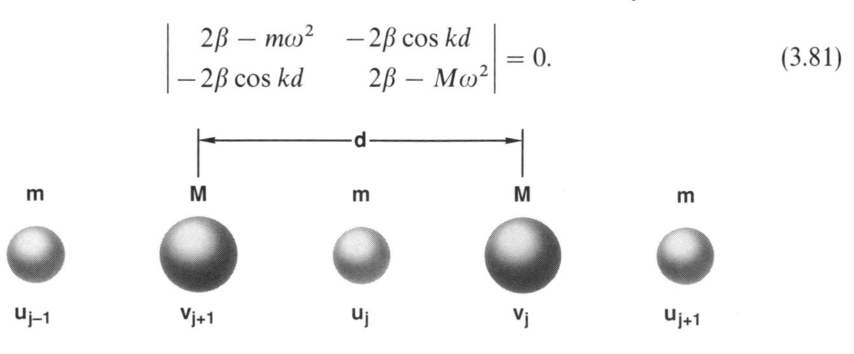

Figure 3.8.

Arrangement of atoms in a one-dimensional diatomic lattice.

Expanding the determinant in the usual way gives

$$(2\beta - m\omega^2)(2\beta - M\omega^2) - (2\beta \cos kd)^2 = 0, \tag{3.82}$$

Solving the quadratic in ω^4, for ω^2, we obtain (using $\sin^2 x + \cos^2 x = 1$)

$$\omega^2 = \beta\left(\frac{1}{m} + \frac{1}{M}\right) \pm \beta\left[\left(\frac{1}{m} + \frac{1}{M}\right)^2 - \frac{4\sin^2 kd}{Mm}\right]^{1/2}. \tag{3.83}$$

This equation clearly has two roots. We can examine the solutions for small k and for the zone boundary at $k = \pi/2d$. At small k the two solution are

$$\omega^2 = 2\beta(1/M + 1/m) \tag{3.84}$$

and

$$\omega^2 = 2\beta k^2 d^2/(M + m), \tag{3.85}$$

the latter obtained by letting $\sin^2 kd \cong k^2 d^2$ and by expanding the square root. At $k = \pi/2d$ we have $\sin kd = 1$, and the two solutions are

$$\omega^2 = 2\beta/M \tag{3.86}$$

and

$$\omega^2 = 2\beta/m. \tag{3.87}$$

For $M > m$ we may sketch the variation of ω with k as in Fig. 3.9, where we have identified the two branches as the optical, Eqs. (3.84) and (3.87), and the acoustical, Eqs. (3.85) and (3.86), branches.

Let us calculate the ratio of the amplitudes A and B from (3.79) for small k by substituting the appropriate values of ω^2. We find that in the optical branch

$$A/B = -m/M, \tag{3.88}$$

showing that the atoms vibrate against each other. The other solution, as $K \to 0$, is

$$A = B, \tag{3.89}$$

showing that the atoms move together as in acoustical vibrations.

As already noted, the heavy and light atoms move out of phase in the optical branch; that is, they oscillate around their equilibrium positions toward and away from each other. In an ionic crystal, where the two atoms are oppositely charged ions, such a motion will set up an oscillating dipole moment. This oscillation, which is of long wavelength, can interact strongly with light of the same wavelength. For

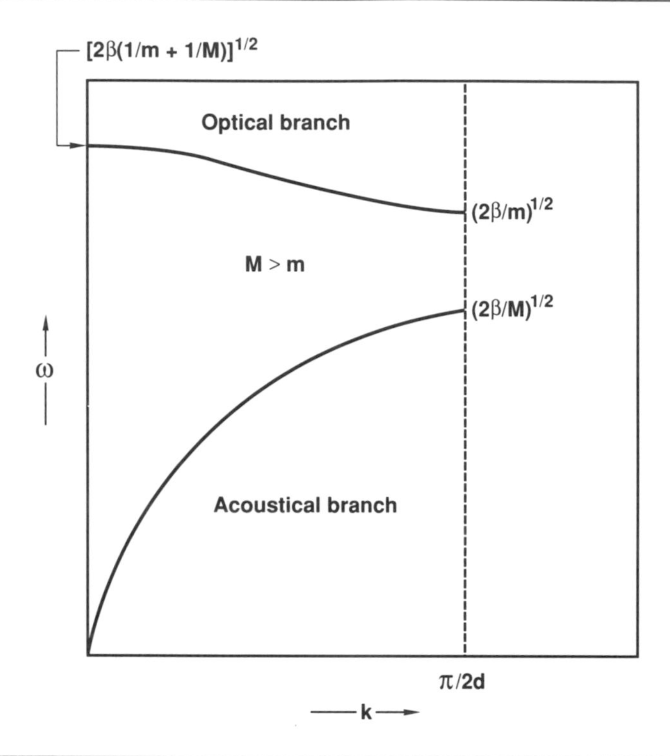

Figure 3.9.

Frequency versus wave number for a diatomic linear lattice.

ions in crystals this is in the infrared region, for example, at 50 to 100 μm for absorption and reflection in the alkali halides.

Chapter 3 Exercises

1. Using the classical partition function

$$Z = \sum_i \exp\left(-\frac{\varepsilon_i}{kT}\right):$$

a. Show that the fraction of particles f_i occupying energy state ε_i is given by

$$f_i = \frac{\exp(-\varepsilon_i/kT)}{Z}.$$

[Hint: Use the Boltzmann law $N_i = K \exp(-\varepsilon_i/kT)$.]

b. Derive the following thermodynamic functions in terms of Z:

$$F = kT \ln Z,$$

$$S = k \ln Z + \frac{kT}{Z}\left(\frac{\partial Z}{\partial T}\right)_v,$$

$$E = \frac{kT^2}{Z}\left(\frac{\partial Z}{\partial T}\right)_v,$$

$$C_V = T\left(\frac{\partial^2(kT \ln Z)}{\partial T^2}\right)_v.$$

2. Discuss the implications of the results of problem 8 in Chapter 2 in terms of the vibrational properties of solids.

3. Analyze the following heat capacity data for copper to obtain the electronic contribution.

T (K)	$10^2 C_P$ (cal g-atom^{-1} deg^{-1})	T (K)	$10^2 C_P$ (cal g-atom^{-1} deg^{-1})
0.418	0.0070	3.020	0.0809
0.517	0.0087	3.498	0.1065
0.787	0.0136	3.579	0.1109
1.252	0.0229	4.023	0.1407
1.452	0.0275	4.606	0.1867
1.968	0.0410	5.301	0.2567
2.603	0.0621	6.013	0.3464

4. Estimate the change in the thermal energy of 100 g copper (atomic weight = 63.5, $\Theta_D = 348$ K) when it is cooled from (a) 300 to 4 K, (b) 78 to 4 K, and (c) 20 to 4 K.

5. M. Born proposed to "cut off" the vibrational modes of an isotropic solid so that both the longitudinal and transverse modes have the same minimum wave length. (Recall that the Debye cut off leaves the maximum frequencies equal.) Hence,

$$\lambda_{min} = (4\pi V/3N)^{1/3},$$

and thus

$$\nu_l = C_l(3N/4\pi V)^{1/3} \quad \text{and} \quad \nu_t = C_t(3N/4\pi V)^{1/3}.$$

Using the Born scheme, show that

$$C_V = R[\phi_D(\Theta_l/T) + 2\phi_D(\phi_t/T)],$$

where ϕ_D is given by (3.60). Also make a schematic graph of $N(\nu)$ versus ν based on the Born approximation. (See Fig. 3.2 for the Debye equivalent.)

6. The anharmonic potential is given by (3.73). Show that the anharmonic contribution to the heat capacity is given approximately by

$$C_A \cong k_B\left[1 + \left(\frac{3C}{2A^2} + \frac{15B^2}{8A^2}\right)k_B T\right].$$

You can assume that $A[x^2]$ is the dominating term in (3.73) because of the small amplitude of the oscillations, where $[x]$ is the mean value of the displacement.

7. Using the harmonic approximation, derive the formula for the vibrational energy levels for a hydrogen atom attached to a much heavier atom. How are the levels changed if the hydrogen is replaced by deuterium?

8. Suppose that the Einstein model is valid in giving the vibrational energy of a solid that has a high-temperature polymorph A, and a low-temperature one, B. Show that at the transition temperature T_{tr}, where A and B are in equilibrium,

$$T_{tr} = \frac{E_A - E_B}{k \ln(\nu_A/\nu_B)},$$

where ν_A and ν_B are the Einstein characteristic frequencies.

9. For diamond $\Theta_D = 2000$ K, its density is 3.51 g/cm^3 and its interatomic spacing is 1.54 Å. Calculate the velocity of sound in diamond. Estimate the dominant phonon wavelength at 300 K, and calculate the frequency of lattice vibration to which this corresponds.

Additional Reading

J. C. Slater, *Introduction to Chemical Physics*, Chaps. XIII, XIV, McGraw-Hill, New York, 1939. A classic text in the field of solid state science, very clearly written.

L. A. Girifalco, *Statistical Physics of Materials*, Chaps. 2, 3, Wiley, New York, 1973. Contains Einstein and Debye functions in Appendix VII and VIII.

D. C. Wallace, *Thermodynamics of Crystals*, Chap. 1, Wiley, New York, 1972.

L. Brillouin, *Wave Propagation in Periodic Structures*, McGraw-Hill, New York 1946. An advanced treatise for those who wish to pursue this subject in detail.

G. Burns, *Solid State Physics*, Chaps. 12, 13, Academic Press, New York 1985. A very detailed, but understandable and well-written, introductory treatment of vibrations in solids.

Chapter 4

Potential Functions

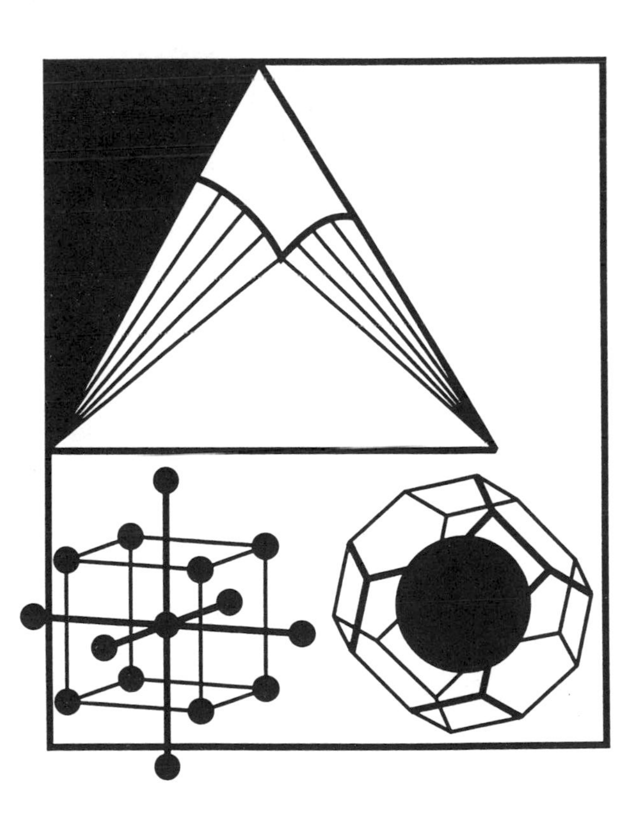

Chapter 4

Contents

The push and pull do nearly cancel leaving naught, but therein lies the calculation that binds us to our reckoning.

—Anonymous Elizabethan writer

Potential Functions

Potential functions describe the strength of the attraction or repulsion between atoms or ions as a function of their distance of separation. At equilibrium the interatomic spacings of molecules and solids result from the precise cancellation of these opposing forces. Although the functional dependence of the potential upon the interatomic distances varies greatly, depending upon the nature of the chemical bonds, it is always the consequence of electrostatic interactions. Undoubtedly, the best known of all the potential functions is contained in Coulomb's law, which states that unlike charges attract and like ones repel with a force proportional to the product of the charges divided by the square of their distance of separation.

Although such functions can be simple and exact for a static isolated pair, calculating the strength of the collective interaction of a statistically significant number of particles becomes a formidable problem in all but the simplest cases. Nevertheless, in principle, if the cohesive energy is known as a function of the interatomic spacings, then the thermodynamic properties relevant to that ensemble are calculable. Also, if the local deformations are uniform throughout the solid, as is the case for hydrostatic deformation, the problem is much simplified. Contrariwise, inhomogeneous local strains, as are sometimes produced by crystal defects, constitute a much more difficult class of problems.

Nowadays, the enormous computational power of the large computers partially compensates for these manifest difficulties. However, most, if not all, such computations contain one or more model-dependent approximations. Frequently, a central force model is used in which a representative average atom interacts with an isotropic average potential resulting from the remainder of the ensemble. Nevertheless, empirical potential functions derived from measured values of the compressibility allow such calculations some fair measure of success. The potential ϕ is a scalar defined by

$$d\phi = -\mathbf{f} \cdot d\mathbf{r}, \tag{4.1}$$

where **f** is the force that depends upon the type of electrostatic interaction, and the vector **r**, in terms of its directional cosines, is given by

$$\mathbf{r} = x \cos \theta_x + y \cos \theta_y + z \cos \theta_z.$$

Because we discuss only pairwise interactions, and the coordinates can always be rotated so as to be collinear with a major axis, the following exposition can be confined to one dimension. Hence, (4.1) reduces to

$$d\phi = -f\, dx. \tag{4.2}$$

The potential is given then by the integral of (4.2) and, thus, has the dimensions of energy. The force is determined by the nature of the interaction of a test charge, itself possessing a unit charge, with an electrostatic field. From this we see that **f** basically depends upon how the electrostatic field varies from point to point in the space about its source. For example, the field source could be as simple as a single spherical ion (e.g., singly charged alkali metal atom) or the spatially complicated distribution of the outer electrons of a large organic molecule.

Because the electrostatic field is a conservative force field, the work done in moving the unit test charge between any two given points is the same, regardless of the path taken. In this regard the work has the same property as a thermodynamic state function.

We now turn to a brief review of two particle interactions, and develop afterwards the formalism associated with the application of generalized potential functions to statistical ensembles.

1. Isolated Two-Body Systems

A. Ion–Ion Interactions

Coulomb's law states that unlike charges attract and like ones repel with a force directly in proportion to their product and inversely proportional to the square of their distance of separation. In vacuo, for single ions, this force is given by

$$f = (z_A e)(z_B e)/r^2, \tag{4.3}$$

where z_A and z_B are the respective number of charges (i.e., oxidation states) of ions A and B, e is the electronic charge, and r is the mutual distance of separation. The force can have either sign, depending on the sign of the charges, and is negative (i.e., attractive) when the charges are of opposite sign. If the charges are not in a vacuum but in a dielectric medium, they become partially screened by the polarization that they themselves induce. Corrected for this effect, (4.3) becomes

$$f = \frac{1}{D} \frac{z_A z_B e^2}{r^2}, \tag{4.4}$$

where the dielectric constant D, therefore, depends upon the polarizability of the medium.

The potential, or potential energy, is given by

$$\int_{\phi^0}^{\phi} d\phi = \int_{r_1}^{r_2} f\,dr = -z_A z_B e^2 \int_{r_1}^{r_2} \frac{dr}{r^2} \tag{4.5}$$

$$= z_A z_A e^2 \left(\frac{1}{r_2} - \frac{1}{r_1} \right). \tag{4.6}$$

The standard state ϕ° is usually taken to be zero, which corresponds to infinite separation of the charges; hence,

$$\phi = z_A z_B e^2 / r_2, \tag{4.7}$$

which is negative and, hence, attractive for unlike charges.

Coulombic interactions are additive because the principle of superposition applies, and the resultant electrostatic field at any point in space is simply the sum of all the individually generated fields. This becomes a simplifying consideration when we compute the internal energy for ionic crystals.

The force acting on a unit charge is called the electrostatic field. The strength of this field F produced by a single spherical ion is given by

$$F = ze/r^2. \tag{4.8}$$

With this definition it follows that

$$\phi = -\int F\,dr. \tag{4.9}$$

B. Ion–Permanent Dipole Interaction

Two charges of equal magnitude, ze, but of opposite sign and more or less rigidly separated by a fixed distance d constitute a permanent dipole, μ, defined by

$$\mu \equiv zed. \tag{4.9}$$

Because of the varying charge distribution within a molecule, z can be fractional. Familiar examples of molecular dipoles are H_2O, HCl, and CH_3Cl, whose polar character is the direct result of the difference in the electronegativities of the component atoms.

The coulombic interaction between a dipole and a single ion is simply the sum of the independent interactions. The configuration and notation necessary for this essentially geometrical derivation of the equation for the potential energy are shown in Fig. 4.1. The potential energy is the result of the separate interactions of the net charges on A and C with the charge e_B on the test ion B. Thus,

$$\phi = -zee_B/AB + zee_B/CB. \tag{4.11}$$

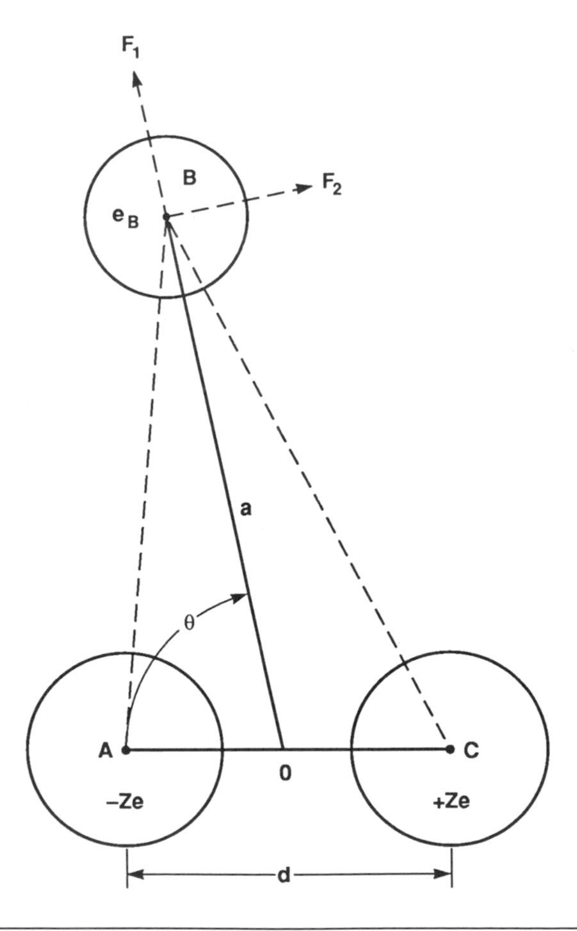

Figure 4.1.

The dipole *zed* interacts with the ion B at a mean distance a from the midpoint of the dipole axis.

Labeling the altitude $OB = a$, we can write

$$\begin{aligned} AB &= (a^2 - ad\cos\theta + d^2/4)^{1/2} \\ &= a(1 - (d/a)\cos\theta + d^2/4a^2)^{1/2} \end{aligned} \tag{4.12}$$

and

$$\begin{aligned} CB &= (a^2 + ad\cos\theta + d^2/4)^{1/2} \\ &= a(1 + (d/a)\cos\theta + d^2/4a^2)^{1/2}, \end{aligned}$$

and substituting in (4.11) for the lengths AB and CB leads to

$$\phi = -\frac{zee_B}{a}\left[\left(1 - \frac{d}{a}\cos\theta + \frac{d^2}{4a^2}\right)^{-1/2} - \left(1 + \frac{d}{a}\cos\theta + \frac{d^2}{4a^2}\right)^{-1/2}\right]. \tag{4.13}$$

Because the separation d is generally but the length of a single chemical bond,

$$((d/a)\cos\theta \pm d^2/4a^2) \ll 1,$$

and, as a result, we can do a binomial expansion of (4.13). Using the formula for small x, we have

$$(1 + x)^{-1/2} = 1 - x/2 + x^2/8,$$

which gives finally

$$\phi = -\frac{zee_B}{a^2}(d\cos\theta)\left(1 - \frac{3}{8}\frac{d^2}{a^2}\right). \tag{4.14}$$

When $(d/a)^2$ is small enough to be neglected,

$$\phi \simeq -(\mu e_B/a^2)\cos\theta. \tag{4.15}$$

We also note that at a distance a the strength of the electrical field F arising from the ion B is e_B/a^2, so the potential energy contained by the ion–dipole coupling is

$$\phi = -F\mu\cos\theta. \tag{4.16}$$

We shall make use of this equation later in the treatment of induced dipoles.

C. Permanent Dipole Field

We calculate the field generated by a permanent dipole by first taking the derivative of (4.9), obtaining

$$d\phi/dr = -F. \tag{4.17}$$

Applying this to (4.15) and the configuration shown in Fig. 4.1, but replacing e_B by a unit test charge e, gives

$$\left(\frac{\partial \phi}{\partial a}\right)_\theta = -\frac{2e\mu \cos \theta}{a^3} = eF_1, \tag{4.18}$$

where F_1 is the radial component of the field. The tangential component is then given by

$$\frac{1}{a}\left(\frac{\partial \phi}{\partial \theta}\right)_a = -\frac{e\mu \sin \theta}{a^3} = eF_2. \tag{4.19}$$

The total field, or force upon the unit charge at distance a, is then derived from

$$F^2 = F_1^2 + F_2^2. \tag{4.20}$$

Solving (4.18) and (4.19) for F_1 and F_2 and substituting into (4.20), we arrive at

$$F^2 = (\mu^2/a^6)(1 + 3 \cos^2 \theta). \tag{4.21}$$

The average value of $\cos^2 \theta$ taken over 4π is 1/3. Hence,

$$\overline{F^2} = 2\mu^2/a^6. \tag{4.22}$$

D. Induced Dipolar Interactions

The localized charge distribution of an atom or molecule suffers a displacement in response to the presence of an external electrostatic field. The source of such a field can be the oppositely charged plates of a capacitor, or it can derive from the proximity of other molecules having permanent dipoles. The extent of the electronic displacement reflects the degree of polarizability of the atom or molecule, which, in turn, reflects how tightly bound such electrons are. This displacement induces a dipole moment that, unlike a permanent dipole, vanishes in the absence of an external field.

In moderate fields the induced moment is proportional to the applied field. Consequently,

$$\mu = zex = \alpha F, \tag{4.23}$$

where the constant of proportionality, α, designates the polarizability. The moment, as always, is given by the charge times the distance of separation. Here the charges on the induced dipole refer to the equal amounts of positive and negative charge displaced by the external field, and x is the distance between the centers of charge density. Because the local surplus charge need not be an integral multiple of the charge on a single electron, z can be fractional. The work expended in separating the effective charges ze to a distance x is found from

$$
\begin{aligned}
W &= \int_0^x Fze\,dx = \int_0^x \frac{\mu}{\alpha} ze\,dx \\
&= z^2e^2 \int_0^x \frac{x}{\alpha}\,dx = \frac{1}{2} z^2 \frac{e^2x^2}{\alpha} \\
&= \frac{1}{2}\frac{\mu^2}{\alpha} = \frac{1}{2}\alpha F^2.
\end{aligned}
\tag{4.24}
$$

The potential energy contained in the interaction between an electrostatic field and a dipole has already been derived, and for charges along a single line of centers (4.16) simplifies to

$$\phi = -F\mu = -\alpha F^2. \tag{4.25}$$

Now, the total potential energy contained by this system is given by (4.25) plus the work that was necessary to induce the dipole. This implies that the standard, or zero-energy state, for this reaction is an ion at infinite separation from an unpolarized molecule. Notice that we have calculated the work necessary to create spontaneously the dipole in the absence of an external field and then the work done in bringing the ion from infinity to the distance a. The result is the same as for the more realistic process in which the ion approaches the intended dipole along the line of centers gradually displacing the charge in a continuous manner. Because the potential behaves as a state function, the work is uniquely determined by the initial and final states of the system. Thus, combining (4.24) and (4.25) leads to

$$\phi = -\alpha F^2/2. \tag{4.26}$$

Back-substituting ze/a^2 for F gives

$$\phi = -\alpha(ze)^2/2a^4, \tag{4.27}$$

which is the potential for an ion–molecule interaction.

Also, we can apply (4.27) to the interaction of a polarizable molecule A with another, B, that has a permanent dipole. If the polarizability of the first is α_A, the potential is given by combining (4.25) with (4.22) to obtain

$$\phi_{AB} = -\alpha_A F_B^2/2 = -\alpha_A \mu_B^2/a^6. \tag{4.28}$$

If the interaction is between like molecules, both of which are polarizable but at the same time have permanent dipoles, the potential is simply the sum of two terms, such as (4.28), viz.,

$$\phi_{AA} = -2\alpha_A \mu_A^2/a^6. \tag{4.29}$$

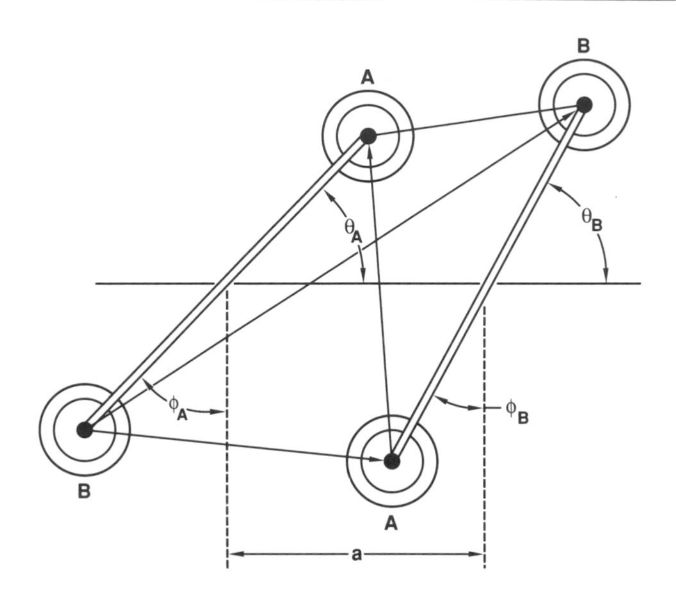

Figure 4.2.

Two identical dipoles, A–B, interact via the vector sum of the coulombic forces. The dashed lines are perpendicular to the solid line of centers.

E. Dipole–Dipole Interaction

The interaction between two permanent dipoles, connected by the heavy lines, is just the vector sum of the four coulombic interactions shown in Fig. 4.2, where a is the distance between the centers of the dipoles. The derivation of the expression for the potential energy of this interaction is quite involved,[1] but as it introduces no new physical principles we will not present it here. The final result of this calculation is with the notation of Fig. 4.2,

$$\phi = -\frac{\mu_A \mu_B}{a^3}\left[2\cos\theta_A \cos\theta_B - \sin\theta_A \sin\theta_B \cos(\phi_A - \phi_B)\right]. \tag{4.30}$$

F. Thermal Effects

A statistical ensemble of polar molecules in a polarizing field will exhibit an average moment reflecting the degree of thermal agitation and the strength of the dipole–field

[1] J. O. Hirschfelder, C. F. Curtis, and R. B. Bird, *Molecular Theory of Gases and Liquids*, pp. 848–851, Wiley, New York, 1965.

coupling. The following derivation of the temperature dependence of the bulk moment of the ensemble omits the dipole–dipole interaction and attributes the entire electrostatic energy to the dipole–field interaction. The energy for this, as previously derived, is

$$\phi = -F\mu \cos \theta. \tag{4.31}$$

Now, according to the Boltzmann equation,

$$N \exp\left[\frac{-E}{kT}\right] d\omega = N \exp\left[\frac{F\mu \cos \theta}{kT}\right] d\omega, \tag{4.32}$$

where N is the total number of dipoles and $d\omega$ is the solid angle within which, at a given temperature, a certain fraction of all dipoles are oriented. The partially ordered array of moments is depicted by Fig. 4.3, which illustrates a net moment caused by the deviation from random orientation.

The average value of the ensemble moment $\bar{\mu}_b$ in the direction of the applied field is given by

$$\bar{\mu}_b = N \int \exp \frac{F\mu \cos \theta}{kT} \mu \cos \theta \, d\omega \Big/ N \int \exp \frac{F\mu \cos \theta}{kT} \, d\omega. \tag{4.33}$$

By setting $F\mu/kT = x$, $\cos \theta = y$, and $d\omega = 2\pi \sin \theta \, d\theta = 2\pi dy$, we can carry out

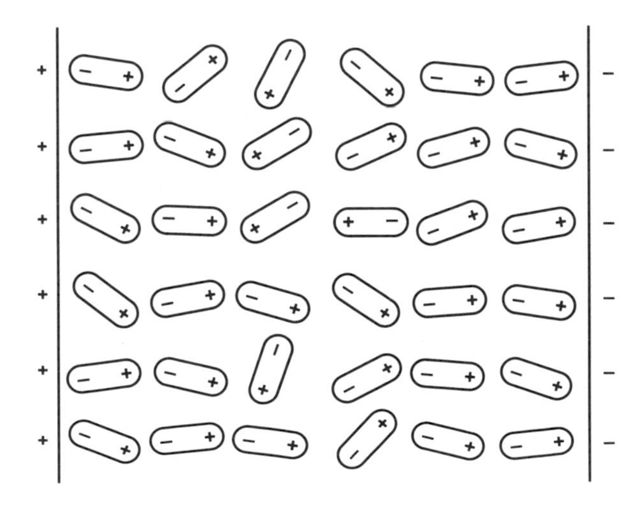

Figure 4.3.

An array of partially aligned dipoles randomized, to a degree, by thermal agitation.

the integration over all solid angles, so that now

$$\frac{\bar{\mu}_b}{\mu} = \frac{\int_{-1}^{+1} e^{xy} y\, dy}{\int_{-1}^{+1} e^{xy}\, dy} = \frac{d \ln}{dx} \int_{-1}^{+1} e^{xy}\, dy$$

$$= \frac{d}{dx} \ln(e^x - e^{-x}) - \frac{d}{dx} \ln x$$

$$= \coth x - \frac{1}{x} \equiv L(x). \tag{4.34}$$

This defines $L(x)$, the Langevin function, named after its first proponent. We note, in passing, that $L(x)$ applies equally well to magnetic moments and, in fact, was first derived to describe the temperature dependence of paramagnetic susceptibilities. $L(x)$ is shown graphically in Fig. 4.4. Note that at high temperature $x \to 0$, $L(x) \to 0$, and, therefore, $\bar{\mu}_B \to 0$. This is the expected behavior for complete randomization. At the other extreme, as $x \to \infty$, $L(x) \to 1$ and $\bar{\mu}_B \to \mu$, the condition of complete alignment along the field. When x is small, but not negligible, $L(x) \cong x/3$, as is easily shown by expanding $L(x)$ up to the cubic terms. Thus, at low fields,

$$\bar{\mu}_B \cong F\mu^2/3kT.$$

The linear approach to $x = 0$ (dashed line in Fig. 4.4) conforms to the essentially

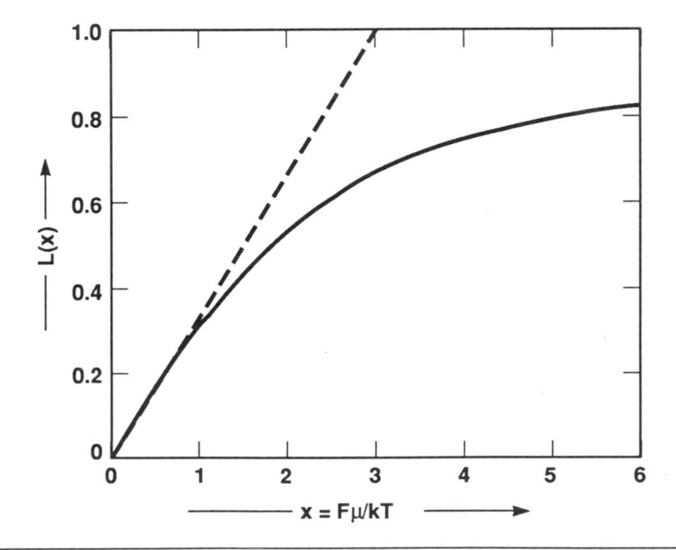

Figure 4.4.

The Langevin function $L(x)$ plotted vs. the interaction energy of dipoles with an applied electrostatic field.

exponential character of the Boltzmann distribution function, which also dictates the rather rapid convergence to saturation with increasing x.

2. Generalized Potential Functions

We have shown that potential functions of all sorts are the sums of all the coulombic interactions. Simple two-body systems allow exact potential functions to be derived from the basic laws of electrostatics. However, the geometric complexities presented by most condensed systems do not permit such a simple treatment, with the exception of elementary ionic crystals, a topic treated in Chapter 5. It is, therefore, more practical to institute an empirical method using experimentally determined quantities, such as enthalpies of dissociation and bulk compressibilities, to calculate the interaction strengths. However, we must first develop equations that will transform these measurable quantities into the interaction potential.

A. The Mie Equation

The graphical form of a general interatomic potential is shown in Fig. 4.5. The steep segment of the curve, commencing at the minimum and rising with decreasing r, reflects the repulsive interaction of the positive ion cores. One should imagine this potential to arise from the repulsion of the partially screened positively charged nuclei and from the interpenetrating negatively charged electron clouds. The kinetic energy of the electrons increases strongly as the atoms or ions are compressed from their equilibrium positions, and the electrons are promoted to higher energy levels because of the exclusion principle. The segment of the curve describing the attractive interactions rises much more gradually from the minimum with increasing r. That portion of the potential that closely circumscribes the minimum itself, and that is nearly parabolic, is the so-called elastic region within which Hooke's law is obeyed. This is easily derivable from elementary mechanics. The change in potential energy due to a small displacement ε of an atom from its equilibrium position, E_{pot}, is given by [see (3.25)]

$$E_{pot} = C\varepsilon^2/2 \tag{4.35}$$

and the force F acting on the system at any instant is given by [see (3.24)]

$$F = -dE_{pot}/d\varepsilon = -C\varepsilon, \tag{4.36}$$

which is Hooke's law. In accord with Fig. 4.5, the displacement is given by

$$\varepsilon = r - r_\varepsilon.$$

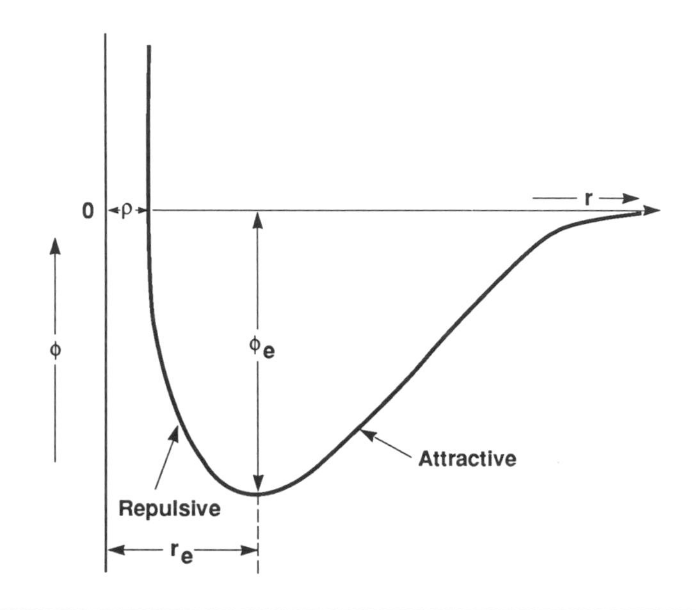

Figure 4.5.

The general form of a pair potential function. r_e is the equilibrium interatomic separation, and ϕ_e is the dissociation energy.

To develop the most general sort of function, we arbitrarily divide the potential into two components, repulsive for $r < r_e$ and attractive for $r > r_e$:

$$\phi = \phi_r + \phi_a. \tag{4.37}$$

We have shown in the previous section that all electrostatic interactions vary according to the inverse distance of separation raised to various powers. Therefore, (4.37) is written in a form first proposed by Mie,[2] a somewhat oversimplified representation, as

$$\phi = Ar^{-n} - Br^{-m}, \tag{4.38}$$

with $n > m$, where A, B, n, and m depend upon the chemical bonding of the system and are to be determined experimentally. The first term is for the repulsive, and the second accounts for the attractive, interactions. Referring to Fig. 4.5, at the minimum in ϕ, that is, at r_e, we have

$$\left(\frac{\partial \phi}{\partial r}\right)_{r_e} = 0 = -nAr^{-(n+1)} + mBr^{-(m+1)}, \tag{4.39}$$

[2] Mie, *Ann. Phys.* **11**, 657 (1903).

and, hence,

$$nA/mB = r_e^{n-m}. \tag{4.40}$$

Substituting (4.40) into (4.38) gives us

$$\phi(r_e) = Br_e^{-m}(m/n - 1) = Ar_e^{-n}(1 - n/m). \tag{4.41}$$

Note that ϕ_e is inherently negative since it represents the maximum negative value of ϕ at the bottom of the well, as shown in Fig. 4.5. The negative of ϕ_e is the dissociation energy ϕ_d, a measurable quantity.

Making use of (4.41) to express A and B in terms of ϕ_e, leads to

$$B = \frac{n\phi_e r_e^m}{m - n}, \qquad A = \frac{m\phi_e r_e^n}{m - n}. \tag{4.42}$$

Substituting (4.42) into (4.38), we arrive at

$$\phi = \frac{\phi_e}{m - n}\left[m\left(\frac{r_e}{r}\right)^n - n\left(\frac{r_e}{r}\right)^m\right] \tag{4.43}$$

as the working form of Mie's equation, in which ϕ_e can be obtained from the temperature dependence of the equilibrium constant covering the dissociation, r and r_e from x-ray or electron diffraction, and n and m from compressibilities or spectral data.

There are several types of potential functions bearing the name of their inventor, which merely prove to be special cases of the Mie equation. Three of traditional importance are the Morse, Lennard–Jones, and Born functions. It should be emphasized that all are intended to deal only with isolated molecules, and further modification is generally required to adapt them to condensed systems.

B. The Morse Potential

The dissociation energy ϕ_d, as already noted, corresponds to

$$\phi_d = \phi_{r=\infty} - \phi_{r=r_e} = -\phi_e,$$

having taken infinite separation as the standard state. The difference between ϕ and ϕ_e is given by

$$\begin{aligned}\Delta\phi = \phi - \phi_e &= \frac{\phi_e}{m - n}\left[m\left(\frac{r_e}{r}\right)^n - n\left(\frac{r_e}{r}\right)^m\right] - \phi_e \\ &= \frac{\phi_e}{m - n}\left[m\left(\frac{r_e}{r}\right)^n - n\left(\frac{r_e}{r}\right)^m - (m - n)\right],\end{aligned}$$

or

$$\Delta\phi = \frac{\phi_d}{n - m}\left[n\left(1 - \left(\frac{r_e}{r}\right)^m\right) - m\left(1 - \left(\frac{r_e}{r}\right)^n\right)\right]. \tag{4.44}$$

Notice that with these definitions and the thermodynamic convention that assigns a positive value to energy absorbed by the system, $\phi_d > 0$.

In order to see how the dissociation energy varies with small atomic displacements, we let

$$x = (r - r_e)/r_e$$

so that

$$r_e/r = 1/(1 + x).$$

Using these relations to rewrite (4.44), we arrive at

$$\Delta\phi = \frac{\phi_d}{n-m}\{n[1 - (1 + x)^{-m}] - m[1 - (1 + x)^{-n}]\}. \tag{4.45}$$

When x is small, $(1 + x)^{-m}$ is sufficiently well approximated by preserving only the first two terms of the expansion, which are identical to the first two terms of the expansion of e^{mx}. According to this approximation,

$$(1 + x)^{-m} = e^{-mx},$$

and substituting this into (4.45) yields

$$\Delta\phi = \frac{\phi_d}{n - m}[n(1 - e^{-mx}) - m(1 - e^{-nx})]. \tag{4.46}$$

The Morse potential is the special case resulting from $n/m = 2$ and leads to

$$\Delta\phi = \frac{\phi_d}{m}[2m(1 - e^{-mx}) - m(1 - e^{-2mx})]$$

$$= \phi_d(1 - e^{mx})^2 = \phi_e[1 - e^{-m(r - r_e)/r_e}]^2, \tag{4.47}$$

which is the Morse equation.

C. Oscillator Frequency in a Mie Potential

Returning to (4.45), we perform a binomial expansion to second order, obtaining

$$\Delta\phi = \frac{\phi_\mathrm{d}}{n-m}\left\{n\left[1-\left(1-mx+\frac{m(m+1)x^2}{2}\right)\right]\right.$$
$$\left.-m\left[1-\left(1-nx+\frac{n(n+1)x^2}{2}\right)\right]\right\}. \tag{4.48}$$

Collecting terms and simplifying leads to

$$\Delta\phi = \frac{1}{2}\frac{mn\phi_\mathrm{d}}{r_\mathrm{e}^2}(r-r_\mathrm{e})^2. \tag{4.49}$$

Thus, truncation of the expansion after the quadratic term leaves us, predictably, with the harmonic approximation so that the displacement energy is proportional to the square of the displacement.

Recall that the frequency of a simple harmonic oscillator is given by [see (3.68)]

$$\nu = (1/2\pi)\sqrt{C/M}, \tag{4.50}$$

where C is the force constant, or elastic stiffness, and M is the oscillator mass. We can determine C from (4.49) by using

$$\Delta\phi = \frac{1}{2}\frac{mn\phi_\mathrm{d}}{r_\mathrm{e}^2}(r-r_\mathrm{e})^2 = \frac{1}{2}C(r-r_\mathrm{e})^2. \tag{4.51}$$

Substituting into (4.50) and defining μ as the reduced mass of the system gives

$$\nu = \frac{1}{2\pi r_\mathrm{e}}\left(\frac{mn\phi_\mathrm{d}}{\mu}\right)^{1/2}. \tag{4.52}$$

If the displacement of the vibrating atom or molecule of mass M is considered only in relation to its z nearest neighbors, the reduced mass is then given by

$$\mu = zM/(z+1). \tag{4.53}$$

D. The Lennard–Jones Potential

Looking at Fig. 4.5, we see that ρ defines the distance of closest approach for two bonded atoms for which the interaction potential is zero. From (4.43) we may write, at $\phi = 0$,

$$m(r_\mathrm{e}/\rho)^n = n(r_\mathrm{e}/\rho)^m. \tag{4.54}$$

Solving for r_e gives

$$r_e = \rho(n/m)^{1/(n-m)}. \tag{4.55}$$

Substituting back into (4.43) for r_e results in, with $\phi_d = -\phi_e$,

$$\phi = \frac{\phi_d}{n-m}\left(\frac{n^n}{m^m}\right)^{1/(n-m)}\left[\left(\frac{\rho}{r}\right)^n - \left(\frac{\rho}{r}\right)^m\right]. \tag{4.56}$$

When $n = 12$ and $m = 6$, values that have been found empirically to frequently give good results, then

$$\frac{1}{n-m}\left(\frac{n^n}{m^m}\right)^{1/(n-m)} = 4,$$

and (4.56) becomes

$$\phi = 4\phi_d[(\rho/r)^{12} - (\rho/r)^6]. \tag{4.57}$$

Its often-used equivalent form, obtained by letting $n = 12$ and $m = 6$ in (4.43), is

$$\phi = \phi_e[(r_e/r)^{12} - 2(r_e/r)^6].$$

This equation is frequently used and is referred to either as the Lennard–Jones or the 6-12 potential. It has found its greatest success in describing rare gas crystals that, at low temperatures, are held together by the dipole-induced van der Waals bonds. Unfortunately, this potential function does not work especially well for metals.

Another popular potential sets $n = 9$ and $m = 6$ and, accordingly, is referred to as the 9-6 potential.

E. The Born–Mayer Potential

A special case of the Mie potential is obtained by setting $m = 1$ in (4.38). This is a very important case because the $1/r$ dependence of the attractive part of the potential represents simple Coulomb attraction and is the basis for the cohesive energy of ionic molecules and crystals, as discussed in detail in Chapter 5. In a modern modification, the inverse power repulsive term is replaced by an exponential repulsion. Thus, for a pair of positive and negative ions, the so-called Born–Mayer potential is given by

$$\phi = -ze^2/r + Ae^{-br}. \tag{4.58}$$

This simple function has been surprisingly successful in describing the properties of alkali halide molecules and crystals. Equilibrium is obtained from

$$\left(\frac{d\phi}{dr}\right)_{r_e} = 0 = +\frac{ze^2}{r_e^2} - Abe^{-br_e}, \tag{4.59}$$

and solving for A and substituting in (4.58) gives

$$\phi(r_e) = (ze^2/r_e^2)(1/b - 1). \tag{4.60}$$

There are several modifications of the Born–Mayer potential designed to improve agreement with experiment, which include polarization interactions and also eliminate some of the difficulties as $r \to 0$. We mention only the shell model[3] here, since it will be used in the next section. The physical basis for this modification is that the repulsive interaction is determined primarily by the same part of the electron distribution that displaces during polarization. This suggests that the repulsive interaction is a function of both the internuclear separation and the dipole moments on the ions. The simplest representation of this effect is the introduction of an effective interionic separation in the expression for the repulsive energy, and making it proportional to the dipole moment $\boldsymbol{\mu}$ yields

$$\mathbf{r}^{\text{eff}} = \mathbf{r}_{ij} + \frac{\boldsymbol{\mu}_i}{Q_i} - \frac{\boldsymbol{\mu}_j}{Q_j}, \tag{4.61}$$

where $\mathbf{r}_{ij}$ is the vector from nucleus i to nucleus j and the Q's are the shell model parameters. Applying this approximation to the energy of interaction of a pair of ions, and including the polarization terms, leads to

$$E(r) = -\frac{Z^2e^2}{r} - \frac{Z(\mu_+ + \mu_-)e}{r^2} - \frac{2\mu_+\mu_-}{r^3} + \frac{\mu_+^2}{2\alpha_+} + \frac{\mu_-^2}{2\alpha_-} + A\exp\left[-b\left(r + \frac{\mu_+}{Q_+} - \frac{\mu_-}{Q_-}\right)\right]. \tag{4.62}$$

In (4.62) the first term is the Coulomb attractive interaction, and the last term is the modified repulsive interaction. The second term is the ion–dipole interaction, and the third one is the dipole–dipole term. The last two terms represent the induced dipole interaction, also referred to as the self-energy of polarization, where the $\alpha_\pm$ are the polarizabilities of the free ions. All of these interactions have been discussed in Section 1 of this chapter. Q_i and Q_j, the shell model parameters, have dimensions

[3] See Appendix E.

of charge and may be thought of as shell charges, but are treated as adjustable parameters.

The equilibrium of the system is determined by

$$\frac{\partial E}{\partial r} = 0, \qquad \frac{\partial E}{\partial \mu_+} = 0, \qquad \frac{\partial E}{\partial \mu} = 0.$$

The application of (4.62) to molecules and crystals, evaluation of the equilibrium conditions, and determination of the various parameters is discussed in Section 3C.

3. Some Applications of the Potential Functions

Potential functions have been applied to a variety of solid state phenomena. At the empirical level, they are useful for extrapolation and interpolation as well as for relating different physical properties to each other and for systematizing large bodies of experimental information. At the more theoretical level, there has been a great deal of progress in recent years in molecular dynamic simulations of solids and solid state processes. These can only be done when interaction potentials are available. These may, of course, be constructed based on quantum mechanics, but often at a heavy price in complexity. The simple potential functions described in this chapter have often been found adequate to explain a variety of solid state phenomena and can yield accurate numerical values for the thermodynamic properties of ionic crystals. We discuss in this section some applications of these potential functions.

Although, thus far, we have presented potential functions exclusively in terms of pairwise interactions between isolated ions, atoms, and molecules, we can apply functions of similar form to condensed phases by assuming that the internal energy (cohesive energy) per atom or ion is given by

$$E = -\frac{1}{2}\sum_i \phi(r_i),$$

where r_i is the distance between interacting ions and ϕ is the potential. This assumption is implicit in the treatment of condensed phases that follows.

A. Experimental Methods of Determining m and n

We will present two methods that, in principle, can be used to determine the product mn and the sum $m + n$ from a knowledge of the compressibility and its pressure dependence. Let us begin by deriving

$$\frac{\partial^2 E}{\partial V^2} = (\chi V)^{-1}, \tag{4.63}$$

where E is the internal energy and χ is the isothermal compressibility. Because the Helmholtz free energy is given by

$$F = E - TS, \tag{4.64}$$

$F = E$ at absolute zero or, in the notation developed in Chapter 2, $F(T = 0) = E_0$, where E_0 is the collective bond energy of the ensemble at absolute zero and the zero-point vibrational energy has been neglected. Consequently,

$$\left(\frac{\partial F}{\partial V}\right)_{T=0} = -P = \left(\frac{\partial E_0}{\partial V}\right)_{T=0} \tag{4.65}$$

and

$$\frac{\partial^2 F}{\partial V^2} = -\left(\frac{\partial P}{\partial V}\right)_{T=0} = \frac{\partial^2 E_0}{\partial V^2}$$

or

$$\frac{\partial^2 E_0}{\partial V^2} = (\chi V)^{-1}. \tag{4.63}$$

Considering only the pairwise potential so that we can make use, ultimately, of the Mie equation, we write for a single atom of volume Ω,

$$\Omega = \alpha r^3, \tag{4.66}$$

where r has the meaning of some effective or average interatomic spacing and α is a geometric constant dependent upon the crystal structure. By taking ϕ_0 as the electrostatic bond energy per atom at $T = 0$, we can write, by analogy to (4.63),

$$\frac{d^2\phi_0}{d\Omega^2} = \frac{1}{3\alpha r^2}\frac{d}{dr}\left(\frac{d\phi_0}{d\Omega}\right) \tag{4.67}$$

and finally obtain

$$\chi^{-1} = \frac{1}{9\Omega}\left[r^2\frac{d^2\phi_0}{dr^2} - 2r\frac{d\phi_0}{dr}\right]. \tag{4.68}$$

The difference between ϕ at 0 K (i.e., ϕ_0) and ϕ is that the latter includes the vibrational energy of the condensed phase and the kinetic energy of the separated particles in the reference or standard state. For solids these corrections are small, but one should be aware of them. Thus, the differentiation in (4.67) and (4.68) is carried out on ϕ of (4.43), and ϕ_0 per molecule is finaly equated to E_0 per mole, and Ω with V. Thus,

$$\chi^{-1} = -mn\phi_0/9\Omega = -mnE_0/9V. \tag{4.69}$$

Needless to say, (4.69) works best for relatively small values of the compressibilities. Table 4.1 lists the values of mn along with the other parameters in (4.69) for a variety of different substances. For ionic materials ($m = 1$), see (4.58), and, thus, the mn column directly gives n, the exponent for the repulsive interaction.

A second method determines the sum $m + n$ from the pressure dependence of the compressibility. Commencing by taking the first and second derivatives of ϕ, as given by (4.43) with respect to r, and substituting into (4.63), we arrive after a certain amount of algebra, at

$$\chi^{-1} = \frac{mn\phi_0}{9\Omega(m-n)}\left[(n+3)\left(\frac{r_e}{r}\right)^n - (m+3)\left(\frac{r_e}{r}\right)^m\right]. \tag{4.70}$$

To determine the variation of χ with pressure, we take the derivative

$$\frac{d\chi^{-1}}{dP} = \frac{d\chi^{-1}}{d\Omega}\frac{d\Omega}{dP} = -\chi\Omega\frac{d\chi^{-1}}{d\Omega} = -\frac{\chi r}{3}\frac{d\chi^{-1}}{dr}. \tag{4.71}$$

Carrying out the differentiation after substituting αr^3 for Ω in (4.70) and then substituting the results in (4.71) gives

$$\frac{d\chi^{-1}}{dP} = -\frac{1}{3}\left[\frac{-(n+3)^2(r_e/r)^n + (m+3)^2(r_e/r)^m}{(n+3)(r_e/r)^n - (m+3)(r_e/r)^m}\right].$$

Letting, as before, $r_e/r \to 1$ gives

$$\frac{d\chi^{-1}}{dP} = \frac{1}{3}(n + m + 6). \tag{4.72}$$

Table 4.1. The product mn for selected solids.[a]

Substance	T(K)	V (ccs/mole)	$\chi \times 10^{12}$ (cm^2/dyne)	$-E_0$ (cal/mole)	mn
C	273.1	3.418	0.16	107,400	27.0
LiF	273.1	9.818	1.53	238,00	5.80
KCl	273.1	37.43	5.64	163,000	8.75
RbBr	273.1	49.23	7.95	152,000	8.76
He	2.71	27.7	7840	14.26	53.3
Hg	0	14.035	3.55	15,327	55.5
Hg	273.1	14.7556	3.946	15,142	53.1

[a] Adapted from E. A. Moelwyn-Hughes, *Physical Chemistry*, p. 323, Pergamon Press, 1961.

The values for m and n, determined by these and other methods, yield fairly good potential functions for liquids, including liquid Hg, for which, upon comparison with experiment, $m = 6$ and $n = 9$ were quite clearly established. The $m = 6$ value is to be expected if the attractive interaction originates in dispersion effects. For other liquids, m and n vary widely. For crystalline metals such potentials are usually not very satisfactory. However, the specialized potential as embodied in the Morse equation (4.47) has been quite successful in describing the basic properties of a number of metals.[4] It has also been found quite satisfactory for molecular dynamic simulations of the equation of state of body-centered cubic iron.

B. The Lennard–Jones Potential and the Rare Gas Crystals

As stated in Section 2D, the 6-12 Lennard–Jones potential, given by (4.57), quite accurately describes the properties of the rare gas crystals. However, once a suitable potential is available, or has been chosen, there is a variety of theoretical techniques available for calculating the equation of state and related properties. Here our discussion is limited to the Lennard–Jones potential in combination with self-

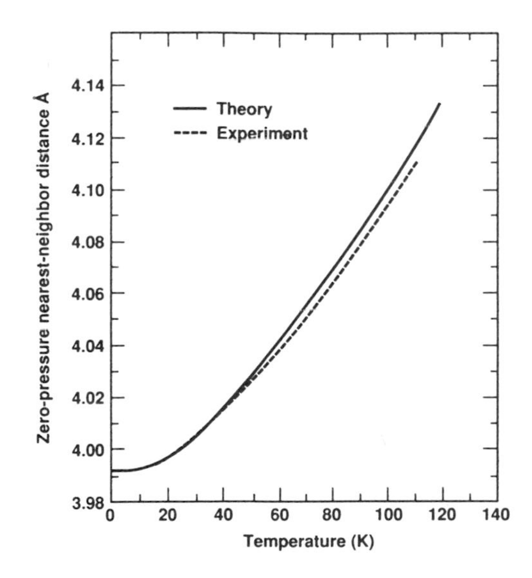

Figure 4.6.

A comparison of the zero-pressure nearest-neighbor distance for Kr with theory. [After A. Paskin, A. M. Llois de Kreiner, K. Shukla, D. O. Welch, and G. J. Dienes, *Phys. Rev.*, **25B**, 1297 (1982).]

[4] See, for example, L. A. Girifalco and V. G. Weizer, *Phys. Rev.* **114**, 687 (1959).

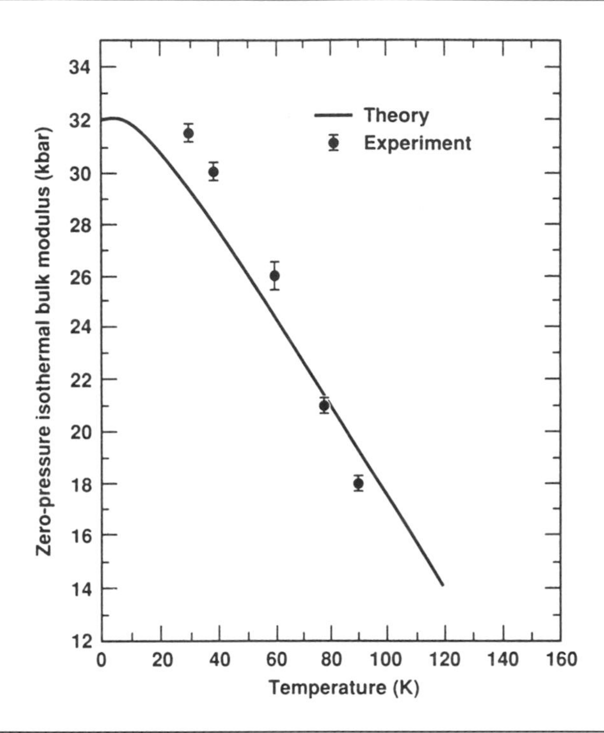

Figure 4.7.

A comparison of the zero-pressure isothermal bulk modulus for Kr with theory. (After A. Paskin *et al.*, loc. cit.)

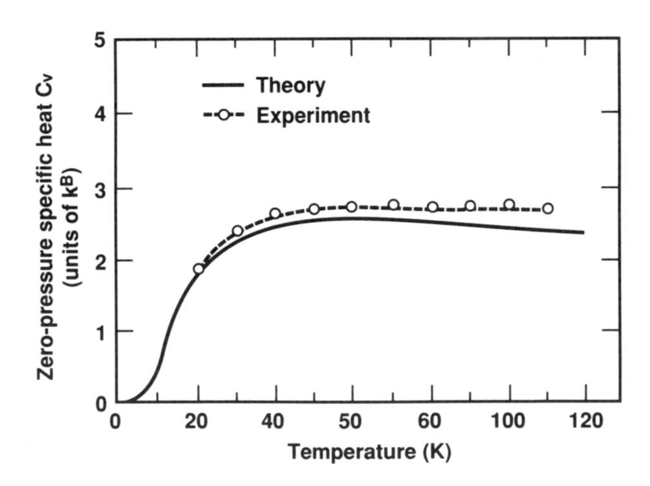

Figure 4.8.

A comparison of the zero-pressure specific heat C_v for Kr with theory. (After A. Paskin *et al.*, loc. cit.) (k_B is Boltzmann's constant).

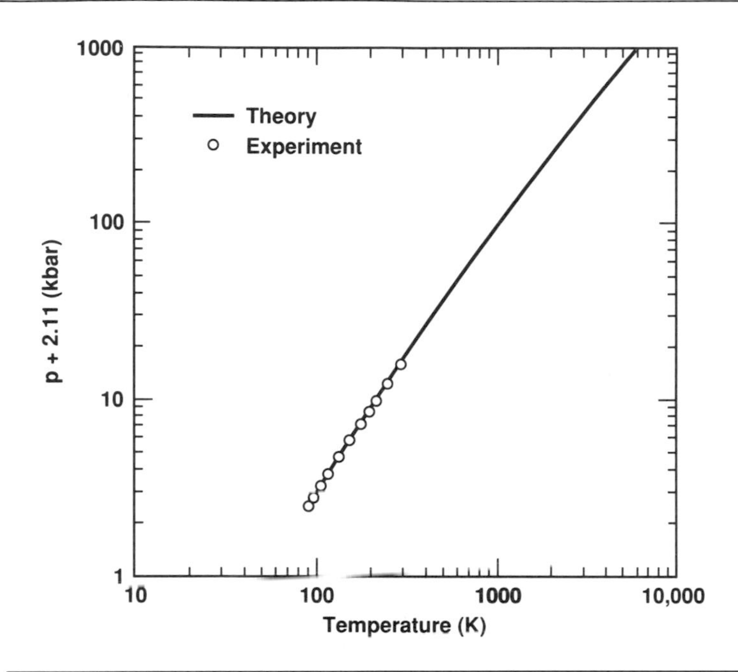

Figure 4.9.

The pressure along the melting curve for Ar. The pressure plus 2.11 kbar is plotted vs. the melting temperature. (After A. Paskin *et al.*, loc. cit.)

consistent phonon calculations, which gives excellent results for rare gas crystals. Figure 4.6 shows the representative behavior of the zero-pressure nearest-neighbor distance as a function of temperature. The behavior of the bulk modulus is shown in Fig. 4.7 and that of the specific heat in Fig. 4.8. One can also extrapolate, by this and similar techniques, over very wide ranges of the variables inaccessible to current experimentation. This is illustrated in Fig. 4.9, where the pressure versus melting temperature plot for argon is shown along the melting line over a very wide range.

C. Ionically Bonded Molecules and Crystals

A somewhat different illustration of the use of potential functions is discussed in this section. The most stable configuration of alkali halide clusters can be calculated from

$$E(r) = -\frac{Z^2e^2}{r} - \frac{Z(\mu_+ + \mu_-)e}{r^2} - \frac{2\mu_+\mu_-}{r^3} + \frac{\mu_+^2}{2\alpha_+} + \frac{\mu_-^2}{2\alpha_-} + A\exp\left[-b\left(r + \frac{\mu_+}{Q_+} - \frac{\mu_-}{Q_-}\right)\right]. \tag{4.62}$$

This equation was discussed in some detail in Section 2E, and its relation to the "polarization catastrophe" is described in Appendix E. The dertermination of the various parameters that appear in (4.62) are derived from the equilibrium conditions

$$\frac{\partial E}{\partial \mu_+} = 0, \qquad \frac{\partial E}{\partial \mu_-} = 0, \qquad \frac{\partial E}{\partial r} = 0$$

and can be used to determine μ_+ and μ_- in terms of the α's (see Appendix E) and also to evaluate the relation between the cohesive energy and the equilibrium interionic separation r_e. The constants A and b are specific to a given ionic salt and are usually determined from measurements of the cohesive energy, lattice parameters, and compressibility.

The structure and properties of ionic crystals are discussed in detail in Chapter 5, and it is shown there that ionic radii can be assigned to a given positive and negative ion independent of the nature of the ionic molecule or crystal. For the present discussion we use a consistent set of values of A and b, and r_+ and r_-, as determined for all the alkali halides by Tosi and Fumi and listed in Table E.1 of Appendix E. These values guarantee very close agreement with the experimental cohesive energies. The free-ion polarizabilities, α_i, and the shell model parameters, Q_i, are determined from molecular data, such as binding energies, vibrational frequencies, and dipople moments. The best values of α_i and Q_i are obtained by fitting a very large number of measurements for all the alkali halide crystals and monomer molecules. Just as in the case of the ionic radii, α_+, α_-, Q_+, and Q_- values are found for all the alkali and halide ions that are characteristic of a given ion, independent of the salt, that describe the crystalline and molecular properties with good accuracy. These parameters are also listed in Table E.1 of Appendix E.

Since all the characteristic interaction parameters have been assigned, one can now use (4.62) with no adjustable parameters to calculate the properties and equilibrium configuration of dimer and trimer molecules and larger clusters of the alkali halides. A computerized iteration technique tried successive molecular structures in order to arrive at the one with minimum free energy. The equilibrium configuration of the dimer molecules was found to be a planar rhombus with equal interionic distances. The distortion arises from the unequal ionic radii. Large halide ions, for example, push the corresponding corners of the square apart and decrease the metal–halogen–metal angle, θ, as illustrated in Fig. 4.10 in full agreement with experiment. For trimers a "noncubic" configuration, viz., a perfect planar hexagon, is found to be the most stable configuration. Experimentally, it is known only that the trimer is a highly symmetrical planar molecule. It is of interest to note that qualitatively correct molecular structures are predicted even by the simple Born–Mayer potential (4.58)—

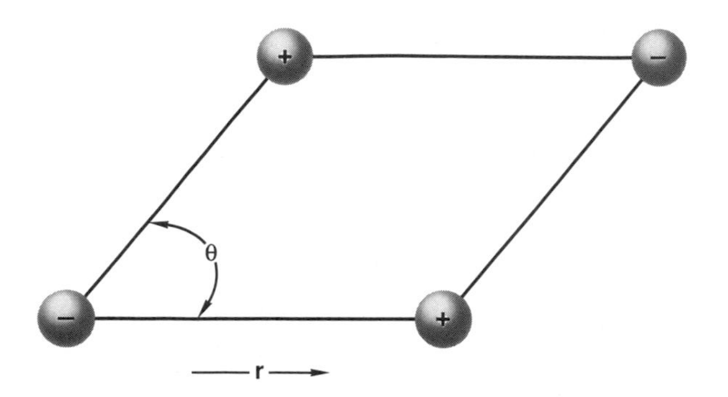

Figure 4.10.

The equilibrium configuration for four ions (two molecules).

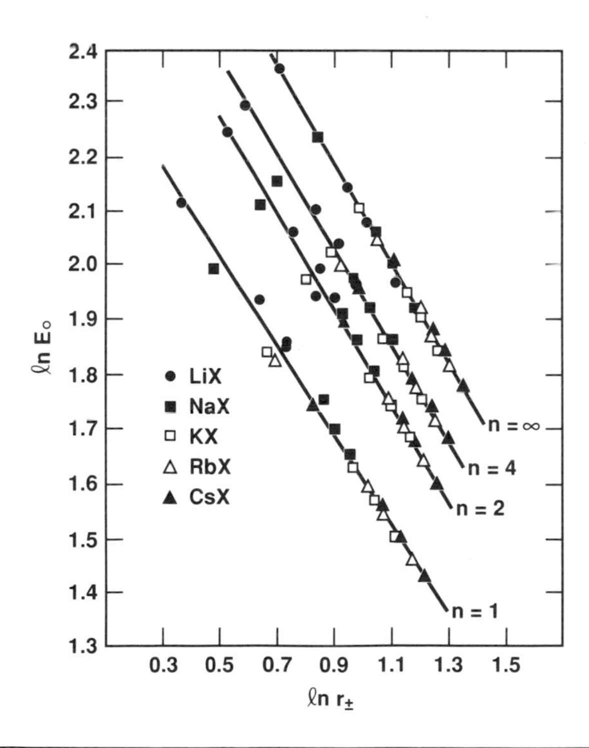

Figure 4.11.

Calculated variation of the binding energy per ion pair E_0 (in ev), with the equilibrium near-neighbor spacing $r_{\pm}$ (in Å), for alkali halide clusters $(MX)_n$. [After D. O. Welch, O. W. Lazareth, G. J. Dienes, and R. D. Hatcher, *J. Chem. Phys*, **68**, 2159 (1978).]

that is, without using the shell model. However, this potential predicts θ values in poor agreement with experiment.

The next higher cluster contains four molecules, $n = 4$. As might be expected by now, the equilibrium configuration is the distorted cube, in complete analogy to the rhombus of Fig. 4.10. The configuration for $n = 6$ is again noncubic, namely a double hexagon.

One can obviously proceed now and build up larger and larger clusters, and the physical properties vary in a nicely systematic way with increasing n. The variation in binding energy, for example, varies as illustrated in Fig. 4.11, as a function of the equilibrium interionic spacing $r_{\pm}$ for all the alkali halides, showing very neat progression toward the crystal values as n approaches infinity ($n \to \infty$). Other properties vary in a similar way with increasing n. By the time n reaches 44, one is dealing with a cubic crystal whose properties approach that of an infinite crystal.

Chapter 4 Exercises

1. Two charges of magnitude 100 and -50 esu are located in the XY plane at (0, 0) and (2 cm, 0).
a. Find a point on the X-axis where the electrical force is zero.
b. Find the potential in part (a) at any point in the XY plane. Is the zero point determined in (a) a stable or unstable equilibrium? (Construct approximate equipotential surfaces.)

2. In the general pair potential function (Fig. 4.5) the force is zero at $r = r_e$ and at $r = \infty$. Find the value of the maximum force for the (a) Morse potential, (b) Lennard–Jones potential, and (c) Mie potential.

3. An interesting extension of the Mie equation involves the addition of a second attractive interaction. Thus,

$$\phi = Ar^{-n} - Br^{-m} - Cr^{-q}.$$

Show that in this case

$$\phi_e = (B/r_e^m)\left(\frac{m}{n} - 1\right) + (C/r_e^q)\left(\frac{q}{n} - 1\right).$$

4. What is the configuration of minimum (greatest) potential energy for a single positive ion interacting with a simple linear dipole?

5. For the Mie potential, discuss the influence of increasing the attractive interaction by increasing B (with A, n, and m unchanged) on

(1) equilibrium separation r_e, and
(2) cohesive energy, ϕ_e.

Calculate r'_e/r_e and ϕ'_e/ϕ_e for $B'/B = 2$ for various values of n and m. Discuss the sensitivity to the exponent of the attractive potential.

6. Two adjacent potential wells with an energy barrier between them play an important role in solid state transitions (see Chapters 8 and 14). Construct such double well systems with two intersecting quadratic potentials and investigate:
a. The energy barrier location and height for two identical potentials as a function of the separation between the two minima.
b. As in a., but for two wells of different depth.
c. As in a., but for two wells of different steepness.

7. Derive analytical expressions for the energy barrier location and height for two general intersecting quadratic potentials and apply them quantitatively to parts a., b., and c. of Exercise 6. Use the general quadratic form $y = a(x - b)^2 + c$.

8. Replace the quadratic potentials of Exercise 6 with two identical Lennard–Jones potentials with the attractive parts intersecting to form the energy barrier. Discuss the location, the height, and the limiting height of the energy barrier as the two minima are moved apart.

Additional Reading

E. A. Moelwyn-Hughes, *Physical Chemistry*, Chap. VII, Pergamon Press, New York (1964). A very thorough treatment of the subject presented in a clear, understandable style.

H. R. Glyde and M. L. Klein, *Crit. Rev. Solid State Sci.*, **2**, 181 (1971). A review of self-consistent phonon and related formalisms.

T. P. Martin, "Alkali Halide Clusters and Microcrystals," *Phys. Rep.*, **95**, 167–199 (1983).

Chapter 5

Ionic Crystals

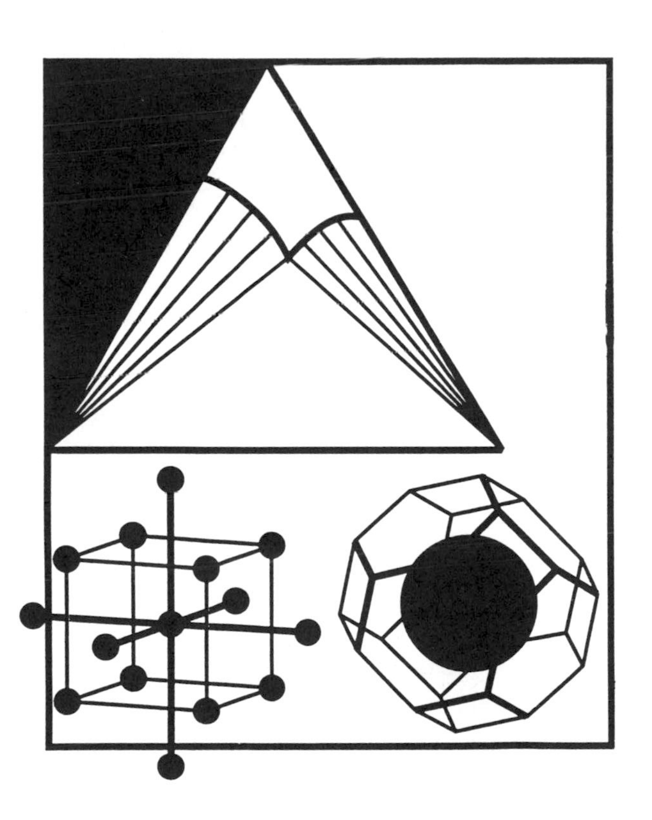

Chapter 5

Contents

Some circumstantial evidence is very strong, as when you find a trout in the milk.

—Henry D. Thoreau, *Journal* (November 11, 1850)

Ionic Crystals

The thermodynamic stability of a crystal with purely ionic bonds is greatest for that particular structure that maximizes the coulombic attractive forces between anions and cations while minimizing the repulsive interactions between ions of like charge. The sum of these potentials that minimizes the free energy clearly must embody the geometry of the crystal structure. In turn, it can be shown that the type of crystal structure is dictated by the relative size of the constituent anions and cations. Although most so-called ionic compounds have bonds that are a mixture of ionic and covalent character, one can frequently assume these compounds to be strictly ionic when predicting the expected crystal structure. Consequently, the principles and calculations that follow are much more widely applicable than just to the alkali metal halides, which are the commonly employed pedagogical examples. It is perhaps appropriate to remark that alkali metal halides, with the exception of CsCl, CsBr, and CsI, as well as the oxides and sulfides of the alkaline earths and the silver halides, excepting the iodide, possess the rock salt structure, which may help to explain why it is the preferred prototypic structure for theoretical consideration.

Before we examine the basis for defining ionic sizes and their influence on crystal structure, we will elucidate a method, first proposed by Evjen[1], of doing the summation of the coulombic potentials for simple crystal lattices.

[1] H. M. Evjen, *Phys. Rev.*, **39**, 680 (1932). There are several ways to evaluate the Madelung constant, but Evjen's is probably the most straightforward and revealing of the physics involved.

1. The Madelung Constant and Lattice Energies

In what follows, we will assume that all ions are point charges located on rational lattice sites and that each pair of ions interacts with the point charge Coulomb potential,

$$\phi = z_i z_j e^2/r_{ij}, \tag{5.1}$$

where e is the electronic charge, z_i and z_j represent the valence of ions i and j, and r_{ij} is the distance separating the two ions. The individual terms in (5.1) are positive or negative, depending on the signs of the ionic charges z, with opposite charges giving a contribution that is attractive and hence has a negative sign. We might indeed anticipate that the sum over all pairs, which gives the total static Coulomb energy of the crystal, will be dominated by the attractive nearest-neighbor positive and negative ion interactions and will be negative. If the net value of the potential energy were positive, the crystal would not exist. It would dissociate into its constituent ions. Upon summation, the total Coulomb potential energy is usually expressed as

$$\phi = -Mz^2e^2/r, \tag{5.2}$$

where M is a constant depending only on the crystal structure; e is the electronic charge; and r is the distance between nearest neighbors of opposite sign, which is generally the distance of closest approach and is the sum of the ionic radii. The calculation of ionic radii will be discussed later in this chapter. The constant M is called the Madelung constant,[2] which can be calculated in several ways.

A. The Evjen Solution

This method, illustrated here for the sodium chloride structure, sums the coulombic potential between a designated ion and all of its neighbors in the successive ith shells. One allows i to increase from unity, which represents the nearest neighboring shell, to whatever ultimate value produces the desired accuracy, or exhausts the patience of the researcher. Figure 5.1 shows a few of the distances used in the summation.

If the distance between nearest neighbors is designated by r, then the distance of the ith shell from the origin is given by

$$r_i = r\sqrt{n_1^2 + n_2^2 + n_3^2}, \tag{5.3}$$

[2] E. Madelung, *Phys. Z.* **19**, 524 (1918).

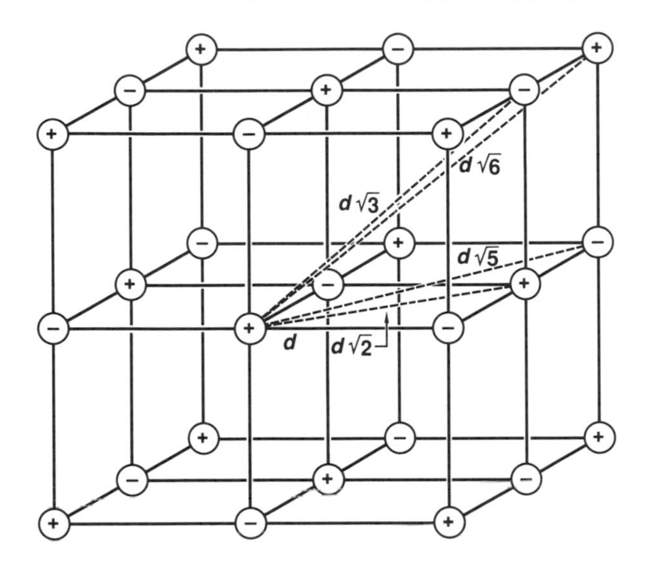

Figure 5.1.

Illustrating the distances between a specified ion and its neighbors in the first, second, third, fifth, and sixth shells.

where n_1, n_2, and n_3 are the coordinates of a single ion in the ith shell, the reference ion being located at the origin. However, since all ions in the ith shell are the same distance from the origin, we need only multiply (5.3) by the corresponding number of ions in the ith shell to account for all of their interactions with the one at 0 0 0. Table 5.1 lists all distances out to and including the ninth shell. A minus sign precedes those values that are for interactions between ions of opposite charge. Consequently,

Table 5.1. Madelung constant terms for NaCl.

$n_1 n_2 n_3$	Number of ions (N_i)	$(n_1^2 + n_2^2 + n_3^2)^{1/2}$	$N_i(n_1^2 + n_2^2 + n_3^2)^{-1/2}$
1 0 0	6	$\sqrt{1}$	−6.000
1 1 0	12	$\sqrt{2}$	8.485
1 1 1	8	$\sqrt{3}$	−4.620
2 0 0	6	$\sqrt{4}$	3.000
2 1 0	24	$\sqrt{5}$	−10.730
2 1 1	24	$\sqrt{6}$	9.800
2 2 0	12	$\sqrt{8}$	−4.244
2 2 1	24	$\sqrt{9}$	−8.000
2 2 2	8	$\sqrt{12}$	2.310

we now can write

$$\phi_i = \frac{z^2e^2}{r} N_i(n_1^2 + n_2^2 + n_3^2)^{-1/2}, \tag{5.4}$$

where N_i is the number of ions with coordinates n_1, n_2, n_3, and where the sign is positive if $n_1 + n_2 + n_3$ is even and negative if it is odd. The standard state is taken to be the ions at infinite separation, which at 0K is the state of zero energy.

The total potential, or coulombic potential energy, is now given by adding the individual values of ϕ_i out to the ith shell. Unfortunately, this series shows no tendency to converge. However, by including only that fraction of each ion that lies within each shell, convergence is achieved quite rapidly, and it is this artifice that makes this method work. The scheme is explained pictorially in Fig. 5.2, which shows the first, second, and third neighboring shells labeled with their appropriate distances in units of r, viz. 1, $\sqrt{2}$ and $\sqrt{3}$. Also drawn to scale are examples of ions with coordinates, 2 2 2, 2 1 1, and 2 0 0. Now ions occupying faces of the first coordination

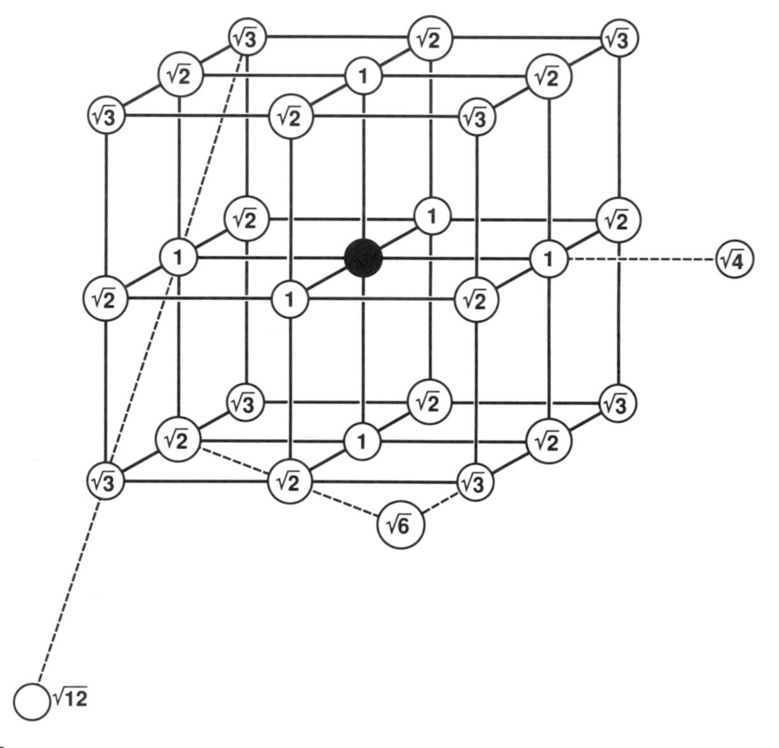

Figure 5.2.

The first coordination shell plus examples from other shells for the NaCl structure.

cube—that is, those with coordinates 1 0 0—are considered to have one-half their volumes within the first shell. In like manner, those on the edges of the 1 1 0 type have one-fourth their volume within the first shell, and those at the corners, in the 1 1 1 positions, contribute one-eighth.

The sum of the first three geometric terms is

$$\frac{-6.000}{2} + \frac{8.485}{4} - \frac{4.620}{8} = -1.457.$$

The second coordination shell includes the first completely, but contains only fractional volumes of the ions in the second shell. The sum now becomes

$$-6.000 + 8.485 - 4.620 + \frac{3.000}{2} - \frac{10.730}{2} + \frac{9.800}{2}$$

$$+ \frac{4.244}{4} - \frac{8.000}{4} + \frac{2.310}{8} = -1.750 \tag{5.5}$$

Expanding the dimensions of the shell leads utimately to convergence on the value of $M = -1.747558$. This is directly substituted into (5.2) and multiplied by N, which is the number of NaCl molecules in the crystal, to obtain the electrostatic binding energy

$$\phi_a = -1.747558 N z^2 e^2/r. \tag{5.6}$$

The multiplicative factor becomes N rather than $2N$ because our calculation is based upon the total number of bonds, not the total number of ions.

The method we have just described encounters some difficulty when naively applied

Table 5.2. Values of the Madelung constant for various structures.

Zinc blende (ZnS)	1.63806
Wurtzite (ZnS)	1.64132
Sodium chloride	1.747558
Cesium chloride	1.762670
Cuprite (Cu_2O)	4.3224
β-Quartz (SiO_2)	4.4394
High quartz (SiO_2)	4.4633
Cadmium iodide	4.71
Anatase (TiO_2)	4.800
Rutile (TiO_2)	4.816
Fluorite (CaF_2)	5.03878
Antifluorite	5.03878
Corundum (Al_2O_3)	25.0312

to the CsCl structure. Depending upon the charge of the central ion, one arrives at two different limits for the convergence. This arises because the boundary surfaces of the succeeding shells are not electrically neutral but oscillate in their residual charge. It has been proven that the correct value is the average of these two numbers. This problem is avoided if the calculation is based upon the rhombohedral lattice for which the CsCl structure has neutral boundary planes.

Values of the Madelung constant for some of the more common ionic crystal structures are listed in Table 5.2.

B. The Ewald Solution

Other methods of summation are mentioned for the sake of completeness. Madelung.[3] in his original work employed a method of relying on a summation over the reciprocal lattice. He transformed the electrostatic potential of a neutralized Bravais lattice of unit point charges, at a point outside the lattice, into a rapidly converging series over the reciprocal lattice. A very important modifiction and improvement on this idea was carried out by Ewald.[4] The important point about the Ewald method is that the Coulomb potential so calculated by a sophisticated mathematical technique is valid at any point in the crystal and not only at the lattice sites. This is clearly important for any calculation where ions have moved off their lattice sites (e.g., imperfections, color centers, etc.).

2. The Repulsive Potential

A. The Inverse Power Law

The net electrostatic potential that we have just calculated is attractive, i.e., negative for all values of the interionic separation distances. Now we must account for the repulsive interaction that must necessarily exist in order for there to be a minimum in the potential energy at r_0, the equilibrium ion separation distance (see, for example, Fig. 4.5). The repulsive forces derive from the interpenetration of electron clouds that surround anions and cations alike. As neighboring ions are pushed closer together, their electron orbitals begin to overlap, and if these are already filled with electrons

[3] E. Madelung loc. cit.

[4] P. P. Ewald, *Ann. Phys.*, **64**, 253 (1921).

they are forced into higher energy states as a consequence of the Pauli exclusion principle. It is this increase in energy that accounts for the compressibilities of solids.

Because there is repulsion between all ions, regardless of their charge, there are three characteristic potentials, viz., ϕ_{++}, ϕ_{--}, and ϕ_{+-}. Continuing with NaCl for our sample calculation, we can use the distances between ions of like and unlike charges already computed for the first two neighboring cells and listed in Table 5.1. Compressibility mesurements show that repulsive potentials increase steeply with diminishing interionic separation. As pointed out in Chapter 4, all electrostatic potentials can be represented as a function of the reciprocal distance between charges raised to some power. In other words, the repulsive energy can be written as[5]

$$\phi(r) = \phi/r^n, \tag{5.7}$$

where n is expected to be relatively large. A general function describing the total noncoulombic repulsive potential ϕ_r is expressed by

$$\phi_r = r^{-n}\left[\frac{6\phi_{+-}}{(\sqrt{1})^n} + \frac{12\phi_{++}}{(\sqrt{2})^n} + \frac{8\phi_{+-}}{(\sqrt{3})^n} + \cdots\right] \tag{5.8}$$

Experimental data show that $n \simeq 9$, and, by assuming this value, (5.8) converges to

$$\phi_r = r^{-9}(6.075\phi_{+-} + 0.550\phi_{++}), \tag{5.9}$$

where the central reference ion is chosen to be a cation; an analogous equation describes the situation when it is an anion. Therefore, the total repulsive energy for a crystal of N NaCl molecules is obtained by multiplying each of these two expressions by $N/2$, the number of cations or anions, and adding them together, leading to

$$\phi_r = r^{-9}N[6.075\phi_{+-} + 0.550(\phi_{++} + \phi_{--})/2]. \tag{5.10}$$

The term in brackets is a constant for each specific crystal type and, multiplied by N, can be abbreviated as A, so that (5.10) becomes

$$\phi_r = A/r^n. \tag{5.11}$$

The attractive and repulsive potentials, (5.6) and (5.11), combine to give the total lattice energy

$$\phi_a + \phi_r = E_0 = -1.748N\frac{z^2e^2}{r} + \frac{A}{r^n}, \tag{5.12}$$

where the subscript zero on E indicates $T = 0\,\text{K}$. Under equilibrium conditions,

[5] The following treatment of repulsive potentials for NaCl follows that of Slater, *Introduction to Chemical Physics*, Chap. 23, McGraw-Hill, New York, (1939).

E_0, is at its minimum at $r = r_e$, leading to

$$\frac{dE_0}{dr} = 0 = 1.748\,\frac{Nz^2e^2}{r^2} - \frac{nA}{r^{n+1}}, \tag{5.13}$$

and solving for A, we obtain

$$A = \frac{1.748}{n}\,Nz^2e^2r_e^{n-1}. \tag{5.14}$$

Taking the value of A from (5.14) and substituting back into (5.12) yields

$$\Delta E_0 = -1.748\,\frac{Nz^2e^2}{r_e}\left[\frac{r_e}{r} - \frac{1}{n}\left(\frac{r}{r_e}\right)^n\right], \tag{5.15}$$

where ΔE_0 is the electrostatic lattice energy as a function of the ionic separation distance at 0 K.

We can relate our model-dependent expression for the internal energy to the earlier derived thermodynamic equations of state as follows: at 0 K, (3.23) beomes

$$\Delta E_0 = E_{00} + V_0\left[P_0\,\frac{V_0 - V}{V_0} + \frac{1}{2}\,P_1\left(\frac{V_0 - V}{V_0}\right)^2 + \frac{1}{3}\,P_2\left(\frac{V_0 - V}{V_0}\right)^3 + \cdots\right], \tag{5.16}$$

where E_{00} is the electrostatic lattice energy at 0 K for the case of equilibrium separation at $P = 0$, (and hence $V = V_0$).

In order to make (5.16) compatible with (5.15), we must express V and V_0 in terms of r_e. Imagine that each ion is to be circumscribed by a cube of volume cr^3, where c is a geometric constant that is characteristic of the specific crystal structure. The important feature of this device is that it allows us an expression to be derived in terms of r that directly reflects the interionic separation and that, in turn, is governed by the balance between the attractive and repulsive forces as prescribed by (5.15). Clearly, this dimension could be the ionic radius or any quantity that is a measure of the interionic distance, providing (5.15) is expressed in the same units. Thus, in the present case $V_0 = Ncr_e^3$, and similarly, $V = Ncr^3$.

$$\frac{V_0 - V}{V_0} = \frac{r_e^3 - r^3}{r_e^3} = 3\left(\frac{r_e - r}{r_e}\right) - 3\left(\frac{r_e - r}{r_e}\right)^2 + \left(\frac{r_e - r}{r_e}\right)^3. \tag{5.17}$$

Substituting into (5.16) and truncating the series after inclusion of the cubic term leads to

$$\triangle E_0 = E_{00} + Ncr_e^3\left[\frac{9}{2}\,P_1^\circ\left(\frac{r_e - r}{r_e}\right)^2 - 9(P_1^\circ - P_2^\circ)\left(\frac{r_e - r}{r_e}\right)^3\right]. \tag{5.18}$$

In order to give (5.15) the same form as (5.18), we perform a binomial expansion of the former after first having done some algebraic manipulation. The term in brackets in (5.15) can be rewritten as

$$\frac{r_e}{r} - \frac{1}{n}\left(\frac{r_e}{r}\right)^n = \frac{r_e - r}{r} + 1 - \frac{1}{n}\left(\frac{r_e - r}{r} + 1\right)^n. \tag{5.19}$$

The last term in parentheses in (5.19) is of the form $(1 + x)^n$ and can be expanded accordingly. Substituting the resulting series into (5.15) and discarding all terms beyond the cubic results in.

$$\Delta E_0 = -1.748\frac{z^2e^2}{r_e}\left\{\left(1 - \frac{1}{n}\right) + \left(\frac{n+1}{2} - 1\right)\left(\frac{r_e - r}{r_e}\right)^2 + \left[\frac{(n+1)(n+2)}{6} - 1\right]\left(\frac{r_e - r}{r_e}\right)^3 \cdots\right\}. \tag{5.20}$$

By comparing (5.20) termwise with (5.18), we can establish the following equivalencies:

$$E_{00} = -1.748N\frac{z^2e^2}{r_e}\left(1 - \frac{1}{n}\right), \tag{5.21}$$

$$P_1^\circ = \frac{1.748}{18}\frac{z^2e^2}{r_e^4}(n-1), \tag{5.22}$$

$$P_2^\circ = \frac{1.748}{108}\frac{z^2e^2}{r_e^4}(n-1)(n+10). \tag{5.23}$$

The calculated values for P_1° and P_2° vary considerably in their agreement with their experimental counterparts for the alkali halides. However, in even the worst cases agreement is generally better than 30%.

B. The Born Method and Lattice Energies

Two major functional forms have been used to represent the repulsive interactions. The inverse power potential can be thought of as some sort of average eletrostatic potential, since all vary as an inverse power of the distance (see Section 2A in Chapter 4). The other commonly used form is the Born–Mayer exponential repulsion. This is the more modern one, justifiable by quantum mechanical arguments.

Here we show how to obtain the characteristic parameters of these repulsive interaction functions from experimental data, specifically from cohesive energies and compressibilities. For NaCl then, using the inverse power law, we have, with only six nearest neighbors (i.e., for one unit cell)[6]

$$\phi = -Me^2/r + 6A/r^n, \tag{5.24}$$

where M is the Madelung constant and A and n are the repulsive interaction parameters to be determined. A and n are specific to a given ionic compound, although one value of n may be approximately valid for a set of homologous salts such as the alkail halides. Application of the equilibrium condition $\partial\phi/\partial r = 0$ gives

$$6A = (Me^2/n)(r_e)^{n-1}, \tag{5.25}$$

where r_e is the equilibrium interionic distance. Substitution into (5.24) yields

$$\phi_e = -(Me^2/r_e)(1 - 1/n), \tag{5.26}$$

which gives n in terms of the cohesive energy ϕ_e.

Next, we use the Born–Landé equation, derived in Chapter 4, to relate the compressibility χ to the cohesive energy. This relation, at equilibrium when $\partial\phi/\partial r = 0$, is

$$B \equiv \frac{1}{\chi} = \frac{1}{9V} r^2 \left(\frac{d^2\phi}{dr^2}\right), \tag{5.27}$$

evaluated at $r = r_e$. Here B is the bulk modulus, and V is the molar volume. Carrying out the differentiation gives, for the bulk modulus, two equivalent expressions:

$$B = \frac{Me^2}{18r_e^4}(n - 1) = \frac{6nA}{18r_e^{n+3}}(n - 1). \tag{5.28}$$

From (5.26) and (5.28), n and A can be evaluated from experimental values of ϕ_e and B.

Treating the Born–Mayer exponential form the same way, starting with

$$\phi = -Me^2/r + 6A\exp(-br), \tag{5.29}$$

where A and b are now the Born–Mayer repulsive parameters and are specific to a given ionic compound, although again a single value of b may be a reasonable approximation for the alkali halides. [A in (4.38) and (4.39) is numerically not the same as in (5.29).] Equilibrium now gives

$$Me^2/r_e^2 = 6Ab\exp(-br_e), \tag{5.30}$$

---

[6] This approximation is partially justified by the short-range nature of the repulsive interaction.

and, therefore,

$$\phi_e = -(Me^2/r_e)(1 - 1/br_e), \tag{5.31}$$

a relation connecting the cohesive energy, the lattice parameter, and the repulsive interaction constant b (in the literature the notation $\rho = 1/b$ is often used). Following the prescription of (5.27), we obtain for the bulk modulus B the two equivalent expressions

$$B = \frac{Me^2}{18r_e^4}\left(\frac{b}{r_e} - 2\right) = \frac{6Abr_e^2 \exp(-br_e)(b/r_e - 2)}{18r_e^4}. \tag{5.32}$$

As before, A and b can be evaluated from experimental values of ϕ_e and B by means of (5.31) and (5.32).

The lattice energy may be obtained from the Born equations by combining measured values of the compressibility and Eqs. (5.26) and (4.69) and equating ϕ_e to E_0, which is to say that at 0 K all the internal energy is electrostatic with the exception of the zero-point vibrational energy, which is very small and, hence, neglected.

$$\chi^{-1} = -mnE_0/9V. \tag{4.69}$$

For NaCl, $m = 1$ and (4.69) reduces to

$$\chi^{-1} = -nE_0/9V_0, \tag{5.33}$$

where V_0 is the molecular volume equal to $2r_e^3$. Now taking (5.26) with (5.33), we obtain

$$n = 1 + 18r_e^4/\chi Az^2e^2. \tag{5.34}$$

Eliminating n from (5.26) by substitution of (5.34) finally yields for the lattice energy

$$E_0 = -\frac{Az^2e^2}{r_e}\left(1 + \frac{\chi Az^2e^2}{18r_e^4}\right)^{-1}. \tag{5.35}$$

The accuracy of (5.35) in predicting the lattice energies of the alkali halides is evident from the agreement between the calculated and the experimental results obtained from the experimental Born–Haber cycle, as shown in Table 5.3.

Before leaving the Born model, it is interesting to note along with Moelwyn-Hughes[7] that widely dissimilar elements even in differing physical states can have nearly equal repulsive potentials if they are isoelectronic. An example of this is solid KCl for which $\phi_r = 6.55 \times 10^{-81}$ erg-cm as calculated from (5.35) and (5.11), taking

[7] E. A. Moelwyn-Hughes, *Physical Chemistry*, Pergamon Press, 1961.

Table 5.3. Lattice energies of the alkali halides.[a]

Halide	$r_e \times 10^8$ (cm)	$\chi \times 10^{12}$ (dynes/cm^2)$^{-1}$	n	$-E_0$ (kcal/mole)[b] Equation (5.35)	Experimental
LiF	2.014	1.53	5.86	238	238
LiCl	2.570	3.48	6.66	191	192
LiBr	2.746	4.28	7.00	180	182
LiI	3.010	7.2	6.15	161	170
NaF	2.330	(1.90)	(8.00)	217	214
NaCl	2.849	4.16	8.16	178	179
NaBr	2.982	5.09	8.02	169	171
NaI	3.236	7.1	7.98	156	160
KF	2.679	3.3	8.05	189	189
KCl	3.149	5.64	8.87	163	163
KBr	3.304	6.66	9.08	155	156
KI	3.538	8.54	9.29	146	148
RbF	2.815	(3.64)	(8.80)	178	181
RbCl	3.286	7.4	8.12	154	158
RbBr	3.434	7.95	8.72	152	151
RbI	3.663	9.58	9.49	141	143
CsF	3.005	(3.07)	(13.0)	177	172
CsCl	3.559	5.9	13.1	148	142
CsBr	3.716	7.0	13.2	142	142
CsI	3.952	9.3	12.7	133	135

[a] Taken from E. A. Moelwyn-Hughes, *Physical Chemistry*, p. 557, Pergamon Press, 1961. Based upon calculations by M. Born and J. Mayer, *Z. Phys.*, **75**, 1 (1932) and J. Sherman, *Chem. Rev.*, **11**, 93 (1932).

[b] Equation (5.35) has been multiplied by Avogadro's number to convert to moles.

$n = 9$. The value for ϕ_r is 6.18×10^{-81} erg-cm for a pair of gaseous Ar atoms, which, of course, are isoeletronic with the halide salt.

3. The Born–Haber Cycle

The lattice energy ΔE_0, as calculated in the preceding section, is based upon a standard that has the constituent ions at infinite separation at absolute zero. As a consequence, the kinetic and coulombic energies of the standard state are zero. However, the standard internal energy of formation, ΔE°, from thermochemical measurements more commonly involves the equilibrium phases of the reactants at 1 atm of a pressure and 25°C. Consequently, several steps are needed in order to compare ΔE_0, the electrostatic energy, with the internal energy of formation ΔE_0° at 0 K. The sum of these ingredient energy values is called the Born–Haber cycle, after those first responsible for its use, and it is illustrated for NaCl in Fig. 5.3.

The various energy terms are enumerated as follows:

ΔE_0° The standard internal energy of formation for NaCl(*s*)
ΔE_D The internal energy of dissociation of $Cl_2(g)$
ΔE_v The internal energy of vaporization of elemental Na(*s*)
ΔE_E The internal energy associated with the electron affinity of Cl(*g*)
ΔE_I The ionization energy of Na(*g*)
ΔE_0 Lattice energy

Values of ΔE_D can be obtained from molecular spectra, ΔE_v from the temperature dependence of the equilibrium values of the vapor pressure, and ΔE_I from atomic spectra. The electron affinity ΔE_E is a more elusive quantity and prior to 1970 had been measured only for H, F, Cl, Br, I, C, O_2, and S. Now, values exist for several elements as a result of new spectroscopic techniques.

Although the energy terms are listed as internal energies, they could equally well be taken as enthalpies becuase the calculation is for 0 K. The generally good agreement between the theoretically calculated lattice energies and those obtained from the use of thermochemical measurements in the Born–Haber cycle is apparent in the last two columns of Table 5.3.

In order to convert the electron affinities in Table 5.4 to enthalpies or internal

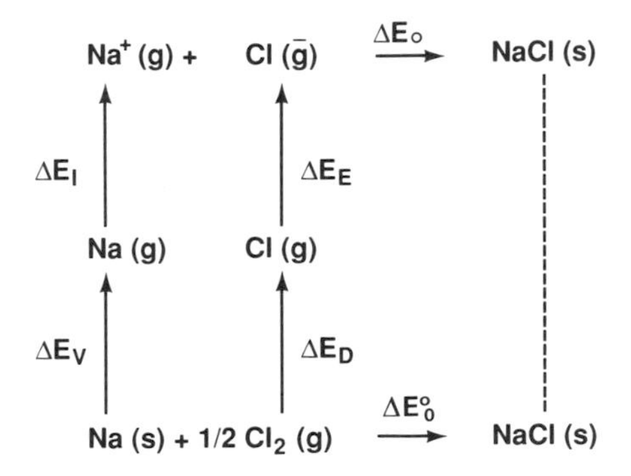

Figure 5.3.

The Born–Haber Cycle.

Table 5.4. Atomic electron affinities.[a]

Z	Symbol	Electron Affinity (eV)[b]	Z	Symbol	Electron Affinity (eV)[b]
1	H	0.754	34	Se	2.02
2	He	(−0.22)	35	Br	3.363
3	Li	0.620	36	Kr	(−0.40)
4	Be	(−2.5)	37	Rb	0.4859
5	B	0.86	38	Sr	(−1.74)
6	C	1.270	42	Mo	1.0
7	N	0.0	47	Ag	1.303
8	O	1.465	49	In	0.35
9	F	3.339	50	Sn	1.25
10	Ne	(−0.30)	51	Sb	1.05
11	Na	0.548	52	Te	1.90
12	Mg	(−2.4)	53	I	3.061
13	Al	(0.52)	54	Xe	(−0.42)
14	Si	1.24	55	Cs	0.472
15	P	0.77	56	Ba	(−0.54)
16	S	2.077	73	Ta	0.8
17	Cl	3.614	74	W	0.5
18	Ar	(−0.36)	75	Re	0.15
19	K	0.501	78	Pt	2.128
20	Ca	(−1.62)	79	Au	2.3086
22	Ti	0.391	81	Tl	0.5
23	V	0.937	82	Pb	1.05
24	Cr	0.66	83	Bi	1.05
26	Fe	0.582	84	Po	(1.8)
27	Co	0.936	85	At	(2.8)
28	Ni	1.276	86	Rn	(−0.42)
29	Cu	1.276	87	Fr	0.456)
31	Ga	(0.37)			
32	Ge	1.20			
33	As	0.80			

[a] Taken from T. Moeller, *Inorganic Chemistry*, p. 81, Wiley, New York, 1982.
[b] All values experimental except those in parentheses.

energies, one must change the sign that, for historical reasons, is given as positive for exothermic reactions.

The ionization potentials of the elements are given by Table 5.5. The values that occur just above the dashed horizontal lines are the energies necessary to remove an eletron from an ion with a closed-shell, rare gas structure. Clearly these values are much larger than those necessary to remove other electrons from the same ions or atom.

Table 5.5. Ionization energies of the elements.[a]

Z	Symbol	Outermost Configuration	Ionization Energy (eV) I	II	III	IV	V	VI
1	H	$1s^1$	13.598					
2	He	$1s^2$	24.587	54.416				
3	Li	$2s^1$	5.392	75.638	122.451			
4	Be	$2s^2$	9.322	18.211	153.893			
5	B	$2s^22p^1$	8.298	25.154	37.930			
6	C	$2s^22p^2$	11.260	24.383	47.887			
7	N	$2s^22p^3$	14.534	29.601	47.448			
8	O	$2s^22p^4$	13.618	35.116	54.934			
9	F	$2s^22p^5$	17.422	34.970	62.707			
10	Ne	$2s^22p^6$	21.564	40.962	63.45			
11	Na	$3s^1$	5.139	47.286	71.64			
12	Mg	$3s^2$	7.646	15.035	80.143			
13	Al	$3s^23p^1$	5.986	18.828	28.447			
14	Si	$3s^23p^2$	8.151	16.345	33.492			
15	P	$3s^23p^3$	10.486	19.725	30.18			
16	S	$3s^23p^4$	10.360	23.33	34.83			
17	Cl	$3s^23p^5$	12.967	23.81	39.61			
18	Ar	$3s^23p^6$	15.759	27.629	40.74			
19	K	$4s^1$	4.341	31.625	45.72			
20	Ca	$4s^2$	6.113	11.871	50.908			
21	Sc	$3d^14s^2$	6.54	12.80	24.76			
22	Ti	$3d^24s^2$	6.82	13.58	27.491			
23	V	$3d^34s^2$	6.74	14.65	29.310			
24	Cr	$3d^54s^1$	6.766	15.50	30.96			
25	Mn	$3d^54s^2$	7.435	15.640	33.667			
26	Fe	$3d^64s^2$	7.870	16.18	30.651			
27	Co	$3d^74s^2$	7.86	17.06	33.50	51.3	79.5	
28	Ni	$3d^84s^2$	7.635	18.168	35.17	54.9	75.5	
29	Cu	$3d^{10}4s^1$	7.726	20.292	36.83	55.2	79.9	
30	Zn	$3d^{10}4s^2$	9.394	17.964	39.722	59.4	82.6	
31	Ga	$3d^{10}4s^24p^1$	5.999	20.51	30.71	64		
32	Ge	$3d^{10}4s^24p^2$	7.899	15.934	34.22	45.71	93.5	
33	As	$3d^{10}4s^24p^3$	9.81	18.633	28.351	50.13	62.63	
34	Se	$3d^{10}4s^24p^4$	9.752	21.19	30.820	42.944	68.3	
35	Br	$3d^{10}4s^24p^5$	11.814	21.8	36	47.3	59.7	
36	Kr	$3d^{10}4s^24p^6$	13.999	24.359	36.95	52.5	64.7	
37	Rb	$5s^1$	4.177	27.28	40	52.6	71.0	
38	Sr	$5s^2$	5.695	11.030	43.6	57	71.6	
39	Y	$4d^15s^2$	6.38	12.24	20.52	61.8	77.0	
40	Zr	$4d^25s^2$	6.84	13.13	22.99	34.34	81.5	

(continued on next page)

Table 5.5. *(Continued)*

Z	Symbol	Outermost Configuration	Ionization Energy (eV)					
			I	II	III	IV	V	VI
41	Nb	$4d^35s^2$	6.88	14.32	25.04	38.3	50.55	
42	Mo	$4d^55s^1$	7.099	16.15	27.16	46.4	61.2	
43	Tc	$4d^55s^2$	7.28	15.26	29.54			
44	Ru	$4d^75s^1$	7.37	16.76	28.47			
45	Rh	$4d^85s^1$	7.46	18.08	31.06			
46	Pd	$4d^{10}$	8.34	19.43	32.93			
47	Ag	$4d^{10}5s^1$	7.576	21.49	34.83			
48	Cd	$4d^{10}5s^2$	8.993	16.908	37.48			
49	In	$4d^{10}5s^25p^1$	5.786	18.869	28.03	54		
50	Sn	$4d^{10}5s^25p^2$	7.344	14.632	30.502	40.734	72.28	
51	Sb	$4d^{10}5s^25p^3$	8.641	16.53	25.3	44.2	56	
52	te	$4d^{10}5s^25p^4$	9.909	18.6	27.96	37.41	58.75	
53	I	$4d^{10}5s^25p^5$	10.451	19.131	33			
54	Xe	$4d^{10}5s^25p^6$	12.130	21.21	32.1			
55	Cs	$6s^1$	3.894	23.1				
56	Ba	$6s^2$	5.212	10.004				
57	La	$5d^16s^2$	5.577	11.06	19.175			
58	Ce	$4f^15d^16s^2$	5.47	10.85	20.20	36.72		
59	Pr	$4f^36s^2$	5.42	10.55	21.62	38.95	57.45	
60	Nd	$4f^46s^2$	5.49	10.72				
61	Pm	$4f^56s^2$	5.55	10.90				
62	Sm	$4f^66s^2$	5.63	11.07				
63	Eu	$4f^76s^2$	5.67	11.25				
64	Gd	$4f^75d^16s^2$	5.85	11.52				
65	Tb	$4f^96s^2$	5.85	11.52				
66	Dy	$4f^{10}6s^2$	5.93	11.67				
67	Ho	$4f^{11}6s^2$	6.02	11.80				
68	Er	$4f^{12}6s^2$	6.10	11.93				
69	Tm	$4f^{13}6s^2$	6.18	12.05	23.71			
70	Yb	$4f^{14}6s^2$	6.254	12.17	25.2			
71	Lu	$4f^{14}5d^16s^2$	5.426	13.9				
72	Hf	$4f^{14}5d^26s^2$	7.0	14.9	23.3	33.3		
73	Ta	$4f^{14}5d^36s^2$	7.89					
74	W	$4f^{14}5d^46s^2$	7.98					
75	Re	$4f^{14}5d^56s^2$	7.88					
76	Os	$4f^{14}5d^66s^2$	8.7					
77	Ir	$4f^{14}5d^76s^2$	9.1					
78	Pt	$4f^{14}5d^96s^1$	9.0	18.563				
79	Au	$4f^{14}5d^{10}6s^1$	9.225	20.5				
80	Hg	$4f^{14}5d^{10}6s^2$	10.437	18.756	34.2			
81	Tl	$4f^{14}5d^{10}6s^26p^1$	6.108	20.428	29.83			

Table 5.5. (*Continued*)

Z	Symbol	Outermost Configuration	Ionization Energy (eV) I	II	III	IV	V	VI
82	Pb	$4f^{14}5d^{10}6s^26p^2$	7.416	15.032	31.937	42.32	68.8	
83	Bi	$4f^{14}5d^{10}6s^26p^3$	7.289	16.69	25.56	45.3	56.0	88.3
84	Po	$4f^{14}5d^{10}6s^26p^4$	8.48					
85	At	$4f^{14}5d^{10}6s^26p^5$						
86	Rn	$4f^{14}5d^{10}6s^26p^6$	10.748					
87	Fr	$7s^1$						
88	Ra	$7s^2$	5.279	10.147				
89	Ac	$6d^17s^2$	6.9	12.1				
90	Th	$6d^27s^2$		11.5	20.0	28.8		
91	Pa	$5f^26d^17s^2$						
92	U	$5f^36d^17s^2$						
93	Np	$5f^46d^17s^2$						
94	Pu	$5f^67s^2$	5.8					
95	Am	$5f^77s^2$	6.0					
96	Bk	$5f^76d^17s^2$						

[a] Adapted from T. Moeler, loc. cit.

4. The Kapustinskii Equations

The Kapustinskii[8] equation can be used to estimate lattice energies when the crystal structure is unknown. This lack of information is much less likely to occur now than when this scheme was first proposed in 1933. However, the method does provide additional insight into the calculation of crystal energies, as well as being a viable mechanism for their determination when the Madelung constant is unknown. Although he was perhaps the first to employ the technique next described, there have been numerous additions to and modifications of this approach. (References 8–11 contain exhaustive bibliographies on the subject of lattice energies.)

This approach is a straightforward semiempirical substitution for an undetermined Madelung constant contained in either the Born equation

$$E_0 = \frac{MNz_1z_2e^2}{r_e}\left(1 - \frac{1}{n}\right) \tag{5.36}$$

[8] A. F. Kapustinskii, *A. Phys. Chem.*, **22B**, 257 (1933); *Quart. Rev.*, **10**, 283 (1956).

[9] G. J. Moody and J. D. R. Thomas, *J. Chem. Ed.*, **42**, 204 (1965).

[10] M. F. C. Ladd and W. H. Lee, *Prog. Sol. State Chem.*, **1**, 37 (1964); **2**, 378 (1965); **3**, 265 (1967).

[11] T. C. Waddington, *Adv. Inorganic Chem. Radiochem.*, **1**, 157–211 (1959).

or the Born–Mayer equation

$$E_0 = \frac{MNz_1z_2E^2}{r_e}\left(1 - \frac{\rho}{r_e}\right), \tag{5.37}$$

where ρ is found empirically to be nearly constant for most crystals with a value of 0.345. A new constant α is now defined by

$$\alpha \equiv \frac{M}{n_i/2}, \tag{5.38}$$

where n_i is the number of ions per molecule. The Madelung constant is taken as 1.7476 (note that earlier work frequently uses the incorrect value of 1.745), the value for NaCl, for all crystals regardless of their actual structure, and r_e, the equilibrium distance of separation between anion–cation nearest-neighbor pairs, is set equal to the sum of their Pauling or Goldschmidt radii;[12] i.e., $r_e = r_+ + r_-$ (see Table 5.12).

These substitutions effectively reduce all crystals to the rock salt structure with sixfold coordination for all ions. Although at first these would appear to be drastic approximations, one can see from Fig. 5.11 that the energy difference between the CsCl, NaCl, and ZnS structures is less than 10% over most of the range of radius ratios. Thus, in view of the apparent insensitivity of the energy to crystal structure, one would hope that accounting for the energy by simple pairwise interactions might furnish a reasonable approximation. However, this is not always the case, and spurious results are not uncommon.

To proceed, the numerical factors in (5.36) and (5.37) are collected into a single constant, so that taking $n = 9$, $\rho = 0.354$, $M = 1.748n_i/2$, $Ne^2 = 329.7$ kcal/Å, and $r = r_+ + r_-$ reduces (5.36) and (5.37) to

$$E_0\ (\text{kcal/mole}) = \frac{256.1 n_i z_+ z_-}{r_+ + r_-} \tag{5.39}$$

and

$$E_0\ (\text{kcal/mole}) = \frac{288.2 n_i z_+ z_-}{r_+ + r_-}\left[1 - \frac{0.345}{r_+ + r_-}\right]. \tag{5.40}$$

The success of these approximate calculations can be best judged from an examination of Tables 5.6 and 5.7.

A useful application of the Kapustinskii equations is the determination of the possible stability or instability of a hypothetical compound. To be sure, the only

[12] Section 5.5 explains the derivation of these radii, which are based upon the sixfold coordination of the rock salt structure.

Table 5.6. Lattice energies (kcal/mole) of the alkali metal halides.

Salt	Born–Haber Cycle	Kapustinskii
LiF	241.2	227.7
LiCl	198.2	192.1
LiBr	188.5	189.5
LiI	175.4	170.4
NaF	216.0	211.5
NaCl	183.8	179.9
NaBr	175.9	175.5
NaI	164.5	161.0
KF	191.5	188.5
KCl	166.8	162.7
KBr	160.7	161.3
KI	151.0	146.8
RbF	183.6	181.7
RbCl	162.0	158.2
RbBr	155.2	149.7
RbI	146.5	141.0
CsF	171.0	170.4
CsCl	153.2	149.4
CsBr	148.3	143.9
CsI	140.3	134.7

Table 5.7. Lattice energies (kcal/mole) of some transition metal(II) chalcogenides.

		Calculated Values Kapustinskii Eqs.	
MCh	Born–Haber Cycle	Eq. (5.40)	Eq. (5.39)
MnO	911	931	880
FeO	937	953	899
CoO	954	967	908
NiO	974	976	914
ZnO	964	957	901
ZnS (sph)	865	794	771
ZnS (wur)	861		
MnS	801	776	757
MnSe	790	737	724
ZnSe	863	753	737

Table 5.8. The enthalpies of formation (kcal/mole) of some hypothetical dihalides and the corresponding oxides.[a]

	Salt				
M	MF_2	MCl_2	MBr_2	MI_2	MO
Li		+1061	+1095	+1133	+1037
Na	+403	+513	+533	+580	+522
K	+104	+204	+233	+267	+243
Rb	+39	+131	+158		+188
Cs	−30	+51	+76	+113	+123
Al	−185	−65	−35	+2	−55
Cu	−127	−52	−34	−5	−37
Ag	−49	+23	+40	+67	+55

[a] T. Waddington, p. 217, loc. cit.

rigorous criterion of stability is the sign of the free energy of formation. Nevertheless, approximate values of the internal energies as derived from the Kapustinskii or similar equations, coupled with estimates of the entropy of formation, gives a fair degree of predictability. Such estimates of the enthalpies of formation are given in Table 5.8.

5. Entropies of Solid Compounds

A method for estimating the entropies of solid compounds was proposed by Latimer[13] well over half a century ago. For compounds whose specific heat has reached the classical Dulong–Petit limit, viz. $C_v = 3R$ cal/mole, he initially assumed that the contribution of each ith constituent to the total entropy of the compound is given by the empirical relation[14]

$$S_i^\circ(298\ \text{K}) = \frac{3}{2} R \ln A - 0.94,$$

where A is the atomic weight. This function yields values that are slightly too low for large atomic weights, and vice versa for the lower ones. Consequently, the curve described by the empirical entropy equation was appropriately (and arbitrarily)

[13] W. M. Latimer, *J. Am. Chem. Soc.*, **43**, 818 (1921).

[14] W. M. Latimer, *The Oxidation States of the Elements and Their Potentials in Aqueous Solutions*, App. III, Prentice-Hall, Englewood Cliffs, NJ 1952.

raised and lowered at its extremities to agree with the measured values derived from simple salts. This resulted in the compilation of entropies of the elements in Table 5.9 based solely on atomic weight.

The entropies of formation of most common compounds have now been measured, and there is no need to use estimated values. Nevertheless, it is instrutive to observe the relative excellence of the predicted values based upon such a simple concept.

The entropies of the anions are estimated from the averages of the measured values, which are grouped according to the charge on the cation. That is to say, the value of the metallic ion is taken as the constant given in Table 5.9. This value, weighted by its stoichiometric proportion, is subtracted from the measured entropy of formation of a compound and the remainder assigned to the anion(s). This being done for several compounds, average values per anion are calculated within each category as determined by the oxidation state of the cation, thus creating a "universal" scale. Thus, Latimer calculates average values of 22.0 cal/mole-deg for the sulfate in M_2SO_4, 17.2 cal/mole-deg for MSO_4 and 13.7 cal/mole-deg for $M_2(SO_4)_3$. In this fashion, Table 5.10 is created. Thus, the entropy dependence upon the oxidation state of the cation is assigned completely to the anion. The relative constancy of the anion entropy with varying cations and their varying charge states justifies the usefulness of this scheme. As compounds become less ionic and thus more covalent, their true entropy values increasingly deviate from the Latimer predictions.

An example calculation of the free energy of formation of AuF follows: a rock salt structure is assumed for AuF (as AgF has this structure), and in any case the Kapustinskii equations assume this structure. Pauling's ionic radii (Table 5.12), $r(Au^+) = 1.37$ Å and $r(F^-) = 1.36$ Å, allow calculation of the lattice enthalpy. Thus, the Born–Haber cycle yields for the total enthalpy of formation:

	ΔH
$Au(s) \rightarrow Au(g)$	+88
$Au(g) \rightarrow Au^+(g) + e$	+214
$(1/2)F_2(g) \rightarrow F(g)$	+19
$F(g) + e \rightarrow F^-(g)$	−84
$Au^+(g) + F^-(g) \rightarrow AuF(s)$	+186
$Au(s) + (1/2)F_2(g) \rightarrow AuF(s)$	+51 kcal/mole

The entropy of solid AuF is obtained from adding the values given in Table 5.9 and 5.10, yielding $S^\circ_f(AuF) = 20.8$ cal/mol-deg. The measured entropies of the pure elements are $S^\circ(Au) = 11.4$ cal/mol-deg and $S^\circ(F_2/2 = 24.3$ cal/mol-deg. Hence,

$$\Delta G^\circ = \Delta H^\circ - T\,\Delta S^\circ = 51 - 298 \times 10^{-3}(-14.9) = 56 \text{ kcal/mol.}$$

Thus, it is predicated that AuF is very unstable, in agreement with fact.

Table 5.9. Entropy contribution of metals in solid compounds at 25°C (cal/mole-deg).

Li	Be													
3.5	4.3													
Na	Mg	Al												
7.5	7.6	8.0												
K	Ca	Sc	Ti	V	Cr	Mn	Fe	Co	Ni	Cu	Zn	Ga	Ge	As
9.2	9.3	9.7	9.8	10.1	10.2	10.3	10.4	10.6	10.5	10.8	10.9	11.2	11.3	11.45
Rb	Sr	Y	Zr	Nb	Mo	Tc	Ru	Rh	Pd	Ag	Cd	In	Sn	Sb
11.9	12.0	12.0	12.1	12.2	12.3	—	12.5	12.5	12.7	12.8	12.9	13.0	13.1	13.2
Cs	Ba	La	Hf	Ta	W	Re	Os	Ir	Pt	Au	Hg	Tl	Pb	Bi
13.6	13.7	13.8	14.8	14.9	15.0	15.0	15.1	15.2	15.2	15.3	15.4	15.4	15.5	15.6

Table 5.10. Entropy contributions of negative ions in solid compounds at 25°C (cal/mole-deg).

	Charge on Cation			
	+1	+2	+3	+4
F^-	5.5	4.7	4.0	5.0
Cl^-	10.0	8.1	6.9	8.1
Br^-	13.0	10.9	9	10
I^-	14.6	13.6	12.5	13.0
O^{2-}	2.4	0.5	0.5	1.0
S^{2-}	8.2	5.0	1.3	2.5
OH^-	5.0	4.5	3.0	
CN^-	7.2	6		
NO_2^-	17.8	15		
NO_3^-	21.7	17.7	15	14
CO_3^{2-}	15.2	11.4	8	
HCO_3^-	17.4	13	10	
ClO_3^-	24.9	20		
BrO_3^-	26.5	22.9	19	
IO_3^-	25.5	22.9	19	
MnO_4^-	31.8	28		
ClO_4^-	26.0	22		
SO_3^{2-}	19	14.9	11	
SO_4^{2-}	22	17.2	13.7	12
CrO_4^{2-}	26.2	21		
$C_2O_4^{2-}$	22	17.7	14	
PO_4^{2-}	24	17.0	12	

6. The Size of Ions

The preceding sections have explained how the lattice energy of simple ionic structures depends only upon the ionic charge combined with the interionic distances of separation. The calculation of the lattice energy is predicated upon a specific crystal structure, the one thermodynamically most stable. Given particular ionic species, it is plausible to believe that the distances of separation in a stable crystal reflect their effective sizes. Hence, we come upon the rather old concept that the structure of maximum stability is decided by the sizes of the component ions. However, establishing the criteria for a scale of *effective* ionic radii proves to be somewhat indirect, although the actual internuclear distance is directly available from x-ray diffraction.

The structural basis for calculating ionic radii is sixfold coordination (e.g., the NaCl structure). Corrections for four-and eightfold coordination are then applied to the initial values. In like fashion, following Pauling,[15] univalent radii are first calculated and afterwards correted for their multiple charge states.

A. Empirical Ionic Crystal Radii

Values for the empirical ionic crystal radii are obtained from either the direct measurement of the internuclear distance[16] by x-ray diffraction or from the molar refraction volumes of the ions. The latter technique seems only to have been influential in the early history of ionic radii calculations;[17] the results served as the basis for further refinements. The idea that ionic radii may be constant, and hence additive, becomes obvious when, for example, one considers the near constancy of the differences in the internuclear distances, Δr, between the alkali metal halides, as shown in Table 5.11. Table 5.11a demonstrates the relative constancy of the anion sizes. The average values of $\overline{\Delta r}_{BrCl} = 0.15_8 \pm 0.01$ Å and $\overline{\Delta r}_{ClF} = 0.51 \pm 0.04$ Å show a remarkable degree of invariance, considering the crudeness of this model.

However, this simple scheme cannot be extended per se to more complex systems without incurring significant error. It is perhaps disturbing that even CsF, the only Cs salt with the NaCl structure, agrees less well with $\overline{\Delta r}_c$ than do the bcc CsCl, CsBr, and CsI. This clearly points up the importance of other factors not included in this first approximate theory. The departure of CsF from the mean is not surprising, since Cs^+ is the largest of the metal ions and F^- the smallest and most electronegative, and, furthermore, our theory completely ignores polarization.

---

[15] Linus Pauling, *Nature of the Chemical Bond*, Chap. 13, Cornell University Press, Ithaca, NY, 1960.

[16] Knowledge of the cation–anion distances alone is not sufficient for a determination of the individual ionic radii [see H. Wondratschek, *Amer. Mineral.*, **72**, 82 (1987)].

[17] J. A. Wasastjerna, *Soc. Sci. Fenn. Comm. Phys. Math.*, **38**, 1 (1923).

Table 5.11. The internuclear distances (Å) and their differences in alkali halide salts.

(a)

Anion	Counterion				
	Li^+	Na^+	K^+	Rb^+	Cs^+
Br^-	2.75	2.97	3.29	3.44	3.71
Δr (Br–Cl)	0.18	0.16	0.15	0.15	0.15
Cl	2.57	2.81	3.14	3.27	3.56
Δr (Cl–F)	0.56	0.50	0.48	0.45	0.56
F^-	2.01	2.31	2.66	2.82	3.00

(b)

Cation	Counterion			
	F^-	Cl^-	Br^-	I^-
Li^+	2.01	2.57	2.75	3.00
Δr (Na–Li)	0.28	0.24	0.22	0.23
Na^+	2.29	2.81	2.97	3.23
Δr (K–Na)	0.37	0.33	0.32	0.30
K^+	2.66	3.14	3.29	3.53
Δr (Rb–K)	0.16	0.13	0.15	0.13
Rb^+.	2.82	3.27	3.44	3.66
Δr (Cs–Rb)	0.18	0.29	0.27	0.29
Cs^+	3.00	3.56	3.71	3.95

Nevertheless, by assuming a value for the ionic radius of a single ion and applying the difference scheme as just outlined, one can generate a set of self-consistent ionic radii. This was done by Goldschmidt,[18] who assumed values of 1.33 Å for F^- and 1.32 Å for O^{2-}, based upon the optical refraction measurements of Wasastjerna.[19] The figure for O^{2-} was amended to 1.40 Å, which gives better agreement with the measured values. We will not report his values here, but a selected number are tabulated along with those obtained by other methods in Table 5.12.

Another quite different method of evaluating the ionic radii was proposed very early in this game by Landé.[20] It is based upon the (correct) premise that in certain

---

[18] V. M. Goldschmidt, *Geochemische Verteilungsgesetze der Elemente, Schft. Norske-Videnskaps-Acad, Oslo I Mat.-Naturv. Kl.* (1926).

[19] J. Wasastjerna, loc. cit.

Table 5.12. Selected values for some ionic radii[a] with six-fold coordination.

Li^+	(P)	0.68	Be^{2+}	(P)	0.35	Ti^{4+}	(G)	0.64
Na^+	(P)	0.97	Mg^{2+}	(S)	0.72	Cr^{2+}	(S)	0.73, 0.82
K^+	(P, G)	1.33	Ca^{2+}	(S)	1.16	Mn^{2+}	(S)	0.67, 0.82
Rb^+	(G, S)	1.49	Ba^{2+}	(S)	1.36	Mn^{3+}	(S)	0.58, 0.65
Cs^+	(S)	1.70				Mn^{4+}	(S)	0.54
Fe^{2+}	(S)	0.61, 0.77	Cu^+	(S)	0.73	F^-	(P, G, S)	1.33
Fe^{3+}	(S)	0.55, 0.66	Ag^+	(S)	1.15	Cl^-	(P)	1.81
Co^{2+}	(S)	0.65, 0.74	Au^+	(P)	1.37	Br^-	(P)	1.95
Co^{3+}	(S)	0.53, 0.61	Th^{4+}	(P)	1.02	I^-	(P, G)	2.20
			U^{3+}	(S)	1.06	O^{2-}	(P, S)	1.40
Ni^{2+}	(P)	0.69	U^{4+}	(P)	0.95	S^{2-}	(P)	1.84
Ni^{3+}	(S)	0.56, 0.60	U^{5+}	(S)	0.76			
			U^{6+}	(S)	0.75			

[a] All radii are in Ångstrom units. Ions showing two values of r have their low-spin states listed first. S = Shannon and Prewitt; P = Pauling and Ahrens; G = Goldschmidt.

crystals the ions are in actual contact. His choice for the reference structure was LiI. A (100) plane with the ions drawn correctly to scale is shown in Fig. 5.4. The absolute magnitude of the iodide ion is taken as one-half the iodide–iodide internuclear distance as determined by x-ray diffraction.

Also, the similarity of the anion–cation separations in the pair Mg–S, $d = 2.60$ Å, and Mn–S, $d = 2.59$ Å, and in another pair Mg–Se, $d = 2.73$ Å, and Mn–Se, $d = 2.73$ Å, suggests that the sulfide ions and selenium ions are in contact. This is inferred not only from the fact that Mg^{2+} and Mn^{2+} are much smaller than either S^{2-} or Se^{2-}, but also because Mn^{2+} is much smaller than Mg^{2+}. Hence, the interionic distance between cation and anion is really just the interval between an anion center and the center of the adjacent anion cage, if the cation is located at the center of the anion cage.

B. Semitheoretical Ionic Radii

Semitheoretical values for the radii of ions in crystals have been produced by several authors, but the discussion by Pauling[21] remains the most widely quoted. His

[20] A. Landé *A. Phys.* **1**, 191 (1920).

[21] L. Pauling, loc. cit.

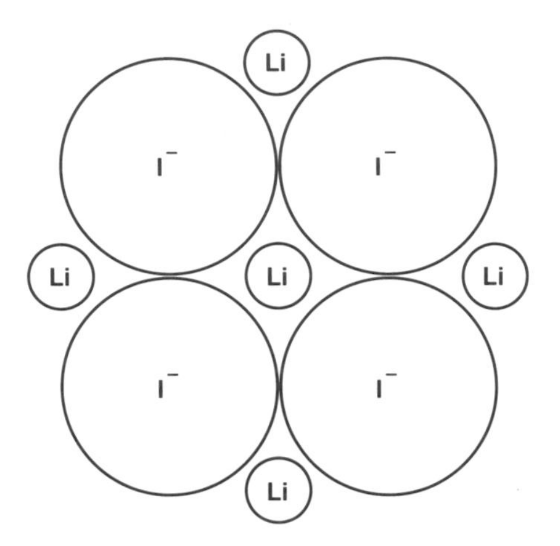

Figure 5.4.

A (100) face of LiI showing the contact between I^- ions as is necessary for the Landé model.

calculations are founded on NaF, KCl, RbBr, and CsI, which he designated as standard crystals. The components that make up each of these salts are isoelectronic with one another, having the closed shell structures of Ne, Ar, Kr, and Xe, respectively. All of these compounds have radius ratios $r_c/r_a \simeq 0.75$, where c and a refer to cation and anion. It is important that this ratio be significantly larger than 0.414, at which point all ions are in contact. Below this value the anions contact each other and double repulsion occurs—i.e., not only the normal coulombic repulsion of like (nuclear) charges but pronounced repulsion between the interpenetrating electron shells. The geometry portraying the relative sizes of the alkali halides is shown in Fig. 5.5. All the standard compounds except CsI have the rock salt structure. By reducing the interionic distance by 2.3%, the latter is corrected for the structural difference. As with the empirical methods previously described, the success of the following scheme is to be judged by its agreement with measured internuclear distances.

The fundamental formula is

$$r_i = C_n/Z_{\text{eff}} = C_n/(Z - S), \tag{5.41}$$

which states that the magnitude of the ionic radius r_i is inversely proportional to the effective charge Z_{eff} on the outermost shell of electrons. This is an intuitively attractive idea. The effective charge is then the formal nuclear charge Z corrected for the shielding supplied by the intervening electrons. Pauling's original method for calculating the screening or shielding factor S was partly theoretical and partly based

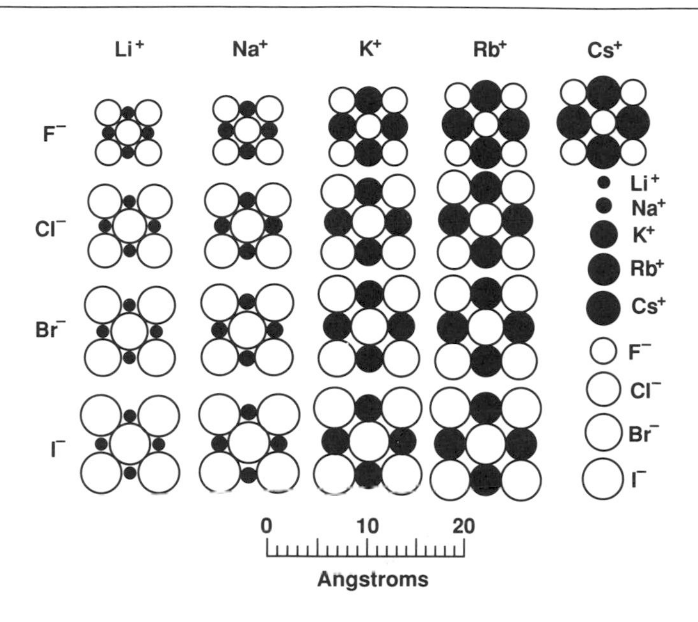

Figure 5.5.

The (100) face of the alkali halides with the NaCl structure, all ions drawn to scale. (Reprinted from Linus Pauling: *The Nature of the Chemical Bond*, 3rd edition. © 1939, 1940, 3rd edition © 1960 by Cornell University. Used by permission of the publisher, Cornell University Press.)

upon measurements of molar refraction and x-ray–produced electron densities. More frequently cited, perhaps mostly because of their explicit and compact formulation, are the rules for calculating S proposed by Slater.[22] His numerical values are based on only the radial part of the wave function, and his rules may be paraphrased as follows:

1. All electrons with principal quantum numbers n greater than that in the shell under consideration are neglected, although when the outermost shell is the one under consideration, as is most frequently the case, all inner shells must be accounted for in the calculation.
2. Electrons having the same shielding capacity are grouped according to the following scheme: $1s;2s2p;3s3p;3d;4s4p;4d;5s5p;5d\ldots.$
3. The shielding constants for s and p electrons are:
   0.35 for each in the shell of interest,
   0.85 for each in the next inner shell,
   1.00 for all nearer the nucleus.

[22] J. C. Slater, *Phys. Rev.*, **36**, 57 (1930); **34**, 1293 (1929).

4. The values of S for d and f electrons are:
   0.35 for the shell being calculated,
   1.00 for all d and f electrons of smaller principal quantum number.

The quantity C_n in (5.41) is an empirically derived constant that depends on the quantum number of the outermost shell.

An example calculation applied to KCl is as follows: the observed interionic distance d is 3.147 Å; hence,

$$r_{K^+} + r_{Cl^-} = d = 3.147. \tag{5.42}$$

The electrons of this isoelectronic pair are divided according to the Slater rules into three groups: $1s^2;2s^22p^6;3s^23p^6$. Therefore,

$$S = 2(1.00) + 8(0.85) + 8(0.35) = 11.60.$$

Accordingly,

$$\begin{aligned} r_{K^+} &= \frac{C_n}{19 - 11.60} = \frac{C_n}{7.40}, \\ r_{Cl^-} &= \frac{C_n}{17 - 11.60} = \frac{C_n}{5.40}. \end{aligned} \tag{5.43}$$

Solving (5.42) and (5.43) results in $C_n = 9.83$. Reintroducing the numerical values of Z_{eff} and C_n into (5.41), we calculate $r_{K^+} = 1.33$ Å and $r_{Cl^-} = 1.82$ Å.

Multivalent ions can be incorporated into this scheme as follows. Pauling has termed the ionic radii calculated in the following manner crystal radii, r(crystal). Rewriting (5.12) as

$$E_0 = -Mz^2e^2/r + A/r^n,$$

we again perform the differentiation of E_0 with respect to r and, assuming A is known, solve for r_e, finding that

$$r_e = (nA/Mz^2e^2)^{1/(n-1)}. \tag{5.44}$$

For ions with $z = 1$, i.e., the singly charged oxidation state,

$$r_{e1} = (nA/Me^2)^{1/(n-1)}. \tag{5.45}$$

Consequently, multiply charged ions, $r^{\pm x}$, are determined from the ratio of r_x/r_{e1}, leading to

$$r(\text{crystal}) = r_1(1/z^2)^{1/(n-1)}. \tag{5.46}$$

Thus, the crystal radii are related to the univalent radii through the assumption that n and A are the same for multivalent ions as for the univalent oxidation state.

There are several compilations[23] of ionic radii, most of which differ slightly, if at all, in their individual magnitudes. Among the prominent names associated with these calculations are Goldschmidt,[24,25] Zachariasen,[26] Shannon and Prewitt,[27] and, of course, Pauling,[28] whose results have been modified and amplified by Ahrens.[29] Because coordination number as well as spin state (i.e., low or high) influence ionic size, a comprehensive tabulation requires several printed pages, and the interested reader is referred to Refs. 24, 27 and 29 or to the Landolf–Börnstein Tabellen 1:4 (1950). Some selected representative values are listed in Table 5.12. The variations in ionic radii between the low and high magnetic spin states are about 0.1 Å, and this subject will be discussed in greater detail in the section of ligand fields.

The variations in the ionic radii as a function of coordination number are about 0.3 Å. It has been suggested[30,31] that, by defining sixfold coordination to be the standard state, one can correct for other coordinations by multiplying the ionic radii by the following factors.

Coordination Number	Correction Factor
12	1.12
8	1.03
6	1.00
4	0.94

These correction factors are experimentally derived from the measured values of selected anions and cations in compounds of varying coordination. By a process of iteration, initial values are refined until their consistency in reproducing interionic distances is deemed satisfactory.

Pauling[32] has suggested an approximate theoretical correction for changing the

---

[23] F. D. Bloss, *Crystallography and Crystal Chemistry*, pp, 206–216, Holt, Rinehart and Winston, 1971. This is an especially comprehensive list that includes several oxidation states and coordination numbers for all the elements, when known, and compares values from several authors.

[24] V. M. Goldschmidt, *Geochem. Oxford Univ. Press* (1954).

[25] V. M. Goldschmidt, T. Barth, G. Lund, and W. H. Zachariasen, *Skr, Norske Vidensk. Akad. I, Mat.-Nat. Kl* #2.

[26] W. H. Zachariasen, *Z. Krist*, **80** 137 (1931).

[27] R. D. Shannon and C. T. Prewitt, *Acta. Cryst.*, **B25**, 925 (1969).

[28] L. Pauling, loc. cit.

[29] L. H. Ahrens, *Geochem, et Cosmochim. Acta*, **2**, 155 (1952)

[30] R. D. Shannon and C. T. Prewitt, loc. cit.

[31] G. S. Zhadanov, *Crystal Physics*, Academic Press, New York, 1965.

[32] L. Pauling, loc. cit.

coordination number, which has the form

$$\frac{R_{\mathrm{II}}}{R_{\mathrm{I}}}=\left[\frac{B_{\mathrm{II}}}{B_{\mathrm{I}}}\cdot\frac{M_{\mathrm{I}}}{M_{\mathrm{II}}}\right]^{-(n-1)}. \tag{5.47}$$

The symbols have their usual meaning; i.e., R_{I} and R_{II} are the sums of the cation and anion radii, B_{I} and B_{II} are the repulsive constants defined by (5.24). The M's are the respective Madelung constants. His example calculation is the conversion of the standard NaCl structure with six anion–cation contacts to the fluorite structure, which has 8. The numerical values of B_{I} for the NaCl structure and B_{II} for the fluorite are assumed to differ only by the factor $B_{\mathrm{II}}/B_{\mathrm{I}} = 8/6$, reflecting the change in the number of points of tangency. To convert the sum of the univalent (NaCl structure) Ca^{2+} and F^- radii to their true values in the CaF_2 structure, n is taken equal to 8 and $A(\mathrm{NaCl})/A(\mathrm{CaF_2}) = 1.7476/5.0388$, which ultimately leads to $R_{\mathrm{II}}/R_{\mathrm{I}} = 0.894$.

7. Ionic Packing and Crystal Structure

Because ionic bonds derive from an inherently spherical potential, in contrast to covalent bonds, they are nondirectional. Consequently, the geometry of ionic crystal structures is governed by the balance between the coulombic attractive and repulsive forces in combination with the interionic electron repulsions. The strength of these forces naturally depends upon the distances between the ions. This line of reasoning directly leads us to suppose that the structure of ionic crystals is controlled by the size of the ions. Indeed, this turns out to be the case for most simple binary salts. The greatest degree of stability is achieved by maximizing the number of contacts between anions and cations and, of course, at the same time maximizing the separations between ions of like charge. Hence, stated most tersely, the thermodynamic stability depends primarily upon the radius ratio r_a/r_c.

The limit of stability defined in terms of the radius ratio is simply summarized as follows: a given three-dimensional array is stable so long as r_a/r_c does not permit contact between like ions. When contact does occur, the ions rearrange to a different geometry that once again precludes contact. It is convenient to commence our calculation of these ranges of stability based upon lattices derived from the close-packed fcc and hcp structures as described in Chapter 1. These structures are shown once again in Fig. 5.6. It is seen that each atom in the close-packed structures has 12 nearest neighbors. Ionic crystals are seldom close-packed because of the differing sizes of anions and cations. However, when all ions are of the same size, and thus there are 12 points of common tangency between each ion and its neighbors, the radius ratio r_c/r_a equals unity. When this ratio falls below unity, the screening or

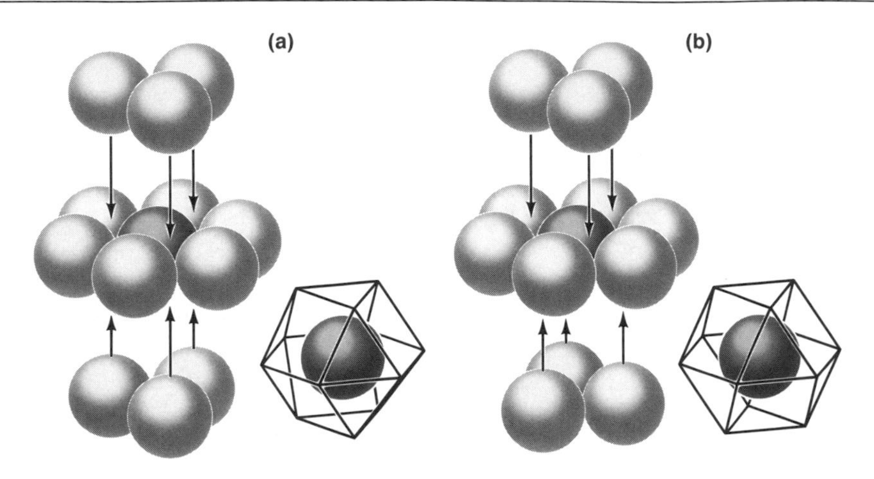

Figure 5.6.

(a) The hcp arrangement with the 12-fold coordination polyhedron; (b) the same for the fcc structure. (Taken from F. O. Bloss, *Crystallography and Crystal Chemistry*, Holt, Rinehart and Winston, p. 232, 1971.)

neutralization of like charges from one another is reduced and the CsCl structure becomes more stable than the close-packed ones.

The CsCl structure is based upon the bcc lattice and has eightfold coordination. When all ions are in contact, as shown in Fig. 5.7, a cube diagonal that is a [111] line connects the centers of two corner ions with that of the body-centered ion, all three lying in a contacting chain. The length of this line is given by

$$(2r_a + 2r_c)^2 = 12r_a^2, \tag{5.48}$$

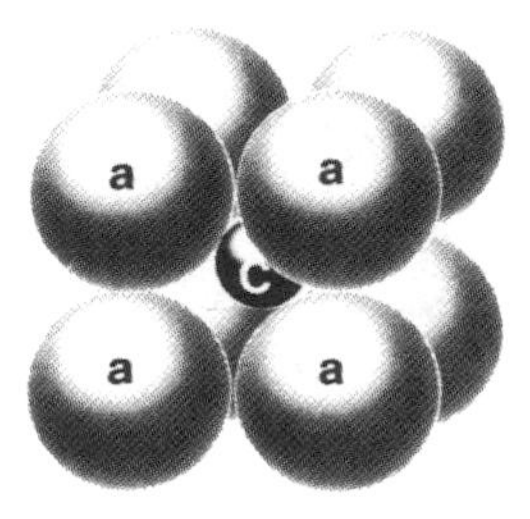

Figure 5.7.

The CsCl structure with all ions in contact.

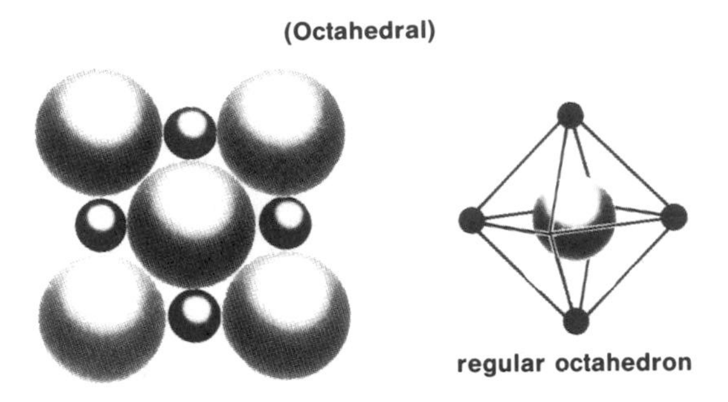

Figure 5.8.

The (100) face of the NaCl structure with all ions in contact and the octahedral coordination polyhedron made by the six ligands located at its corners.

from which we obtain $r_c/r_a = 0.732$. Consequently, cubic crystals with radius ratios in the range of 0.732 to 1.00 fall within the CsCl limits of stability.

The octahedral coordination to six nearest neighbors of opposite charge is typified by NaCl. When all ions are in contact, the {100} faces resemble Fig. 5.8. The sixfold coordination can be visualized by placing an imaginary black sphere in front and another behind the large open central circle in Fig. 5.8. Connecting the centers of the six surrounding spheres will create the octahedron also shown in Fig. 5.8. Since $r(Cl^-) > rNa^+$, we assign the large circles the role of representing the anion, and the smaller darker ones the cation. Then the diagonal [100] is given by

$$(4r_a)^2 = 2(2r_a + 2r_c)^2, \tag{5.49}$$

which, when solved for the radius ratio, yields

$$r_c/r_a = 0.414. \tag{5.50}$$

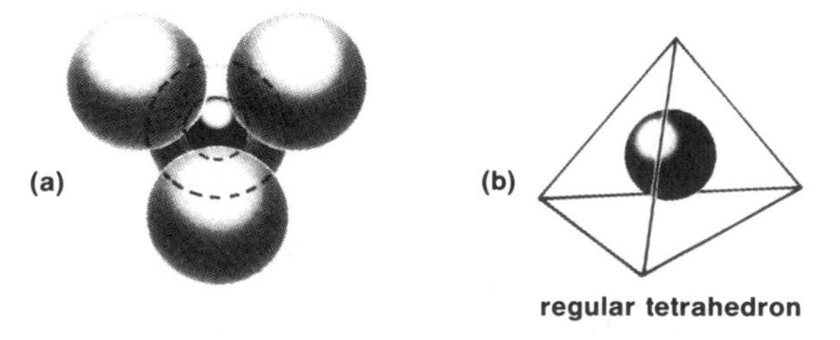

Figure 5.9.

Tetrahedral coordination provides stability for compounds where $0.22 < r_c/r_a < 0.41$. (a) The four larger anions embrace the smaller cation, the dashed circle represents an anion resting on and above the cation; (b) the coordination tetrahedron.

It can be shown that when $r_c/r_a \gtrsim 0.414$ the anion–anion repulsion is less for tetrahedral coordination than for octahedral because of the reduction from 6 to 4 in the number of anions surrounding each cation. The tetrahedral coordination is illustrated in Fig. 5.9. The triangular and linear coordination configurations are shown in Fig. 5.10.

The critical radius ratio for triangular coordination is $r_c/r_a = 0.22$, and this remains the favored configuration until it reaches 0.155, below which the linear arrangement is favored. The number of unlike ions surrounding a given ion and the coordination configurations are summarized as follows:

Packing configurations of ionic crystals.

r_+/r_-	Maximum No. of Ligands	Coordination Configuration	Examples
~1	12	Close-packed	$KAl_3Si_3O_{10}(OH)_2$–mica
0.732–1	8	Cubic CsCl	
0.414–0.732	6	Octahedral or square	NaCl
0.22–0.414	4	Tetrahedral	Si
0.15–0.22	2	Triangular	B_2O_3
<0.15	—	Linear	

Although we have referred to the radius ratio as the radius of the cation divided by that of the anion, the packing rules are unaltered if this ratio is reversed (i.e., r_a/r_c).

Although these simple calculations can help one to visualize the primary interactions between ions and also provide a fair degree of predictability as to the expected structure, there are numerous exceptions to these rules. The electrostatic energies for eightfold, sixfold, and fourfold coordination are plotted versus the radius ratio r_+/r_- in Fig. 5.11. (We again remind ourselves that this graph remains unaltered if r_+/r_- is inverted to r_-/r_+, provided that $r_+ > r_-$.)

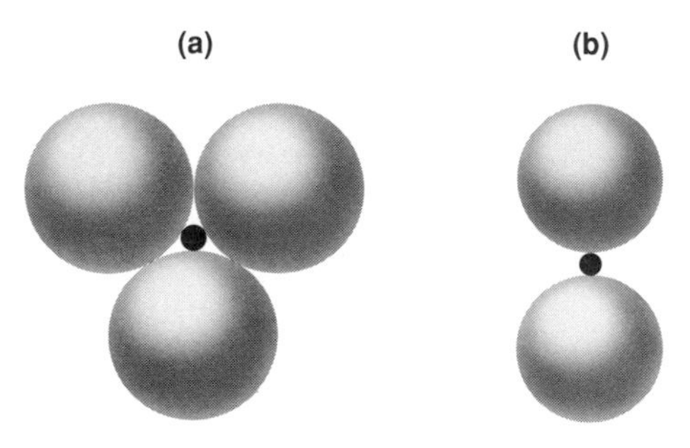

Figure 5.10.

(a) Triangular coordination $0.15 < r_c/r_a < 0.22$; (b) linear coordination $r_c/r_a < 0.15$.

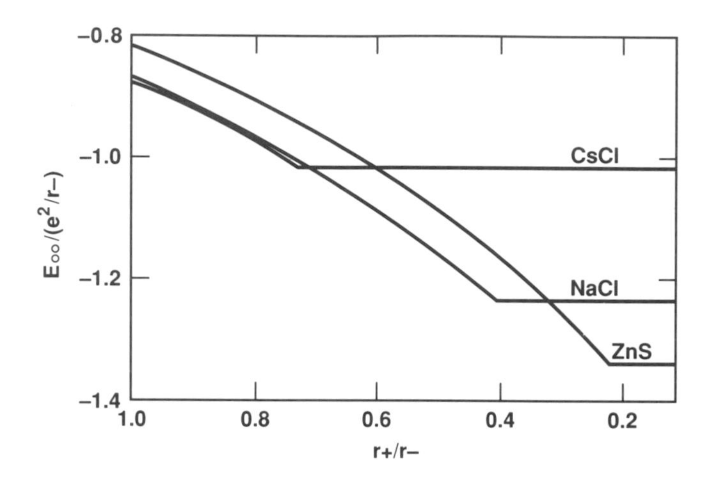

Figure 5.11.

The electrostatic energy of the eight-, six-, and fourfold coordinated structures as a function of radius ratio.

The small difference in energy between the CsCl and NaCl structures for the range $0.7 \gtrsim r_+/r_- \lesssim 1$ is clearly revealed in this graph. Also, at the crossover points and in their immediate vicinity the energies are equal or similar. Only CsCl, CsBr, and CsI have the body-centered lattice, although a comparison of the values of r_+/r_- as listed in Table 5.13 would indicate KF, RbF, RbCl, and RbBr should have this structure as well. The charged hard-sphere model has neglected differences in the electronegativities, which give rise to polarization as well as induced dipolar van der Waals forces. Although these are small contributions compared with the coulombic interactions, they can reverse the predictions of the simple electrostatic model when the energy differences between different structures are small.

Table 5.13. Values of r_+/r_- for the alkali halides.

	Li^+	Na^+	K^+	Rb^+	Cs^+
F^-	0.51	0.73	1.00	1.12	1.28
Cl^-	0.38	0.54	0.74	0.82	0.94
Br^-	0.35	0.50	0.68	0.76	0.87
I^-	0.31	0.44	0.60	0.68	0.77

8. Complex Ionic Crystals—Pauling's Rules

The preceding section has been devoted to a concept sometimes described as Pauling's first rule, viz. that the radius ratio of cation to anion determines the coordination polyhedron and thus the ligancy of the fiducial ion.

Pauling has called the second rule the electrostatic valence rule. This so-called rule is intuitively appealing in that it requires maximum thermodynamic stability for structures in which the charge on the anion or cation is completely neutralized by its nearest neighbors. Needless to say, if electric neutrality required the participation of ions in more distant shells, the total number of bonds per ion would include coulombic potentials weakened by the greater distance of separation. Or, stated more concisely, ionic bonds between nearest neighbors are the strongest. The strengths of the individual bonds are divided pro rata among the v_i ligands, so that if all bonding is to adjacent ions and $-ze$ is the electric charge on the anion and ze the charge on the cation;

$$\zeta = \sum_i \frac{z_i}{v_i} = \sum_i s_i, \tag{5.51}$$

where ξ is the total bond strength, so s_i is the strength of the individually participating bonds. In the words of Pauling, "(we) postulate that in a stable ionic structure the valence of each anion, with changed sign, is exactly or nearly equal to the sum of

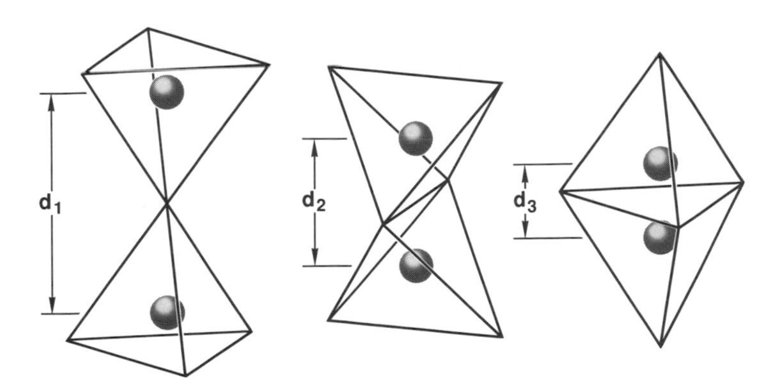

Figure 5.12.

The distances d_1:d_2:d_3 are in relation 1:0.58:0.33, thus increasing the cationic repulsive potential.

the strengths of the electrostatic bonds to it from the adjacent cations."[33] Consequently, the MgO (periclase) structure with six oxide ions coordinated to each magnesium ion has $s = 1/3$, and aluminum oxide with sixfold coordination has $s = 1/2$. If ions in more distant shells are required for neutralization, then different bond strengths must be assigned for each.

Pauling's third rule states that the sharing of edges or faces of coordination polyhedra decreases the stability of the structure, and this decrement is enhanced by cations of large valence and a small number of ligands. If one continues to consider the cases wherein the cation resides within the coordination polyhedron of the larger anions, it is then the cation–cation repulsion terms that cause the diminution of stability. The decreasing cation separation distance in the tetrahedral coordination in going from corner sharing to face sharing is demonstrated in Fig. 5.12. The separation between cations shows a relative decrease from unity for shared corners to 0.58 for shared edges to 0.33 for shared faces. The corresponding shrinkage is less for octahedra, being 0.71 and 0.58.

9. Electronegativity: Ionic Bonds

Electronegativity is defined as the power of an atom in a molecule to attract an electron to itself. Fluorine is the most electronegative of all the elements, and cesium is the most electropositive, as it gives up its $6s^1$ electron most easily. Although all are directly a function of the nuclear attraction for electrons, a list follows that defines the various potentials commonly employed in chemistry.

$M(g) \rightarrow M^+(g) + e^-(g)$	Ionization potential
$M(g) + e^-(g) \rightarrow M^-$	Electron affinity
$M(\text{soln.}) \rightarrow M^+(\text{soln.}) + e^-$	Oxidation or electrode potential
$M - X \rightarrow M^+ - X^-$	Electronegativity

A chemical bond can vary in the symmetry of its charge distribution from one that is purely ionic, such as CsF, to the opposite extreme of strict covalence, such as between the carbon atoms in diamond. Strictly speaking, all bonds between unlike atoms will demonstrate some degree of polarity, reflecting their difference in electronegativity.

A relative measure of the degree of ionicity of a chemical bond can be obtained from a comparison of the calculated value of the dipole moment, assuming the bond

---

[33] L. Pauling, p. 548, loc. cit.

Table 5.14. Calculated (er_0) and measured (μ) values of the electric dipole moments of the hydrogen halides in debye units (D).

	r_0 (Å)	er_0 (D)	μ (D)[a]	μ/er_0
HF	0.92	4.42	1.98	0.45
HCl	1.28	6.07	1.03	0.17
HBr	1.43	6.82	0.79	0.12
HI	1.62	7.74	0.38	0.05

[a] Experimental values.

to be entirely ionic, with the measured value. The results of such calculations are shown in Table 5.14 for the hydrogen halides. It is clear that even HF, the most ionic of the hydrogen halides, does not exhibit a dipole consistent with the complete transfer of an electron from the hydrogen to the fluorine. As the electron affinity of the halogen decreases with increasing atomic number, the polarity of the H—X bond decreases.

A. Ionic Bonds

A quantitative measure of the ionic character of a chemical bond can be obtained if one can assume that the enthalpy of bond formation can be divided between the covalent and ionic contributions; i.e., for the A—B bond one is permitted to write

$$\Delta H_{AB} = \Delta H_{\text{covalent}} + \Delta H_{\text{ionic}}. \tag{5.52}$$

There is some quantum mechanical justification for this assumption as the best wave function, and thus the one that provides the greatest stability for a bond between unlike atoms will be a combination of covalent and ionic wave functions, which for an arbitrary diatomic A—B molecule can be written

$$\Psi = C_1\psi_{AB} + C_2\psi_{A^+B^-} + C_3\psi_{A^-B^+}.$$

When A and B differ greatly in electronegativity, the least probable resonance hybrid (i.e., $\psi_{A^+B^-}$ or $\psi_{A^-B^+}$) is omitted from the total wave function.

Pauling has proposed two non–quantum mechanical methods of estimating ΔH_{ionic} based upon thermochemical data. The first proposal assumes

$$\Delta H_{\text{ionic}} = \Delta H_{AB} - \Delta H_{\text{covalent}} = \Delta H_{AB} - (1/2)(\Delta H_{AA} + \Delta H_{BB}). \tag{5.53}$$

In words, (5.53) states that the covalent contribution to the A—B bond is the arithmetic mean of the covalent bonds of the pure constituents, and, hence, the enthalpy of the ionic contribution is the difference between the total enthalpy of formation and the covalent fraction.

A second approximation, which appears more in accord with experimental data when ΔH_{AA} and ΔH_{BB} differ significantly, makes use of the constituent bond energies as the geometric mean. Thus,

$$\Delta H'_{\text{ionic}} = \Delta H_{AB} - (\Delta H_{AA} \cdot \Delta H_{BB})^{1/2}. \tag{5.54}$$

It is found that neither ΔH_{ionic} nor $\Delta H'_{\text{ionic}}$ is additive. This is to say, that they are not unique values that remain strictly invariant regardless of the nature of the compound. However, it was discovered that $(\Delta H'_{\text{ionic}})^{1/2}$ as calculated with (5.54) does approximately satisfy an additivity relation that allows us to write

$$\Delta H'_{\text{ionic}} = k(x_A - x_B)^2, \tag{5.55}$$

where x_A and x_B are thus defined as the electronegativities. An arbitrary value of $x_F = 4.0$ for fluorine establishes the base number to which all other electronegativity values are referenced. The empirically determined value of k is found to be 30 kcal. Now (5.54) can be expressed in the form

$$\Delta H_{AB} = (\Delta H_{AA} \cdot \Delta H_{BB})^{1/2} + 30(x_A - x_B)^2 \tag{5.56}$$

Table 5.15. The complete electronegativity scale.[a]

Li	Be	B											C	N	O	F
1.0	1.5	2.0											2.5	3.0	3.5	4.0
Na	Mg	Al											Si	P	S	Cl
0.9	1.2	1.5											1.8	2.1	2.5	3.0
K	Ca	Sc	Ti	V	Cr	Mn	Fe	Co	Ni	Cu	Zn	Ga	Ge	As	Se	Br
0.8	1.0	1.3	1.5	1.6	1.6	1.5	1.8	1.8	1.8	1.9	1.6	1.6	1.8	2.0	2.4	2.8
Rb	Sr	Y	Zr	Nb	Mo	Tc	Ru	Rh	Pd	Ag	Cd	In	Sn	Sb	Te	I
0.8	1.0	1.2	1.4	1.6	1.8	1.9	2.2	2.2	2.2	1.9	1.7	1.7	1.8	1.9	2.1	2.5
Cs	Ba	La–Lu	Hf	Ta	W	Re	Os	Ir	Pt	Au	Hg	Tl	Pb	Bi	Po	At
0.7	0.9	1.1–1.2	1.3	1.5	1.7	1.9	2.2	2.2	2.2	2.4	1.9	1.8	1.8	1.9	2.0	2.2
Fr	Ra	Ac	Th	Pa	U	Np–No										
0.7	0.9	1.1	1.3	1.5	1.7	1.3										

[a] The values given in the table refer to the common oxidation states of the elements. For some elements variation of the electronegativity with oxidation number is observed; for example, Fe^{II} 1.8, Fe^{III} 1.9; Cu^{I} 1.9, Cu^{II} 2.0; Sn^{II} 1.8, Sn^{IV} 1.9. For other elements see W. Gordy and W. J. O. Thomas, *J. Chem. Phys.*, **24**, 439 (1956). (Reprinted from Linus Pauling: *The Nature of the Chemical Bond*, 3rd Edition. © 1939, 1940, 3rd edition © 1960 by Cornell University. Used by permission of the publisher, Cornell University Press).

Table 5.16. Comparison of electronegativity with average of ionization energy and electron affinity.

	Ionization Energy	Electron Affinity	Sum/125	x (Pauling)
F	403.3	83.5	3.90	4.0
Cl	300.3	87.3	3.10	3.0
Br	274.6	82.0	2.86	2.8
I	242.2	75.7	2.54	2.5
H	315.0	17.8	2.66	2.1
Li	125.8	0	1.01	1.0
Na	120.0	0	0.96	0.9
K	101.6	0	0.81	0.8
Rb	97.8	0	0.78	0.8
Cs	91.3	0	0.73	0.7

Reproduced from Pauling, p. 96, loc. cit.

and

$$x_A - x_B = (1/30)[\Delta H_{AB} - (\Delta H_{AA} \cdot \Delta H_{BB})^{1/2}]^{1/2}. \tag{5.57}$$

Table 5.15 reproduces Pauling's "complete electronegativity scale."

A somewhat simpler scheme for calculating electronegativities has been proposed by Mulliken[34], in which he simply sums the electron affinity and the ionization potential of the atom under consideration. These values are in good agreement with Pauling's, with the exception of hydrogen, as shown in Table 5.16.

B. Electronegativities and Crystal Structure

It is apparent from elementary chemical reasoning that chemical bonds become increasingly ionic in character as the difference in the electronegativities of the ligands increases. Conversely, bonds between atoms with similar or equal values of their electronegativities tend to be covalent or metallic in nature. In both the ionic and covalent cases the greatest stability is achieved when the atoms or ions can complete their outer shell of electrons to attain a rare gas electron configuration. This can be accomplished by sharing covalent electrons or by the gaining or losing of electrons to form ionic bonds. When covalent bonds are formed, tetrahedrally coordinated compounds generally result, as exemplified by the sp^3 bonding of group IVB. Ionic

[34] R. S. Mulliken, *J. Chem. Phys.*, **2**, 782 (1934); **3**, 573 (1935).

compounds are most often octahedral or cubic coordination. Thus one anticipates a direct link between electronegativity and crystal structure.

The electronegativity differences in a sample of over 100 different compounds are plotted against the average value of the principal quantum number, i.e. $\bar{n} = (n_A + n_B)/2$, in Fig. 5.13. The plot clearly reveals two distinct populations, with the octahedrally coordinated compounds occupying the upper region of the graph, and the tetrahedrally coordinated the lower. There are but eight exceptions in which seven fourfold compounds are in the nominally sixfold region and one sixfold coordinated compound that is displaced.

The relative number of octahedral compounds increases with increasing values of $\bar{n}$. This trend is rationalized by recalling that the energy required to delocalize or completely remove an electron decreases with increasing atomic number within each group. Hence, it becomes energetically more favorable to involve d and f electrons in the bonding with increasing $\bar{n}$ with a resultant decrease in the stability of the sp^3 bonds.

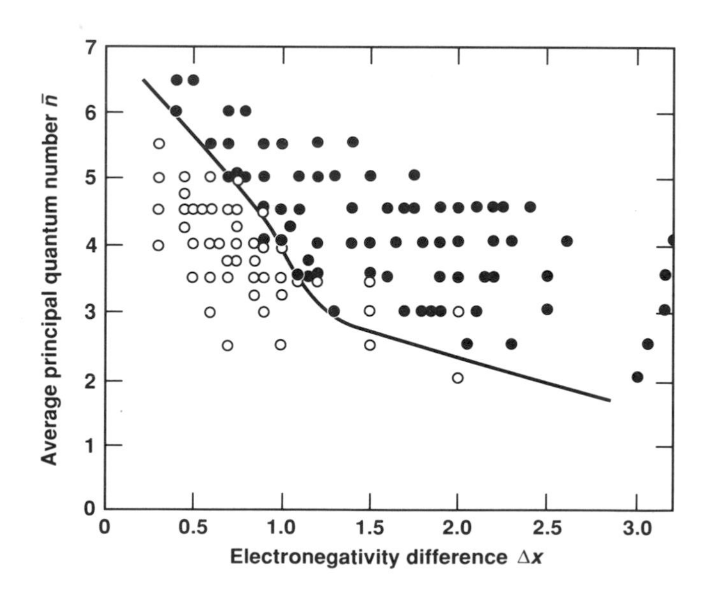

Figure 5.13.

The separation of fourfold and sixfold coordinated structures, according to E. Mooser and W. B. Pearson, *Acta Cryst.* **12**, 1015 (1959). The abscissa is Pauling's electronegativity difference; the ordinate is the average principal quantum number of the valence electrons. [Reproduced from J. C. Phillips, *Rev. Mod. Phys.*, **42**, 317 (1970).]

10. Some Important Ionic Structures

Having gotten this far into Chapter 5, even the casual reader might conclude that sodium chloride is the most important of all the inorganic compounds. Although this opinion is shared in common by cooks and textbook writers, compounds having commerically useful magnetic, electrical or mechanical properties generally possess crystal structures more complicated than that of rock salt. Two examples are discussed here in order to illustrate how physical and chemical properties can reflect the complexities of the underlying atomic order.

A. The Perovskite Structure

The naturally occurring mineral perovskite has the formula $CaTiO_3$ and is named for a 19th-century Russian mineralogist, Count L. A. Perovski. When occurring naturally, it is primarily igneous in origin, although it is also occasionally found in metamorphic formations, particularly in the boundaries between magmatic intrusions and impure limestones where contact metamorphism often produces Ce- or Nb-rich perovskites. The compounds $(Mg, Fe)SiO_3$ and $CaSiO_3$ transform to the perovskite structure at very high pressures. It has been proposed that they are the major components of the earth's mantle.

The general formula for this mineral is ABX_3, where A and B are cations and X is the anion. The idealized cubic crystal structure is shown in Fig. 5.14. The larger of the two cations occupies the central position, and the smaller one the cube corners. The latter is octahedrally coordinated to the anions that occupy the $\frac{1}{2}$ 0 0 positions. This is better visualized with the help of Fig. 5.15.

A great many elements are able to form ideal or modified perovskites. The latter are slightly distorted from the ideal cubic form so that adjacent cells have slight differences in symmetry. As a result, the true unit cell consists of a cluster of distorted cubes.

The range in relative sizes spanned by the ionic radii that can be accommodated by the pervoskite structure can be specified as follows: initially the larger cation, A, is taken to be exactly the same as X, the anion. Then, in terms of ionic radii,

$$r_A + r_X = (r_B + r_X)\sqrt{2}$$

which can be derived from purely geometrical considerations. However, the structure tolerates a finite range of values that then can be expressed by

$$r_A + r_X = t(r_B + r_X)\sqrt{2},$$

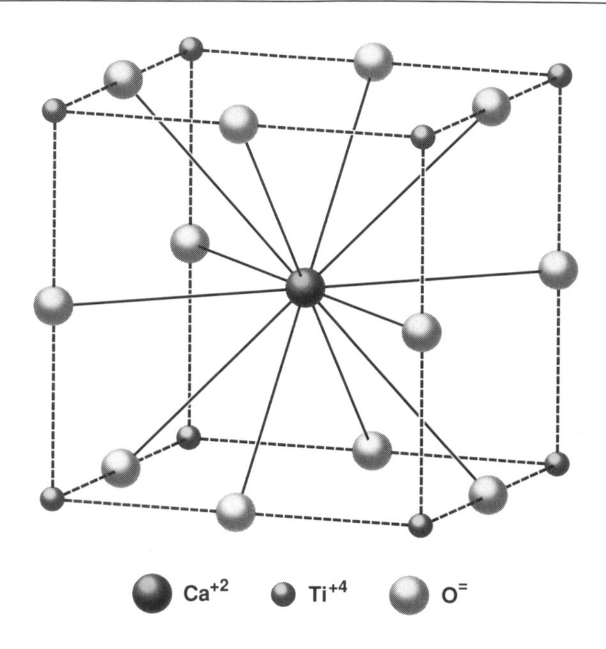

Figure 5.14.

The idealized cubic unit cell of $CaTiO_3$ (perovskite).

where, as a rule, the tolerance factor t is $0.8 \leq t \leq 1.0$. If $0.8 \leq t \leq 0.89$, the unit cell is deformed from ideal cubicity. When this is the case, i.e., the A ion is too small, then the principal axes of the cation octahedra are tilted with respect to their neighbors, as shown in Fig. 5.16. This results in puckered networks of linked B and X ions that are the basis for one of the unusual electrical properties of some perovskites.

Over 50 different metallic ions are accepted by the perovskite structure, which only requires that the positive and negative charges be of equal number. This is clearly illustrated by the chemically disparate constituents in the following list of perovskites

$A^{+n}B^{+m}X_3$	$n{:}m$
$NaWO_3$	1:5
$CaSnO_3$	2:4
$YAlO_3$	3:3
$CsCdBr_3$	1:2

An even more striking indifference to the association of a particular charge state with an A and B lattice site is exemplified by such perovskites as $K_xLa_{(2/3-x/3)}TiO_3$, where singly charged potassium ions share the same A-type lattice

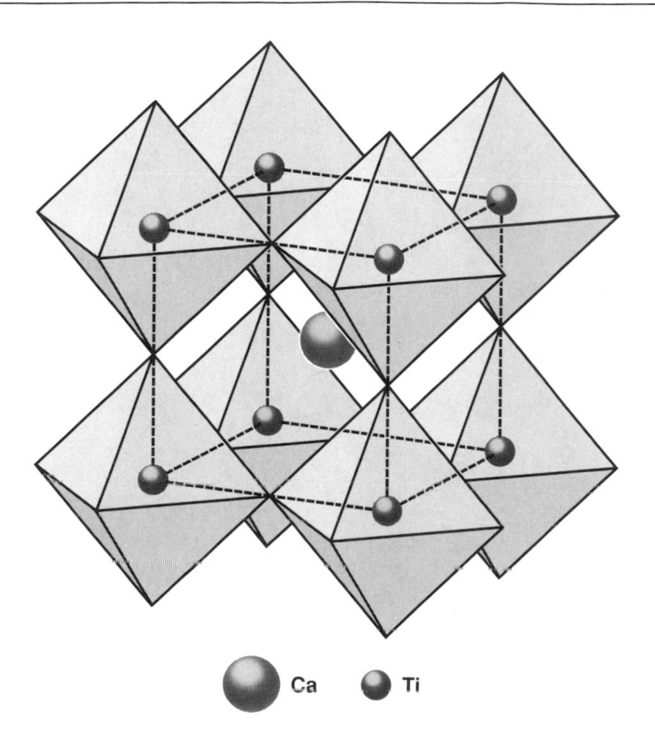

Figure 5.15.

The ideal structure of perovskite, showing the octahedral coordination of the Ti^{4+} ions to the O^{2-}. The oxide ions are located at the points of contact between adjacent octahedra.

sites with triply charged lanthanum. Almost all known perovskites are oxides, followed by a much smaller number of fluorides, sulfides, and other halides, as well as a few selenides.

B. Piezoelectric Perovskites

Some crystals can maintain a permanent electric dipole because the center of the positive charge does not coincide with that of the negative one. In such cases the macroscopic single crystal will behave as a giant dipole and has a positive end and a negative end; such crystals are ferroelectric. When a ferroelectric crystal is placed in an electrostatic field, the response of the crystal's dipole causes a distortion of the lattice. Removal of the field causes the strain to vanish, and the structure relaxes to its equilibrium zero-field dimensions. The previously mentioned small distortions of the cubic structure, which are exhibited by so many perovskites, are the source of

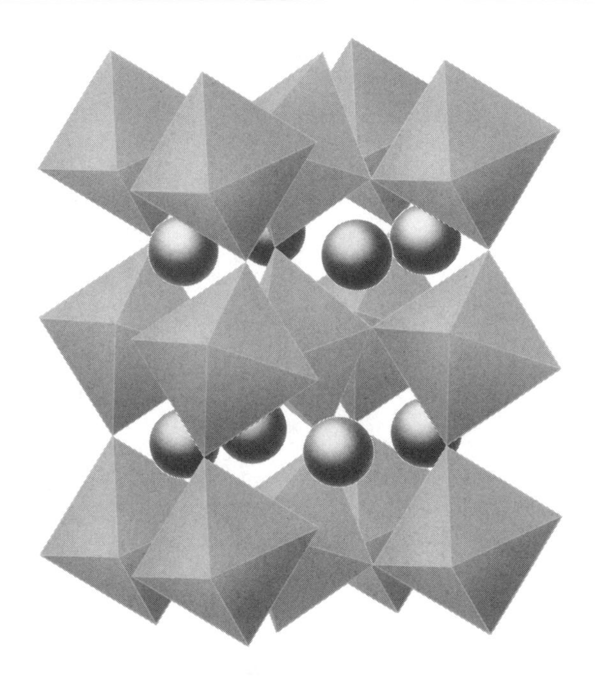

Figure 5.16.

Illustrating the distorted perovskite structure that gives rise to the puckered configuration.

their crystal dipole moments and, thus, the cause of their piezoelectric behavior. At temperatures somewhat above a characteristic critical temperature, thermal vibrations nullify these inherent distortions and the compound is no longer ferroelectric. This is completely analogous to the temperature-dependent transition of a ferromagnet that becomes a paramagnet above its Curie temperature. The crystal distortions giving rise to piezoelectric behavior are shown sequentially in Fig. 5.17, with the most widely used piezoelectric substance, barium titanate ($BaTiO_3$) as the representative example.

It should be understood that the transition from the cubic to the distorted one is not sharply defined as it is in first-order phase transitions (e.g., melting, allotropic transitions, boiling, etc.). The ferroelectric to nonferroelectric (paraelectric state) transition belongs to the category of transitions known as second-order ones, which includes magnetism, order–disorder, and all the transitions collectively known as cooperative phenomena. The transition temperatures of some representative perovskites are listed in Table 5.17.

Piezoelectrics find many applications since they can be used to transform electrical oscillations into mechanical vibrations, and vice versa. Components of pressure gauges, electrical relays, and audio apparatus are representative of their many

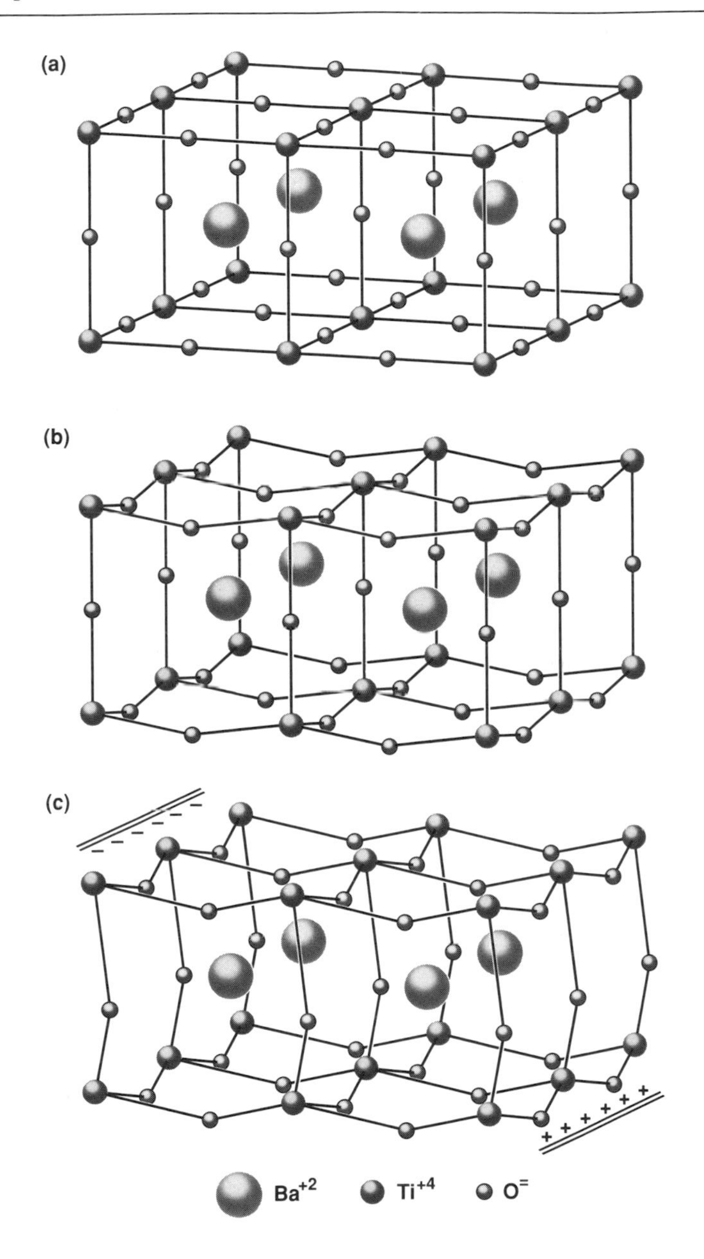

Figure 5.17.

The barium titanate structure. The ions shown are not to scale, since the O^{2-} is in actuality much larger than Ti^{4+}: (a) the perfect cubic structure as it exists at $T \gg 393$ K; (b) the distorted zero-field structure existing at $T < 393$ K; (c) the structure with enhanced deformation caused by an electrostatic field at $T < 393$ K.

Table 5.17. Ferroelectric transition temperatures, T_c, for some representative perovskites.

Compound	T_c (K)
$BaTiO_3$	393
WO_3	223
$KNbO_3$	712
$PbTiO_3$	763

applications. In particular the so-called PZT perovskites, which encompass a continuous series of solid solutions between the end members $PbTiO_3$ and $PbZrO_3$, can generate strong voltages up to 100 V in response to very slight compressions.

C. Perovskite Superconductors

The dc electrical resistivity vanishes in the superconducting state below a critical temperature. Oxides having the perovskite or related structures can become higher-temperature superconductors than any other class of compounds. In the two years preceding the writing of this text, the highest critical temperature, T_c, then known for any compound has risen from 23 to 122 K, and there are unconfirmed reports of even higher values. The commercial potential for superconductors that do not require cooling with liquid He or H_2 refrigerants can scarcely be overestimated. The rise in superconducting transition temperatures with the date of their respective discovery is displayed in Fig. 5.18.

At the present writing there is good reason to expect the discovery of new compounds with even higher T_c values than the current record holder.

Probably the most-studied class of perovskites to date has the formula $YBa_2Cu_3O_{7-x}$, where $0 \geq x \leq 1$. This clearly is not a multiple of ABO_3 but is an oxygen-deficient modification of $A_3B_3O_9$ that would consist of three basic cubic cells as shown in Fig. 5.19. In the slang of the trade, compounds of this sort have been named "1–2–3", since this is the ratio of the metal ions in the formula. The unit cell is shown in Fig. 5.19.

Charge neutrality is maintained by varying the relative concentrations of Cu^+:Cu^{2+}:Cu^{3+} in response to the varying O^{2-} concentration. The better superconductor is the oxygen-rich member. It is generally assumed that, in passing from six O^{2-} per formula unit to seven, the additional oxide ions occupy the unfilled terminal positions of the 1–2–3 unit cell rather than the number 2 or central cube. This is attributed to the increase in Cu^{3+} with increasing O^{2-}. The cupric ion is

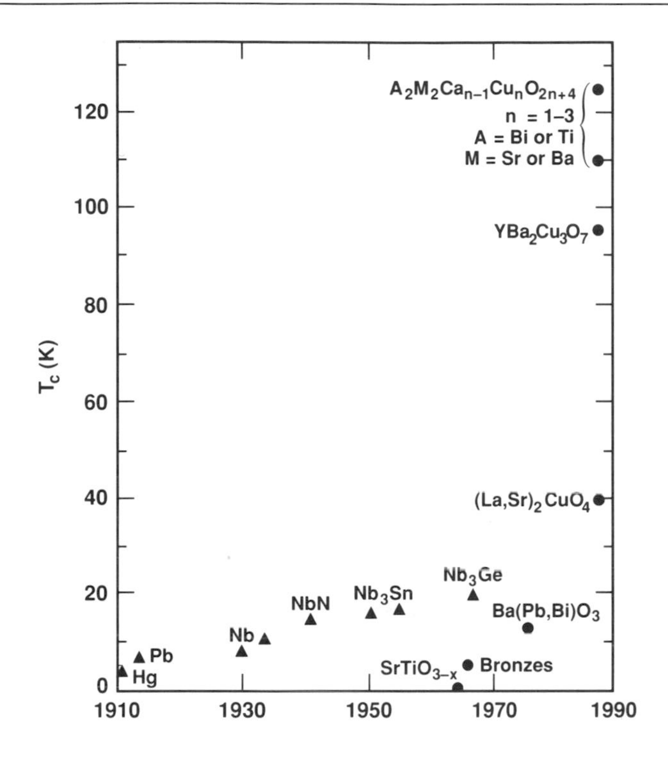

Figure 5.18.

The rise in superconducting temperatures with the date of their discovery. [After A. W. Sleight, *Science*, **242**, 1519, 1520 (1988). Copyright 1988 by the AAAS.]

considered to supply the charge carrier. A generalized phase diagram representing the electrical and magnetic states of several compositions of oxide superconductors is shown in Fig. 5.20.

At present, there is no theory that fully explains the superconductivity of inorganic compounds. Nevertheless, there are some interesting correlations to be made between T_c and composition. It has been pointed out that increasing the Cu^{3+} concentration raises the critical temperature of $YBa_2Cu_3O_{7-x}$, since it represents an increase in the charge-carrying entity. In addition, increasing the Cu^{3+}/Cu^{2+} shortens the average Cu—O bond length, which has been found by independent methods to raise the T_c of the $La_{2-x}A_xCuO_4$ group of superconductors. For example, substitution of ions of decreasing radius (e.g., Ba, Sr, Ca, and Na) causes a corresponding decrease in the Cu—O distance and a proportionate rise in T_c. Furthermore, compression of these compounds also increases their transition temperatures.

Another example of the sensitivity of oxide superconductors to variation in the

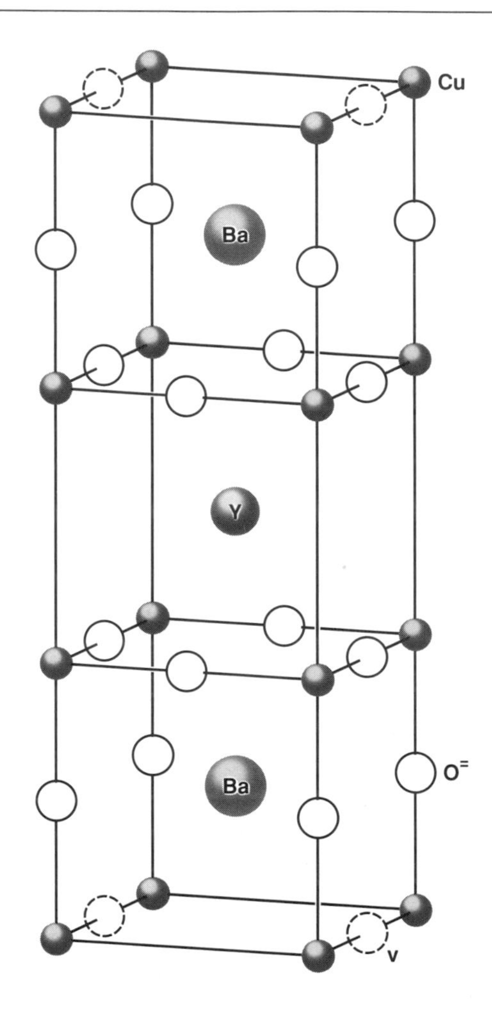

Figure 5.19.

The 1–2–3 structure has three cubic units. O^{2-} ions are absent from the vertical edges of the Y cell. They are also missing from the terminal horizontal planes $YBa_2Cu_3O_6$, but there are two O^{2-}, shown in dashed circles, in $YBa_2Cu_3O_7$.

composition is shown by the substitution of Bi^{4+} for Pb^{4+} in the simple ABO_3 structure presented in Fig. 5.21.

It is probably essential that inorganic superconductors have metallic, or at least semiconductive, properties. Insulators with tightly bound localized electrons cannot conduct electricity in either a normal or a superconducting state because the energy required to excite the outer electrons to the level of the conduction band is simply too great.

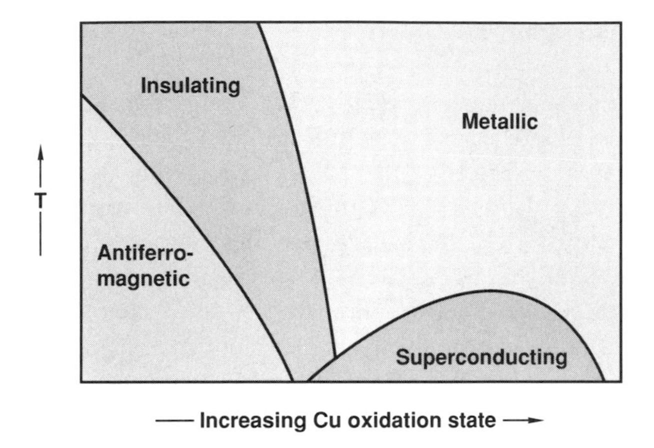

Figure 5.20.

The electrical and magnetic states for systems such as $YBa_2Cu_3O_{6+x}$, $Bi_2Sr_{3-x}Y_xCu_2O_8$, and $La_{2-x}A_xCuO_4$ (where A, for example, can be Ba, Sr, Ca, or Na). (After A. W. Sleight, p. 1525, loc. cit., Copyright 1988 by the AAAS.)

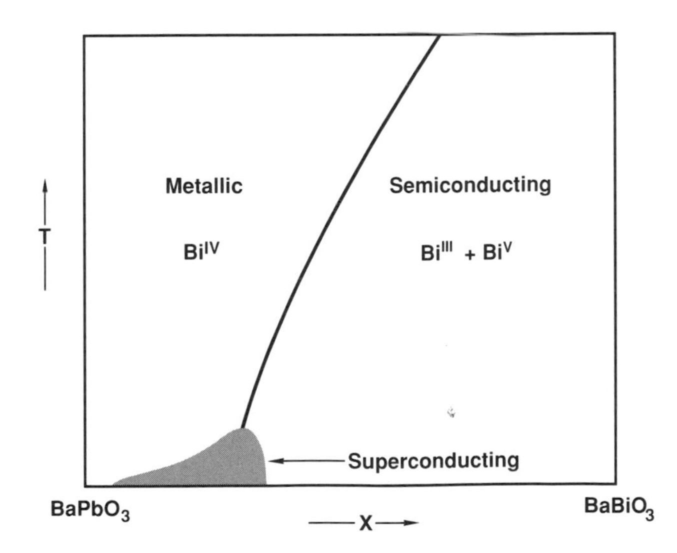

Figure 5.21.

$BaPbO_3$ and $BaBiO_3$ form a continuous series of solid solutions, yet only a limited range of compositions becomes superconducting. (After A. W. Sleight, loc. cit., Copyright 1988 by the AAAS.)

D. The Spinel Structure

Naturally occurring spinel minerals are found as minor constituents of both igneous and metamorphic rocks. The prototypic mineral after which the structure is named has the idealized formula $MgAl_2O_4$. In nature, other elements in varying amounts will substiture for Mg^{2+} and Al^{3+}. The most renowned member of the spinel group is doubtless magnetite, Fe_3O_4, the lodestone, man's first useful magnet. The general formula for spinel is AB_2X_4, where A and B are metallic cations in the +2 and +3 oxidation states, respectively, and can equally well be written as $(AX)(B_2X_3)$. Most spinels are oxides, although a much smaller number of sulfides, selinides, and

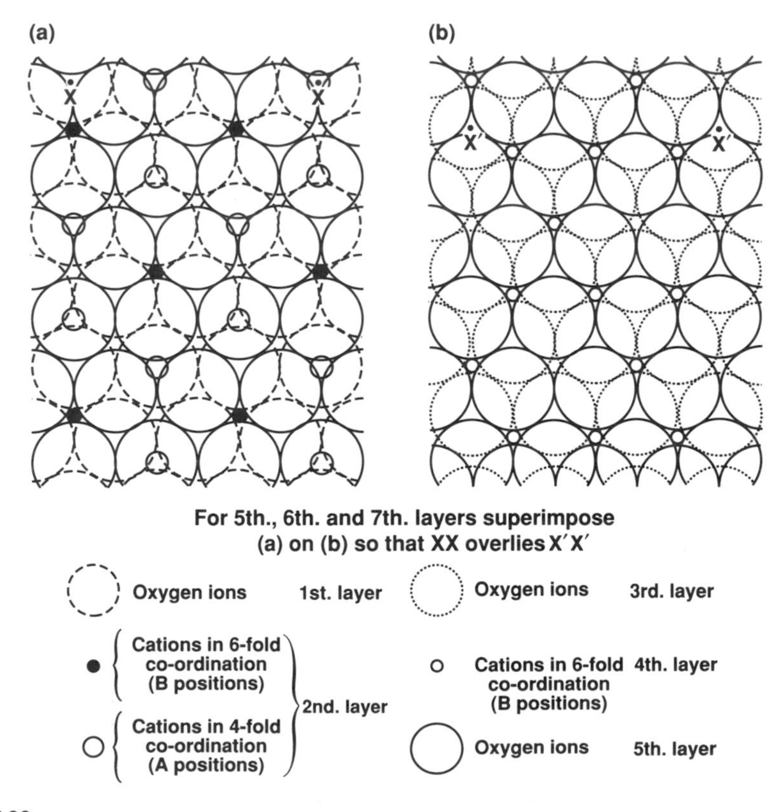

Figure 5.22.

The structure of spinel minerals projected along the (triad) z-axis. The close-packed planes of the O^{2-} ions, [110] and [$\bar{1}$10], alternate with those of the cations. The planes containing the fourfold coordinated ions alternate with those having ions in the sixfold positions. [Taken from W. A. Deer, R. A. Howie, and J. Zussman, *Rock-Forming Minerals*, Wiley, vol. 5, p. 58, 1962, as adapted from G. D. Nicholls, *Adv. Phys.*, **4**, 113 (1955).]

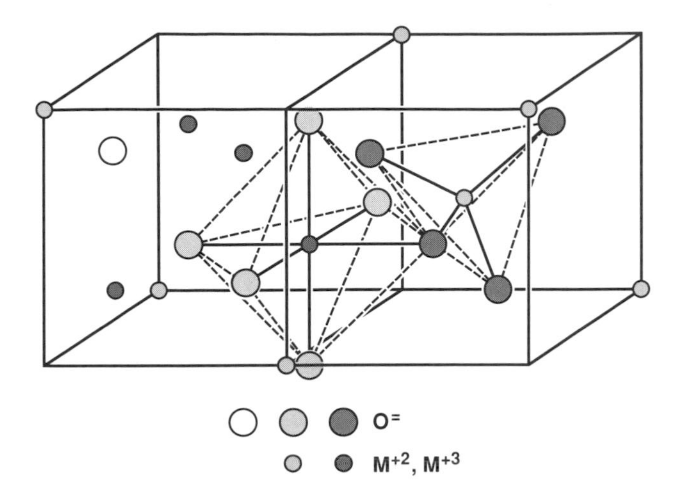

Figure 5.23.

Two octants of the spinel structure. Note that the octahedrally coordinated cation has one O^{2-} ligand in common with the tetrahedrally coordinated ion.

tellurides also exist. Stability decreases with decreasing electronegativity, making the oxides the most stable.

The spinel crystal structure is frequently described by saying that it is a close-packed face-centered cubic configuration of X^{2-} ions with the cations filling one-eighth of the tetrahedral and one-half of the octahedral interstices. A schematic diagram of such an arrangement is shown in Fig. 5.22.

An expanded view of two of the eight cubic units that make up the unit cell is shown in Fig. 5.23.

When the tetrahedrally coordinated A sites are occupied by M^{2+}, and the octahedral B sites by M^{3+} the structure is called a normal spinel. If the A sites are occupied by M^{3+} and the B by a randomly disposed mixture of M^{2+} and M^{3+}, the structure is known as an *inverse* spinel.[35] The oxidation states of the cations are not restricted necessarily to +2 and +3. They can also be formed of M^{4+}, M^{2+}, and even M^{+}, where the quadrivalent ion occupies the smaller A site and the lower oxidation state the B site.

Because ionic radii decrease with increasing oxidation number, the more highly charged cations tend to be smaller and hence tend to go into the A sites. However, charge neutrality can be achieved in ways other than by varying the composition. In particular, spinels tend to incorporate vacant lattice sites into their structure as

[35] This inversion is explainable in terms of ligand field theory and will be discussed in Chapter 6.

a method for preserving electroneutrality. Gamma hematite, γ-Fe_2O_3, can possess as many as 2 and 2/3 vacancies per unit cell. A formula[36] more revealing of the actual structure is written as Fe^{3+} $[\square_{1/3}Fe^{+3}_{5/3}]$, where the symbol $\square$ stands for the vacant lattice site.

E. Ferrites

Spinels that have the general formula MM'_2X_4, where M can be divalent Fe, Ni, Mn, Zn, Cu, Co, but M′ must be Fe^{3+} and X^{2-} = O, S, Se, Te, are called ferrites. These materials have magnetic properties that make them useful for a variety of electrical and electronic applications. A few examples of their uses are as cores in inductors and transformers, gyrators in microwave circuitry, and formerly as memory elements in electronic computers.

In order to illustrate the relation between the magnetic properties and the composition and crystal structure, we take magnetite[36] as our first example. We can estimate the maximum value of the magnetic moment of the Fe_3O_4 formula unit, based upon a free-ion approximation, as follows:

$$\mu = gJ\mu_\beta, \tag{5.58}$$

where the Landé g factor is taken equal to 2.00, its value for a free electron, and μ_β gives the equation its units in terms of the Bohr magneton. The total electron spin J is calculated in the following way: the electronic configuration of iron consists of an inner core with the Ar structure and outer shells holding $3d^6 4s^2$ electrons. Consequently, Fe^{2+} has six 3d electrons distributed among the five d orbitals, which leaves four of them unpaired in accordance with the Hund principle of maximum multiplicity for the ground state. Each electron has a spin of $\frac{1}{2}$, so (5.58) predicts $\mu = 4\mu_\beta$ for each Fe^{2+}. In a similar manner, one derives $\mu = 5\mu_\beta$ for each Fe^{3+}. Thus, Fe_3O_4 should have a total moment of $14\mu_\beta$ based upon the free-ion approximation. However, the measured value is $\cong 4\mu_\beta$. Although the free-ion approximation is admittedly crude, this discrepancy is so great that another explanation is required.

A true ferromagnet such as Fe or Ni has all its moments aligned in the same direction within a single magnetic domain when the temperature is below T_c, the Curie temperature. However, this is not the only way in which the atomic moments can align themselves with respect to one another. All arrangements other than the collinear one, with each spin preserving the same sense for its north and south poles,

---

[36] In speaking about nonstoichiometric compounds, it is preferable to refer to them by their generic name rather than by the stoichiometric formula unless this composition is specifically intended.

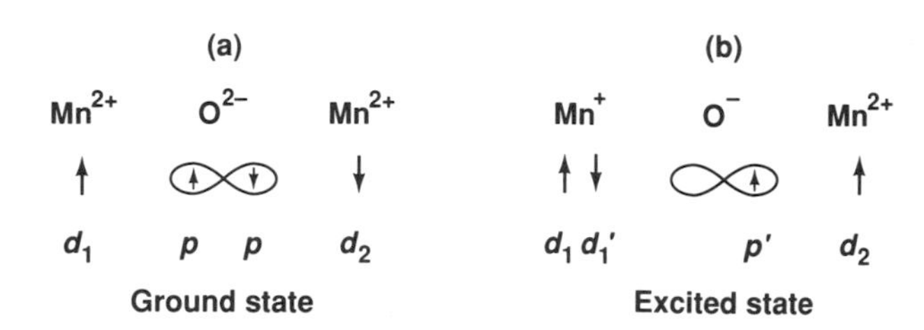

Figure 5.24.

Superexchange coupling leading to antiferromagnetic spin order in MnO. (After A. H. Morrish, *The Physical Principles of Magnetism*, Wiley, New York, p. 465, 1965.)

necessarily leads to a diminution of the bulk moment as a result of internal cancellation. When this cancellation is complete, the compound is said to be antiferromagnetic and there are innumerable arrangements of the atomic moments that can create this condition. However, in some cases, of which magnetite is one, there is but a partial cancellation, leaving the substance with a residual moment. Such compounds belong to a class termed ferrimagnets. The magnetic coupling between the atomic moments of adjacent atoms in simple substances, such as Fe, is called direct exchange coupling. Such a direct interaction is clearly impossible in the spinel structure because all the magnetic cations are separated from each other by oxide ions. Consequently, the spin coupling is indirect, proceeding via the O^{2-} bridge. This particular method of indirect exchange is called superexchange and was first proposed by Kramers.[37–39]

Leaving spinels for the moment, we can most simply illustrate superexchange using antiferromagnetic MnO as our example. Nearest-neighboring Mn^{2+} ions are separated by only 4.43 Å, and direct exchange is expected to be ferromagnetic. What is found instead is antiferromagnetic coupling between ions more distantly separated by an intervening O^{2-}. The mechanism for superexchange is shown in Fig. 5.24 and is seen to involve a shift in the spin density of O^{2-}. Also shown in this figure are the excited states shown as electron transfers from O^{2-} to Mn^{2+}. This configuration, if complete, nullifies the moment on the manganese ion, but the remaining Mn^{2+} remains coupled to the oxygen. The strongly overlapping p orbitals of the O^{2-} with the collinear adjacent *d* orbitals of the adjacent Mn^{2+} outweigh the strength of the direct exchange between nearest-neighboring Mn^{2+} ions and the indirect exchange between them that would be forced to proceed through a right angle as shown by Fig. 5.25.

[37] H. A. Kramers, *Physica*, **1**, 182 (1934)
[38] P. W. Anderson, *Phys. Rev*, **79**, 350 (1950).
[39] H. G. Van Vleck, *J. Phys. Radium*, **12**, 262 (1951).

O^{2-} Mn^{2+}

Mn^{2+}

Figure 5.25.

The interactions between nearest-neighboring Mn^{2+} that are not bridged by a common collinear O^{2-}. (After Morrish, p. 466, loc. cit.)

It is interesting to note that the superexchange phenomenon requires extended wave functions, implying a significant covalent character to the bonding in MnO. Nevertheless, the enthalpy of formation based upon a strictly ionic model gives good agreement with the experimental value (see Table 5.7).

It was pointed out in Fig. 5.23 that a common O^{2-} ion links the tetrahedrally coordinated cations to the octahedrally coordinated ones. Although there are also oxide bridges between A sites and between B sites, the A—B configuration is the most favorable one for superexchange as described by Fig. 5.26. When the cations are not linked in a direct linear configuration by the O^{2-}, the optimal strength for the exchange interaction becomes a compromise between the bridging angle and the bond lengths. It turns out that the Fig. 5.26a configuration is the favored one since (b) and (d) each have one very long bond and (c) and (e) have the least propitious angles.

Returning to our example, Fe_3O_4, we can see how the moment of the Fe^{2+} coupled antiferromagnetically with an equal number of Fe^{3+} ions will reduce the total moment from $14\mu_\beta$ to $6\mu_\beta$. It is not unexpected that our oversimplified free-ion

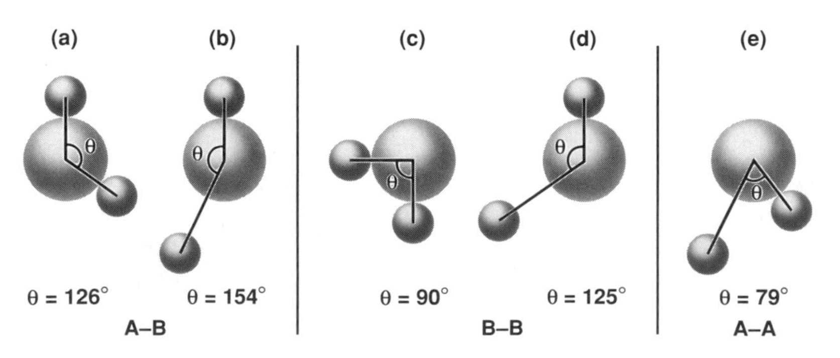

Figure 5.26.

All possible O^{2-} linkages between cations are shown, with distances drawn to scale. [After Morrish, p. 504, loc. cit. Taken from E. W. Gorter, *Phillips Res. Rpt.*, **9**, 295 (1954).]

Table 5.18. The magnetic ordering temperature T_o as a function of the fraction of A sites, η, occupied by Mg^{2+}.

η	T_o (C°)
0.27	325
0.23	331
0.19	339
0.18	355
0.16	372
0.09	410

From Landolt–Bornstein, **4/3**, 217 (1970).

model produces a value that disagrees with the measured one. In fact, the discrepancy, such as it is, can be explained, although it would carry us too far afield to do so here. The ferrimagnetic spin arrangement has, of course, been amply confirmed by neutron diffraction measurements.

As expected, the magnetic ordering temperature depends upon the distribution of the ions between the tetrahedral and octahedral sites even when the overall composition is a constant. This is recorded for the $MgFe_2O_4$ system in Table 5.18. The observation that the ordering temperature rises with the increasing fraction of Fe^{3+}—Fe^{3+} conjoined in the A—B linkage is again consistent with the preceding argument contending that this configuration provides the strongest magnetic coupling.

F. Superionic Conductors

Fast ionic conductors are a diverse group of ionic compounds embracing a variety of crystal structures and compositions. What they have in common is their ability to carry an electric current, in contrast to most ionic crystals, which are insulators. The specific conductivities of this group are much larger than typical semiconductors, though considerably less than those of true metals. As the name superionic implies, the electrical charge carriers are ions rather than electrons or positive holes in the valence band. Consequently, their high conductivity reflects the rapid rate of diffusion of the charge-bearing ion. The range in conductivity spanned by these materials, with a few examples specifically noted, is compared in a diagrammatic fashion with that of the electron conductivities of more conventional substances in Fig. 5.27.

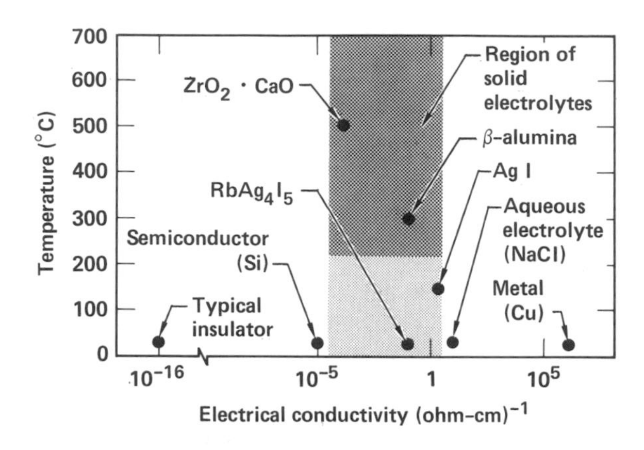

Figure 5.27.

Electrical conductivities of several common substances and representative solid electrolytes are shown at temperatures where the materials might be used. β-Alumina is the sodium form, in which Na^+ is the mobile species. In silver iodide, Ag^+ is responsible for the electrical conductivity, as it is in $RbAg_4I_5$. [After Shriver and Farrington, *C&E News*, **63**, 42 (1985).]

The essential requirements for a good ionic conductor are that it have

1. A high concentration of charge carriers
2. A high concentration of vacancies or interstitial ions
3. A low enthalpy of activation, $\Delta H_m^\ddagger$, for ion hopping

These requirements are met in different ways, but a generally common feature of fast ionic conductors is the possession of somewhat continuous open paths for interstitial diffusion. Frequently, an order–disorder transformation occurs that increases the availability of additional lattice sites for diffusion at elevated temperatures. This mechanism is, in particular, typical of the cubic silver-based ionic conductors. Here the effective vacancy concentration is increased in the disordered phase, and the conductivity is further augmented by the relatively open passageways inherent in these crystal structures.

The ability of fast ionic conductors to serve as high-temperature electrolytes promises a wide variety of commercial applications.

G. β and β″-Alumina

Two of the most widely studied superionic conductors are called β- and β″-alumina for historical reasons, although they are really aluminates. They have a modified spinel structure, with 50 of the 58 ions in the unit cell having the same configuration as in the spinel structure. The general formula for Na–β-alumina is $Na_{1+x}Al_{11}O_{17+0.5x}$, where the excess O^{2-} occupies interstitial lattice sites. A typical composition for this compound is $Na_{1.2}Al_{11}O_{17.1}$. The general formula for Na–β''-alumina is $Na_{1+x}M_xAl_{11-x}O_{17}$. The cation M is divalent, and by substituting for Al^{3+} in the tetrahedral sites it compensates for the excess positive charge supplied by the Na^+ ions, thus allowing the oxide concentration to remain constant. Commonly, M is Mg^{2+}, Ni^{2+}, or Zn^{2+}, although a host of others will substitute as well.

The fundamental structural lattice unit is shown in Fig. 5.28 for β-alumina. The

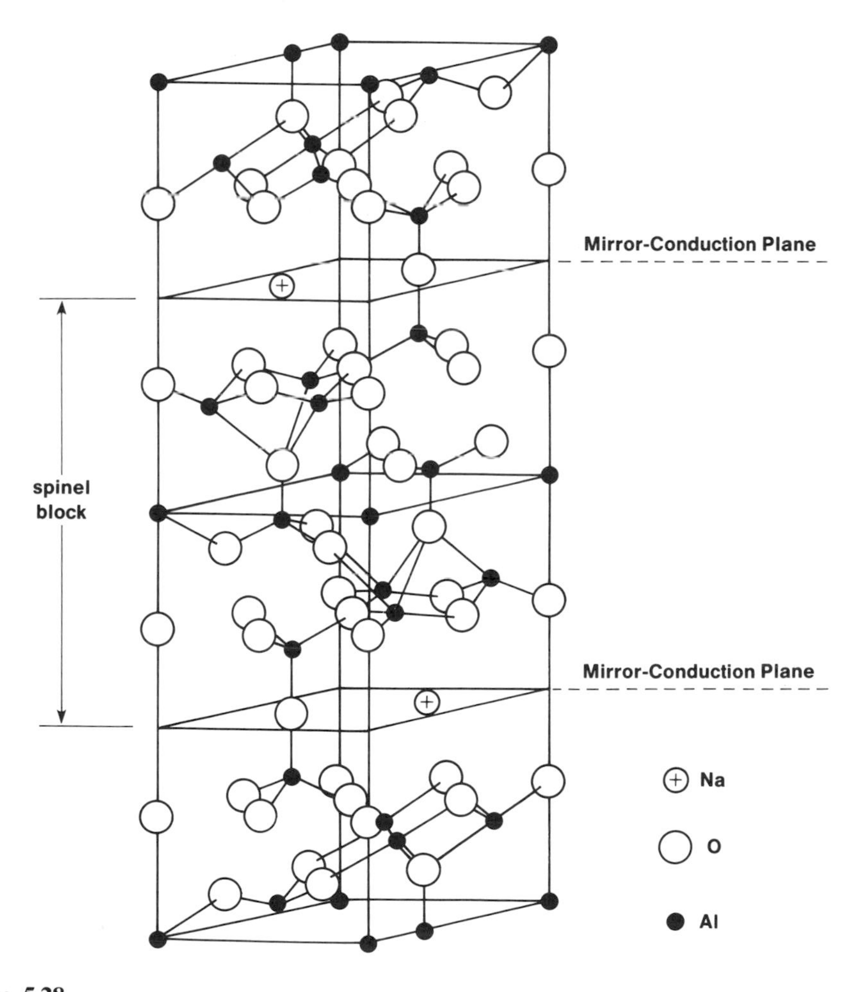

Figure 5.28.

Illustrating the relation between β-alumina and the spinel structure.

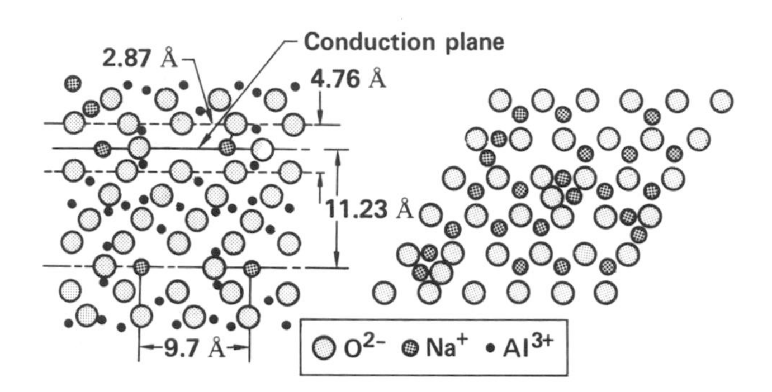

Figure 5.29.

The structure of sodium β-alumina is seen left as tightly packed Al—O spinel-type blocks held together by Al—O—Al bonds. The interblock region, defined by the midpoints of the oxygen atoms in these bridging bonds, is the conduction plane in which sodium ions diffuse rapidly in two dimensions around the bridging oxygens. A 90° rotation out of the plane of the page gives the view on the right showing (looking down on) the conduction plane. Note how sparsely populated this plane is with Na^+ ions. There are no Al^{3+} ions in this plane.

Na^+ charge carriers diffuse within the 4.76-Å-wide band that encompasses the mirror plane, as shown in Fig. 5.29, zigzagging around the columns of oxide ions. The β''-alumina phase, though still containing spinel blocks, has the blocks rotated 120° and is a more complicated structure and more difficult to portray in two dimensions. It does provide a more open structure for the diffusion of the sodium ions.

Only a few of the superionic conductors, such as the aluminates, are sufficiently refractory to allow their use significantly above room temperature. The stability of these compounds is in no small measure due to the stability of their predominately spinel-like structures.

Chapter 5 Exercises

1. Calculate the Madelung constant for a linear chain of alternating positive and negative charges.

2. a. Calculate the Madelung constant M, as defined by (5.2), for the infinite planar array shown below, to four significant figures. (Start with the central negative ion.)

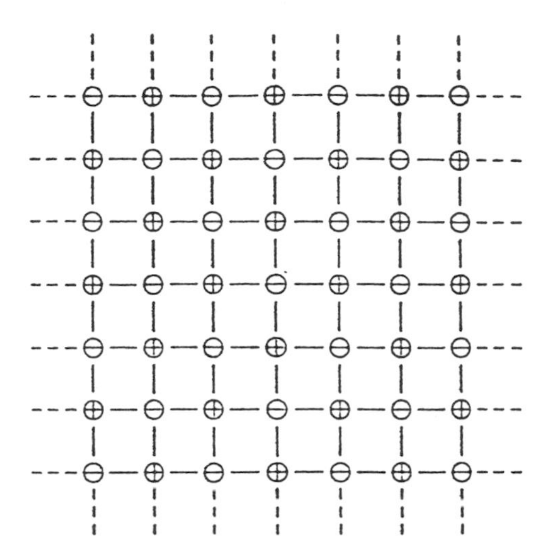

b. Calculate the total lattice energy of the array, assuming coulombic interactions between unlike ions and a repulsive potential given by $\phi_r = A/r^n$.

3. The compressibility of NaCl is 3.3×10^{12} cm^2-$dyne^{-1}$, and the equilibrium spacing between Na^+ and Cl^- is 2.81 Å. Calculate n in (5.34).

4. The Evjen method sums over expanding concentric shells that are electrically neutral. The potential converges more rapidly with this procedure than shells that bear a charge. Draw the first neutral shell for the CsCl structure, giving the coordinates of the relevant ions.

5. Estimate the lattice energy of the hypothetical compound Xe^+F^-.

6. Estimate the lattice energy of zinc blende when $n = 9$. Given that the equilibrium constant for $Cl_2 \rightleftarrows 2Cl$ is 4.8×10^{-6} at 600 K and is 1.04×10^{-10} at 800 K and enthalpy of vaporization of Li (metal) $\rightarrow$ Li (gas) is 38 kcal, estimate the standard internal energy, ΔE°_0, for the formation of 1 mole of LiCl (solid) at 0 K, using the Born–Haber cycle.

7. Discuss and compare the Pauling and Mulliken schemes for calculating electronegativities. Why do the values obtained by these two very different methods show such good agreement?

8. Assume CaS has the rock salt structure. Commencing with (5.12), calculate the equilibrium Ca–S separation, r_e, the compressibility, and the lattice energy, given that $n = 9.4$ for NaCl.

Additional Reading

Most of the additional reading references relevant to (1) Madelung constants, (2) repulsive potentials, (3) lattice energies, and (4) the Born–Haber cycle have been cited in this chapter. However, for convenience we collect the most important ones here with a brief description of their content.

M. F. C. Ladd and W. H. Lee, *Prog. Solid State Chem.*, **1**, 37–82 (1964); **2**, 378 (1965). Cites 147 references and is perhaps the most comprehensive bibliography ever assembled on these subjects.

T. C. Waddington, *Adv. Inorg. Chem. Radiochem.*, **1**, 157–221 (1959). 135 references present the results of calculations involving dozens of compounds.

W. E. Dasent, *Nonexistent Compounds*, Marcel Dekker, New York (1965). As the whimsical title suggests, this is a very imaginative little book dealing with a variety of ways in which to estimate the stability of compounds that are of low stability.

E. A. Moelwyn-Hughes, *Physical Chemistry*, Pergamon Press, Chap. 13 (1961). A very clear, concise presentation of the fundamental equations used to calculate lattice energies.

J. C. Slater, *Introduction to Chemical Physics*, McGraw-Hill, Chap. 23 (1939). Gives a detailed example of how to calculate Madelung constants and lattice energies and integrates these results with empirical equations of state. Probably the best presentation from the standpoint of pedagogy.

L. Pauling, *The Nature of the Chemical Bond*, Cornell University Press, Ithaca, NY (1960). Although occasionally rambling and without a clearly organized plan of presentation, it nevertheless remains the seminal work on these topics as well as many others. It also is instructive to observe how the Pauling intuition is used to explain and solve chemical problems.

F. Donald Bloss, *Crystallography and Crystal Chemistry, An Introduction*, Holt, Rinehart and Winston, Chaps. 8, 9 (1971). This is an exceptionally well-written introductory presentation to the subject of ionic radii and crystal structure, with emphasis on the latter as befits a crystallographer.

F. G. Fumi and M. P. Tosi give a somewhat more advanced treatment of ionic radii and repulsive potentials in *J. Phys. Chem. Sol.*, **31**, 21 (1964).

Chapter 6

Quantum Mechanical Principles and the Covalent Bond

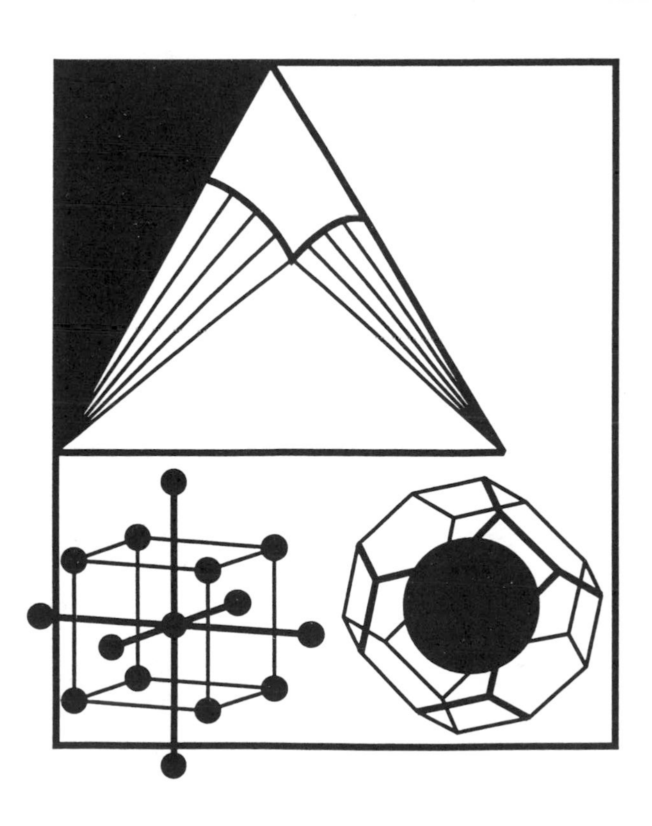

Chapter 6

Contents

Schrödinger's cat's a mystery cat, he illustrates the laws;
The complicated things he does have no apparent cause;
He baffles the determinist and drives him to despair
For when they try to pin him down—*the quantum cat's not there*!

—"Mr. Eliot's Guide to Quantum Theory", John Lowell, *Physics Today*, April 1989

Quantum Mechanical Principles and the Covalent Bond

Except in simple cases, quantum mechanics generally fails to deliver quantitative values of the chemical and physical properties of matter. Nevertheless, some understanding of elementary quantum mechanics is mandatory in order to explain at least qualitatively atomic and molecular behavior. Also, in a qualitative sense, its judicious application can offer a substantial degree of predictability for which there is no alternative method. The material presented in this chapter is intended chiefly to furnish a ready reference for terms, concepts, and equations that are to be used in later portions of the book. Those aspects of the subject that do not directly or indirectly pertain to chemical bonding or to electric and magnetic properties have been purposefully excluded.

1. The Schrödinger Equation

We will not probe very deeply into the foundations of quantum mechanics. Nevertheless, it is appropriate to reflect briefly upon the origins and significance of the Schrödinger equation. The solution of this equation in some suitable form underlies all nonrelativistic wave mechanics.

The starting point for the imputation of wavelike behavior to particulate matter is the de Broglie relation

$$\lambda = h/mv = h/p, \tag{6.1}$$

where λ is the wavelength, h is Planck's constant, and p is the momentum of the particle. This equation, by analogy, can be based on the same expression derived earlier for the phonon. The quantization of electromagnetic radiation requires that the energy be specified by $E = h\nu$, ν being the frequency. Also according to the special theory of relativity $E = mc^2$, where m is the relativistic mass of the photon and c is the speed of light. As a result, we arrive at (6.1) as

$$\lambda = c/\nu = h/mc = h/p. \tag{6.2}$$

A further substantiation of this idea is found in the development of the Bohr atom. In his famous model, Bohr proposed that the electrons of the hydrogen atom were to be found in quantized orbitals such that the circumference of a stable orbital must equal an integral number of wavelengths as given by the de Broglie equation.[1] This requirement implied that the bound electrons were, in fact, to be regarded as standing waves so that $n\lambda = 2\pi r$, where n is an integer and r is the radius of the orbit. Substituting this into (6.1) gives

$$r = n\hbar/p, \tag{6.3}$$

where $\hbar$ is the common designation for $h/2\pi$. According to (6.3), the atomic electrons can exist only in discrete orbitals, the radii of which are a function only of their momenta. The Bohr model was experimentally justified in that it quantitatively reproduced the absorption and emission spectra of atomic hydrogen. Later, the experimentally observed diffraction of electron beams was the direct proof of the wave properties of electrons.

One must presume that Schrödinger reasoned that if electrons can be treated as waves, they must perforce obey the fundamental equation for wave motion. In its most elementary form, this is simply the equation of motion of a vibrating string and led to a semiclassical derivation of the one-dimensional Schrödinger equation as follows:

$$\frac{\partial^2 u}{\partial x^2} = \frac{1}{v^2}\frac{\partial^2 u}{\partial t^2}. \tag{6.4}$$

This is the equation for a vibrating string and is hence identical to (3.28), where u is the local transverse displacement, v the phase velocity, and t the time. There are several ways of expressing the solutions to (6.4) for standing waves (see Appendix D). For example,

$$u = A \sin 2\pi(x/\lambda - \nu t) = \psi(x) \sin 2\pi\nu t, \tag{6.5}$$

[1] In actuality, the Bohr atom preceded the de Broglie hypothesis, and it was de Broglie who first noted that his equation could explain Bohr's standing waves.

where, for the moment, ψ is some unspecified function of x only. Taking the second derivative of the last term on the r.h.s. of (6.5) with respect to x and t and back-substituting into (6.4) leads us to

$$\frac{d^2\psi}{dx^2} + \frac{4\pi^2\nu^2}{v^2}\psi = 0. \tag{6.6}$$

In this way the partial differential equation is reduced to an ordinary one that no longer includes time as a variable.

Equation (6.6) is now modified as follows: the total energy of the system, E, is the sum of the kinetic energy $p^2/2m$ and potential energy U; thus,

$$E = p^2/2m + U \tag{6.7}$$

or

$$p = [2m(E - U)]^{1/2}. \tag{6.8}$$

Applying the relation $\nu^2/v^2 = \lambda^{-2}$ to the de Broglie relation between λ and p and substituting this result into (6.6) leads to

$$\frac{d^2\psi}{dx^2} + \frac{8\pi^2 m}{h^2}(E - U)\psi = 0, \tag{6.9}$$

which is the time-independent Schrödinger equation in one dimension. The exact nature of $\psi(x)$, the wave function, is specified by the physical (i.e., electric or magnetic) circumstances surrounding the electron. We will deal with this topic later.

A more rapid, though less insightful, derivation of (6.9) is to begin by rearranging (6.7) to appear as follows:

$$-p^2/2m + \mathrm{E} - U(x) = 0. \tag{6.10}$$

One then substitutes

$$\frac{\hbar}{i}\frac{d}{dx}$$

for p, thus obtaining the operator equation

$$\frac{\hbar^2}{2m}\frac{d^2}{dx^2} + [E - U(x)] = 0. \tag{6.11}$$

Since our ultimate goal is to obtain an equation for standing waves that will be time-independent, we let (6.11) operate on an arbitrary function of x only, viz. $\psi(x)$, to obtain

$$\frac{\hbar^2}{2m}\frac{d^2\psi}{dx^2} + [E - U(x)]\psi = 0. \tag{6.12}$$

After rearrangement this is identical to (6.9). It turns out that substituting $\hbar/i$ for the linear momentum into the classical Hamiltonian representations of the energy converts the classical equation into a quantum mechanical Hamiltonian operator (see Appendix F).

In three dimensions the second derivative in (6.9) is replaced by the Laplacian

$$\nabla^2 = \frac{\partial^2}{\partial x^2} + \frac{\partial^2}{\partial y^2} + \frac{\partial^2}{\partial z^2} \tag{6.13}$$

and so leads to

$$\nabla^2\psi + \frac{8\pi^2 m}{h^2}(E - U)\psi = 0. \tag{6.14}$$

The Hamiltonian operator in three dimensions is now defined by

$$\hat{H} \equiv -\frac{h^2}{8\pi^2 m}\nabla^2 + U. \tag{6.15}$$

Written in its briefest form, the time-independent Schrödinger equation becomes

$$\hat{H}\psi = E\psi. \tag{6.16}$$

When particles obeying the wave equation are constrained within a well-defined volume, the energy becomes quantized and is no longer continuous but assumes only discrete values. Each quantized value of E corresponds to a specific ψ and is referred to as the *eigenvalue* E_i associated with its particular *eigenfunction* ψ_i. The eigenvalues are a direct consequence of the requirement that all allowable eigenfunctions must represent standing waves. Thus, intermediate values of λ that are incommensurate with the relation $n\lambda/2 = L$, where L is the length of the confining one-dimensional potential energy well, are forbidden. The mathematics describing wave motion in three dimensions are more complicated, but the principle remains the same. In classical terms, one would say that the forbidden wavelengths are self-damping.

It should be understood that the confining volume leading to standing waves effectively is provided by the total electrostatic potential supplied by the nuclei of atoms or ions in combination with their associated electrons.

A. The Free-Electron Gas

A realistic approximation of a free-electron gas is represented by the electrons in the conduction band of a metal. They are presumed to be screened from the positively charged nuclei by the inner core electrons in a way that produces a uniform attractive potential that is invariant throughout the crystal. This being the case, the potential energy, $U(x)$ in (6.12), is conveniently set equal to zero, so (6.12) becomes

$$-\frac{\hbar^2}{2m}\frac{d^2\psi_n}{dx^2} = E_n\psi_n. \tag{6.17}$$

The boundary conditions necessary to maintain a standing wave along a line of length L are

$$\psi_n(0) = 0 \qquad \text{and} \qquad \psi_n(L) = 0, \tag{6.18}$$

which expresses the fact that ψ_n must vanish at the boundaries. The subscript n is appended to ψ and E to denote the series of allowable solutions to (6.17) for a fixed value of L. A periodic function that satisfies (6.17) and fulfills the requisite boundary conditions is

$$\psi_n = a\sin(2\pi x/\lambda) = a\sin(n\pi x/L). \tag{6.19}$$

Thus, $L = n\lambda/2$, where n assumes integral values 1, 2, 3, ... That this is a solution may be verified by differentiation, and a more thorough discussion of this and similar equations is given in connection with (3.38) and Appendix D.

Taking the second derivative of ψ_n in (6.19) yields

$$\frac{d^2\psi_n}{dx^2} = -a\left(\frac{n\pi}{L}\right)^2\sin\left(\frac{n\pi x}{L}\right) = -\left(\frac{n\pi}{L}\right)^2\psi_n. \tag{6.20}$$

Substituting for ψ_n and its second derivative in (6.17) leads to

$$E_n = \frac{\hbar^2}{2m}\left(\frac{n\pi}{L}\right)^2. \tag{6.21}$$

Thus, for a fixed value of the length, E_n is a function solely of the quantum number n. The first four energy levels in units of $\hbar^2/2m$ are shown by the horizontal lines in Fig. 6.1. The corresponding wave functions are superimposed at each energy level. Figure 6.1 illustrates a general principle, viz. that the wave function producing the lowest energy level has no nodes—i.e., $\psi_1 = 0$ only at the boundaries of $x = 0$ and L. Each successively higher energy level produces an additional node so that ψ_2 has one node, ψ_3 has two, and so on.

Each energy level shown in Fig. 6.1 can hold two electrons with opposing spins. The levels fill sequentially starting at the lowest level. The Pauli *exclusion principle* states that no two electrons distributed over an identically same volume can have an identical set of quantum numbers. Another way of stating the exclusion principle is to state that the complete wave function for two or more electrons, which is the product of their individual wave functions, must be antisymmetrical. Thus, although

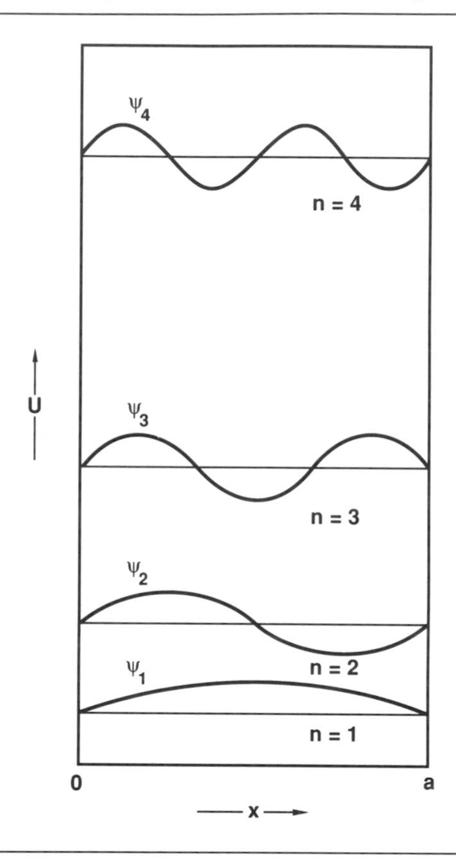

Figure 6.1.

The first four energy levels of a free-electron gas of noninteracting electrons in a potential energy well of infinite depth. The corresponding wave functions are superimposed on each level.

the energy remains unchanged, interchanging the coordinates of the individual wave functions for two electrons causes a change of sign. In other words, if ψ is a function solely of the spatial coordinates and σ stands for the spin coordinates, viz. $\pm 1/2$, then interchange of the electrons, i.e., $\psi \rightarrow \psi$, requires $\sigma \rightarrow -\sigma$ or, if $\psi \rightarrow -\psi$, $\sigma \rightarrow \sigma$. The complete wave function Ψ can always be written as the product $\Psi = \psi\sigma$ so that the interchange of the coordinates of any two electrons in the same quantum state results in $\Psi \rightarrow -\Psi$.

The wave equation has many solutions for each set of boundary conditions. However, only those that represent physical reality are termed acceptable solutions. The mathematical requirements for an acceptable solution are that it be everywhere single-valued, finite, and continuous, and that its gradient be continuous.

Considering the electron as a stationary wave extending from 0 to L implies a distribution of its mass over this distance. However, the particle–wave duality has yet to receive a simple explanation that is generally agreed upon, so it is better to think of the wave function as giving a probability distribution rather than a mass distribution. Thus, the probability of finding the electron at a given position x, where $0 \leq x \leq L$, is then proportional to $\psi_n^2(x)$. When $\psi_n(x)$ is complex, the distribution is given by $\psi_n(x)\psi_n^*(x)$ where ψ_n^* is the complex conjugate of ψ_n, which is formed by replacing i wherever it occurs in ψ with $-i$. The wave function is normally complex, and simply squaring it would result in an imaginary distribution function without physical significance for a real electron. If the wave functions are normalized, that is,

$$\int_0^L \psi_n \psi_n^* \, dx = 1, \tag{6.22}$$

then $\psi_n \cdot \psi_n^*$ can be considered as a probability density function. Thus, if ψ_n is the wave function of an electron, (6.22) states that the probability of finding it between $0 \leq x \leq L$ is unity. Speaking loosely, this procedure converts a density distribution given in arbitrary units to a probability distribution bounded by 0 and 1. This has its analogue in the squaring of the amplitude of an electromagnetic wave in order to obtain its intensity.

Let us now return to (6.22), substitute the complex version of ψ_n of (6.19) into (6.22), and integrate; i.e.,

$$a^2 \int_0^L \psi\psi^* \, dx = 1, \tag{6.23}$$

where a is the normalizing factor to be determined. Setting $a = n\pi x/L$, we can write

$$\sin \alpha = \frac{e^{i\alpha} - e^{-i\alpha}}{2i},$$

and so

$$\sin^2 \alpha = \frac{1}{2} - \frac{1}{2} \frac{e^{2i\alpha} + e^{-2i\alpha}}{2} \tag{6.24}$$

$$= \frac{1}{2} - \frac{1}{2} \cos 2\alpha. \tag{6.25}$$

Because of the symmetry between the sine function and its complex conjugate, $\psi \cdot \psi = \psi^* \cdot \psi$. Substituting (6.25) into (6.22), integrating, and solving for a leads to $a = (2/L)^{1/2}$. Thus, the normalized eigenfunction becomes

$$\psi_n(x) = (2/L)^{1/2} \sin(n\pi x/L). \tag{6.26}$$

B. Particle in a Box

The preceding derivation for a free electron in a linear one-dimensional box is easily expanded to describe the three-dimensional situation. The Schrödinger equation in three dimensions becomes

$$\frac{\partial^2\psi}{\partial x^2} + \frac{\partial^2\psi}{\partial y^2} + \frac{\partial^2\psi}{\partial z^2} + \frac{2m}{\hbar^2}[E - U_x - U_y - U_z]\psi = 0. \tag{6.27}$$

Once again, we specify a uniform potential within the box and set $U_x = U_y = U_z = 0$. It is assumed that we can separate variables, as in the normal mode analysis of vibrations presented in Chapter 3. This leads us to try a solution of the form

$$\psi = X(x)Y(y)Z(z), \tag{6.28}$$

in which $X(x)$, $Y(y)$, and $Z(z)$ are mutually independent. Substituting into (6.27) and dividing by $X(x)Y(y)Z(z)$ yields

$$\begin{aligned} X^{-1}\frac{\partial^2 X}{\partial x^2} + \frac{2m}{\hbar^2}E_x &= 0, \\ Y^{-1}\frac{\partial^2 Y}{\partial y^2} + \frac{2m}{\hbar^2}E_y &= 0, \\ Z^{-1}\frac{\partial^2 Z}{\partial Z^2} + \frac{2m}{\hbar^2}E_z &= 0. \end{aligned} \tag{6.29}$$

Solving these three orthogonal one-dimensional equations is a repetition of the process that lead to (6.26). We impose the following boundary conditions:

$$\begin{aligned} X(0) &= X(a) = 0, \\ Y(0) &= Y(b) = 0, \\ Z(0) &= Z(c) = 0, \end{aligned} \tag{6.30}$$

which defines a rectilinear box of volume *abc*. This corresponds to assuming infinitely high potential barriers at the walls. The three solutions to (6.29) are

$$\begin{aligned} X &= \left(\frac{2}{a}\right)^{1/2} \sin\left(\frac{\sqrt{2mE_x}}{\hbar} x\right), \\ Y &= \left(\frac{2}{b}\right)^{1/2} \sin\left(\frac{\sqrt{2mE_y}}{\hbar} y\right), \\ Z &= \left(\frac{2}{c}\right)^{1/2} \sin\left(\frac{\sqrt{2mE_z}}{\hbar} \frac{z}{c}\right), \end{aligned} \tag{6.31}$$

in which the normalization that gives rise to $(2/a)^{1/2}$, $(2/b)^{1/2}$, and $(2/c)^{1/2}$ is implicit. Since these functions are valid over the entire range (i.e., $0 \leq x \leq a$, $0 \leq y \leq b$, and $0 \leq z \leq c$), we can solve them at their respective boundaries where the sine functions vanish, since their arguments must equal $n\pi$. Thus, at

$$\begin{aligned} x = a, &\qquad n_x\pi = \sqrt{2mE_x}/\hbar, \\ y = b, &\qquad n_y\pi = \sqrt{2mE_y}/\hbar, \\ z = c, &\qquad n_z\pi = \sqrt{2mE_z}/\hbar. \end{aligned} \tag{6.32}$$

Substituting Eqs. (6.32) first into (6.31) and then back into (6.28) yields

$$\psi = \left(\frac{8}{abc}\right)^{1/2} \sin\left(\frac{n_x \pi x}{a}\right) \sin\left(\frac{n_y \pi y}{b}\right) \sin\left(\frac{n_z \pi z}{c}\right), \tag{6.33}$$

which is the proper wave function for particles in a rectilinear three-dimensional box with zero potential energy.

The analogs of this situation in the real world, as mentioned earlier, are delocalized electrons, as in the conduction band of a metal. The actual shapes of real potential energy wells are much more irregular than our elementary rectilinear example. It is just such geometrical complications, as well as the fact that real potentials are finite, that provide the major difficulties for *ab initio* calculations of the energy levels. Nevertheless, the fundamental concepts are accurately portrayed by the elementary model.

C. The Maximum Energy of a Free-Electron Gas

The maximum energy of the occupied states of an idealized electron gas can be estimated without resorting to wave mechanics. First, the uncertainty principle written in three dimensions is

$$\Delta p_x \, \Delta p_y \, \Delta p_z \, \Delta x \, \Delta y \, \Delta z = h^3. \tag{6.34}$$

This gives the volume in phase space occupied by two electrons of opposite spin. Next, the center of an electron cloud is placed at the origin of a set of cartesian coordinates, which are labeled p_x, p_y, and p_z as in Fig. 6.2. This corresponds to the minimum total kinetic energy for the gas since the kinetic energy is equal to $p^2/2m$. The kinetic energy $p^2/2m$ increases with the expansion of the occupancy of the spherical momentum space. As a result, we can write

$$\frac{4}{3}\pi p_{\max}^3 = \frac{N}{2}\frac{h^3}{\Delta V}, \tag{6.35}$$

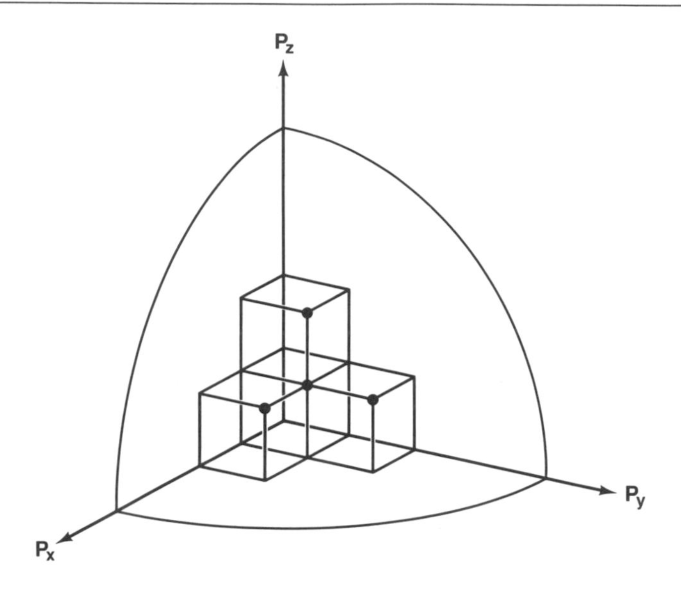

Figure 6.2.

Representation of a free-electron gas in three-dimensional momentum space. Each dot stands for two electrons of opposite spin.

where N is the total number of electrons within the sphere and is divided by 2, acknowledging the two electrons within each hypervolume element ΔV. E and p are related by

$$E_{max} = p_{max}^2/2m$$

and

$$p_{max} = (2mE_{max})^{1/2}. \tag{6.36}$$

Combining (6.35) and (6.36), we arrive at the maximum energy of the electrons:

$$E_{max} = \frac{h^2}{2m}\left(\frac{3N}{8\pi\,\Delta V}\right)^{2/3}. \tag{6.37}$$

This E_{max} term at 0 K corresponds to the Fermi energy of the electron ensemble. For ordinary metals it corresponds to a temperature of $\sim$50,000 K if the electrons were considered to be classical particles with (3/2) kT of kinetic energy. This accounts for the nonclassical behavior of the electronic heat capacity briefly described in Chapter 3, Section 6.

The energy distribution of this free-electron gas differs from the Maxwell–Boltzmann distribution because it is constrained to be within a potential energy well and is required to fill the allowable quantized energy levels sequentially beginning with the lowest. In this sense, the electron gas is not truly free for it must conform to (6.17) and, hence, (6.21), the latter specifying energy values for standing waves.

2. Atomic Wave Functions

A. The Schrödinger Equation in Spherical Polar Coordinates

The construction of wave functions for electrons in their localized orbitals about an atomic nucleus is a logical progression from the solution of the Schrödinger equation for a particle in an artificial box. The constraining force on each particular electron now is supplied by the nucleus, with additional terms coming from the remaining electrons. This is a potential energy well of finite depth, and, of course, the potential varies spatially unlike the constant value for the particle in the box. What follows is perforce a superficial review of this subject. A comprehensive treatment would be too lengthy for a text on solid state chemistry, and, furthermore, this material is well covered in nearly every book on elementary quantum mechanics. An equally, if not more, important reason is that multi-electron wave functions useful for calculations of chemical properties, particularly for condensed phases, cannot be evaluated analytically. With this apology, we now briefly review the origins of hydrogenic wave functions and quantum numbers.

The first step in applying the Schrödinger equation to the hydrogen atom is

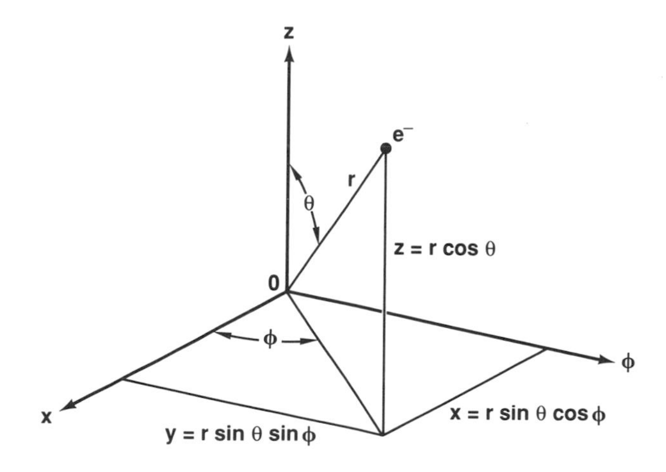

Figure 6.3.

Spherical polar coordinates.

to transform (6.14) from cartesian to spherical polar coordinates; the coordinate symbols for this transformation are defined in Fig. 6.3. Consequently, the coordinate transformations are

$$x = r \sin\theta \cos\phi,$$
$$y = r \sin\theta \sin\phi, \tag{6.38}$$
$$z = r \cos\theta.$$

Further, for this two-body problem the mass of the entire system is expressed in terms of the reduced mass

$$\mu = m_n m_e/(m_n + m_e),$$

where m_n and m_e are the masses of the nucleus and electron, respectively. Now substituting (6.38) and μ into (6.14) we obtain, after some manipulation,

$$\frac{1}{r^2}\frac{\partial}{\partial r}\left(r^2\frac{\partial\psi}{\partial r}\right) + \frac{1}{r^2\sin^2\theta}\frac{\partial^2\psi}{\partial\phi} + \frac{1}{r^2\sin^2\theta}\frac{\partial}{\partial\theta}\left(\sin\theta\frac{\partial\psi}{\partial\theta}\right) + \frac{8\pi^2\mu}{h^2}\left(E + \frac{Ze^2}{r}\right)\psi = 0, \tag{6.39}$$

where r is the distance between the electron and the nucleus, and where $U(r)$, the potential, has been made explicit by the substitution of Ze^2/r, the coulombic attraction between the electron and its nucleus. The wave function in polar coordinates is $\psi = \psi(r, \theta, \phi)$ and can be written as $\psi = R(r)\Theta(\theta)\Phi(\phi)$ with the variables separated. Substituting this last expression into (6.39), dividing by ψ, and multiplying by $r^2 \sin^2\theta$ gives

$$\frac{\sin^2\theta}{R}\frac{d}{dr}\left(r^2\frac{\partial R}{\partial r}\right) + \frac{1}{\Phi}\frac{d^2\Phi}{d\phi^2} + \frac{\sin\theta}{\theta}\frac{d}{d\theta}\left(\sin\theta\frac{\partial\Theta}{\partial\theta}\right) + \frac{2\mu}{\hbar^2}\left(E + \frac{Ze^2}{r}\right)r^2\sin^2\theta = 0. \tag{6.40}$$

Since the Φ term in (6.40) depends solely on ϕ and because r, θ, and ϕ are independent variables, we can set this term equal to a constant:

$$\frac{1}{\Phi}\frac{d^2\Phi}{d\phi^2} = -m^2. \tag{6.41}$$

By substituting this result back into (6.40), we obtain

$$\frac{\sin^2\theta}{R}\frac{d}{dr}\left(r^2\frac{\partial R}{\partial r}\right)+\frac{\sin\theta}{\theta}\frac{d}{d\theta}\left(\sin\theta\frac{\partial\Theta}{\partial\theta}\right)+\frac{2\mu}{\hbar^2}\left(E+\frac{Ze^2}{r}\right)r^2\sin^2\theta=m^2. \tag{6.42}$$

To continue the process of separating variables, we divide (6.42) by $\sin^2\theta$ and set the θ-dependent terms equal to $-\beta$, which is also a constant.

$$\frac{1}{\theta\sin\theta}\frac{d}{d\theta}\left(\sin\theta\frac{d\Theta}{d\theta}\right)-\frac{m^2}{\sin^2\theta}=-\beta. \tag{6.43}$$

The r-dependent terms then become

$$\frac{1}{R}\frac{d}{dr}\left(r^2\frac{dR}{dr}\right)+\frac{2\mu r^2}{\hbar^2}\left(E+\frac{Ze^2}{r}\right)=\beta. \tag{6.44}$$

The solutions for the so-called R, Φ, and Θ equations are developed fully in standard texts on quantum mechanics. At this juncture it suffices to say that each of the three functions leads in a natural way to a different quantum number. The radial wave function R produces n, the principal quantum number that denotes the electron shell (i.e., $n = 1, 2, 3, \ldots$). The Θ function defines the orbital angular momentum whose quantum number l takes values $l = 0, 1, 2, \ldots, (n-1)$, to which are assigned the familiar symbols s, p, d, f. The numerical value of l gives the number of nodes in the corresponding eigenfunction. The magnetic quantum number m_l comes from the solution of the Φ function and is allowed the values $-l, -l+1, \ldots, l, 0, 1, \ldots, l-1$, l for a total of $2l+1$ separate values.

Solutions of the R functions are tabulated in Table 6.1.

The radial charge distribution function is independent of the angular distribution functions and so is accurately portrayed for all directions. The first six hydrogenic radial wave functions of Table 6.1 are plotted in Fig. 6.4. The charge distribution, which is proportional to r^2R^2, is also shown.

We have not discussed the angular solutions, but pictorial representations of the s, p, and d, charge distributions in three dimensions are portrayed in Fig. 6.5. It must be remembered that these solid-appearing cartoons portray only the angular charge pattern, and they must be combined with the radial R^2 function in order to obtain the spatial distribution.

Table 6.1. Hydrogen radial wave functions.[a]

n	l	m	ψ	Function
1	0	0	$1s$	$\frac{1}{\sqrt{\pi}}\left(\frac{Z}{a_0}\right)^{3/2} e^{-\rho/2}$
2	0	0	$2s$	$\frac{1}{4\sqrt{2\pi}}\left(\frac{Z}{a_0}\right)^{3/2}\left(2-\frac{\rho}{2}\right)e^{-\rho/4}$
2	1	0	$2p_z$	$\frac{1}{4\sqrt{2\pi}}\left(\frac{Z}{a_0}\right)^{3/2}\frac{\rho}{2}e^{-\rho/4}\cos\theta$
2	1	+1	$2p_x$	$\frac{1}{4\sqrt{2\pi}}\left(\frac{Z}{a_0}\right)^{3/2}\frac{\rho}{2}e^{-\rho/4}\sin\theta\cos\phi$
2	1	−1	$2p_y$	$\frac{1}{4\sqrt{2\pi}}\left(\frac{Z}{a_0}\right)^{3/2}\frac{\rho}{2}e^{-\rho/4}\sin\theta\sin\phi$
3	0	0	$3s$	$\frac{1}{81\sqrt{3\pi}}\left(\frac{Z}{a_0}\right)^{3/2}\left(27-9\rho+\frac{\rho^2}{2}\right)e^{-\rho/6}$
3	1	0	$3p_z$	$\frac{\sqrt{2}}{81\sqrt{\pi}}\left(\frac{Z}{a_0}\right)^{3/2}\left(6-\frac{\rho}{2}\right)\frac{\rho}{2}e^{-\rho/6}\cos\theta$
3	1	+1	$3p_x$	$\frac{\sqrt{2}}{81\sqrt{\pi}}\left(\frac{Z}{a_0}\right)^{3/2}\left(6-\frac{\rho}{2}\right)\frac{\rho}{2}e^{-\rho/6}\sin\theta\cos\phi$
3	1	−1	$3p_y$	$\frac{\sqrt{2}}{81\sqrt{\pi}}\left(\frac{Z}{a_0}\right)^{3/2}\left(6-\frac{\rho}{2}\right)\frac{\rho}{2}e^{-\rho/6}\sin\theta\sin\phi$
3	2	0	$3d_{z^2x}$	$\frac{1}{81\sqrt{6\pi}}\left(\frac{Z}{a_0}\right)^{3/2}\frac{\rho^2}{4}e^{-\rho/6}(3\cos^2\theta-1)$
3	2	+1	$3d_{xz}$	$\frac{\sqrt{2}}{81\sqrt{\pi}}\left(\frac{Z}{a_0}\right)^{3/2}\frac{\rho^2}{4}e^{-\rho/6}\sin\theta\cos\theta\cos\phi$
3	2	−1	$3d_{yz}$	$\frac{\sqrt{2}}{81\sqrt{\pi}}\left(\frac{Z}{a_0}\right)^{3/2}\frac{\rho^2}{4}e^{-\rho/6}\sin\theta\cos\theta\sin\phi$
3	2	+2	$3d_{z^2-y^2}$	$\frac{1}{81\sqrt{2\pi}}\left(\frac{Z}{a_0}\right)^{3/2}\frac{\rho^2}{4}e^{-\rho/6}\sin^2\theta\cos 2\phi$
3	2	−2	$3d_{xy}$	$\frac{1}{81\sqrt{2\pi}}\left(\frac{Z}{a_0}\right)^{3/2}\frac{\rho^2}{4}e^{-\rho/6}\sin^2\theta\sin 2\phi$

[a] $\rho = 2\alpha r$, $\alpha = z/na_0$, $a_0 = \hbar/\mu e^2$.

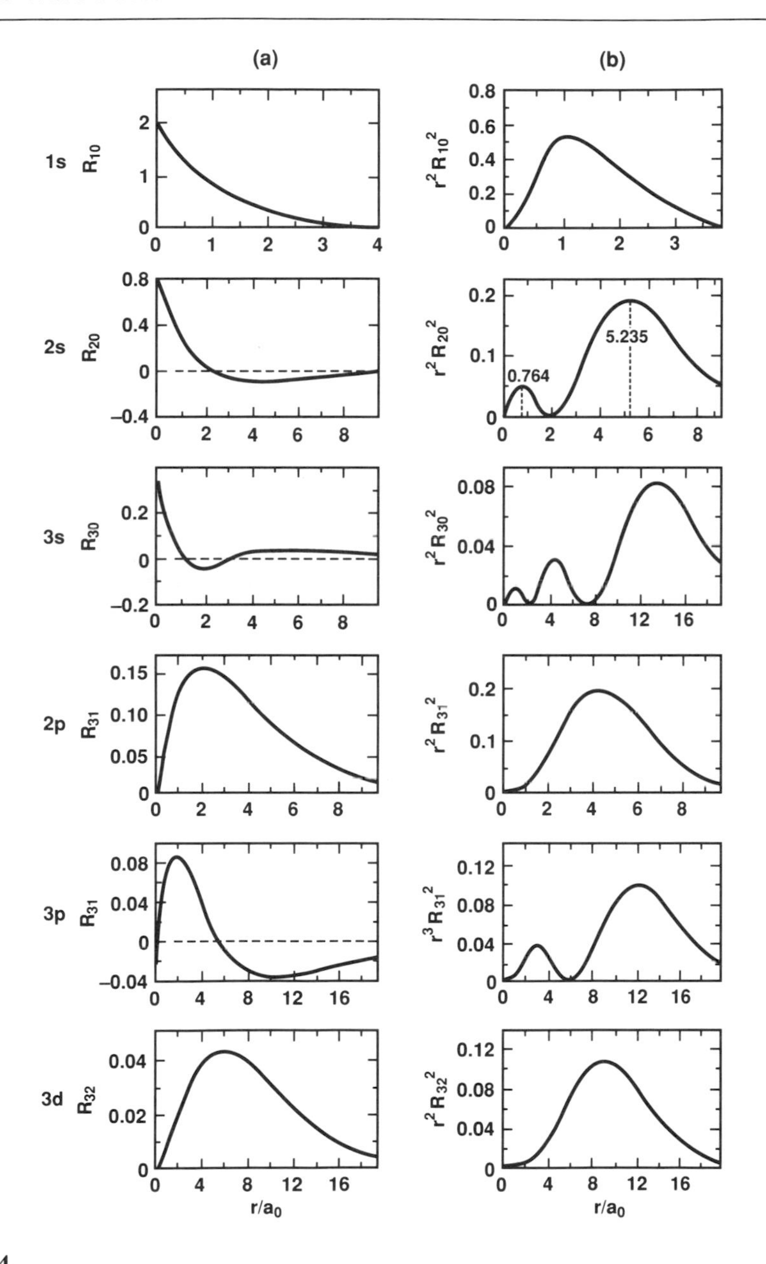

Figure 6.4.

(a) The radial function R for various hydrogenic wave functions; (b) the charge distribution as determined by $|\Psi|^2(4\pi r^2)$, which is the total charge within a shell of thickness dr at a distance r from the origin.

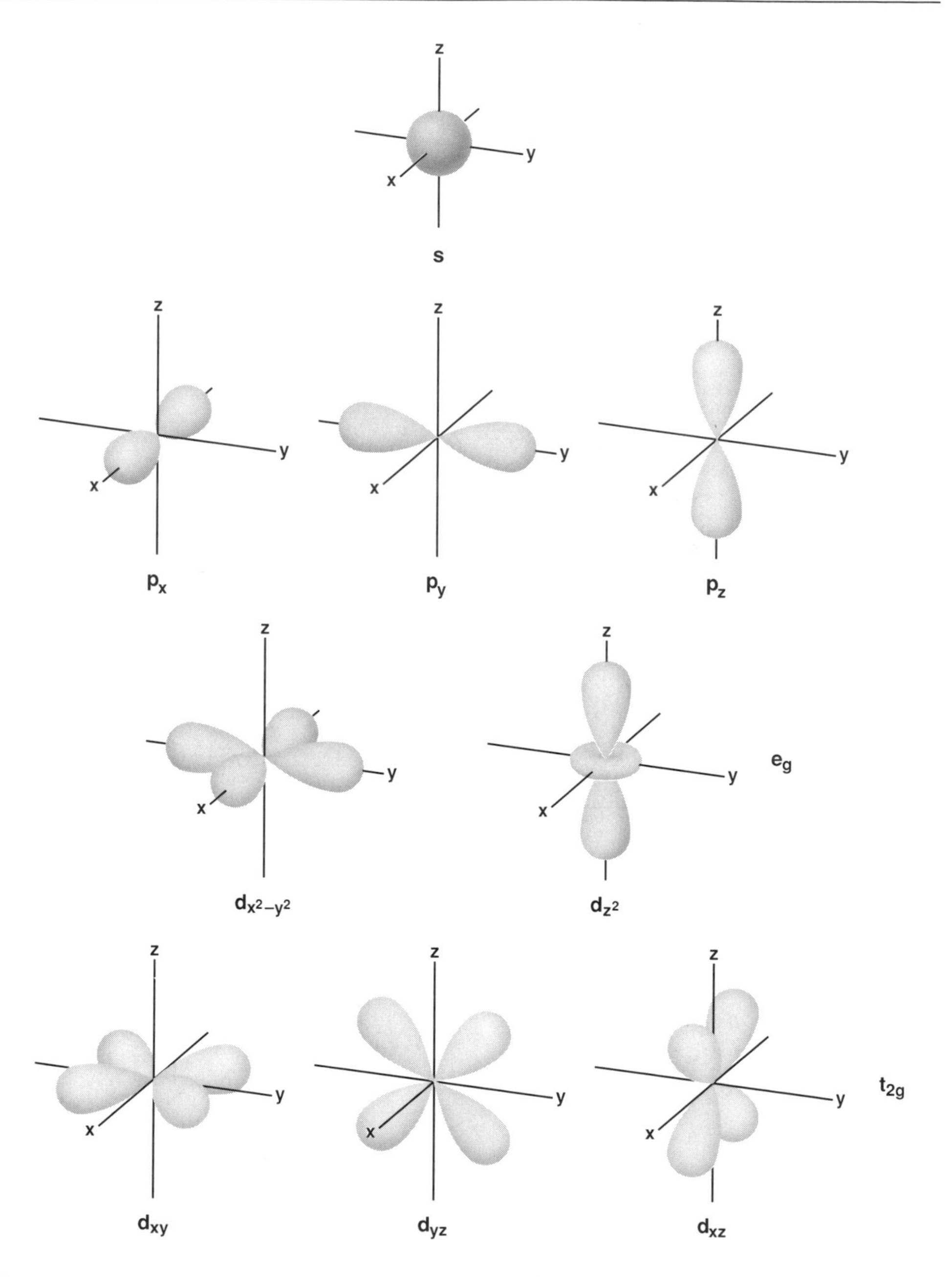

Figure 6.5.

A pictorial representation of the angular charge distribution of hydrogenlike *s*, *p*, *d* wave functions.

B. The Spin Quantum Number

The only remaining quantum number yet to be discussed is the spin quantum number m_s, which has values of $\pm\frac{1}{2}$. The positive value is generally referred to as spin-up, and the negative as spin-down. Of course, the actual direction of the spin is meaningful only with respect to a specified system of coordinates. Most often, the frame of reference is provided by the z-direction of the nuclear magnetic moment or by another electron. The spin angular momentum is restricted to the values $\pm\hbar/2$.

The Pauli exclusion principle requires that no two electrons possess identical quantum numbers. Consequently, if for a pair of electrons the quantum numbers n and l are identical, m_s must be positive for one and negative for the other. The consequences of the spin are twofold. Because m_s has only two values, it limits to 2 the number of electrons described by the same spatial wave function. Second, because the two spins are oppositely directed, the combined wave function is always antisymmetric; that is, an exchange of spatial and spin coordinates between the two electrons requires the wave function to change sign.

3. The Covalent Bond

The basis for the covalent bond was identified in the early part of the 20th century as the sharing of electrons between bonded atoms.[2] Although much oversimplified, it is correct to say that the strength of a covalent bond increases with the amount of shared electric charge or the electron density between the bonded atoms. In the language of wave mechanics, it is said that the greater the overlap of the atomic wave function the stronger is the bond. However, it is incorrect to assume that the molecular wave function simply and accurately reflects the superposition of the unaltered component atomic wave functions. The proximity of the bonding and nonbonding electrons and of the nuclear charges causes the atomic wave functions to become distorted when forming bonds as a result of these many-body interactions. Nevertheless, the atomic wave functions provide a useful starting point for the construction of the total or molecular wave equation.

The universally adopted beginning for the development of the theory of covalent bonding is H_2^+, the hydrogen molecule ion. The potential energy terms are defined as

$u_1 = e^2/r_1$ the attraction of the electron to nucleus 1,

$u_2 = e^2/r_2$ the attraction of the electron to nucleus 2,

$u_{12} = e^2/r_{12}$ the mutual repulsion of the nuclei.

[2] G. N. Lewis, *J.A.C.S.*, **38**, 762 (1916).

The Schrödinger equation for H_2^+ now becomes

$$\nabla^2\psi + \frac{8\pi^2 m}{h^2}\left(E + \frac{e^2}{r_1} + \frac{e^2}{r_2} - \frac{e^2}{r_{12}}\right)\psi = 0.$$

This expression treats the hydrogen nuclei as fixed with respect to the much more rapid electronic motion. This assumption is known as the Born–Oppenheimer approximation, and it allows the kinetic energy of the vibrating nuclei to be disregarded. Since there is no analytical solution for this three-body problem, we must resort to an approximation technique such as the variation method.

A. The Variation Method

A physical interpretation of the variational principle might paraphrase the old cliche that water (in a gravitational potential) seeks its own level. Analogously, electrons confined within an electrostatic potential, if unhindered, will distribute their charge in a pattern that places the system in its lowest possible energy state. Consequently, the calculated wave function that distributes the electronic charge in the way that achieves the lowest energy is regarded as the "true" wave function. As we have no way of knowing, a priori, the true wave function, the best we can do is to apply a trial-and-error approach using our physical intuition as much as possible in the selection of trial wave functions. Such trial functions are plugged into the energy equation with the aim of finding a minimum value for E. The appropriate equation that defines the energy is obtained from (6.16) by first multiplying both sides by ψ, extracting E (which is a constant), and then dividing both sides by ψ^2 and integrating over the entire relevant volume. This leads to

$$E = \int \psi \hat{H} \psi \, d\tau \bigg/ \int \psi^2 \, d\tau, \tag{6.45}$$

which is an expression for the expectation value of E for the chosen wave function. This equation is used rather than (6.16) because the latter will generally be a function of the coordinates varying from point to point without establishing an overall minimum value for E.

Returning now to the H_2^+ molecule, we choose for our first trial wave function a *linear combination* of *atomic orbitals* (LCAO). The simplest trial function is the single-electron molecular wave function given by

$$\psi_{MO} = C_1\psi_1 + C_2\psi_2, \tag{6.46}$$

which represents the electron distributed between the two 1s orbitals of the two protons, as illustrated in Fig. 6.6. C_1 and C_2 are constants to be determined by the

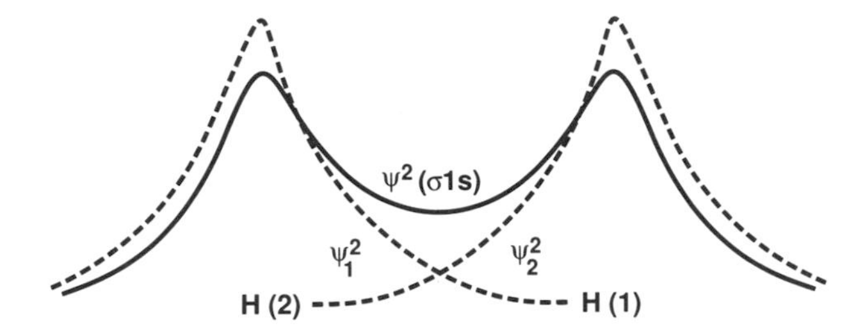

Figure 6.6.

A schematic portrayal of the coalescence of the atomic wave functions ψ_1 and ψ_2 to form a diatomic molecule.

minimization of the energy that maximizes the bond strength. Substituting ψ_{MO}, the molecular wave function from (6.46), into (6.45) gives us

$$E = \frac{C_1^2 \int \psi_1 \hat{H} \psi_1 \, d\tau + 2C_1C_2 \int \psi_1 \hat{H} \psi_2 \, d\tau + C_2^2 \int \psi_2 \hat{H} \psi_2 \, d\tau}{C_1^2 \int \psi_1^2 \, d\tau + 2C_1C_2 \int \psi_1 \psi_2 \, d\tau + C_2 \int \psi_2^2 \, d\tau}. \tag{6.47}$$

The equation becomes less cumbersome if we make the following definitions:

$$\int \psi_i \hat{H} \psi_j \, d\tau \equiv H_{ij} \qquad \text{and} \qquad \int \psi_i \psi_j \, d\tau \equiv S_{ij},$$

where i may or may not equal j. Then (6.47) appears as

$$E = \frac{C_1^2 H_{11} + 2C_1C_2H_{12} + C_2^2 H_{22}}{C_1^2 S_{11} + 2C_1C_2S_{12} + C_2^2 S_{22}}. \tag{6.48}$$

Now, minimizing the energy as given by (6.48) with respect to C_1 and C_2 gives us

$$\frac{\partial E}{\partial C_1} = 0 = C_1(H_{11} - ES_{11}) + C_2(H_{12} - ES_{12}),$$

$$\frac{\partial E}{\partial C_2} = 0 = C_1(H_{12} - ES_{12}) + C_2(H_{22} - ES_{22}). \tag{6.49}$$

The coefficients C_1 and C_2 are without physical significance because the selection of $C_1\psi_1 + C_2\psi_2$ as the molecular wave function is entirely arbitrary and ψ_{MO} could have any number of terms. Solving directly for C_1 and C_2 leads only to the trivial solution $C_1 = C_2 = 0$. Nontrivial solutions to (6.49) require the determinant of (6.50) to vanish, so that

$$\begin{vmatrix} H_{11} - ES_{11} & H_{12} - ES_{12} \\ H_{12} - ES_{12} & H_{22} - ES_{22} \end{vmatrix} = 0. \tag{6.50}$$

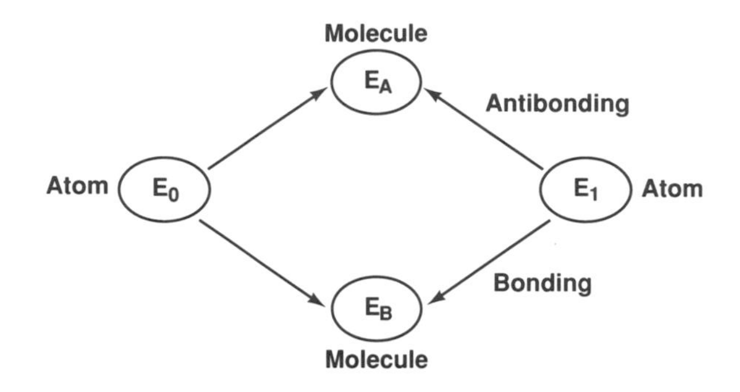

Figure 6.7.

Schematic of the formation of a diatomic molecule with bonding and antibonding molecular orbitals.

Equations of the form of (6.50) are called *secular* equations.[3] Setting $S_{11} = S_{22} = 1$, because of normalization of the wave functions, and $H_{11} = H_{22}$, because of symmetry, and relabeling S_{12} simply as S, we obtain the two solutions to the quadratic equation (6.50) as

$$E_g = \frac{H_{11} + H_{12}}{1 + S}, \qquad E_u = \frac{H_{11} - H_{12}}{1 - S}. \tag{6.51}$$

The subscripts g and u designate the symmetric (German *gerade*) and antisymmetric (*ungerade*) wave functions.

H_{12} stands for the binding energy between the hydrogen atoms that combine to form H_2^+; hence, it is negative. S_{12} is positive but less than 1. Consequently, $E_g < E_u$. Furthermore, since H_{11} is the energy of the neutral hydrogen atom, $E_u > H_{11}$. This is to say that energy would be required in order to promote an electron from its free-atom atomic orbital into the molecular orbital corresponding to E_u. This orbital, therefore, does not lead to the formation of a chemical bond and is appropriately called *antibonding* in contrast to E_g, which is a *bonding* orbital. This result is shown diagrammatically in Fig. 6.7.

Returning now to the molecular wave functions themselves, we substitute E_g and E_u back into (6.49) and hence determine that $C_1 = C_2$ for the former and $C_1 = -C_2$

[3] "From Latin *saeculum*, generation, age. The term secular perturbation was first introduced into celestial mechanics to describe gravitational perturbations with cumulative effects." (According to W. J. Moore, *Physical Chemistry*, Prentice–Hall, 1972.)

for the latter. Thus,

$$\psi_g = C_1(\psi_1 + \psi_2) \qquad \text{(symmetric)}, \tag{6.52}$$

$$\psi_u = C_1(\psi_1 - \psi_2) \qquad \text{(antisymmetric)}. \tag{6.53}$$

Normalizing to unity the individual wave functions, viz.

$$\int \psi_g^2 \, d\tau = 1, \qquad \int \psi_u^2 \, d\tau = 1, \tag{6.54}$$

gives

$$C_1^2 \left[\int \psi_1^2 \, d\tau \pm 2 \int \psi_1 \psi_2 \, d\tau + \int \psi_2^2 \, d\tau \right] = C_1^2(1 \pm 2S + 1) = 1. \tag{6.55}$$

Therefore, $C_1 = (2 \pm S)^{-1/2}$, and the two molecular wave functions become

$$\psi_g = \frac{1}{\sqrt{2 + 2S}} (\psi_1 + \psi_2),$$
$$\psi_u = \frac{1}{\sqrt{2 - 2S}} (\psi_1 - \psi_2). \tag{6.56}$$

We will not perform the intervening algebra,[4] but if we take the Hamiltonian for H_2^+ as

$$\nabla^2 + \frac{8\pi^2 m}{h^2} \left[E + \frac{e^2}{r_1} + \frac{e^2}{r_2} - \frac{e^2}{r_{12}} \right] = 0$$

and solve Eqs. (6.51), we obtain a dissociation energy of 1.77 eV for H_2^+, whereas the experimental value is 2.78 eV. This is not bad agreement, considering the very approximate wave functions that were assumed.

These are some of the general consequences that follow from the variation method:

1. The calculated value of E will always be somewhat greater (less negative) than the true or measured value.
2. The algebraically lowest calculated value of E is closest to the true value.
3. The wave function that yields the lowest calculated value of E most closely approximates the true wave function.

---

[4] See, for example, Moore, pp. 678–679, loc. cit.

B. Molecular Orbitals

The formation of bonding and antibonding orbitals is an inevitable result when atoms combine to construct molecules, whether they be hetero- or homonuclear. We have shown this for H_2^+, but this result is easily cast into a more general form. Let us write the molecular orbital wave function as

$$\psi_{MO} = C_1\psi_1 + C_2\psi_2 = N(\psi_1 + \lambda\psi_2), \tag{6.57}$$

where N is the normalizing factor given by

$$N^{-2} = S_{11} + 2\lambda S_{12} + \lambda^2 S_{22}, \tag{6.58}$$

and if ψ_1 and ψ_2 are individually normalized,

$$N^{-2} = 1 + 2\lambda S_{12} + \lambda^2. \tag{6.59}$$

Here λ is a variable reflecting the polarity of the bond in a heteronuclear molecule, i.e., the asymmetric charge distribution in a two-center bond.

Let $H_{11} \equiv E_1$, which is associated with ψ_1, and, in like fashion, $H_{22} \equiv E_2$ is associated with ψ_2. Thus, E_1 and E_2 are the energies associated with the atomic state. In a like manner, $H_{12} \equiv E_{12}$ and is the bond energy derived from the overlapping atomic wave functions and, in general, is a measure of the bond strength. The secular equations now become

$$\begin{aligned} (E_1 - E) + \lambda(E_{12} - ES_{12}) &= 0, \\ (E_{12} - ES_{12}) + \lambda(E_2 - E) &= 0. \end{aligned} \tag{6.60}$$

Eliminating λ results in

$$(E - E_1)(E - E_2) - (E_{12} - ES_{12})^2 = 0. \tag{6.61}$$

This is quadratic with respect to E, the binding energy, and hence there are two unequal solutions to (6.61) corresponding to

$$E = (E_2 + E_1)/2 - E_{12}S_{12} \pm [(E_2 + E_1 - 2E_{12}S_{12})^2 - 4(1 - S_{12}^2)(E_1E_2 - E_{12}^2)]^{1/2}. \tag{6.62}$$

A stable molecule is formed when E is negative, which occurs when the absolute value of $E_{12}S_{12}$ and the negative of the last term on the right-hand side will exceed the average value of the atomic energy levels. The opposite case then gives rise to the antibonding orbital.

A schematic representation of the formation of molecular orbitals in two dimensions is presented in Fig. 6.8. The symbols for the MO (viz., σ1s, π2p, etc.) list first the molecular orbital and then the atomic ones that contributed to its formation.

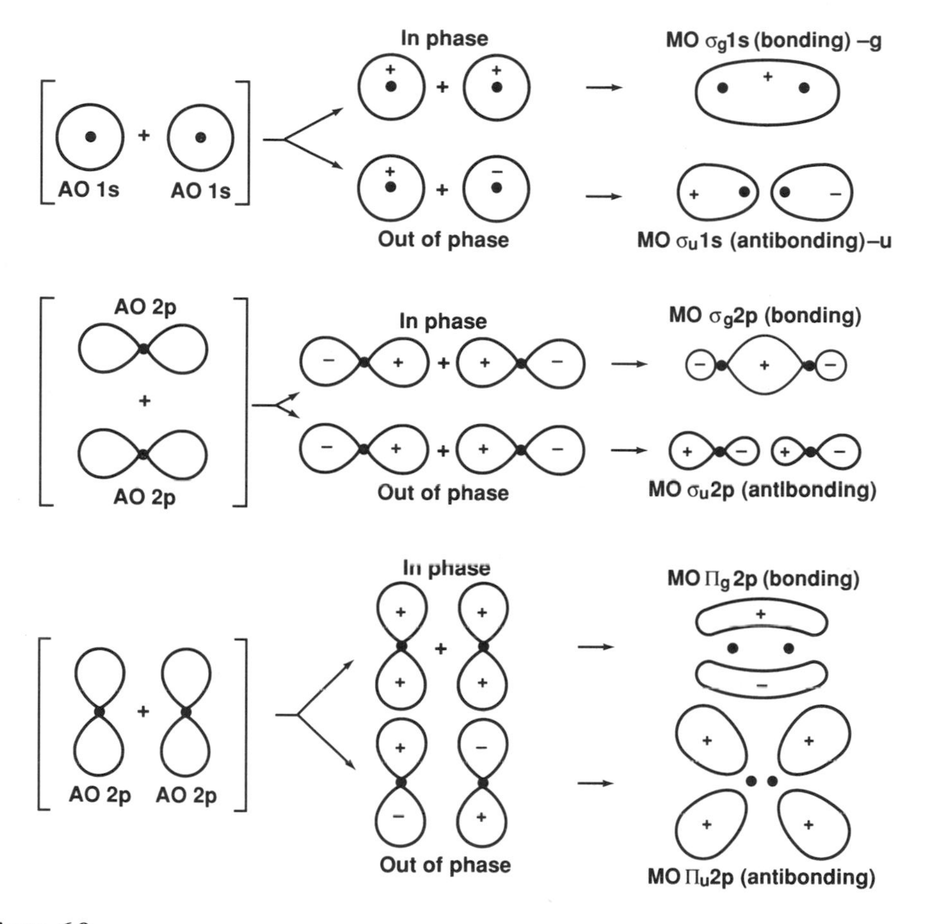

Figure 6.8.

A diagrammatic depiction of the creation of molecular orbitals from atomic ones. Note that the antibonding orbitals minimize the charge between atoms in contrast to the bonding orbitals. Also note that the charge density, ψ^2, is symmetrical about the bond axis for both bonding and antibonding configurations. The wave functions of the bonding orbitals are symmetric with respect to inversion (hence g for *gerade*) and thus reinforce each other. The converse is true for the antisymmetric (*ungerade*) antibonding cases.

The asterisk denotes an antibonding level of higher energy than the atomic contributor.

An important difference between σ and π bonding is illustrated in this figure. The angular momentum about the line of centers for atoms joined by a σ bond is zero, whereas a value of unity is appropriate to the π-bonded molecule. This is entirely analogous to the designations for the s and p atomic orbitals. In σ bonding the electron distribution is symmetrical with respect to the line joining the two nuclei. Thus, there is essentially free rotation around such a bond, since rotating the atoms with respect to each other results in no appreciable change in energy. In π bonding the electron distribution is an asymmetrical rotation of the atoms around the connecting bond and is restricted, since it could not be performed without distortion of the orbitals.

Let us examine ethylene, C_2H_4, as a simple example to illustrate the importance and effects of these considerations. Each carbon atom in this molecule forms three planar trigonal σsp^2 bonds with its three neighbors. Each of these orbitals is filled with two electrons. However, this bonding scheme leaves an unpaired p_z electron on each carbon; these overlap and form a π bond. This leads to a rigidity of the molecule, since rotation around the carbon atoms relative to each other is now constrained, whereas, because of pure σ bonding, hydrogen rotation around the carbon–hydrogen bond is essentially free. The total carbon–carbon bond, consisting of σ and π bonding, is referred to as a double bond.

The sequence of energies for the MO derived from the first row of the periodic

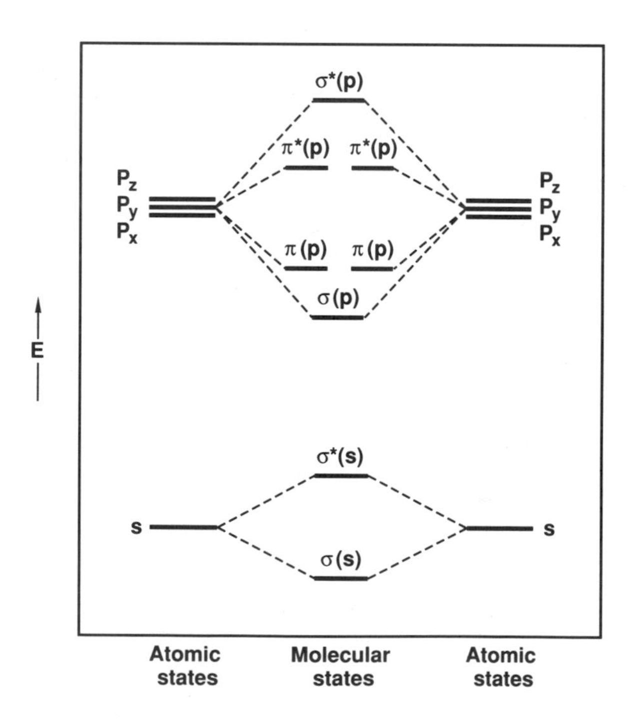

Figure 6.9.

The relative energy levels of the σ and π molecular orbitals.

table is found to be

$$\sigma 1s < \sigma^* 1s < \sigma 2s < \sigma^* 2s < \sigma^2 p < \pi_y 2p$$
$$= \pi_z 2p < \pi_y^* 2p = \pi_z^* 2p < \sigma^* 2p.$$

Reflecting the nomenclature for atomic orbitals, the σ and π MO have total angular momenta of 0 and 1, respectively. This energy level scheme is shown graphically in Fig. 6.9 without regard for the principal atomic quantum number.

The total bond strength of a specific molecule is proportional to the net number of filled bonding orbitals—that is to say, the total number of bonding minus the number of antibonding orbitals. An example is the triply bonded nitrogen molecule shown in Fig. 6.10. The $\sigma 2s$ and $\sigma^* 2s$ bonds cancel one another energetically, leaving one $\sigma 2p_x$ and two $\pi 2p$ bonds. The unfilled excited states, π^* and σ^*, are also shown.

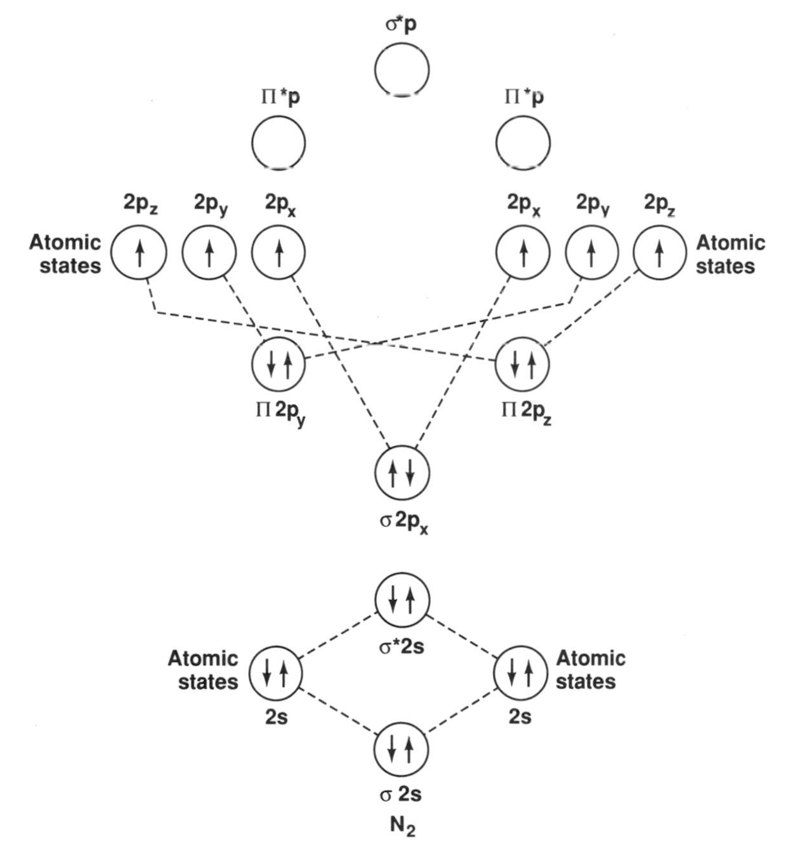

Figure 6.10.

Showing the origin of the triple bond in N_2 in terms of MO formation.

Oxygen also illustrates the interpretive power of molecular orbital theory, as shown in Fig. 6.11. The significance of this example lies in the simplicity with which it reveals the origin of the paramagnetic triplet ground state of doubly bonded O_2. In this example the $\sigma 2s$ and σ^*2s bonds essentially cancel one another, as do the $\sigma 2p$ and the two $2\pi^*2p$. Thus, the stable bond results only from the two $\pi 2p$ molecular orbitals.

The value of $m_s = 1$, which results from the spin alignment of the two unpaired π^*2p electrons, allows m_s to assume values of -1, 0, 1 in the presence of an external magnetic field. Hence, O_2 is in the triplet state. The increase in electron energy derived from promoting the $2p$ atomic electrons to the antibonding π^* orbitals essentially cancels the corresponding decrease obtained from the formation of a single bonding π orbital. This cancellation leaves O_2 a doubly bonded molecule as originally proposed by classical valence theory.

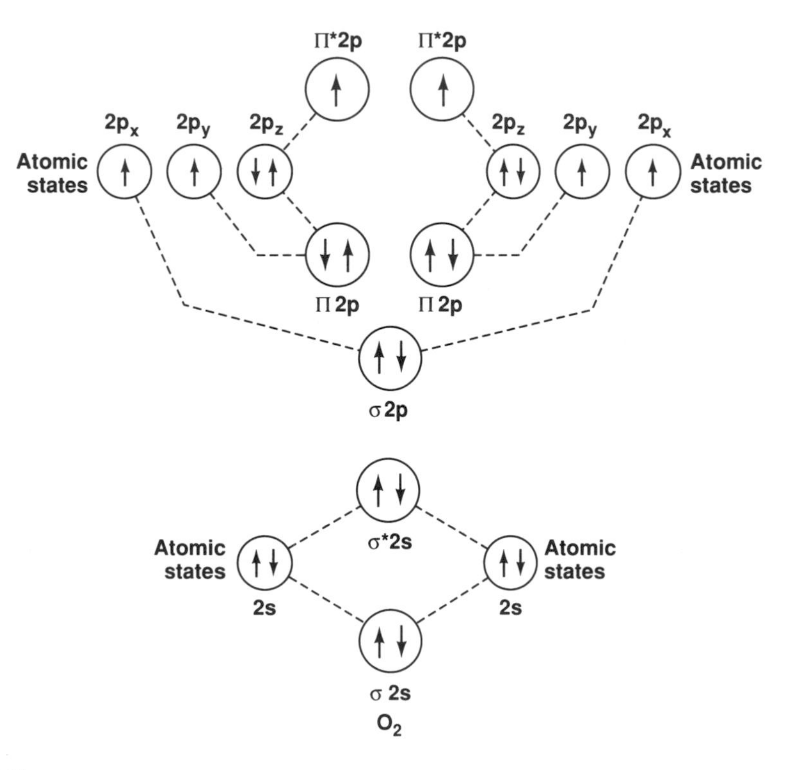

Figure 6.11.

The molecular orbitals of the ground state of O_2. Note the unpaired π^*2p electrons.

C. Mutual Electron Repulsion

As atoms coalesce to form molecules, there is not only a force exerted between the bonding electrons and the screened nuclear charges of their ligands, but also there will be a force of opposite sign originating from the electron–electron interactions. Such interactions are strong enough to completely distort the atomic wave functions from their unperturbed forms that we have thus far assumed. The strengths of the electrostatic repulsive interactions, symbolized by ⇔, decrease in the following order:

[(unbonded electron pair) ⇔ (unbonded electron pair)]
> [(unbonded electron pair ⇔ (bonded electron pair)]
> [(bonded electron pair) ⇔ (bonded electron pair)].

The enhanced diffuseness of the bonding electron cloud, together with the mitigating effect of the positive charges of two rather than one atomic nuclei, furnishes a qualitative explanation of this sequence. Experimental support for this explanation is provided by the change in bond angles observed in the isoelectronic molecules CH_4, NH_3, and H_2O as shown in Fig. 6.12. The methane bonds are all identical. The outer electrons all participate in bonding, and the 109.5° bond angle maximizes the distance between the bonding electrons. The ammonia molecule has one pair of unbonded electrons that repels the bonding ones more strongly than does their mutual interaction, thus shrinking the bond angle from its maximum value. The water molecule with two unbonded pairs of electrons continues this trend by repelling even more strongly the electron clouds of the bonded pairs.

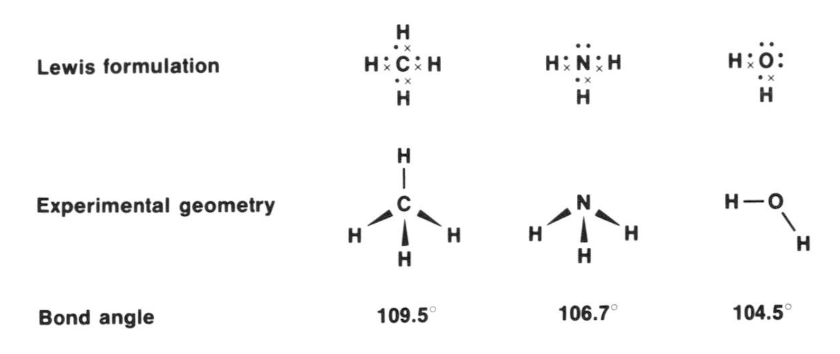

Figure 6.12.

The varying bond angles reflect the varying repulsive potential of the bonded and unbonded electrons. (After J. J. Lagowski, *Modern Inorganic Chemistry*, Marcel Dekker, New York, p. 96, (1973).

D. Hybrid Wave Functions

It has been repeatedly stated that all else being equal the strength of a covalent bond increases with increasing overlap of the bonding wave functions. Stated in more physical terms, the greater the electron density between bonded atoms, the stronger is the bond. The combination of different types of orbitals in order to optimize the bond strength is called hybridization. Taking for our example the carbon atom, the energetics of this process are outlined in Fig. 6.13. The energies required for each of the steps shown in Fig. 6.11 are explained as follows:

E_1 Energy to remove a carbon from graphite, leaving it in its normal $2s^2 2p^2$ state

E_2 Energy to promote a single $2s$ electron in the isolated atom into an empty $2p$ orbital

E_3 Energy to form hybrid orbitals that are a combination of $2s$ and $2p$ wave functions

E_4 Energy released when the carbon atom bonds with other atoms to form a molecule while retaining the hybrid wave functions

The sp^3 atomic state and the atomic sp^3 hybridized state are fictitious, the latter existing only as molecular orbitals. It is the energy liberated by bonding that gives rise to the hybridization.

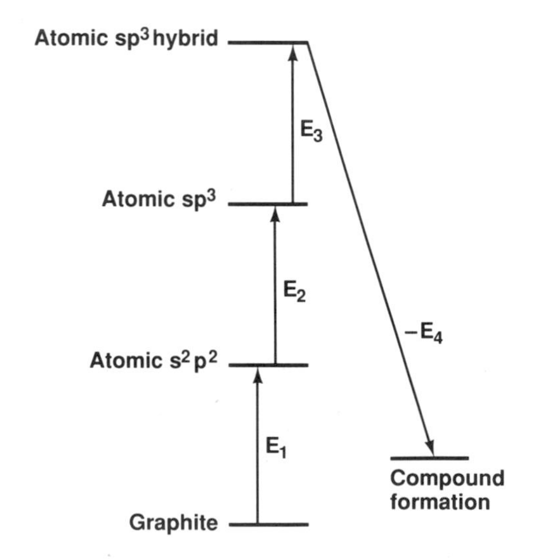

Figure 6.13.

The steps, hypothetical as well as real, giving the energy balance for taking a carbon atom from its standard state, graphite, and forming another carbonaceous bond.

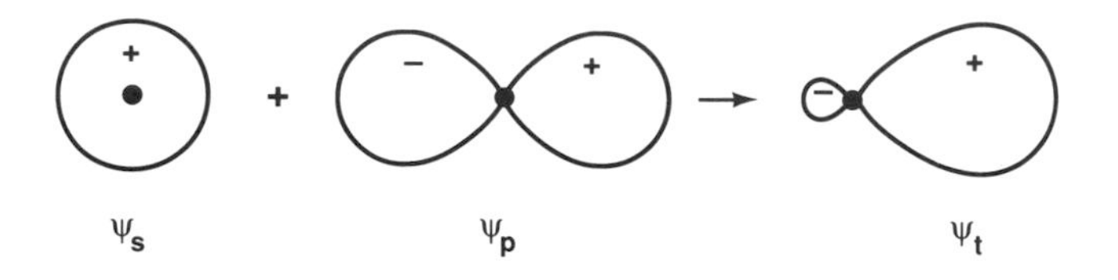

Figure 6.14.

Boundary surfaces in the formation of an sp^3 hybrid atomic orbital ψ_t.

The p wave functions, as previously described, have their maximum charge densities along the x-, y-, and z-axes. The combination of s and p wave functions to make an sp^3 hybrid wave function is depicted in Fig. 6.14. The tetrahedral configuration maximizes the separations of the four hybrid orbitals and enhances the degree of overlap with bonding atoms such as hydrogen in the creation of methane. The geometry of this molecule is conveniently represented by Fig. 6.15.

Following the arguments developed by Pauling[5] and Slater,[6] the relative strength of the sp^3 hybrid bond can be estimated in the following way. It is first assumed that the radial charge distributions of the $2s$ and $2p$ electrons are sufficiently alike so that in this first approximation they are taken to be equal. Because of its spherical charge distribution, the s orbitals can form bonds of equal strengths in any direction. In order to develop a relative scale for the bonding potentials of the various orbitals, an arbitrary value of unity is assigned to the s wave function. Now the angular distribution functions (which we have not derived) for the s and p wave functions are

$$\begin{aligned} \psi_{2s} &= 1, & \psi_{2p_y} &= \sqrt{3}\sin\theta\sin\phi, \\ \psi_{2p_x} &= \sqrt{3}\sin\theta\cos\phi, & \psi_{2p_z} &= \sqrt{3}\cos\theta. \end{aligned} \tag{6.63}$$

The maximum strength of the purely p orbital bonds will lie along the x-, y-, z-axes for ψ_x, ψ_y, and ψ_z, respectively; hence, p bonds are then $\sqrt{3}$ times stronger than s bonds, this being the maximum value of the p wave function. We now ask if there is a combination of s and p wave functions that form a bond stronger than 1.732 (i.e., $\sqrt{3}$), which we are assured exist from chemical evidence. It is simplest to calculate the maximum bond strength in the z-direction and later show how tetrahedral bonds of the same strength are formed at 109°28′ to one another.

---

[5] L. Pauling and J. Sherman, *J.A.C.S.*, **59**, 1450 (1937); also L. Pauling, *Nature of the Chemical Bond*, p. 116, Cornell University Press, Ithaca, NY, 1960.

[6] J. C. Slater, *Phys. Rev.*, **98**, 1093 (1955). Also for another concise, yet clear, explanation of this concept, see H. Eyring, Walter, and Kimball, *Quantum Chemistry*, pp. 221–223, Wiley, New York, 1944.

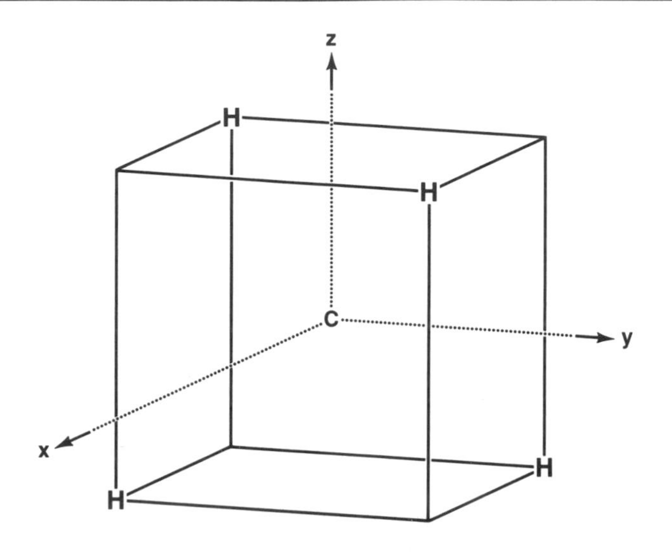

Figure 6.15.

Showing how coordinates are constructed for the methane molecule. The carbon atom is located at the center of a hypothetical cube with hydrogen atoms at opposing corners and rotated 90° about the z-axis with respect to each other.

The tetrahedral bond is first approximated by a linear sum of s and p electrons, to wit,

$$\psi_t = a\psi_s + b\psi_{p_x} + c\psi_{p_y} + d\psi_{p_z}. \tag{6.64}$$

Combining the s wave function with p_z along the z-axis, while neglecting p_x and p_y because of their orthogonality to p_z, leaves us with

$$\psi_{tz} = a\psi_s + d\psi_{p_z}. \tag{6.65}$$

Normalization requires

$$a^2 + d^2 = 1, \tag{6.66}$$

so that (6.65) becomes

$$\psi_{tz} = a + \sqrt{1 - a^2}\sqrt{3}\cos\theta. \tag{6.67}$$

The maximum value in the direction $\theta = 0$ (see Fig. 6.3) is obtained by differentiating ψ_{tz} with respect to a, equating the result to zero, and solving. Thus,

$$\frac{\partial\psi_{tz}}{\partial a} = 0 = \frac{\partial}{\partial a}(a + \sqrt{3 - 3a^2}),$$

which gives $a = 1/2$. Substituting for a in (6.67) leads to

$$\psi_{tz} = \frac{1}{2} + \frac{3}{2}\cos\theta = \frac{1}{2} + \frac{\sqrt{3}}{2}p_z. \tag{6.68}$$

This describes the wave function for the strongest bond in the z-direction. The shape of this orbital has been shown as ψ_t in Fig. 6.14. Since the maximum value of ψ_{tz} occurs at $\theta = 0$, (6.68) gives ψ_{tz} (max) $= 2.0$. This is the figure of merit for the maximum bond strength resulting from the maximum overlap of ψ_s and ψ_p in the tetragonal configuration as well. Because all four tetrahedral bonds are of equivalent strength, they all have strength of 2.0 on an arbitrary scale.

Returning once again to (6.64), we are permitted to write

$$\psi_t = a\psi_s + b(\psi_{p_x} + \psi_{p_y} + \psi_{p_z}) \tag{6.69}$$

with all p orbitals equally weighted. Again normalization requires that

$$a^2 + 3b^2 = 1, \tag{6.70}$$

so that $b = 1/2$ since $a = 1/2$. Along the (1, 1, 1) diagonal, $1/\sqrt{2} = \sin\phi = \cos\phi = (1/\sqrt{2})\cos\theta = 1/\sqrt{3}$ and $\sin\theta = \sqrt{2}/\sqrt{3}$, so $\psi_{p_x} = \psi_{p_y} = \psi_{p_z} = 1$. Consequently, substituting the coefficients for a, b and all the wave functions in (6.64) leads to $\psi_t = 2.0$, as expected. Because this value clearly exceeds the strength of 1.732 for p electron bonding, the tetrahedral hybrid bond is preferred. From considerations of symmetry, we can form the complete set of orthogonal bonding orbitals:

$$\begin{aligned}
\psi_t(1, 1, 1) &= (1/2)(\psi_s + \psi_{p_x} + \psi_{p_y} + \psi_{p_z}),\\
\psi_t(1, -1, -1) &= (1/2)(\psi_s + \psi_{p_x} - \psi_{p_y} - \psi_{p_z}),\\
\psi_t(-1, 1, -1) &= (1/2)(\psi_s - \psi_{p_x} + \psi_{p_y} - \psi_{p_z}),\\
\psi_t(-1, -1, 1) &= (1/2)(\psi_s - \psi_{p_x} - \psi_{p_y} + \psi_{p_z}).
\end{aligned} \tag{6.71}$$

The sp^3 tetrahedral hybridization has been emphasized because, in addition to simple molecules such as CH_4, it is characteristic of the diamond cubic structure. Important elements such as Si and Ge, which are essential to the semiconductor industry, possess this crystal structure. So does diamond, along with elements and compounds of lesser commercial importance.

Other types of hybrid wave functions will be discussed as the subject relates to the specific examples under consideration. However, we will complete this discussion of the hybridization of $2s$ and $2p$ orbitals of carbon by at least mentioning the trigonal and linear molecular orbitals.

The three trigonal orbitals are made up as combinations of ψ_s, ψ_{p_x}, and ψ_{p_y}, and they lie in the xy plane with included angles of 120°. Using the same procedure as previously outlined for the tetrahedral case, we find

$$\begin{aligned}\psi_0 &= (1/\sqrt{3})(\psi_s + \sqrt{2}\psi_{p_x}),\\ \psi_{2\pi/3} &= (1/\sqrt{3})(\psi_s - (1/\sqrt{2})\psi_{p_x} + \sqrt{3/2}\psi_{p_y}),\\ \psi_{4\pi/3} &= (1/\sqrt{3})(\psi_s - (1/\sqrt{2})\psi_{p_x} - \sqrt{3/2}\psi_{p_y}).\end{aligned} \tag{6.72}$$

Here we need not refer to the x- or y-direction because all three orbitals are entirely equivalent. An example of this type of bond is ethylene, illustrated in Fig. 6.16, which shows how this electron configuration leads to the carbon–carbon double bond. One bond is the hybrid just described, and the other is a π bond formed by the remaining p_z electrons. The relative strength of the sp^2 bond is 1.991 and, hence, is almost identical to that for an sp^3 hybrid.

The diagonal or linear bond in its normalized form is given by

$$\psi_{\mathrm{d}} = (1/\sqrt{2})(s + p_x). \tag{6.73}$$

Two additional π bonds are created by the remaining p_y and p_z electrons, thus together forming a triple bond such as that in acetylene. This is shown schematically in Fig. 6.17. The relative strength of the sp bond is 1.932. Thus, a triple bond that includes the two π bonds is nearly three times stronger than a single sp^3 bond. Table 6.2 summarizes the relative strengths of the various types of covalent bonds.

In concluding this topic, we should remind ourselves that although hybrid wave functions are indeed very real, the ones we have created by the LCAO technique are very approximate. What has been demonstrated is the principle of hybridization rather than the creation of realistic wave functions—i.e., that hybrid orbitals will form when the energy necessary to promote electrons into higher energy orbitals is more than offset by the increased stability of the resulting chemical bonds.

E. Valence Bond Theory

Atomic wave functions can be combined to construct chemical bonds in yet another way that is not prescribed by molecular orbital theory. This alternative recipe is the so-called valence bond method, which historically preceded molecular orbital theory. In some respects this approach more nearly resembles the classical chemical description of creating covalent bonds from shared pairs of electrons.

Taking the H_2 molecule as the simplest example, we write the complete molecular wave function as

$$\Psi = C_1\psi_1 \pm C_2\psi_2, \tag{6.74}$$

which represents a linear combination of atomic orbitals, generally abbreviated as LCAO. The constants C_1 and C_2 are the weighting factors to be determined by the

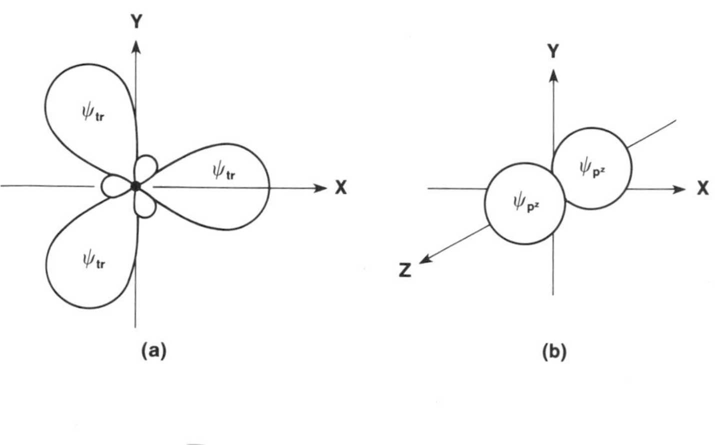

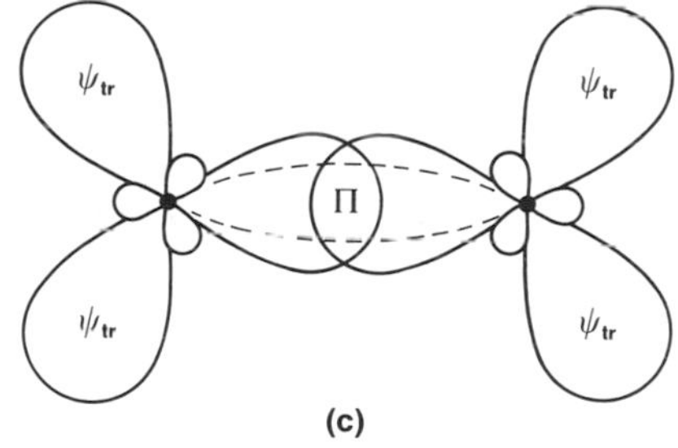

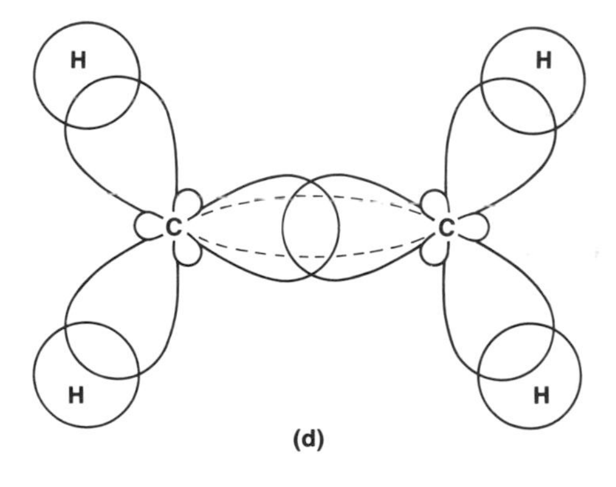

Figure 6.16.

(a) Three trigonal hybrid orbitals; (b) the atomic p_z orbitals that are superposed on (a) but shown separately here for clarity; (c) two carbon atoms doubly bonded via one trigonal and one π molecular orbital. The latter is formed from the p_z atomic orbitals and lies in front of and behind the xy plane and is shown in dashed lines; (d) $H_2C = CH_2$, the compound ethylene.

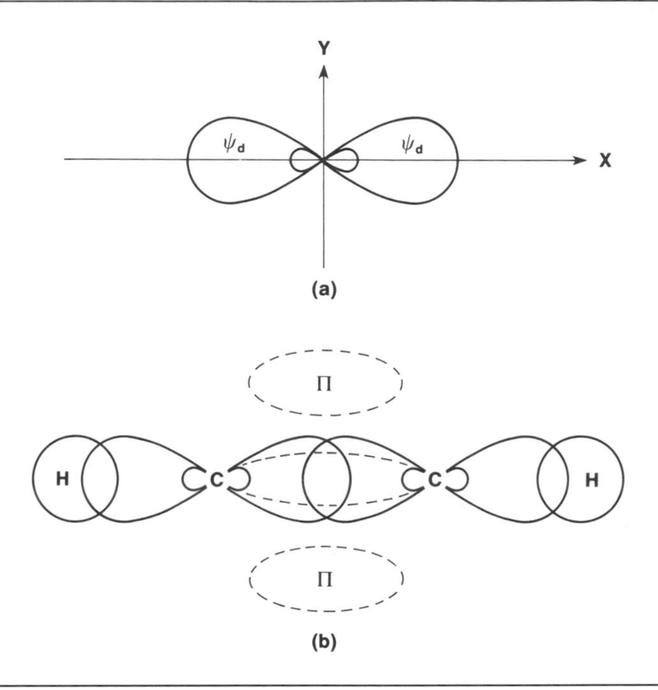

Figure 6.17.

(a) The sp_x hybrid orbital; (b) the acetylene molecule H—C≡C—H. The four π orbitals lie above and below, as well as before and behind, the digonal bond that lies in the plane of the page.

Table 6.2. Relative bond strengths of various atomic and hybrid orbitals.[a]

Wave Function	Relative Maximum Value of Angular Part
s	1
p_x, p_y, p_z	$\sqrt{3}$
d_{z^2}	$\sqrt{5}$
d_{xy}, d_{yz}, d_{xz}	$\sqrt{15/2}$
$d_{x^2-y^2}$	$\sqrt{15/2}$
sp^3	2
dsp^2	2.694
sp^2	1.991
sp	1.932
d^2sp^3	2.923

[a] From J. J. Lagowski, *Modern Inorganic Chemistry*, Marcel Dekker, New York, p. 131, 1973.

variational method. Approximate wave functions[7] that turn out to yield results in fair agreement with the measured values are

$$\begin{aligned} \psi_1 &= \psi_A(1)\psi_B(2) \equiv \phi_1, \\ \psi_2 &= \psi_A(2)\psi_B(1) \equiv \phi_2. \end{aligned} \tag{6.75}$$

The first of Eqs. (6.75) is interpreted to mean that electron 1 is associated with nucleus A and electron 2 with nucleus B. The second equation reverses this order.

From a consideration of the symmetry, it is obvious that $C_1 = C_2$ for homonuclear molecules. Preserving the more general expression, we can formulate the two equations given by (6.74) as follows:

$$\begin{aligned} \Psi_s &= C_1\psi_A(1)\psi_B(2) + C_2\psi_A(2)\psi_B(1), \\ \Psi_a &= C_1\psi_A(1)\psi_B(2) - C_2\psi_A(2)\psi_B(1), \end{aligned} \tag{6.76}$$

where Ψ_s is symmetric and Ψ_a is antisymmetric since it becomes $-\Psi_a$ if the coordinates of electrons 1 and 2 are interchanged. The energies are now given by (6.45):

$$E = \frac{\int (\phi_1 \pm \phi_2)\hat{H}(\phi_1 \pm \phi_2)\, d\tau}{\int (\phi_1 \pm \phi_2)^2\, d\tau}, \tag{6.77}$$

and substituting into the secular equations (6.49) we obtain for the symmetric and antisymmetric configurations that

$$E_s = \frac{H_{11} + H_{12}}{S_{11} + S_{12}}, \qquad E_a = \frac{H_{11} - H_{12}}{S_{11} - S_{12}}. \tag{6.78}$$

Thus far we have neglected the effects of the electron spin, which must be combined with the spatial part of the wave function. Recalling that the complete wave function must be antisymmetric, we are compelled to combine Ψ_s with the antisymmetric spin configuration, and vice versa for Ψ_a. The four possible spin arrangements are listed as follows, where α implies spin-up and β spin-down:

Spin Function	Electron 1	Electron 2
$\alpha(1)\alpha(2)$	$+1/2$	$+1/2$
$\alpha(1)\beta(2)$	$+1/2$	$-1/2$
$\beta(1)\alpha(2)$	$-1/2$	$+1/2$
$\beta(1)\beta(2)$	$-1/2$	$-1/2$

There are four ways of combining these spin states; three of them are symmetric: $\alpha(1)\alpha(2)$, $\beta(1)\beta(2)$, and $\alpha(1)\beta(2) + \alpha(2)\beta(1)$. One is antisymmetric: $\alpha(1)\beta(2) - \alpha(2)\beta(1)$.

[7] There are other simple combinations of atomic wave functions that yield poorer results than (6.74).

The symmetric spin states combine with Ψ_a to create a triplet excited state, $^3\Sigma$, whereas the antisymmetric spin state in combination with Ψ_s creates the singlet ground state, $^1\Sigma$. The relative energies are plotted versus the interatomic separation distance in Fig. 6.18. It is clear that the triplet state is unstable since there is no minimum in the energy curve. The calculated singlet state gives a binding energy of 3.14 eV, which, considering the very approximate nature of (6.76), compares favorably with the experimental value of 4.72 eV. The triplet state can be considered equivalent to the antibonding orbital of the MO treatment, and the singlet state can be considered equivalent to the bonding orbital.

Much more complicated, yet more realistic, wave functions have been employed that yield a calculated value of the energy that is in essentially complete agreement with the measured one.

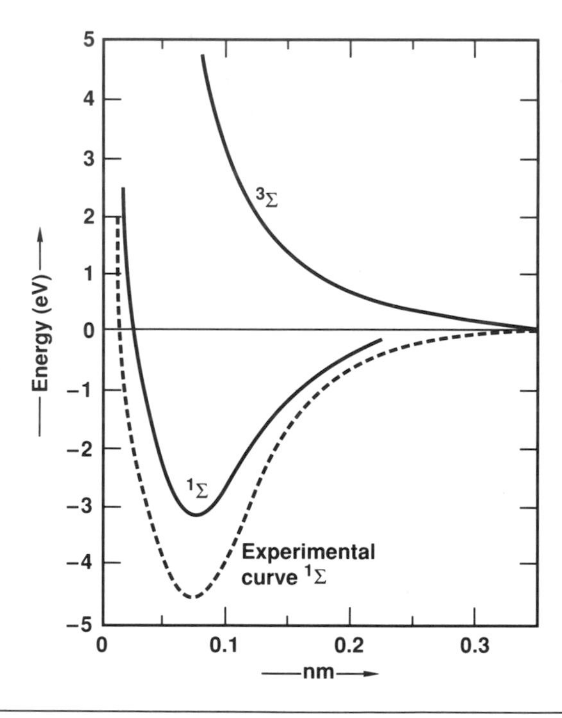

Figure 6.18.

The energies of the calculated triplet and singlet states of H_2 are compared with the experimental singlet state values.

4. Ligand Fields

Ligand field effects stem from the interactions between a central metal ion whose outer electron shell is composed of asymmetric orbitals and its nearest-neighboring negatively charged ligands. The asymmetry requirement limits such effects to ions with d or f outer electron shells. The discussion that follows is restricted to d-electron interactions, which are by far the stronger and applicable to a much greater number of elements. Originally, and still occasionally, referred to as *crystal field effects*, these interactions first were thought to be strictly coulombic and therefore explainable by a point charge model. It is now believed that the covalent character of the interaction is more influential, and, as a consequence, *ligand field* has become the preferred term. Ligand fields make a rather small (~5%) contribution to the overall bond strength and, similarly (~1–4%) to the apparent ionic size. However, they have a more significant impact upon the magnetic and spectroscopic properties and, on occasion, can be the determining factor for the stability of a specific crystal structure.

A. Ligand Field Splitting

In the absence of electrostatic or magnetic fields, the five atomic d orbitals are degenerate, all having the same energy. The charge distribution of the d_{xy}, d_{yz}, and d_{xz} orbitals all have the same symmetry, whereas $d_{x^2-y^2}$ and d_{z^2} are different, as shown in Fig. 6.19. In ligand field parlance in an octahedral field, d_{xy}, d_{yz}, and d_{xz} are

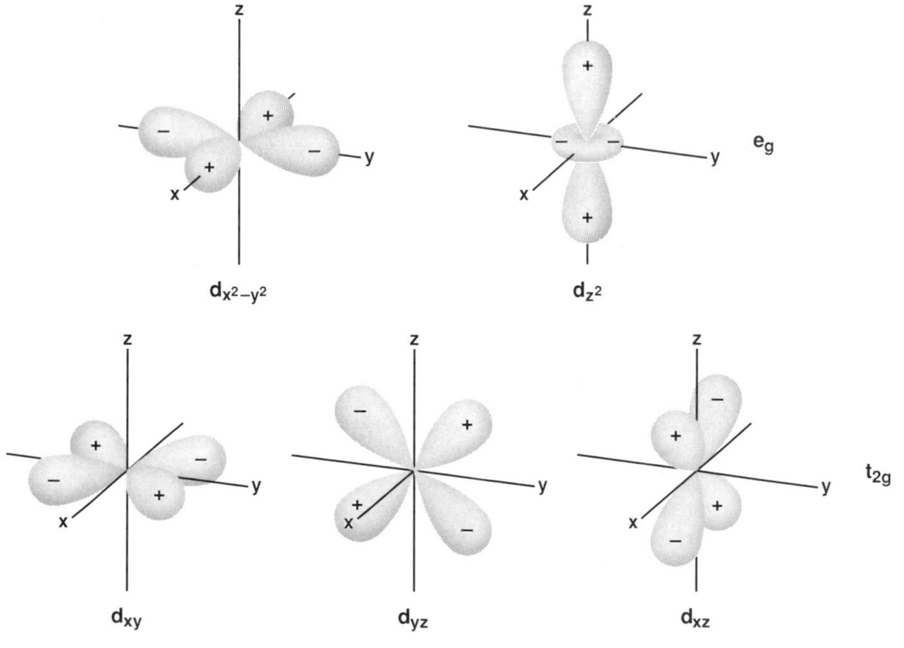

Figure 6.19.

The five d orbitals with the signs of their corresponding wave functions also shown.

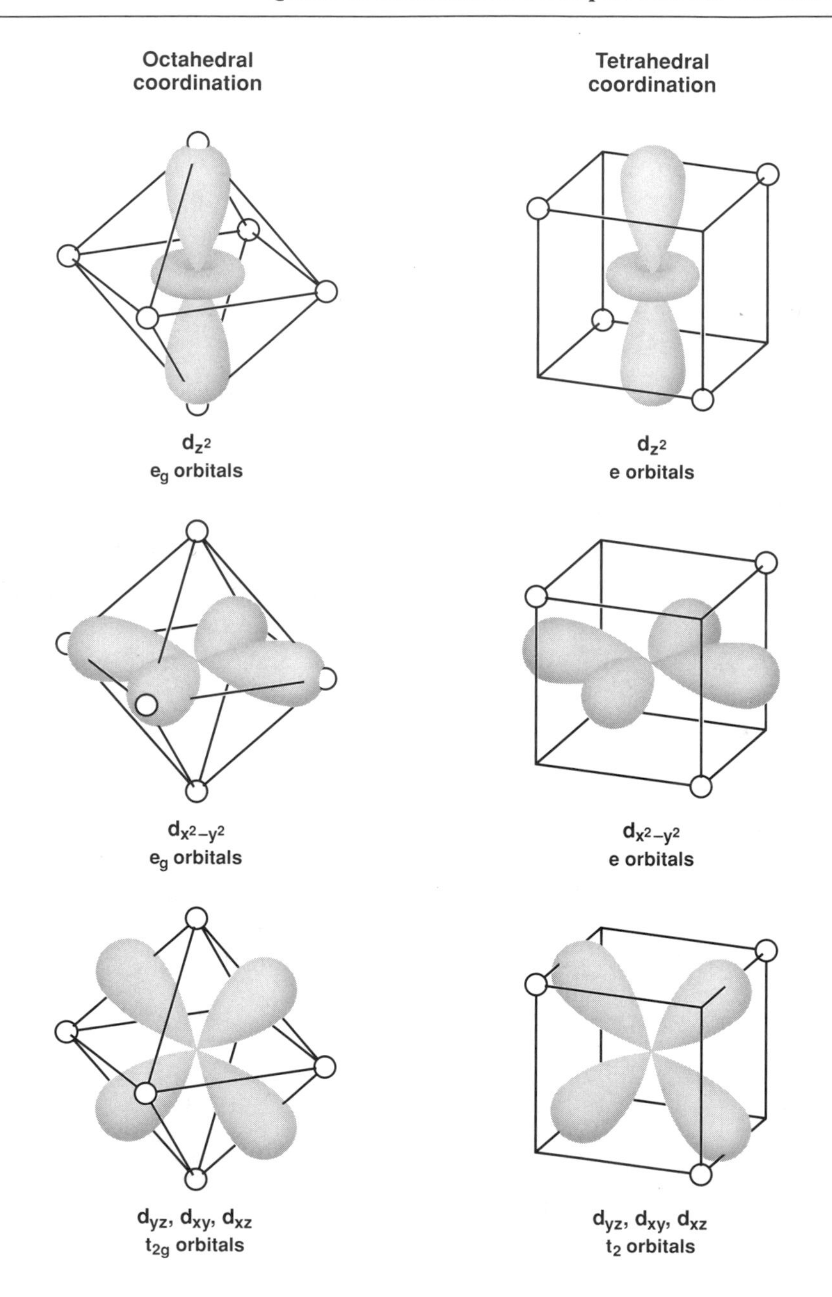

Figure 6.20.

Note that the e_g orbitals are directed at the ligands in octahedral coordination and hence have higher energies than the t_{2g}. The situation is reversed for tetrahedral coordination where the t_2 orbitals lie closer to the ligands.

dubbed t_{2g}, in which t stands for triply degenerate with a twofold axis of symmetry and no change of sign (g for *gerade*, meaning even) upon inversion through the center of the coordinate system. The twofold degenerate orbitals are designated by e_g or simply e, depending upon the coordination. In terms of the simple, though basically incorrect, point charge model, the orbitals most proximate to the ligand charge clouds have the higher energies because of coulomb repulsion. A pictorial demonstration of this for the d orbitals of ions in octahedral and tetrahedral coordination is depicted in Fig. 6.20. The relative values of the splitting energies are given in Fig. 6.21.

The total splitting energies Δ_c, Δ_t, and Δ_o are equal to zero for either an empty or completely filled d shell. This is strictly true only for the point charge model. However, the departure from the ratios of raising and lowering energies given in Fig. 6.18, introduced by covalent interactions, is small enough to be neglected in the first approximation. The actual values of these ratios (2/5:3/5, 2/5:3/5, and 3/5:2/5 for Δ_c, Δ_t, and Δ_o in this order) can be derived as follows: looking back at Fig. 6.20, it is obvious that d_{z^2} and $d_{x^2-y^2}$ are in the high energy configuration with their maximum charge concentration pointed directly at the six ligands in the octahedral configuration. The remaining three wave functions have their lobes intersecting the midpoints of four faces of the octahedron and are of lower energy. Let us now examine Fig. 6.22.

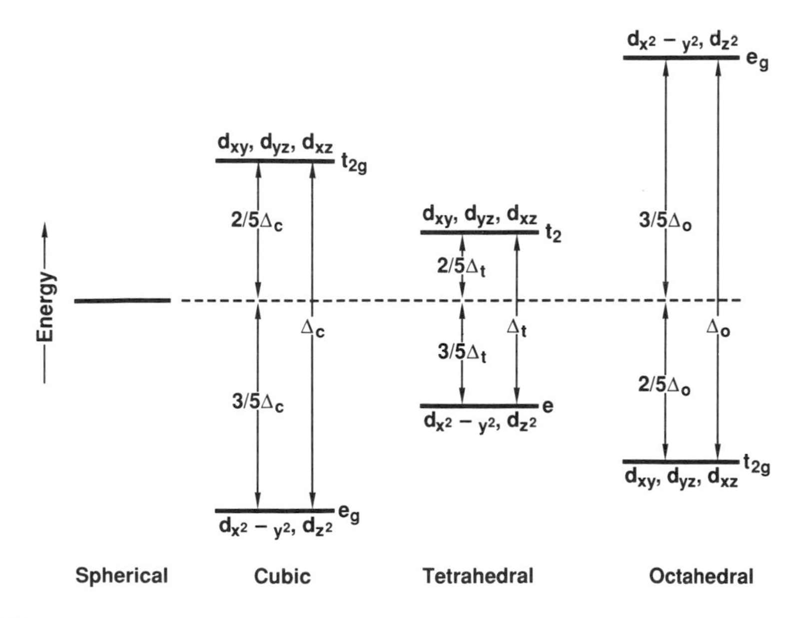

Figure 6.21.

Ligand field splitting of d orbitals in various coordinations.

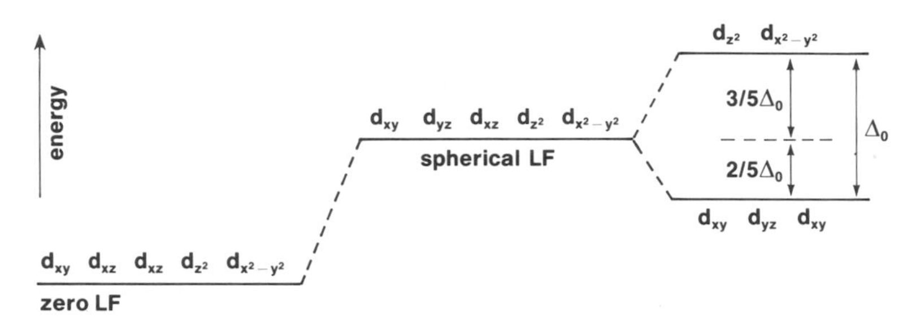

Figure 6.22.

The application of a spherical electrostatic field raises the energy of all orbitals equally. The application of an octahedral field causes splitting.

We see that the introduction of a spherical electrostatic field raises the energy of all orbitals equally. If we distribute the same amount of charge among the six points that define a regular octahedron, the total energy of the electrostatic interaction remains constant. It is thus a matter of simple algebra to show that the raising energy is $(3/5)\Delta_o$ for two orbitals, and the lowering is $(2/5)\Delta_o$ for the remaining three. Analogous arguments give the value of these coefficients for the cubic and tetrahedral geometries as shown in Fig. 6.21.

B. Ligand Field Stabilization Energy (LFSE)

All d shell ions can have their energy lowered by the LFSE, except those that are empty or completely filled. Also the effect is minimal for d^5, the half-filled shell. Experimental evidence of this stabilization energy or, if you will, the increased bond strength is displayed in Fig. 6.23. Here the metal ions are all in the $+2$ oxidation state, and the number of d electrons runs from 0 to 10, with $3d^1$ belonging to Sc^{2+} omitted for lack of data. A rather qualitative argument that is very approximate at best goes as follows: in the absence of ligand field effects lattice energies are assumed to vary in a monotonic manner with increasing atomic number. This is because the oxidation state, crystal structure, and ligand ions remain the same over the sequence of d-electron elements. Further, it is assumed that the d shell fills with constant increments of energy per electron and that the screening of the nuclear charge is a linear function of the increasing number of d electrons. Were this true, which obviously it is not, the lattice energies would be expected to more or less follow the

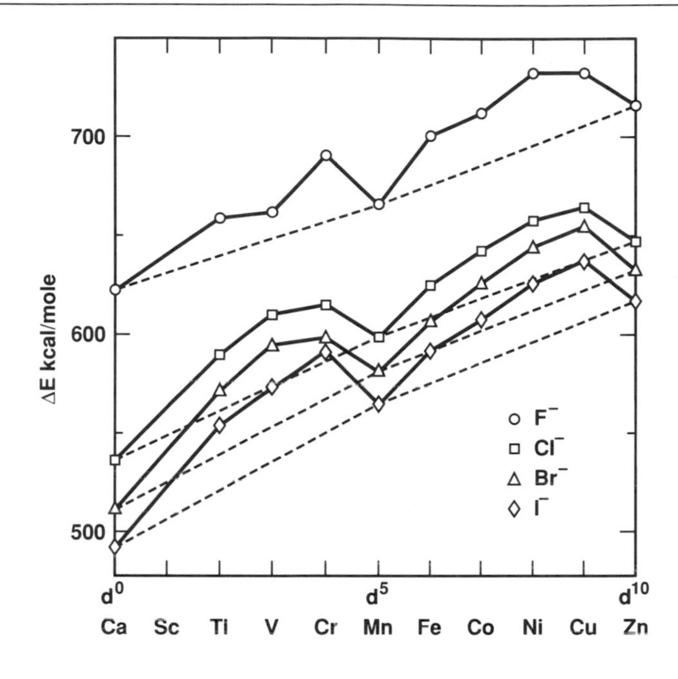

Figure 6.23.

Lattice energies of divalent first row transition metal halides, MX_2, as a function of the number of *d* electrons. (Adapted from B. N. Figgis, *Introduction to Crystal Fields*, Interscience, New York, p. 83, 1966.)

dashed lines in Fig. 6.23. However, the double bow shown by the experimental values reveals the influence of another type of interaction. Also, the energies of hydration resulting from sixfold coordinated hydrated complexes have the same double-bowed pattern with the minimum at Mn^{2+}, as do the lattice energies. This is shown in Fig. 6.24.

Since the only thing the crystalline halides and hydrated complexes have in common is their sixfold (i.e., octahedral) ligand coordination, we can assume that the unusual energy pattern is the result of ligand field effects. In fact, a semiquantitative explanation can be derived as follows: first, let us add *d* electrons uniformly, distributing them pro rata between e_g and t_{2g}. This progression maintains the overall energy constant as prescribed by the dashed line in Fig. 6.23. A more realistic way of filling is to commence with the lower t_g orbitals and then, in observance of Hund's rule, which calls for maximization of the total spin, we proceed to place two unpaired electrons in the e_g before completing the t_{2g}. This sequence we may term "actual,"

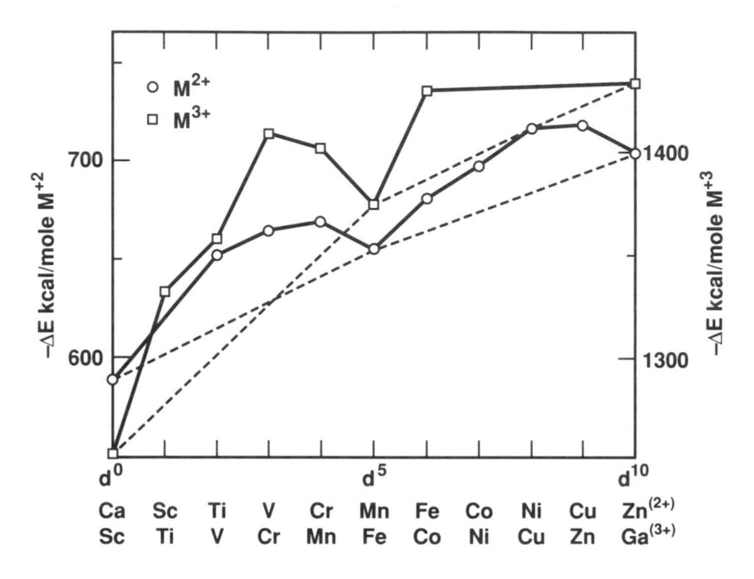

Figure 6.24.

Hydration energies of divalent and trivalent first transition metal ions versus the number of *d* electrons. (Adapted from B. N. Figgis, p. 91, loc. cit.).

in contrast to uniform. Table 6.3 lists these two arrangements and the energy differences. For example, it is seen that for d^2 transferring 4/5 of an electron from the uniform population e_g to construct the actual d^2 population results in an energy loss of $(4/5)\Delta_o$. As consequence of this process, all atomic configurations except d^0, d^5, and d^{10} can derive increased stability from the ligand field interactions. Note also that the relative differences between successive LFSE terms roughly correspond to the empirical double-bow patterns of Figs. 6.23 and 6.24.

It is interesting to note that the ionic separations also reflect the LFSE in a somewhat predictable manner as shown in Fig. 6.25. Here, the pattern substantiates the Pauling rule, viz. that enhanced stability shortens the length of the chemical bond.

Table 6.3. The source of LFSE in an octahedral field.

Total *d* electrons		0	1	2	3	4	5	6	7	8	9
"Uniform" populations	e_g	0	2/5	4/5	6/5	8/5	2	12/5	14/5	16/4	18/5
	t_{2g}	0	3/5	6/5	9/5	12/5	3	18/5	21/5	24/5	37/5
Actual populations	e_g	0	0	0	0	1	2	2	2	2	3
	t_{2g}	0	1	2	3	3	3	4	5	6	6
Stability differences		0	2/5	4/5	6/5	3/5	0	2/5	4/5	6/5	3/5

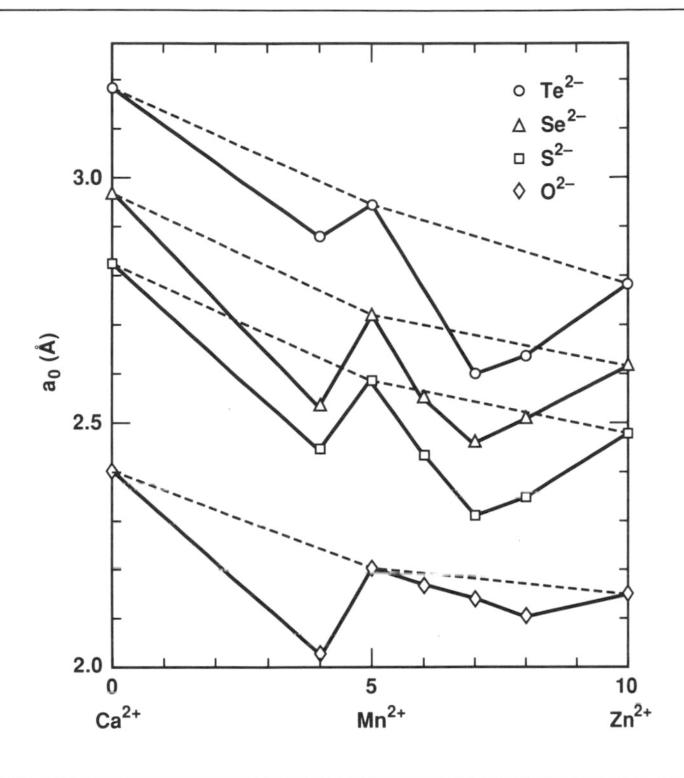

Figure 6.25.

The interionic separations decrease more or less proportional to the increase in the LFSE. (Taken from B. N. Figgis, p. 101, loc. cit.).

C. Low-Spin Versus High-Spin States

Electrons bound to the same atom will tend, whenever possible, to avoid one another in order to minimize the coulombic repulsion energy. According to Hund's empirical rule (which is substantiated by quantum mechanical calculations), the state of maximum multiplicity is generally the most stable. Consequently, if the splitting energies, where Δ_o and Δ_t are for the octahedral and tetrahedral, respectively, are small compared with the pairing energies E_p, electrons will enter singly into the d orbitals sequentially from the lowest to the highest energy levels. To the contrary, if the ligand field splitting is very large, successive electrons will completely fill the lower states with spin pairing before entering the upper energy levels. Examples of both conditions are shown by the two ferric ion complexes in Fig. 6.26 in which $Fe(H_2O)_6^{+3}$ is in the high-spin state and $Fe(CN)_6^{-3}$ is in the low. Additional examples of $3d$ transition elements in low- and high-spin states in tetrahedral coordination are given

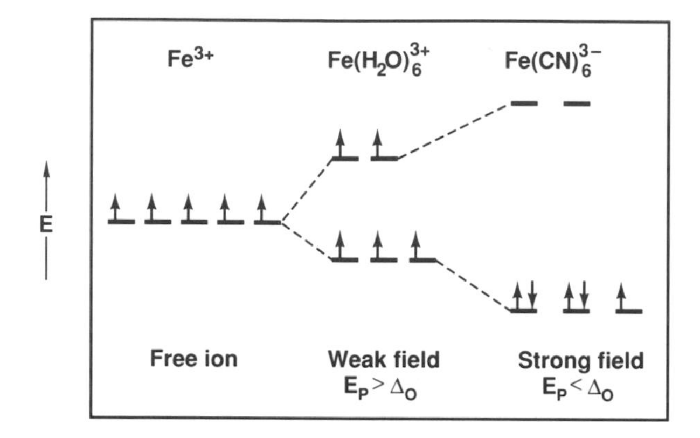

ΔO (Crystal field splitting parameter)

Figure 6.26.

The energy-level diagram for weak- and strong-field Fe^{3+} complexes in an octahedral field. The weak field produces the high spin state. (Adapted from J. J. Lagowski, p. 704, loc. cit.)

in Table 6.4. As would be expected, the spin state has a direct, if not governing, influence on the magnetic behavior of a compound. No sweeping generalizations are possible, and the effects must be considered on a case-by-case basis.

The effective ionic radius is also a function of the spin state and, as anticipated, is rather larger for the high spin configuration. (See Table 5.12 for examples.)

D. The Jahn–Teller Effect

The Jahn–Teller theorem states that if a molecular crystal gives rise to an orbitally degenerate ground state it will distort itself in a way that removes the degeneracy. In other words, if an electron would appear to have a choice between two or more orbitals of equal energy, the coordination cage will deform in a manner that renders the orbitals unequal in energy.

The theoretical justification of the Jahn–Teller theorem lies beyond the scope and level of this text. However, a pictorial representation of the effect with Mn^{3+} as the model is shown in Fig. 6.27. Referring to Fig. 6.24, let us take Mn^{3+} as our example. The high-spin arrangement is $(t_{2g})^3(e_g)^1$. The geometry displayed by Fig. 6.27a lowers the d_{z^2} orbital by decreasing its repulsion with the z-axis ligands relative to the $d_{x^2-y^2}$ orbital, which lies closer to the ligands in the xy plane. In case (b) the t_{2g} and e_g levels are degenerate. In case (c) all t_{2g} levels are lowered, and the levels inverted with

Table 6.4. Ligand field stabilization energies.[a]

Electronic configurations and ligand field stabilization energies of transition metal ions in tetrahedral coordination

Number of 3*d* Electrons	Ion	High-Spin State, $E_p > \Delta_t$: Electronic Configuration e	t_2	Unpaired Electrons	LFSE	Low-Spin State, $E_p < \Delta_t$: Electronic Configuration e	t_2	Unpaired Electrons	LFSE
0	Ca^{2+}, Sc^{3+}, Ti^{4+}			0	0			0	0
1	Ti^{3+}	↑		1	$\frac{3}{5}\Delta_t$	↑		1	$\frac{3}{4}\Delta_t$
2	Ti^{2+}, V^{3+}	↑ ↑		2	$\frac{6}{5}\Delta_t$	↑ ↑		2	$\frac{6}{4}$
3	V^{2+}, Cr^{3+}, Mn^{4+}	↑ ↑	↑	3	$\frac{4}{5}\Delta_t$	↑↓ ↑		1	$\frac{5}{6}\Delta_t$
4	Cr^{2+}, Mn^{3+}	↑ ↑	↑ ↑	4	$\frac{3}{5}\Delta_t$	↑↓ ↑↓		0	$\frac{11}{6}\Delta_t$
5	Mn^{2+}, Fe^{3+}	↑ ↑	↑ ↑ ↑	5	0	↑↓ ↑↓	↑	1	$\frac{10}{6}\Delta_t$
6	Fe^{2+}, Co^{3+}, Ni^{4+}	↑↓ ↑	↑ ↑ ↑	4	$\frac{3}{5}\Delta_t$	↑↓ ↑↓	↑ ↑	2	$\frac{3}{4}\Delta_t$
7	Co^{2+}, Ni^{3+}	↑↓ ↑↓	↑ ↑ ↑	3	$\frac{5}{6}\Delta_t$	↑↓ ↑↓	↑ ↑ ↑	3	$\frac{6}{5}\Delta_t$
8	Ni^{2+}	↑↓ ↑↓	↑↓ ↑ ↑	2	$\frac{4}{5}\Delta_t$	↑↓ ↑↓	↑↓ ↑ ↑	2	$\frac{4}{5}\Delta_t$
9	Cu^{2+}	↑↓ ↑↓	↑↓ ↑↓ ↑	1	$\frac{3}{5}\Delta_t$	↑↓ ↑↓	↑↓ ↑↓ ↑	1	$\frac{3}{5}\Delta_t$
10	Zn^{3+}, Ga^{2+}, Ge^{4+}	↑↓ ↑↓	↑↓ ↑↓ ↑↓	0	0	↑↓ ↑↓	↑↓ ↑↓ ↑↓	0	0

Electronic configurations and ligand field stabilization energies of transition metal ions in octahedral coordination

Number of 3*d* Electrons	Ion	High-Spin State: Electronic Configuration t_{2g}	e_g	Unpaired Electrons	LFSE	Low-Spin State: Electronic Configuration t_{2g}	e_g	Unpaired Electrons	LFSE
0	Ca^{2+}, Sc^{3+}, Ti^{4+}			0	0			0	0
1	Ti^{3+}	↑		1	$\frac{2}{5}\Delta_o$	↑		1	$\frac{2}{5}\Delta_o$
2	Ti^{2+}, V^{3+}	↑ ↑		2	$\frac{4}{5}\Delta_o$	↑ ↑		2	$\frac{4}{5}\Delta_o$
3	V^{2+}, Cr^{3+}, Mn^{4+}	↑ ↑	↑	3	$\frac{6}{5}\Delta_o$	↑ ↑ ↑		3	$\frac{6}{5}\Delta_o$
4	Cr^{2+}, Mn^{3+}	↑ ↑ ↑	↑	4	$\frac{3}{5}\Delta_o$	↑↓ ↑ ↑		2	$\frac{8}{5}\Delta_o$
5	Mn^{2+}, Fe^{3+}	↑ ↑ ↑	↑ ↑	5	0	↑↓ ↑↓ ↑		1	$\frac{10}{5}\Delta_o$
6	Fe^{2+}, Co^{3+}, Ni^{4+}	↑↓ ↑ ↑	↑ ↑	4	$\frac{2}{5}\Delta_o$	↑↓ ↑↓ ↑↓		0	$\frac{12}{5}\Delta_o$
7	Co^{2+}, Ni^{3+}	↑↓ ↑↓ ↑	↑ ↑	3	$\frac{4}{5}\Delta_o$	↑↓ ↑↓ ↑↓	↑	1	$\frac{9}{5}\Delta_o$
8	Ni^{2+}	↑↓ ↑↓ ↑↓	↑ ↑	2	$\frac{6}{5}\Delta_o$	↑↓ ↑↓ ↑↓	↑ ↑	2	$\frac{6}{5}\Delta_o$
9	Cu^{2+}	↑↓ ↑↓ ↑↓	↑↓ ↑	1	$\frac{3}{5}\Delta_o$	↑↓ ↑↓ ↑↓	↑↓ ↑	1	$\frac{3}{5}\Delta_o$
10	Zn^{2+}, Ga^{3+}, Ge^{4+}	↑↓ ↑↓ ↑↓	↑↓ ↑↓	0	0	↑↓ ↑↓ ↑↓	↑↓ ↑↓	0	0

[a] Taken from Roger G. Burns, *Mineralogical Applications of Crystal Field Theory*, Cambridge University Press, 16, 1970.

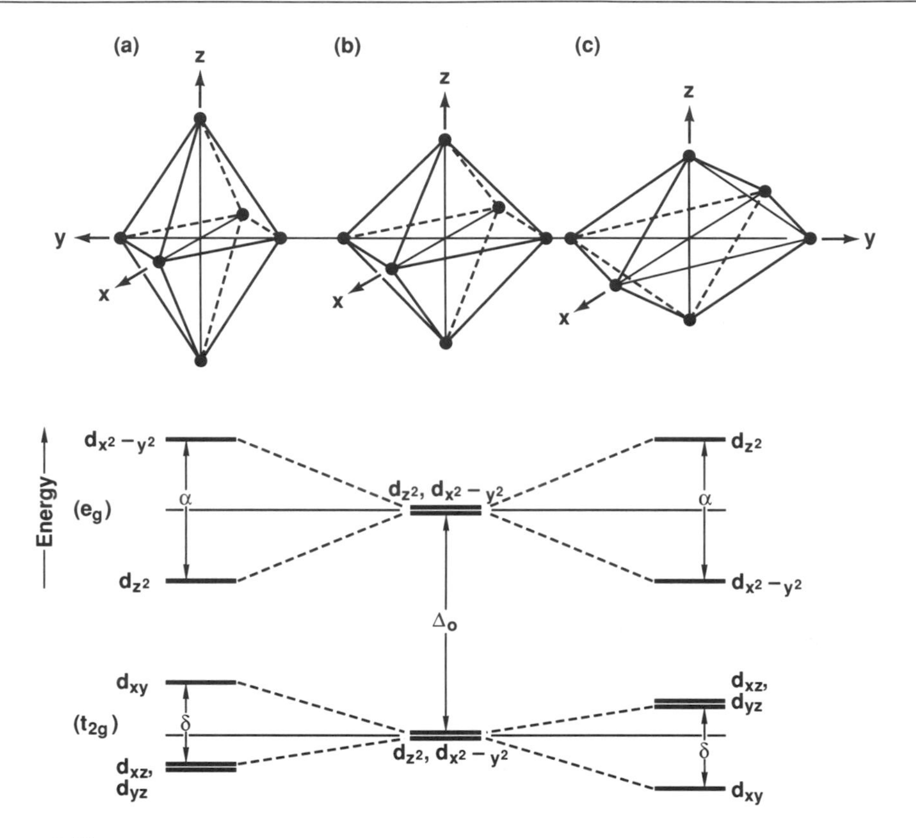

Figure 6.27.

The energy levels of the *d* orbitals of a transition metal ion in octahedral coordination: (a) the octahedron elongated along the *z*-axis; (b) a regular octahedron; (c) the octahedron compressed along the *z*-axis. (After R. G. Burns, *Mineralogical Applications of Crystal Field Theory*, Cambridge University Press, p. 21, 1970.)

respect to case (a). At the same time, as the interionic distances increase in the *xy* plane, the d_{z^2} orbital becomes the highest e_g level. Either of the distorted octahedrons is more stable than the regular one.

In the octahedral coordination, oxides in which the metal ion has d^4, d^9, and low-spin d^7 configurations show large Jahn–Teller distortions. This is because there are an odd number of electrons in the e_g levels. Such ions are Cr^{2+}, Mn^{3+}, Ni^{3+}, and Cu^{2+}. In tetrahedral coordination, ions having d^3, d^4, and d^9 electronic structure are expected to be more stable in distorted sites.

E. Normal and Inverse Spinels and LFSE

In Chapter 5, Section 9D, we discussed the spinel crystal structure, with the general formula $(AO)(B_2O_3)$, where A is divalent and B trivalent. The normal spinel has A^{2+} in the tetrahedral sites and B^{3+} in the octahedral ones (see Fig. 5.23). However, the so-called inverse structure also occurs wherein B^{3+} occupies the tetrahedral positions while the normal B sites are a random mixture of A^{2+} and B^{3+}. The familiar compound magnetite is an inverse spinel. This is rationalized as follows: the transfer of the high-spin d^5 Fe^{3+} to the tetrahedral sites from the octahedral ones does not increase the energy, whereas placing the high-spin d^6 Fe^{2+} in the octahedral holes gives a net gain in the LFSE. A similar situation stabilizes the inverse structure of $NiAl_2O_3$, viz. the extra LFSE arises from the occupation of the B sites by Ni^{2+}.

These inversions are not generally predictable in advance but are usually susceptible to ad hoc explanations, based upon a comparison of the LFSE of the two symmetries as listed in Table 6.5.

Table 6.5. Ligand field stabilization energies in octahedral and tetrahedral ligand fields.

Number of Electrons	LFSE[a]	
	Octahedral	Tetrahedral
0	0	0
1	$(2/5)\Delta_o$	$(3/5)\Delta_t$
2	$(4/5)\Delta_o$	$(6/5)\Delta_t$
3	$(6/5)\Delta_o$	$(4/5)\Delta_t$
4	$(3/5)\Delta_o$	$(2/5)\Delta_t$
5	0	0
6	$(2/5)\Delta_o$	$(3/5)\Delta_t$
7	$(4/5)\Delta_o$	$(6/5)\Delta_t$
8	$(6/5)\Delta_o$	$(4/5)\Delta_t$
9	$(3/5)\Delta_o$	$(2/5)\Delta_t$
10	0	0

[a] Δ_t is generally $\leq \frac{1}{2}\Delta_o$ for a given ion with common ligands.

Chapter 6 Exercises

1. Justify the formulas $\cos\theta = (e^{i\theta} + e^{-i\theta})/2$ and $\sin\theta = (e^{i\theta} - e^{-i\theta})/2i$.
2. Verify that $d^2\psi/dx^2 = -ky$ has the solution $y = A\cos(kx) + B\sin(kx)$, where A and B are arbitrary constants.

3. Show that the wave functions are unchanged by a uniform increase in the potential energy. Do the energy levels change? If so, in what manner?

4. Using the variational method, calculate the ground state energy of hydrogen. Take for the trial function $\psi = e^{-\alpha r}$.

5. Show that if the wave function ψ for a hydrogenic atom is $\psi = (1/\pi a_0^3)^{1/2} e^{-r/a_0}$ the integrated probability density over all space is unity; i.e., the probability of the electron's being somewhere is 1.

6. Show that the value of the potential energy of an electron in the ground state of the hydrogen atom is $-me^4/2\hbar^2$.

7. Describe the bonding in tetrachloroethylene ($Cl_2C = CCl_2$) in terms of molecular orbital theory.

8. Show that e^{ikx} and e^{-ikx} are both solutions to (6.17). Also by imposing the boundary conditions (6.18), show how the exponential solutions lead to (6.25).

9. Verify the uncertainty principle, $\Delta p \; \Delta x \geq h$, for any state of a particle in a one-dimensional box.

10. The orbitals for sp^2 hybridization are

$$\psi_1 = (1/\sqrt{3})s + (\sqrt{2}/\sqrt{3})p_x,$$

$$\psi_2 = (1/\sqrt{3})s - (1/\sqrt{6})p_x + (1/\sqrt{2})p_y,$$

$$\psi_3 = (1/\sqrt{3})s + c_2 p_x + c_3 p_y.$$

Find the constants c_2 and c_3 so as to obtain three normalized orthogonal orbitals.

11. Create a table of LFSE, such as Table 6.3, for d shell electrons in a configuration of cubic symmetry.

12. Use ligand field arguments to predict the coordination number of Mo^{5+}.

Additional Reading

There are so many excellent texts and articles on quantum chemistry that it is impossible to draw up a definitive list of references. Further, the perception of how best to present quantum mechanics varies greatly, which perhaps accounts for the overwhelming number of books dealing with this subject. The present authors have found the following to be most useful.

L. Pauling and E. B. Wilson, Jr., *Introduction to Quantum Mechanics with Applications to Chemistry*, McGraw-Hill, New York (1935).

H. Eyring, J. Walter, and G. E. Kimball, *Quantum Chemistry*, Wiley, New York (1961).
Both of these books are standard references that go far beyond the scope of the present text but are useful because of their completeness and rigor.

C. A. Coulson, *Valence*, Oxford-Clarendon Press (1952). Very readable with a minimum of mathematics. Good presentation of the molecular orbital approach.

H. F. Hameka, *Introduction to Quantum Theory*, Harper and Row, New York (1967). An excellent presentation and explanation of the mathematics necessary to deal with quantum theory.

B. N. Figgis, *Introduction to Crystal Fields*, Interscience, New York (1966). A clear presentation with many examples and correlations.

R. G. Burns, *Mineralogical Applications of Crystal Field Theory*, Cambridge University Press, New York (1970). Dealing as it does with solids, it is more relevant to the present text, though not as general or comprehensive as the preceding reference.

Chapter 7

Covalent Crystals

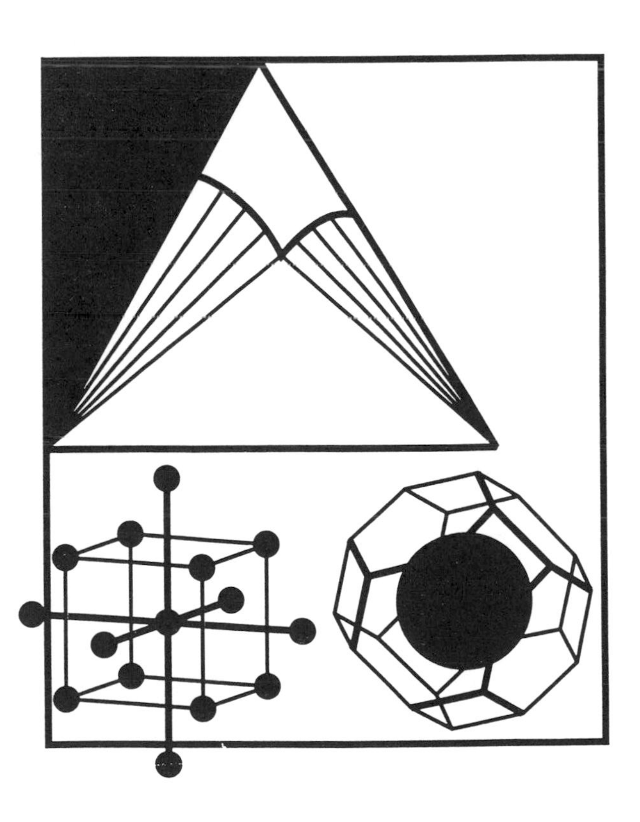

Chapter 7

Contents

"Goodness, what beautiful diamonds."
"Goodness had nothing to do with it, dearie."
—Mae West in *Night after Night*

Covalent Crystals

All covalent bonds between unlike elements retain a certain degree of ionicity that increases with the difference in their electronegativities. Hence, the term "covalent" is more properly interpreted to mean mostly covalent directed bonds in distinction to the simplest ionic bonds, which in the absence of ligand field effects preserve the spherical symmetry of point charges. The basic nature of the covalent bond was developed in the preceding chapter, and it will be illustrated and expanded here by selected examples with diverse chemical properties. This chapter commences with compounds in which the bonding electrons are highly localized, such as in the silicates, and then proceeds into the area of semiconductors and graphite in which they are partially or even completely delocalized. In this manner we will travel from insulators to the final category, which is true metals that are dealt with in Chapter 8.

1. Covalent Radii

It is far more difficult to construct a universal set of self-consistent atomic sizes for covalently bonded atoms than for ionic ones (see Chapter 5, Section 6), not only because the effective atomic size varies in response to multiple bonding but also on account of the complexity and diversity of the myriad crystal structures. Nevertheless, the atomic radii within covalently bonded elemental solids and compounds that have the simple diamond cubic (Fig. 1.21), sphalerite (Fig. 7.1), and wurtzite (Fig. 7.2) structures are more or less constant for each element. In all three structures the cations are tetrahedrally coordinated. The prototypic wurtzite and sphalerite crystals are both ZnS.

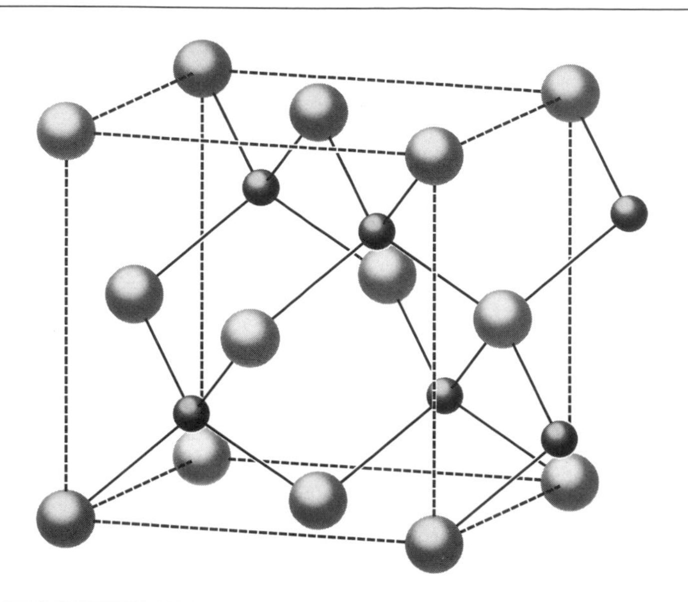

Figure 7.1.

The sphalerite cubic zinc blende crystal structure. The small dark circles represent the metal ions; the larger light ones, the anions.

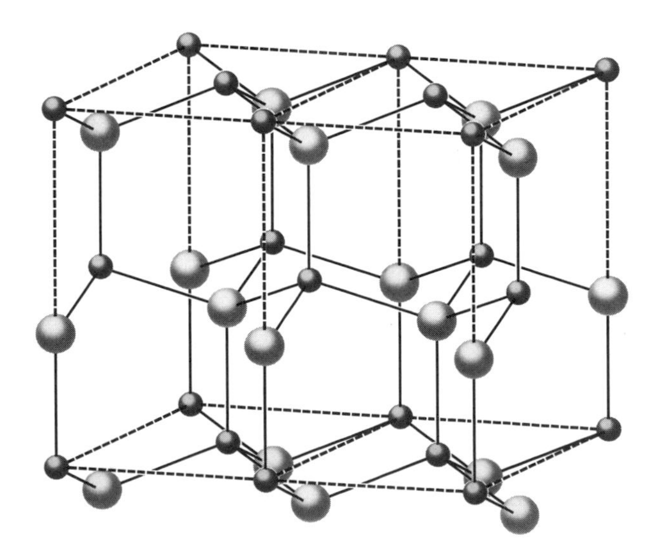

Figure 7.2.

The hexagonal wurtzite crystal structure. The small dark circles represent the cations; the others, the anions.

Table 7.1. Covalent radii for some nonmetallic elements.[1] (Å)

	C	N	O	F
Single-bond radius	0.772	0.70	0.66	0.64
Double-bond radius	0.667			
Triple-bond radius	0.603			
	Si	P	S	Cl
Single-bond radius	1.17	1.10	1.04	0.99
Double-bond radius	1.07	1.00	0.94	0.89
Triple-bond radius	1.00	0.93	0.87	
	Ge	As	Se	Br
Single-bond radius	1.22	1.21	1.17	1.14
Double-bond radius	1.12	1.11	1.07	1.04
	Sn	Sb	Te	I
Single-bond radius	1.40	1.41	1.37	1.33
Double-bond radius	1.30	1.31	1.27	1.23

Table 7.2. Tetrahedral covalent radii.[1] (Å)

	Be	B	C	N	O	F
	1.06	0.88	0.77	0.70	0.66	0.64
	Mg	Al	Si	P	S	Cl
	1.40	1.26	1.17	1.10	1.04	0.99
Cu	Zn	Ga	Ge	As	Se	Br
1.35	1.31	1.26	1.22	1.18	1.14	1.11
Ag	Cd	In	Sn	Sb	Te	I
1.52	1.48	1.44	1.40	1.36	1.32	1.28
	Hg					
	1.48					

Pauling[1] calculated the radii for covalently bonded nonmetallic elements based simply upon the internuclear distances in a representative collection of molecules and crystals. His values are reproduced in Table 7.1. A further compilation based upon the observed interatomic distances in the tetrahedrally coordinated crystals is shown in Table 7.2. A comparison of the calculated and observed interatomic

[1] L. Pauling, *Nature of the Chemical Bond*, 3rd ed., p. 224, Cornell University Press, Ithaca, NY, 1960. (Tables 7.1 and 7.2 used by permission of the publisher, Cornell University Press).

Table 7.3. Comparison of observed and calculated values of tetrahedral interatomic distances.[1]

Sphalerite	Obsvd.	Calcd.	Sphalerite	Obsvd.	Calcd.
CuF	1.99	1.85	AlSb	2.62	2.64
BN	1.58	1.57	GaSb	2.62	2.63
AlP	2.36	2.36	InSb	2.80	2.80
GaP	2.36	2.36	BeTe	2.38	2.41
InP	2.54	2.54	ZnTe	2.63	2.63
BeS	2.10	2.10	CdTe	2.80	2.80
ZnS	2.35	2.35	HgTe	2.80	2.80
CdS	2.52	2.53	CuI	2.63	2.62
HgS	2.52	2.52	AgI	2.80	2.80
CuCl	2.34	2.34	Wurtzite		
SiC	1.94	1.89	AlN	1.96	1.90
AlS	2.44	2.44	GaN	1.96	1.95
GaS	2.44	2.44	InN	2.14	2.15
InS	2.62	2.62	BeO	1.72	1.65
BeSe	2.20	2.20	ZnO	1.97	1.97
ZnSe	2.45	2.45	ZnS	2.35	2.35
CdSe	2.62	2.63	SiC	1.94	1.89
HgSe	2.62	2.63	CdSe	2.62	2.63
CuBr	2.46	2.46	MgTe	2.72	2.76

distances for several compounds having the sphalerite or wurtzite structures is given in Table 7.3.

An alternative scheme for calculating covalent atomic radii has been proposed by van Vechten and Phillips.[2] Their method is based upon the Bohr model of the atom and takes account of the influence of the core electrons, when necessary, calculating their effective charge by applying the Slater rules (Chapter 5, Section 6B). Their results in the main differ only slightly from Pauling's, but are sometimes in better agreement with the measured values.

2. The Silicates

After carbon, silicon is the basis for more compounds than is any other element. The naturally occurring silicate minerals are crystallographically perhaps the most com-

---

[2] J. A. van Vechten and J. C. Phillips, *Phys. Rev.B*, **2**, 2160 (1970). Their method is rather involved, and the interested reader is referred to the original literature.

plex inorganic structures known. The science of crystallography has its origins rooted in mineralogy, and the field is so ancient and prolific that we cannot do it justice in the brief space allotted. On the other hand, silicates present such an intriguing variety of covalently bonded structures that we cannot omit or find a satisfactory substitute for this important field of solid state chemistry.

A. Silicate Structures

The fundamental building block of the silicates is SiO_4^{4-} the orthosilicate anion; a structural representation is shown in Fig. 7.3. The exact nature of the Si—O bond as yet has not been satisfactorily described. Pauling originally proposed that it was $\sim 50\%$ ionic. The basis for this supposition has been questioned, and the role of hybridization has been introduced in its stead to explain qualitatively the stability of SiO_4^{4-}. Also the effect of polarizability has been invoked, but again without yielding numerical results. Diagrammatic representations of silicate crystals generally show the silicon atom to be much smaller than the oxygen. This convention most likely has persisted for historical reasons, stemming from the time when it was believed that the Si—O bond was purely ionic. For the record, $(r_{Si}/r_O)_{ionic} = 0.28$ and $(r_{Si}/r_O)_{Cov} = 1.77$. However, the sum of the ionic radii is very nearly the same as that of the covalent ones (viz. $\sim 5\%$ difference) so that the structural relations remain essentially unchanged. Although the covalent radius of oxygen is smaller than that of Si, it historically has been portrayed as larger, and this practice is so universal that we will continue to follow this tradition in most of our illustrations.

The complexity of the silicate structures derives from the ability of the SiO_4 units to form bonds between themselves via bridging oxygen atoms. In this way one-, two-, and three-dimensional subunits are constructed that can be linked to one another

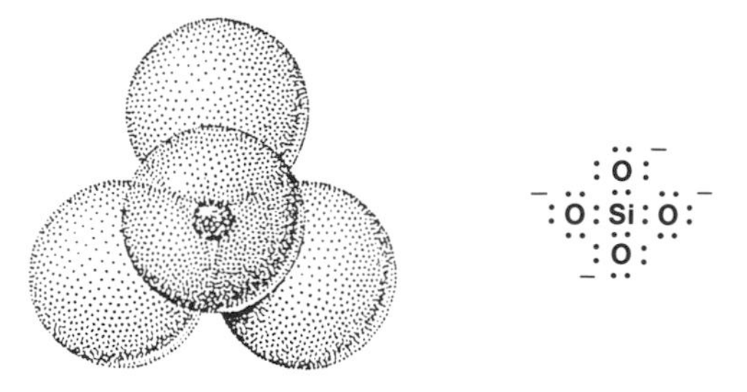

Figure 7.3.

The SiO_4^{4-} ion is a regular tetrahedron. Both the oxygen and silicon have completed their outer shells with eight electrons.

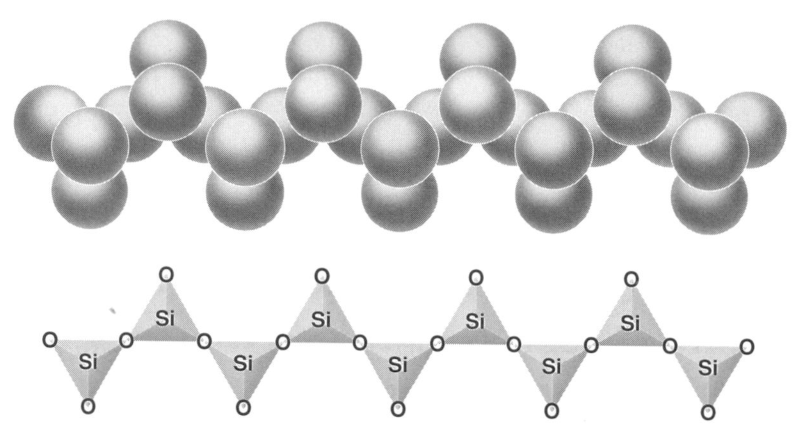

Figure 7.4.

A single chain of SiO_3^{2-}. The singly bonded oxygens, one in the plane of the page and the other behind it, bear a single charge (see Fig. 7.3b) each, and every Si atom is joined by two bridging oxygens.

by intervening metallic ions, such as Al^{3+}, Fe^{2+}, Mg^{2+}, etc. The one-dimensional chain is illustrated in Fig.7.4. Because the bridging oxygen atoms are shared between two silicon atoms, the formula changes from SiO_4 to SiO_3. Examples of the various one-, two-, and three-dimensional structures with the formula unit delineated are shown in Fig. 7.5. A schematic illustration of the way in which elementary SiO_3^{2-} units can combine to form the simplest of the extended structures is shown in Fig. 7.6. An idealized two-dimensional representation showing how silicate chains become linked together through ionic bonds is shown in Fig. 7.7.

The elementary orthosilicate ions, SiO_4^{4-}, are frequently referred to as island silicates. The first minerals to form from a cooling magma are thermodynamically the most stable and most often are neosilicates (see Fig. 7.5), composed of orthosilicate anions, whose structure provides the maximum separation of the Si atoms. This concurs with Pauling's so-called third rule (see Chapter 5, Section 8), which states that the sharing of edges or faces of coordination polyhedra incurs a progressive decrease in stability. Consequently, as the solidification of the molten magma advances, minerals of lower thermodynamic stability are able to form at the lower temperatures, so that the magma becomes increasingly enriched in Si with respect to oxygen and the precipitating minerals follow the trend from top to bottom represented in Fig. 7.5. The sharing of corners, edges, and faces for any given structure can be described by an interesting geometrical relation known as the sharing

Class	Arrangement of SiO_4 tetrahedra (central Si^{4+} not shown)	Unit composition	Mineral example
Nesosilicates	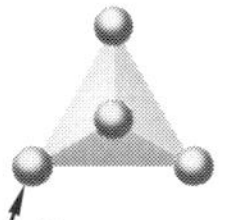 Oxygen	$(SiO_4)^{-4}$	Olivine, $(Mg, Fe)_2SiO_4$
Sorosilicates	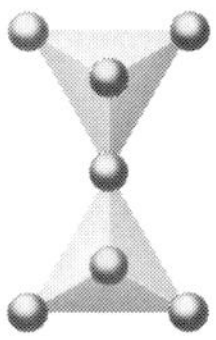	$(Si_2O_7)^{-6}$	Hemimorphite, $Zn_4Si_2O_7(OH) \cdot H_2O$
Cyclosilicates	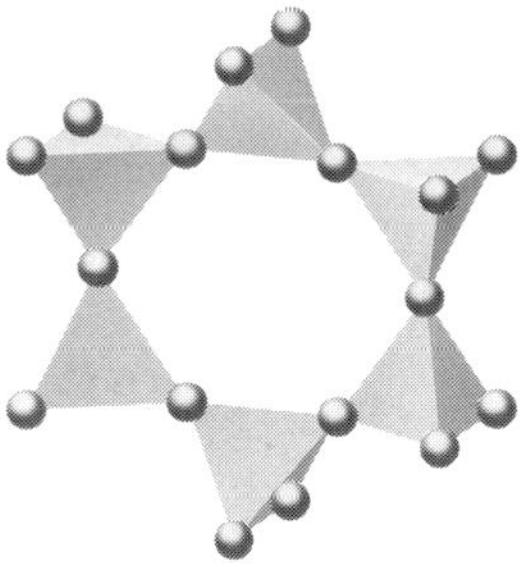	$(Si_6O_{18})^{-12}$	Beryl, $Be_3Al_2Si_6O_{18}$
Inosilicates (single chain)	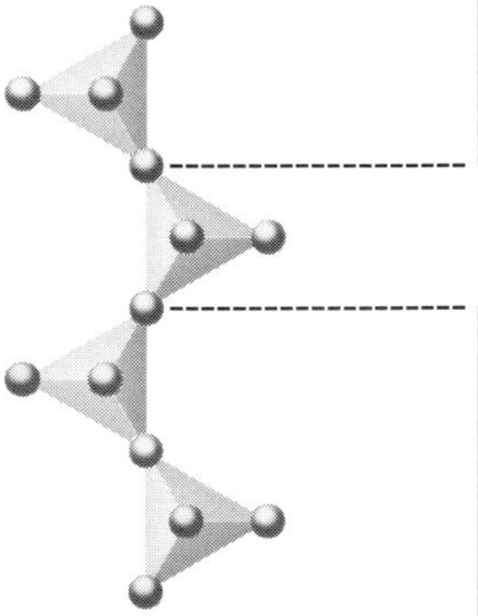	$(SiO_4)^{-4}$	Pyroxene e.g. Enstatite, $MgSiO_3$

Figure 7.5.

The basic structural building blocks of Si_xO_y. (After C. Klein and C. S. Hurlbut, Jr., based upon J. D. Dana, *Manual of Mineralogy*, Wiley, New York, pp. 368–369, 1985).

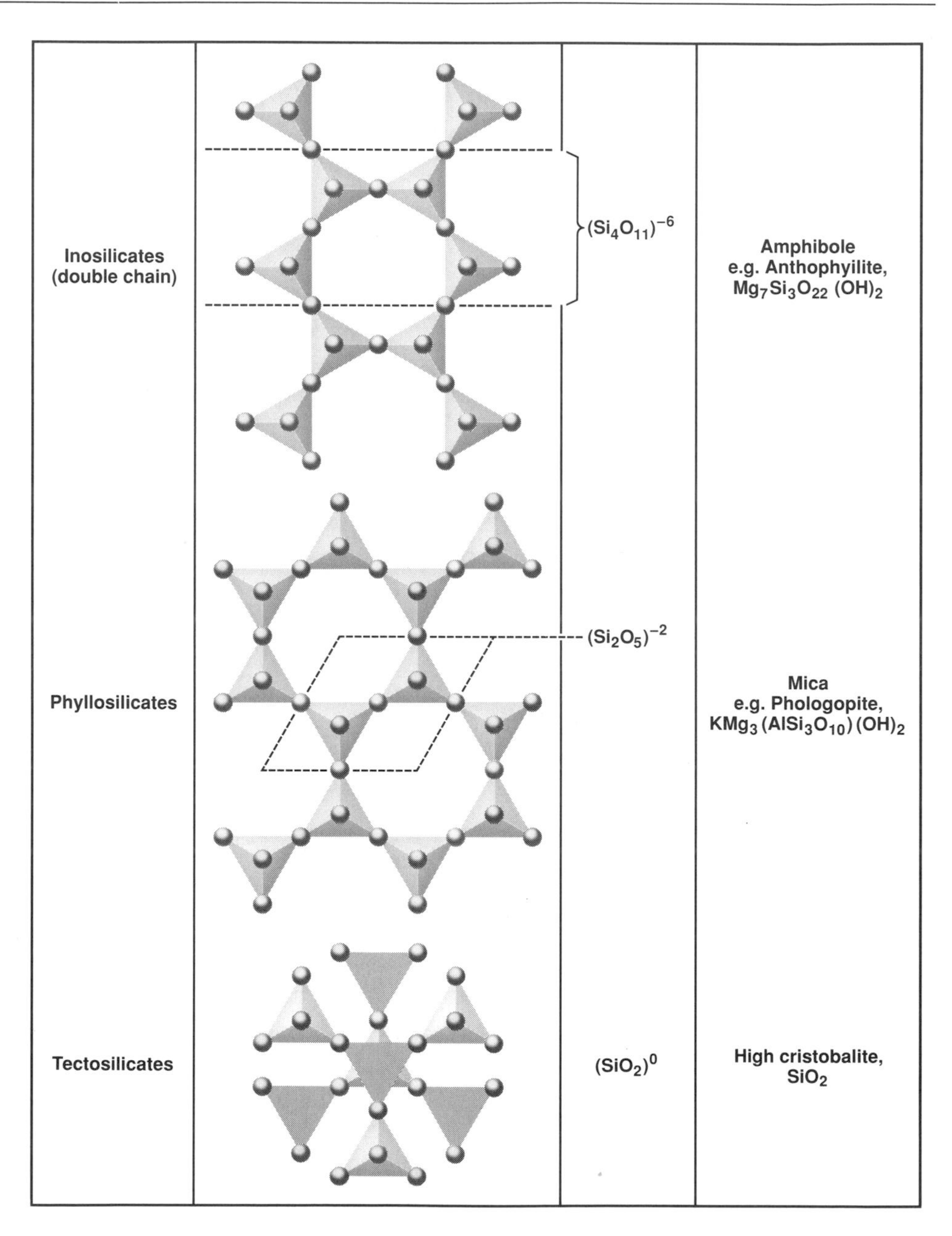

Figure 7.5 (continued)

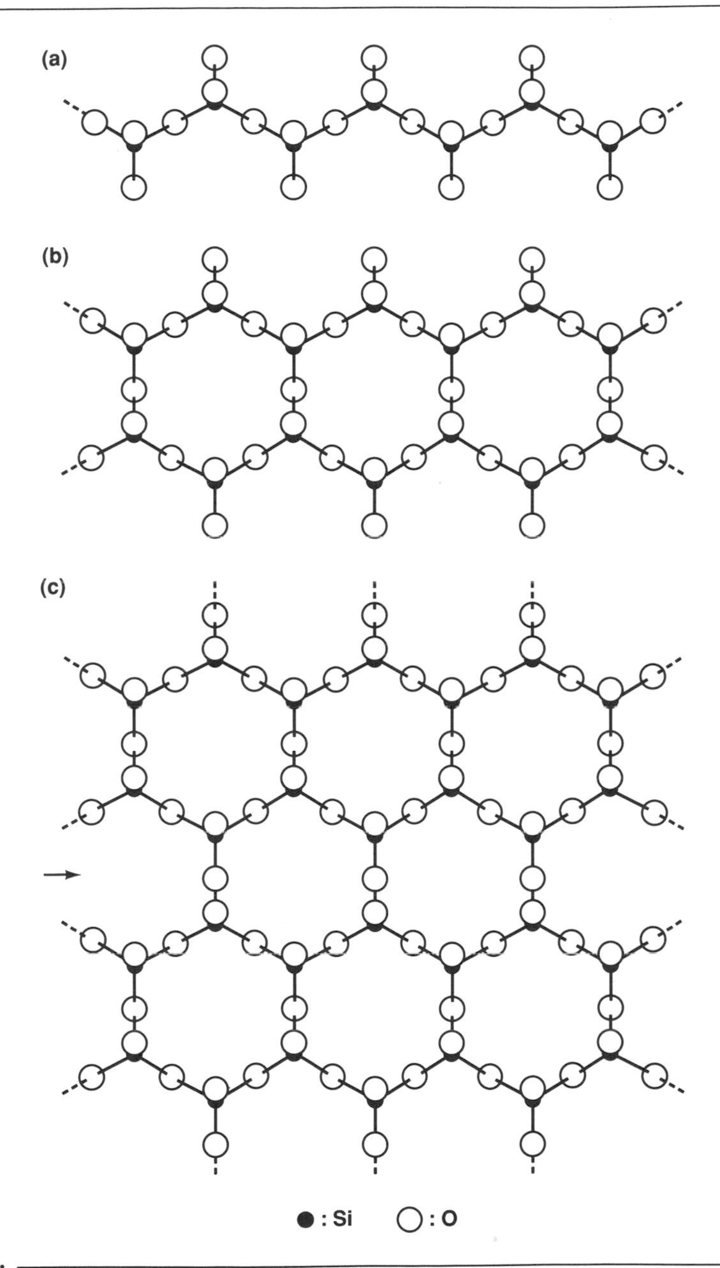

Figure 7.6.

The chain of SiO_3^{2-} units (a) combines to form a double chain (b) and with the addition of further chains become the sheet silicate (c). (After R. C. Evans, *An Introduction to Crystal Chemistry*, Cambridge University Press, p. 246, 1964).

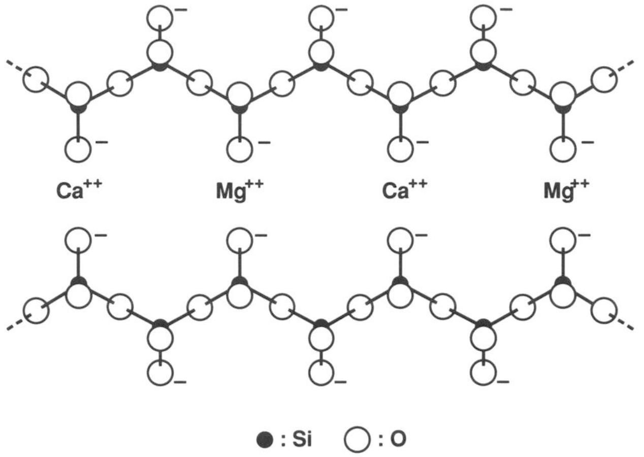

Figure 7.7.

A much-simplified depiction of the bonding of silicate chains in a fictitious crystal illustrating the metal ion bridges between silicate units

coefficient.[3] This is based on the following observation for tetrahedral silicate structures: if n, a positive integer, represents the *minimum* number of tetrahedra sharing a common corner, then the *maximum* number sharing a common corner is $n + 1$. If an anion, generally, O^{2-}, is unshared (i.e., is bonded to but a single Si^{4+}), it is assigned a value of $n = 1$, indicating that these corners are shared by only one tetrahedron. The remaining corners of this same tetrahedron will then be shared by no more than $n + 1$ tetrahedra. Likewise, if the minimum number for the shared anions is 2 (in other words, there are no dangling unshared anions), then the maximum number of tetrahedra that are permitted to share an anion belonging to this tetrahedron is $n + 1 = 3$.

Let us take as an example the mineral tremolite, $Ca_2Mg_5Si_8O_{22}(OH)_2$,[4] a double-chain silicate. The subsequent calculation of the sharing coefficient is based upon the crystallographic unit that exactly contains the unit formula and is shown in Fig. 7.8. Referring to Fig. 7.8, we count 12 unshared oxygens for which $n = 1$ and 20 bridging ones for which $n = 2$. The sharing coefficient S is defined as the average value that is obtained using 12 and 20 as the weighting factors. Thus, the sharing

[3] T. Zoltai, *Am. Min.* **45**, 960 (1960).

[4] The following discussion on sharing coefficients is patterned after F. D. Bloss, *Crystallography and Crystal Chemistry*, Holt, Rinehart and Winston, Chap. 9, 1971.

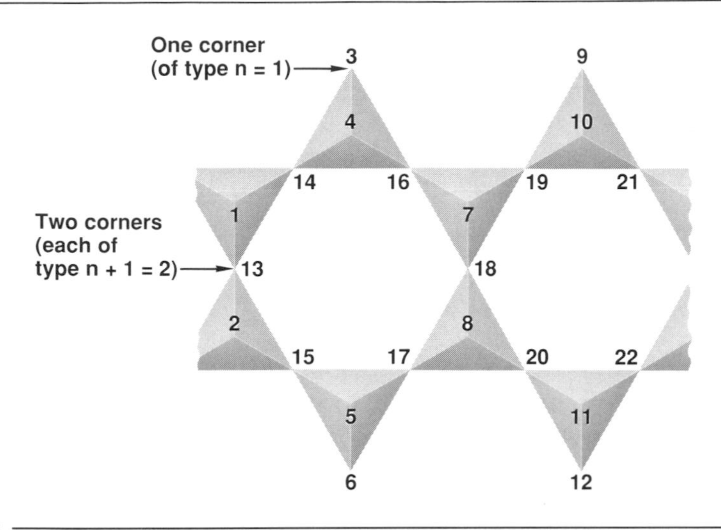

Figure 7.8.

A schematic representation of the SiO_4 tetrahedra in the mineral tremolite, $Ca_2Mg_5Si_8O_{22}(OH)_2$.

coefficient for tremolite is

$$S = (12 \times 1 + 20 \times 2)/32 = 1.625.$$

A general equation for S is derived quite simply as follows: let T stand for the number of tetrahedra per formula unit of which x are joined to anions shared by $n + 1$ tetrahedra. Then

$$S = \frac{x(n + 1) + n(4T - x)}{4T} = \frac{x}{4T} + n. \tag{7.1}$$

For our example the total number of tetrahedral corners per formula unit is $4T = 32$, of which 20 are of the $n + 1$ type. Thus, from the general formula (7.1),

$$S = 1 + 20/32 = 1.625,$$

as obtained before. If A is the number of anions per chemical formula unit, it is also the sum of the number of corners shared between $n + 1$ and n tetrahedra. Thus,

$$A = \frac{x}{n + 1} + \frac{4T - x}{n}. \tag{7.2}$$

From (7.1) and (7.2) one obtains, after some algebra,

$$S = n + (n + 1)(1 - nA/4T). \tag{7.3}$$

One can obtain n from the ratio of the number of available tetrahedral corners to the number of shared corners, $4T/A$, which gives the integer part of the sharing coefficient, i.e., n of (7.3). Thus,

$$n = \text{the integer part of } 4T/A.$$

For our example, $4T/A = 32/22 = 1 + 10/22 = 1 + 0.45$ and, hence, $n = 1$. We may also define the fraction of A anions of type $n + 1$ as A_{n+1} and write

$$4T/A = n + A_{n+1}.$$

For our example, then, $A_{n+1} = 0.45$; i.e., 45% of the A anions (not the corners) are bonded to two tetrahedra and are defined as bridging oxygens. In SiO_2, $n = 2$ and $4T/A = 2$, and therefore $A_{n+1} = 0$. In this structure all four corners are shared, and there are no oxygens bonded to only two tetrahedra.

The sharing coefficient can give helpful guidance in choosing trial structures for minerals whose composition is known. It also quantifies the sequential change in the proportion of (Si)/(O) protrayed in Fig. 7.5.

B. Crystalline SiO_2

There are nine polymorphs of crystalline SiO_2, of which eight are naturally occurring, and all are listed in Table 7.4. All crystalline forms have three-dimensional structures

Table 7.4. Polymorphs of SiO_2.

Name	Symmetry	Min. Crystallization Temperature (°C) for Stable Form at 1 atm.	Inversion to Low Temperature Form at 1 atm.	Specific Gravity
Stishovite[a]	Tetragonal			4.35
Coesite	Monoclinic			3.01
Low (α) quartz	Hexagonal			2.65
High (β) quartz	Hexagonal	574	573	2.53
Keatite (synth.)	Tetragonal			2.50
Low (α) tridymite	Monoclinic or Orthorhombic			2.26
High (β) tridymite	Hexagonal	870	~120–140	2.22
Low (α) cristobalite	Tetragonal			2.32
High (β) cristobalite	Isometric	1470	~268	2.20

[a] Only polymorph with Si in octahedral coordination with oxygen.

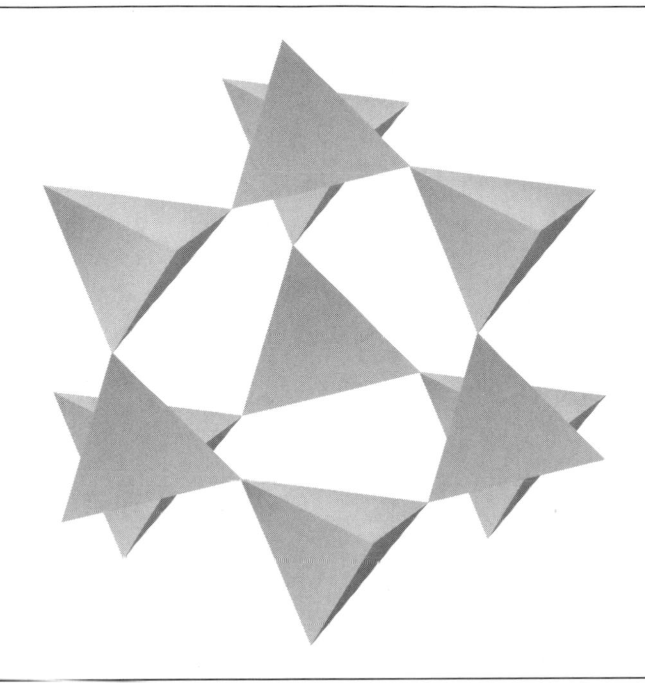

Figure 7.9.

A portion of β-cristobalite projected onto the (111) plane.

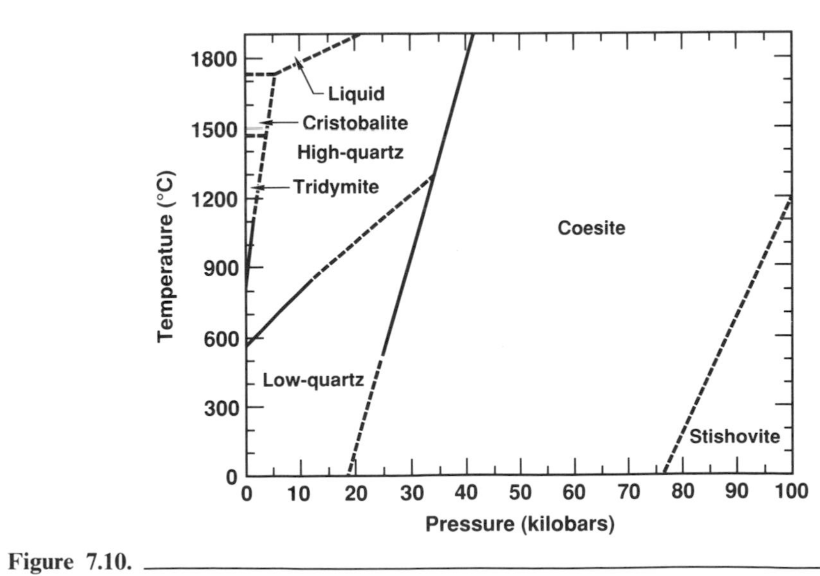

Figure 7.10.

Equilibrium diagram for SiO_2.

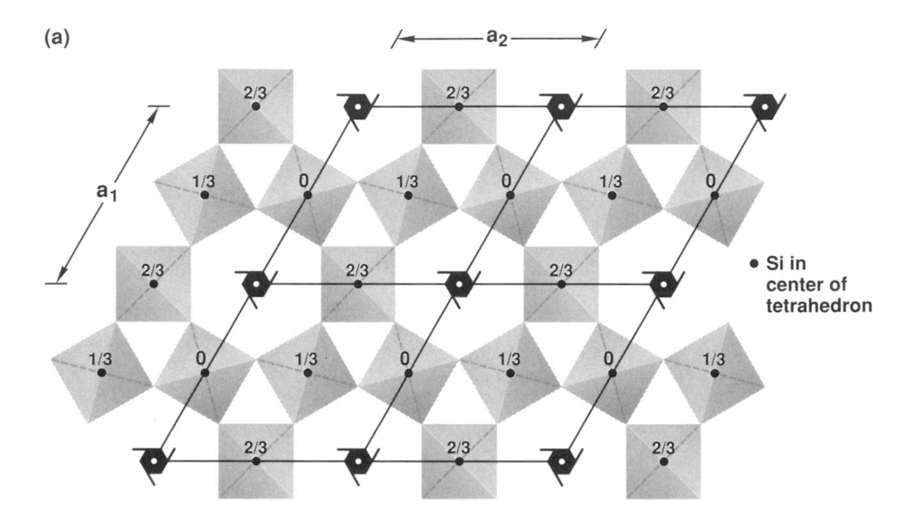

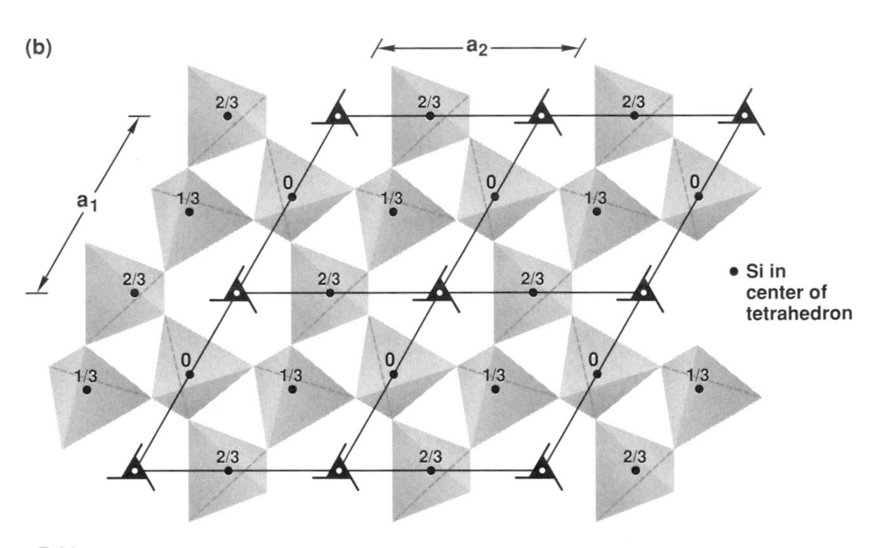

Figure 7.11.

Projections of high (β) and low (α) quartz onto the (0001) plane. The fractional heights of the Si atoms above (0001) are noted in each tetrahedron. The symbols [six-fold symbol] and [three-fold symbol] represent six-fold and three-fold axes of symmetry.

as opposed to the chain or sheet configurations to be discussed later. As such they belong to a category designated as *framework* structures in which all oxygens are shared. Figure 7.9 shows a single example of a framework structure. Figure 7.10 is a somewhat oversimplified phase diagram showing the stability ranges of the various modifications with respect to pressure and temperature. The transformations between the high and low forms of quartz, tridymite, and cristobalite, respectively, involve but minor changes in the atomic positions. However, the transformations between the different polymorphs result in a significant repositioning of the atoms and are very sluggish as a result. The structural relations between α- and β-quartz are illustrated in Fig. 7.11. As recorded in Table 7.3 the densities vary by nearly a factor of 2 in going from stishovite to β-cristobalite. As would be expected, the less dense, more open structures are more capable of incorporating interstitial impurities.

The polymorph stishovite is formed only at very high pressures and, along with coesite, has been found in small amounts at the site of the meteor crater in Arizona. Their creation there has been attributed to the very high pressures and temperatures associated with the original meteor impact.

3. Silicate Minerals

With the exception of pure silica, all silicates are mixtures of covalent and ionic bonds. The Si—O bond is considered to be essentially covalent with, of course, some ionic character due to the difference in electronegativities of Si and O. However, the terminal units of the $(\text{Si—O})_n$ radicals will of necessity carry charges because of the unsaturated bonds. These charges are compensated by the introduction of cations, with the consequent formation of ionic bonds. Thus, silicate minerals in the main are formed from units of interconnected tetrahedra of SiO_4. These units are held together by bridging bonds to other metallic elements, typically Al, Mg, Ca, Fe, Na, and K, although there are examples that include almost every naturally occurring element. Silicates are further complicated by the readiness with which ions of various elements can be interchanged without changing the basic structure. In this respect, silicate minerals are remarkably egalitarian. Aluminium commonly replaces Si at the core of the tetrahedra as well as occurring in a purely ionic site. Anions are also interchangeable, with O^{2-}, OH^-, and F^- commonly exchanging for one another. Only the ionic size seems to govern this type of exchange, and charge balance is maintained by appropriate substitution elsewhere in the lattice, e.g., if Al^{3+} replaces Si^{4+} then charge neutrality can be preserved by replacing an O^{2-} with a F^-, or say a Na^+ by a Ca^{2+}.

In the brief space allotted here, we can only present a superficial introduction to the amazingly complex structural chemistry of silicate compounds. The interested student will find at the end of this chapter a useful bibliography of the standard mineralogical treatises, but is warned that existent data refer mostly to crystallography, although there is a growing literature dealing with thermodynamic and other physical–chemical properties.

A. Neosilicates

Neosilicates, also referred to as orthosilicates, particularly in the older literature, are made up of unconnected SiO_4^{4-} tetrahedra and the necessary metallic cations. The most common naturally occurring example of this type of compound is olivine. Olivine basalts are the most prevalent form of igneous rock found on the surface of our planet. This mineral has a continuously varying composition, with Mg^{2+} and Fe^{2+} occurring in all proportions as described by Fig. 7.12. Members of the forsterite–fayalite series occur as primary crystallization products from Fe, and Mg-rich, silica-poor magmas.

The crystal structure of these minerals is orthorhombic, and the projection of the ideal structure normal to the x-axis is shown in Fig. 7.13.

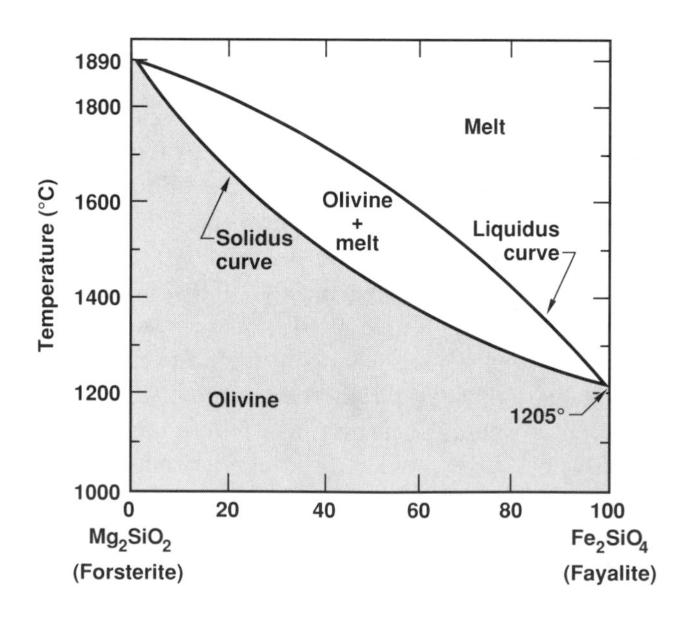

Figure 7.12.

Temperature–composition diagram for the system Mg_2SiO_4–Fe_2SiO_4 at atmospheric pressure.

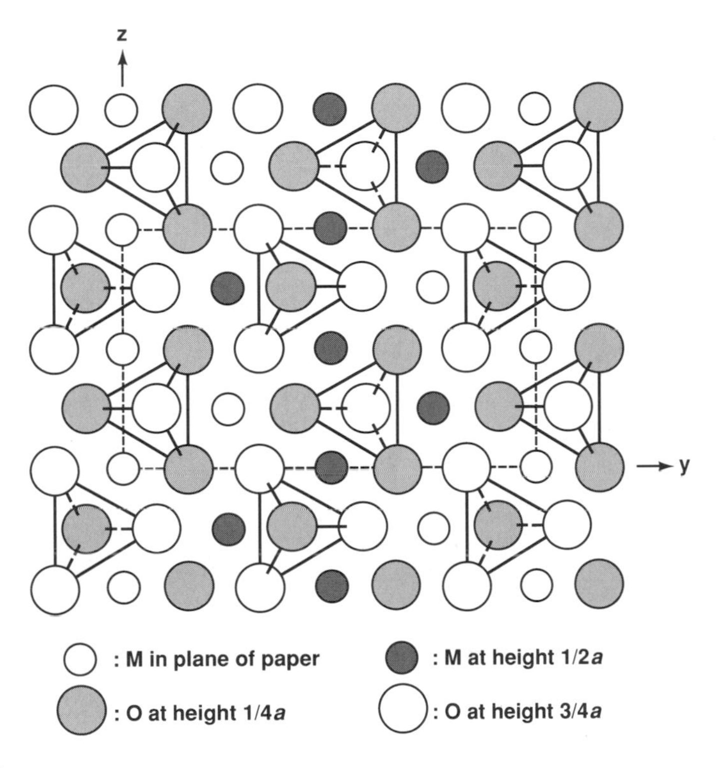

Figure 7.13.

The projection of the idealized structure of forsterite or fayalite along the x-axis. M can be either Mg^{2+} or Fe^{2+} or a mixture of both.

B. Phyllosilicates—Sheet Structures

This class of silicates is based upon an indefinitely extended two-dimensional sheet of linked tetrahedra. These tetrahedra, which are mostly SiO_4, share three of their four oxygens, resulting in (O)/(Si) = 1.5. This is shown in Fig. 7.14.

The phyllosilicates are important constituents of igneous rocks and schists. Talc, kaolinite, and mica in their pure forms historically have been of great commerical significance. Table 7.5 is a compendium of the phyllosilicates. Because of their planar structure these minerals readily cleave along the layer planes, which is one of their useful properties. The soapy or slippery feeling of talc is a consequence of this easy cleavage between planes that are held together only by weak van der Waals forces. By comparing Fig. 7.15, which shows the structure of pyrophyllite and is a representative of the clay minerals, with Fig. 7.16, which shows muscovite, we see that the

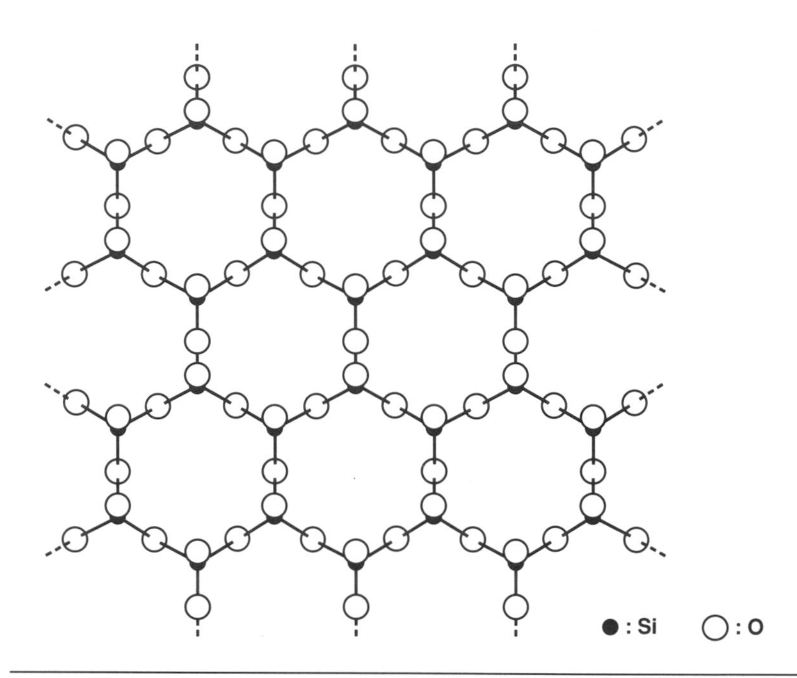

Figure 7.14.

Diagrammatic representation of the SiO_4 sheet configuration, which is the basis for phyllosilicate minerals.

Table 7.5. Phyllosilicates.[a]

Serpentine group	
Antigorite	$Mg_3Si_2O_5(OH)_4$
Chrysotile	
Clay mineral group	
Kaolinite	$Al_2Si_2O_5(OH)_4$
Talc	$Mg_3Si_4O_{10}(OH)_2$
Pyrophyllite	$Al_2Si_4O_{10}(OH)_2$
Mica group	
Muscovite	$KAl_2(AlSi_3O_{10})(OH)_2$
Phlogopite	$KMg_3(AlSi_3O_{10})(OH)_2$
Biotite	$K(Mg, Fe)_3(AlSi_3O_{10})(OH)_2$
Lepidolite	$K(Li, Al)_{2-3}(AlSi_3O_{10})(OH)_2$
Margarite	$CaAl_2(Al_2Si_2O_{10})(OH)_2$
Chlorite group	
Chlorite	$(Mg, Fe)_3(Si, Al)_4O_{10}(OH)_2(Mg, Fe)_3(OH)_6$
and three minerals closely related to the above phyllosilicate groups	
Apopphyllite	$KCa_4(Se_4O_{10})_2F_8H_2O$
Prehnite	$Ca_2Al(AlSi_3O_{10})(OH)_2$
Chrysocolla	$Cu_4H_4Si_4O_{10}(OH)_8$

[a] Taken from C. Klein and C. S. Hurlbut, Jr., *Manual of Mineralogy*, Wiley, New York, p. 426, 1985.

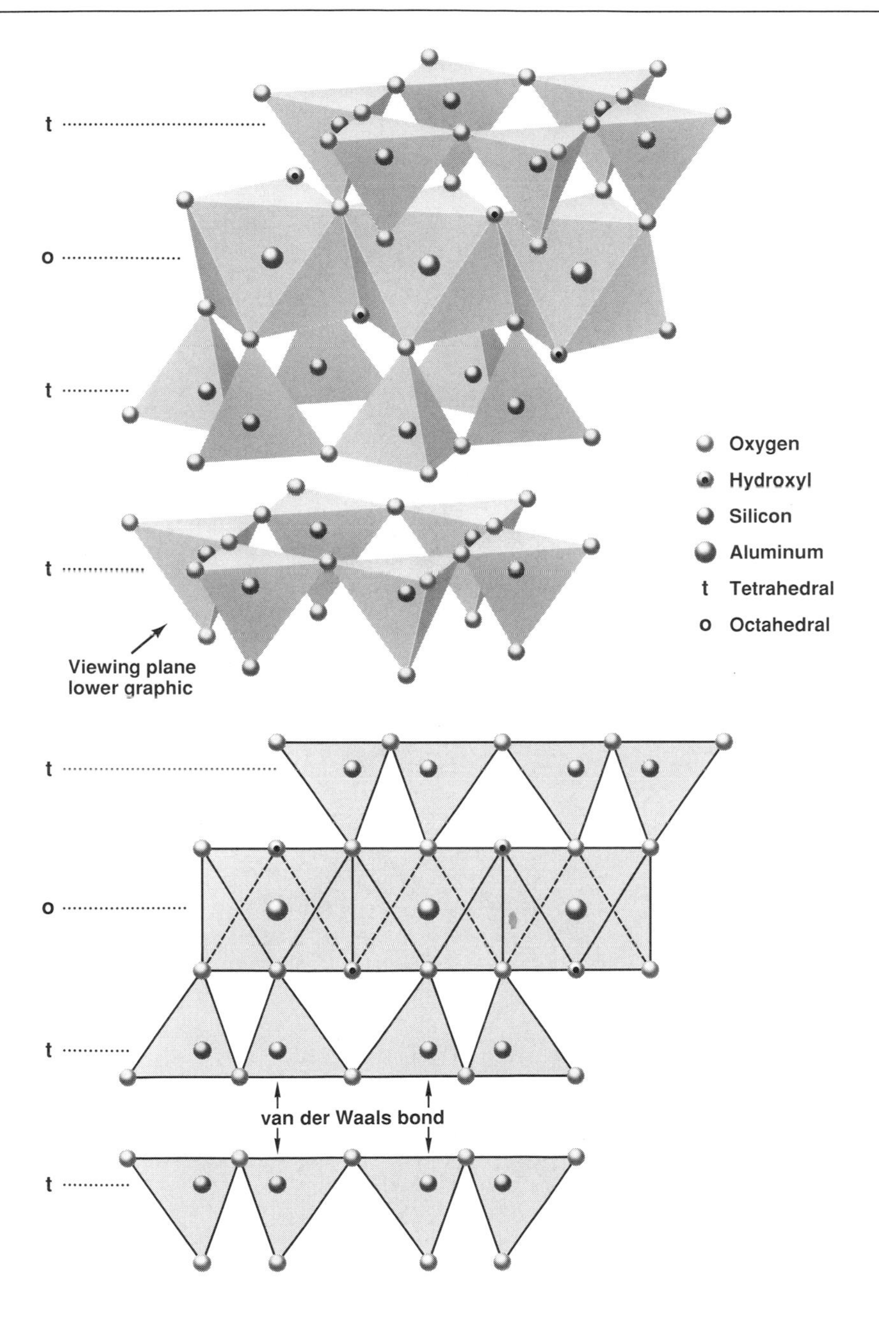

Figure 7.15.

A diagrammatic representation of the clay mineral pyrophyllite. (As presented by Klein and Hurlbut, pp. 420–421, loc. cit.).

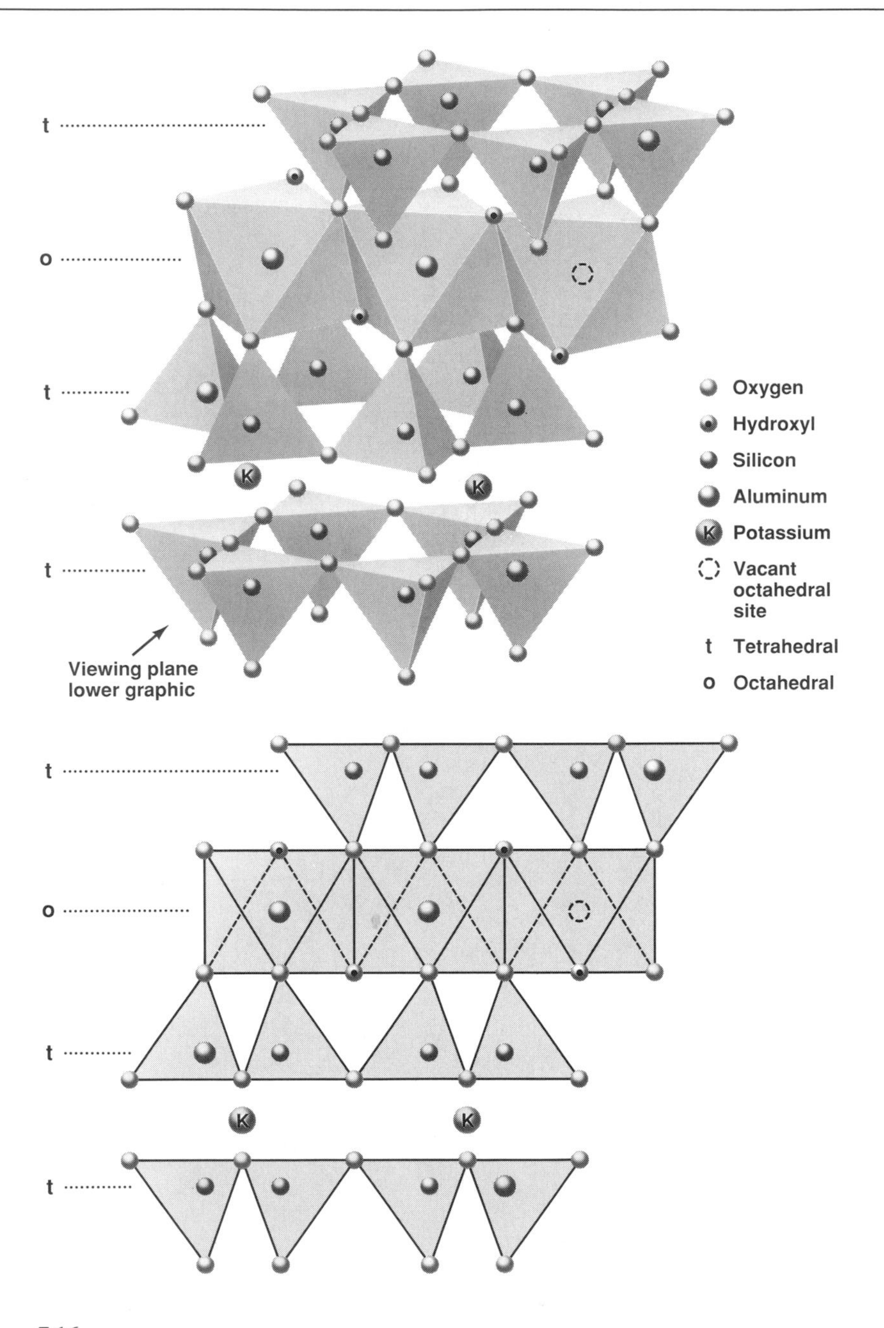

Figure 7.16.

Diagrammatic representation of the mica mineral muscovite. (After Grim 1968, as presented by Klein and Hurlbut, pp. 420–421 loc. cit.)

bonding in the clays is of the weak van der Waals sort, whereas the mica sheets are bonded ionically via K^+ ions. In spite of the ionic bonds between layers, mica readily separates into sheets because the homopolar forces are stronger than the ionic ones.

C. Framework Silicates—Three-Dimensional Structures

All oxygen atoms in the framework (tectosilicate) structure are shared, thus forming three-dimensional chains. The feldspars together with quartz are the mineralogically most important members of this group. Aluminum frequently substitutes for Si, and charge balance is maintained by the addition of interstitial alkali metal or alkaline earth ions. The structure is shown in part in Fig. 7.17, although the bridging oxygen

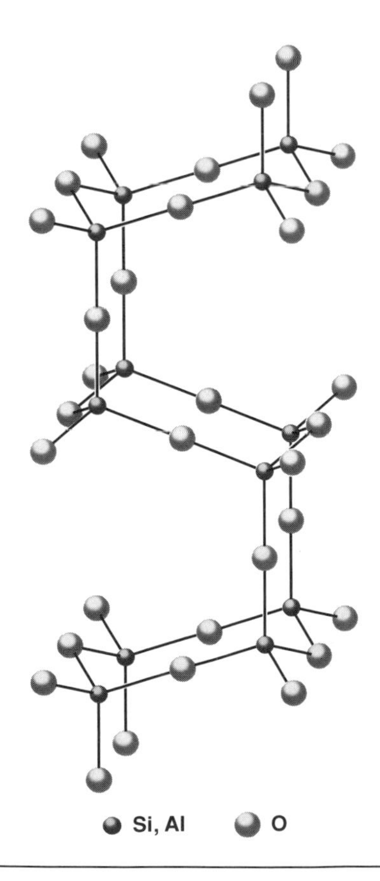

Figure 7.17.

Idealized representation of the silicon–oxygen framework in feldspars.

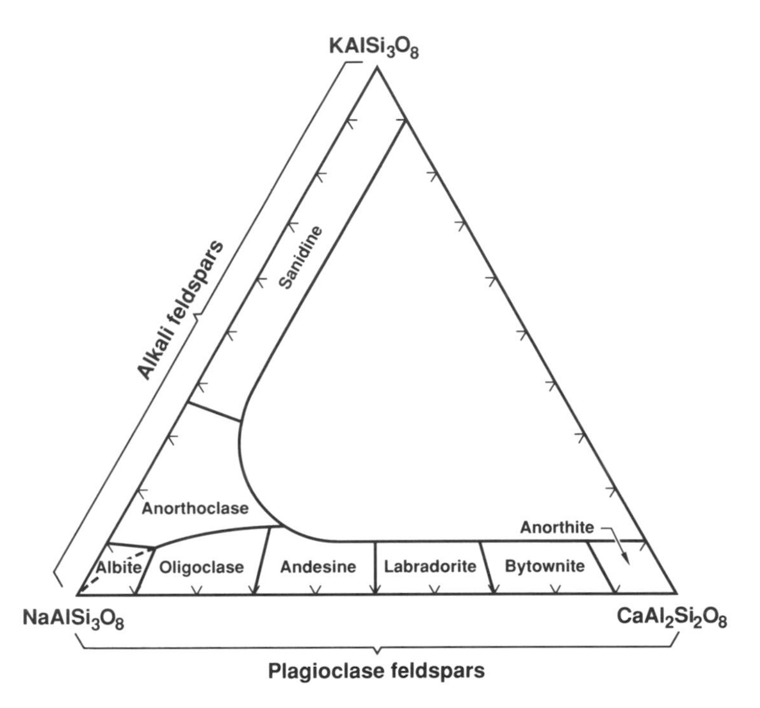

Figure 7.18.
The mineralogical relations between the three feldspar end members $K(AlSi_3O_8)$, orthoclase, $Ca(Al_2Si_2O_8)$, anorthite, and $Na(AlSi_3O_8)$, albite. (After Deer, Howie, and Zussman, *Rock Forming Minerals*, vol. 4, John Wiley and Sons, 1963).

linkages have not been continued in the horizontal direction. The relations between the three end members that encompass the plagioclase and high-temperature alkali feldspars are illustrated in Fig. 7.18. Notice that complete solubility exists between albite and anorthite. This implies a continuous variation of (Al)/(Si) with compensating changes in (Na)/(Ca). The three-dimensional structures often form vitreous, or glassy, phases. These are discussed in Chaper 14, Section 7.

4. Thermodynamics of Semiconduction

The elements C, Si, Ge, and Sn all have diamond cubic allotropes (see Fig. 1.21) and were described in Chapter 1, Section 9A. This structure results from the tetrahedral coordination of the atoms, which is in turn a consequence of the hybrid sp^3 bonds explained in Chapter 6, Section 3D. Similar compounds composed of tetrahedrally coordinated units (e.g., sphalerite and wurtzite) are illustrated at the beginning of this chapter. These, along with other tetrahedrally coordinated compounds, are the basis for most semiconductors and are the subject of this section. The crystallographic relations are described in detail later in the section after our discussion of the principles of semiconduction.

The electrical properties of a semiconductor are intermediate between those of metallic and ionic materials. Metals have their conduction bands partially filled with delocalized electrons that become charge-carrying currents upon application of even a small electrical potential. At the other extreme, purely ionic compounds have an empty conduction band and, as a result, are insulators. At very low temperatures, semiconductors are also insulators, but heating to modest temperatures supplies sufficient energy to promote electrons from the valence into the conduction band, thus changing the semiconductor from an insulator into a conductor. A schematic representation of the energetics relevant to these three cases is shown in Fig. 7.19.

The band gap E_g is defined as the difference in energy between the top of the valence band and the bottom of the conduction band. An electron that has been thermally excited into the conduction band leaves behind an electron hole with a virtual positive charge in the valence band. Both the electron and the hole can serve to carry current, and the relative amounts depend upon their individual mobilities in a semiconductor. Semiconductors in which the current is carried mainly by

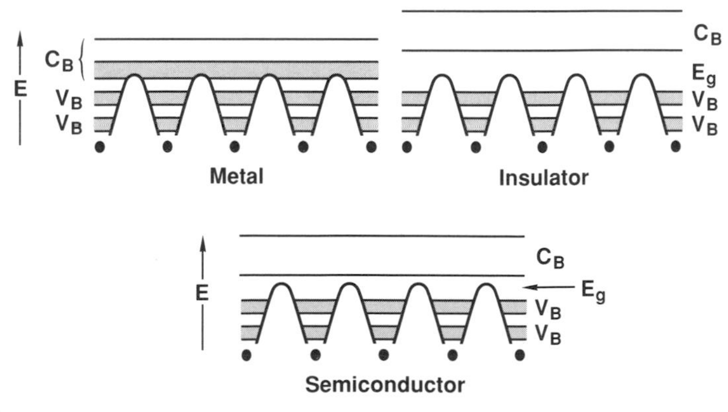

Figure 7.19.

The distribution of electrons in space and energy are schematically represented for linear arrays of atoms or ions (*), showing the differences between conductors, insulators, and semiconductors. The solid circles represent ion cores, V_B stands for the valence band, C_B for the conduction band, and E_g for the band gap energy. Once an electron has sufficient energy to reach the conduction band, it is significantly screened from the local ion core potential.

electrons are called n-type, whereas those in which it is carried mainly by holes are called p-type.

The expression that relates the concentration of conduction electrons to the band gap E_g and to the temperature can be derived for intrinsic semiconductors in a straightforward fashion. Let us first assume both the electron and the hole to be classical particles obeying Boltzmann statistics. Later, we shall see that this approximation is seldom correct, but it nevertheless supplies a convenient starting point for our understanding. We set the total fraction of potential donor sites that donate electrons to the conduction band equal to unity. Then, writing the total partition function (p.f.) for e^- and h^+, but omitting the vibrational contribution while preserving the translational term, we obtain

$$(\text{p.f.})_e = (2pm_e kT/h^2)^{3/2} V e^{-E_g/2kT}, \tag{7.4}$$

$$(\text{p.f.})_h = (2pm_h kT/h^2)^{3/2} V e^{-E_g/2kT}, \tag{7.5}$$

where the formation energy is arbitrarily split equally between the electron and the hole.

Note that m_e and m_h are the *effective masses* of the electron and hole as they both experience an effective potential field within the crystal. Now whether a semiconductor is intrinsic, extrinsic,[5] or partly both, the overall concentration of conduction electrons and holes must obey the law of mass action. Consequently, for electrons in band states, $e^- + h^+ = 0$.

The equilibrium constant K is given by

$$K = (e^-)(h^+), \tag{7.6}$$

and the standard free energy for a reaction that forms 1 mole of conduction electrons and holes is given as

$$\Delta F^\circ = -RT \ln K = -RT \ln(\text{p.f.})_e(\text{p.f.})_h. \tag{7.7}$$

Because of electrical neutrality, the concentrations of e^- and h^+ are given by

$$(e^-) = (h^+) = (2\pi kT/h^2)^{3/2}(m_e m_h)^{3/4} e^{-E_g/2kT}, \tag{7.8}$$

where V has been absorbed in converting e^- and h^+ into units of concentration. Although the foregoing result is valid for classical particles, it generally is not true

---

[5] Intrinsic semiconductors are pure elements or compounds (e.g., Si, Ge, Ga, As). Extrinsic semiconductors contain impurities whose oxidation potentials are such that they can extract electrons from the valence or conduction bands or donate electrons to the conduction band. They act as a stairway in energy space that effectively lowers the band gap energy.

for bound electrons, since they ordinarily do not obey Boltzmann statistics. The Pauli exclusion principle allows only two electrons in each discrete energy state within the continuum of states of a single system. As the continuum of states progressively fills with delocalized electrons from the bottom to the top of the energy band, only those states within an energy range of approximately kT of the top are available for occupancy by additional electrons. Thus, the number of available empty quantum states is not vastly greater than the number of occupying electrons. The derivation of the Fermi–Dirac distribution function, which is appropriate for bound electrons, commences by writing the expression for the total number of a priori arrangements of the n_i electrons among the z_i allowable quantum states for a complexion of energy E_i:

$$\Omega_i = z_i!/n_i(z_i - n_i)!. \tag{7.9}$$

The configurational entropy is given by

$$\Delta S_{\text{config}} = k \ln \frac{z_i!}{n_i!(z_i - n_i)!}, \tag{7.10}$$

and the total free energy of the system is written

$$\Delta F = \sum_i \left(n_i \, \Delta E_i - KT \ln \frac{z_i!}{n_i!(z_i - n_i)!} \right), \tag{7.11}$$

where ΔE_i is the total energy of the ith stationary state of the system referenced to an, as yet, unspecified standard state.

We can now replace the sum by the term that maximizes the entropy and that represents the equilibrium state of the system. Starting with

$$\Delta F = n_i \, \Delta E_i - kT \ln \frac{z_i!}{n_i!(z_i - n_i)!}, \tag{7.12}$$

where the subscript i now denotes the single complexion of the ensemble that minimizes the free energy of the system, and applying Stirling's approximation, one finds

$$\Delta F = n_i \, \Delta E_i - kT[z_i \ln z_i - n_i \ln n_i - (z_i - n_i) \ln(z_i - n_i)]. \tag{7.13}$$

The chemical potential, or partial molar free energy, is obtained by taking the derivative of ΔF with respect to n_i:

$$\frac{\partial \, \Delta F}{\partial n_i} = \mu_i - \mu_i^\circ = \Delta E_i + kT \ln \frac{n_i}{z_i - n_i}. \tag{7.14}$$

We may now drop the subscript, having specified that this particular ith complexion uniquely corresponds to the equilibrium state of the system. Then, taking the antilog of (7.14) and rearranging gives

$$n/z = 1/(1 + e^{-(\Delta\mu - \Delta E)/kT}). \tag{7.15}$$

We note that if $z \gg n$, that is, if the number of allowable quantum states greatly exceeds the number of electrons, (7.14) immediately reduces to the Boltzmann distribution function. Equation (7.15) is the Fermi–Dirac distribution function, and $\Delta\mu$, the *chemical potential* of the electron, is the so-called *Fermi energy* E_F. This equation is further simplied by taking the energy differences from the top of the valence band, which then by definition becomes the standard state. For convenience, we take this value to be zero. Now, for an intrinsic semiconductor, the energy required to form an n–h pair is identical to E_g, the band gap energy. We further assume that z, the density of states, is the same at the top of the valence band and at the bottom of the conduction band. The expressions for electrons and holes are then given by

$$n/z = 1/(1 + e^{-(E_F - E_g)kT}) \tag{7.16}$$

and

$$h/z = 1/(1 + e^{E_F/kT}). \tag{7.17}$$

Because of electrical neutrality, $n = h$ in an intrinsic semiconductor. Thus, we are able to equate (7.4) to (7.5) and solve for E_F:

$$E_F = \frac{1}{2} E_g + \frac{3}{4} kT \ln \frac{m_h}{m_e}. \tag{7.18}$$

However, the last term is very small and is generally neglected, so that the Fermi energy is considered to be located at the center of the energy gap.

The foregoing should not be interpreted to mean that E_F is equal to $E_g/2$ in all cases, because the addition of impurities certainly shifts E_F, but this relation is valid for intrinsic semiconductors and pure stoichiometric compounds. A graphical representation of the electron distribution in the valence and conduction band is shown in Fig. 7.20. The conservation of electrons requires that $-\Delta n_V = +\Delta n_C$. Furthermore, conservation of energy requires

$$-\Delta E_V = \int_V n_e(E)\, dE = \Delta E_C = \int_C n_e(E)\, dE. \tag{7.19}$$

The average energy required to form an n–h pair varies from $(E_C - E_V)$ for the initial electrons removed from the valence band to $(E_C + \Delta E_C) - (E_V - \Delta E_V)$ for the last. If, as is generally the case, $\Delta E_C + \Delta E_V \ll E_C - E_V$, so that ΔE_C and ΔE_V may be neglected (see Fig. 7.20), then the formation of the electron–hole pair is just E_g, of which one-half of this band gap energy is associated with each. Hence, as before $\mu_e = E_g/2$. It should be noted that the foregoing contains the approximation that the

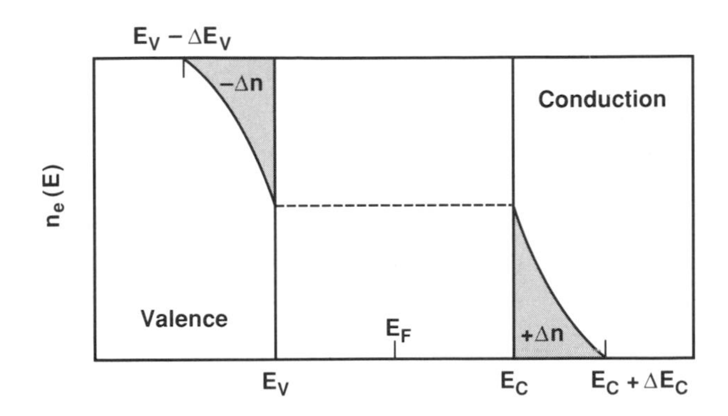

Figure 7.20.

The distribution of electrons in the valence and conduction bands. E_V is the energy at the top of the valence band, E_C is that for the bottom of the conduction band. The conservation of electrons requires $-\Delta n_V$ in the valence band to equal $+\Delta n_C$ in the conduction band. This schematic representation greatly exaggerates the magnitudes of ΔE_C and ΔE_V.

energy level spacings of the valence and conduction bands are of similar magnitude, or, in other words, the density of states is nearly equal in each band.

We next turn our attention to the underlying physics of band theory before returning to a discussion of specific semiconductor materials. This will also lay the groundwork for our study of metallic solids that follows in Chapter 8.

5. Electrons in a Periodic Potential

In Chapter 6 we derived the quantum mechanical description of free electrons in a constant potential energy well, the so-called Sommerfeld free-electron model. However, a realistic model of the conduction electrons in a crystalline solid must take into account the spatially varying potential that reflects the periodicity of the atomic or ionic configuration. It is the periodicity of the potential that causes the energy levels to separate into discrete bands, each of them containing multiple closely spaced energy levels. This has been illustrated for the three major mechanisms of electron conduction in Fig. 7.19.

The discrete energy levels that make up the various bands are themselves a consequence of the Pauli exclusion principle. When the atoms composing a crystal are sufficiently separated to be noninteractive, the Pauli principle assigns two electrons to each localized atomic orbital. As the constituent atoms coalesce into a crystal, the bonding interatomc atomic orbitals overlap as though a macromolecule were being formed. The Pauli principle then continues to apply to these crystal orbitals that are discrete, though closely separated and, as always, contain two electrons of opposing spins. Just as was the case in the formation of molecules where a molecular orbital is formed from each of the overlapping atomic orbitals, even so there is a one-to-one correspondence between the energy levels of the crystal orbitals and the atomic ones. This is illustrated schematically in Fig. 7.21. The inner atomic orbitals do not overlap with those of their neighbors and thus do not participate in the band structure and are termed "nonbonding." A more explicit diagram explaining the origin of the energy bands is shown in Fig. 7.22.

Although the example of band formation by Mg_2Sn relies on molecular orbital theory, bands are also formed by nearly free, nonlocalized electrons for which the MO model is totally inappropriate. In order to see how this comes about, we must quantify our model.

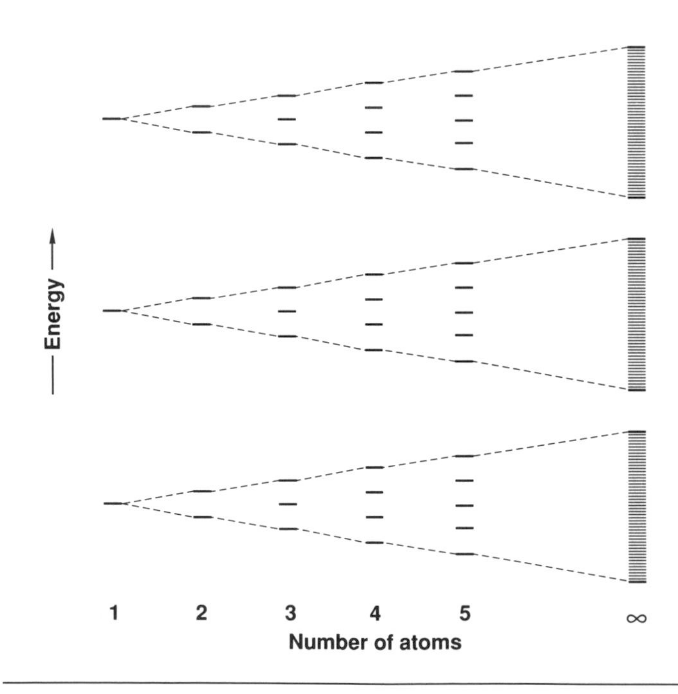

Figure 7.21.

A schematic illustration of the formation of energy bands of closely spaced energy levels. Each atomic orbital gives rise to a single level (crystal orbital) that can hold a maximum of two electrons with opposite spin directions. (After C. A. Coulson, *Valence*, Oxford University Press, p. 280, 1952, by permission of the Oxford University Press).

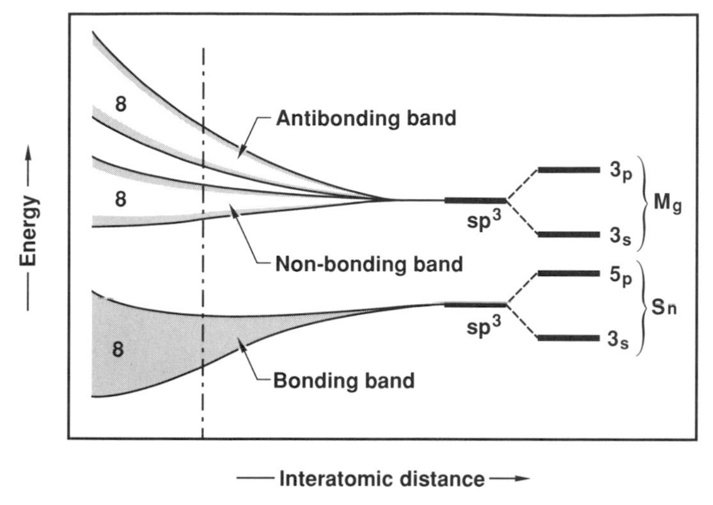

Figure 7.22.

Band structure of magnesium stannide, Mg_2Sn. Each of the four sp^3 bonds creates a crystal orbital that contains two electrons and, hence, results in eight states per band per formula unit.

A. Bloch Functions

Here we seek to find the appropriate expressions that define the energies of electrons under the influence of a periodic potential. Thus we need a solution to the Schrödinger equation where V, formerly a constant, now varies spatially in a periodic fashion. As usual, we begin with a one-dimensional model that is a simple linear chain of atoms separated by distances a_0. Thus, the problem is to solve

$$\frac{d^2\psi}{dx^2} + \frac{2m}{\hbar^2}[E - V(x)]\psi = 0, \tag{7.20}$$

where also

$$V(x) = V(x + a_0). \tag{7.21}$$

Let there be N atoms in the chain. Then, a translation of the wave function by $R_n = na_0$, where $0 \leq n \leq N$ leaves the wave function invariant. Let $\hat{R}$ be an operator that translates $\psi(x)$ and operates between the limits of 0 and a_0. Since all values of $\psi(x)$ are included between these limits, it would be superfluous to extend $\hat{R}$ beyond this range.

Now the eigenfunction can be written[6]

$$\hat{R}\psi(x) = \lambda\psi(x). \tag{7.22}$$

However, by definition,

$$\hat{R}x = x + a_0. \tag{7.23}$$

Therefore,

$$\hat{R}\psi(x) = \psi(x + a_0), \tag{7.24}$$

and, accordingly,

$$\psi(x + a_0) = \lambda\psi(x). \tag{7.25}$$

It can be reasoned that $\psi(x)$ is an exponential function of x since the addition of a_0 to x is equivalent to multiplying $\psi(x)$ by the constant λ. For the same reason,

$$\psi(x + na_0) = \lambda^n\psi(x). \tag{7.26}$$

If we impose periodic boundary conditions such that

$$\psi(x) = \psi(x + Na) \tag{7.27}$$

(this is often done by imagining the ends of the chain of atoms to be connected so as to form a continuous necklace), then, because of (7.26) and (7.27), and the exponential form of $\psi(x)$, we can deduce that

$$\lambda = e^{2\pi in/N}. \tag{7.28}$$

It is now customary to define a variable k, which is a wavenumber, by the relation

$$k = 2\pi n/Na_0, \tag{7.29}$$

where n assumes only integral values (viz. $\pm 1, \pm 2, \ldots$; sometimes the 2π is omitted from the definition and remains explicit in the exponent). The periodic wave function then becomes

$$\psi(x + na_0) = e^{ika_0}\psi(x), \tag{7.30}$$

since it is immaterial where one commences the translation R_n, as long as the period is a_0. A more general form of (7.30) can be written as

$$\psi_k(x) = e^{ika}U_k(x), \tag{7.31}$$

[6] This is derived in Appendix (F) and verified later in this section.

where $U_k(x)$ is any function also having the periodicity of the lattice. Equations such as (7.31) define the so-called Bloch functions. For our purpose $U_k(x)$, sometimes called the cell function, is the fully defined electron wave function that is confined to $0 \leq x \leq a_0$ in the one-dimensional case. Hence, $U(x) = U(x + na)$.

We can now verify (7.26). For the sake of brevity, let us define the indefinite translation by $R_n = na_0$. Then

$$\begin{aligned}\hat{R}(x) &= \exp[ik(x + R_n)]U(x + R_n) \\ &= \exp(ikR_n)\exp(ikx)U(x) \\ &= \exp(ikR_n)\psi(x) = \lambda^n\psi(x) \\ &= \exp \psi(x + R_n),\end{aligned} \tag{7.32}$$

which is in agreement with (7.26).

Looking back at (7.29), it is obvious that if the atoms are at infinite separation, i.e., $a_0 \to \infty$, then $k \to 0$. The cell functions then become the free-atom atomic wave functions, as indicated schematically in Fig. 7.21. The significance of Bloch's function is that it shows the proper form for the solution of the Schödinger equation with a periodic potential. Although it does not give us the actual wave functions, it is indispensible in developing ways of approximating them.

As stated before, each of the discrete overlapping atomic orbitals gives rise to an energy band, and each atom in the crystal contributes a single energy state to each of these bands. This distribution is diagrammed for a fictitious elemental crystal in Fig. 7.23. The outer valence electrons have greater mutual overlap of their wave functions, which produces a narrower band than the less interacting inner shells. This is duly illustrated in Fig. 7.23.

B. Energy Gaps—The Nearly Free Electron Model

In the nearly free electron model we assume the periodic potential to cause only a weak perturbation of the motion of the electrons. It will be shown that energy gaps result from this model. The free, though confined, electron gas of the Sommerfeld model has quantized energy levels defined by (6.21). This equation can be rearranged to become

$$E_n = \frac{\hbar^2}{2m}\left(\frac{2n\pi}{L}\right) = \frac{\hbar^2}{2m}\left(\frac{2n\pi}{Na_0}\right) = \frac{\hbar^2 k_n^2}{2m}. \tag{7.33}$$

Because of the close spacing between energy levels, k_n varies in a quasi-continuous fashion and is usually treated as a continuous variable, as shown in Fig. 7.24. Because

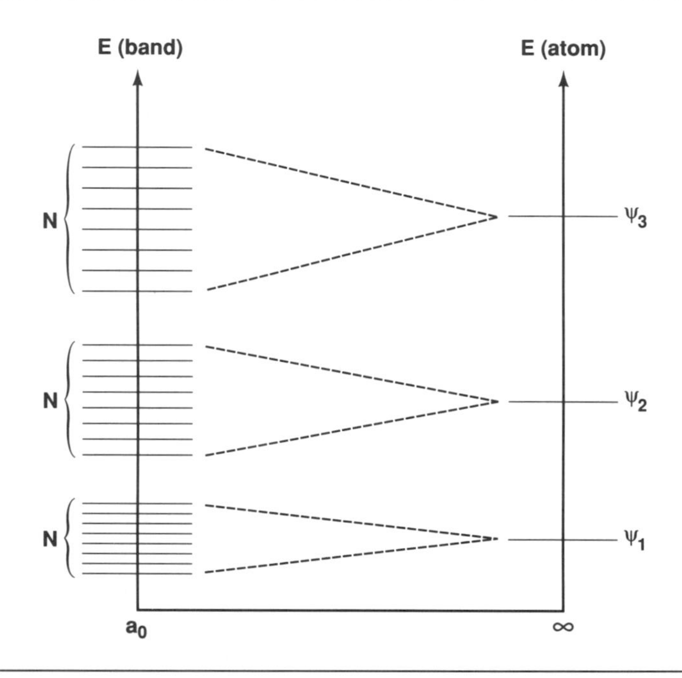

Figure 7.23.

Schematic illustration of the broadening of the localized atomic levels E_a of the isolated atoms into separated bands E_b at the interatomic distance a_0. Each of the N atomic orbitals contributes a single band state.

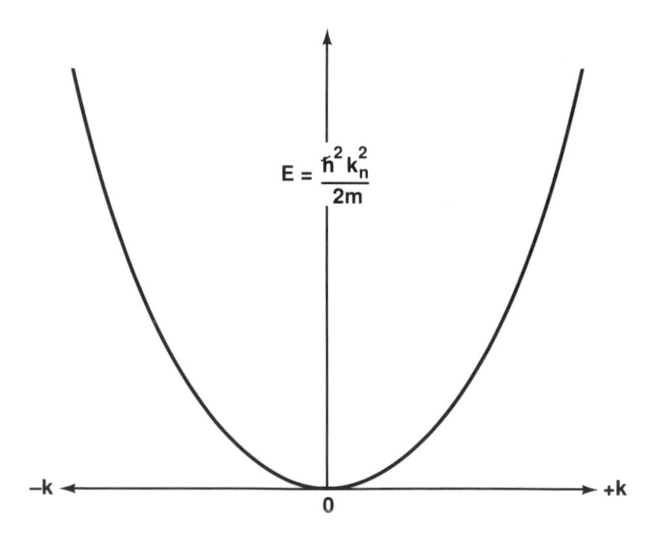

Figure 7.24.

A graph of energy *vs.* wavenumber for a free-electron gas.

$E \propto k_n^2$, the energy is the same for $\pm k_n$ for equal values of n. This degeneracy persists even when the electron energy is perturbed by a periodically varying potential.

It can be seen from its definition, (7.29), that k_n has repetitive intervals proportional to a_0^{-1}, and, as a result, E_n mirrors this periodicity. It is next important to recognize that for a chain of N atoms the total range of k_n is $0 \leq k_n \leq 2\pi/a_0$. This becomes apparent when one writes out the sequence of permitted values for $n = 0, 1, 2, \ldots$

$$k_n = \frac{2\pi n}{Na_0} = 0, \frac{2\pi}{Na_0}, \frac{4\pi}{Na_0}, \ldots, \frac{2\pi(N/2 - 1)}{Na_0}. \tag{7.34}$$

Thus, in k-space, as it is generally called, all values of the energy for each band are contained within the first zone, which runs from zero to $2\pi/a_0$.

The complete wave function can be written as

$$\psi = c_1\psi_k(x) + c_2\psi_{-k}(x), \tag{7.35}$$

where the wave functions are themselves given by

$$\psi_k(x) = L^{-1/2}e^{2\pi ikx}, \qquad \psi_{-k}(x) = L^{-1/2}e^{-2\pi ikx}.$$

These two functions account for the fact that a standing wave may have either a maximum or minimum value at the zone boundary (see Fig. 7.25). These represent plane waves moving in opposite directions. Because the nearly free electrons share the same space (length), viz. $L = Na_0$, the overlap functions S_{ij} are orthonormal.

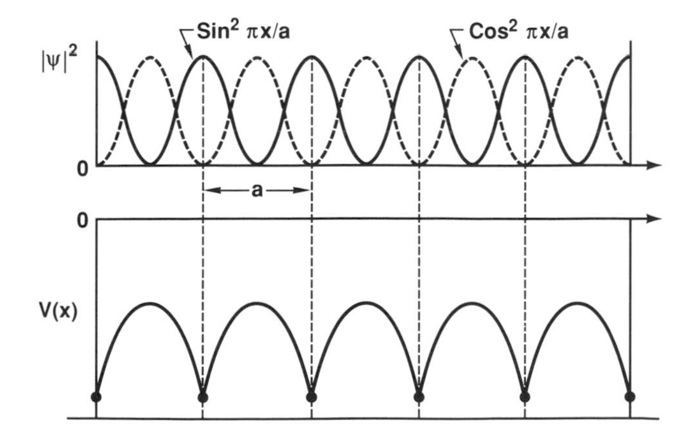

Figure 7.25.

The electron density is proportional to $|\Psi|^2$ and varies in a periodic fashion, with its maxima centered on or midway between the ion cores in this one-dimensional lattice.

That is, as a result of the Pauli principle,

$$S_{ij} = \int \psi_i \psi_j^* \, d\tau = \delta_{ij}, \tag{7.36}$$

where

$$\delta_{ij} = 1 \qquad \text{for } i = j,$$
$$\phantom{\delta_{ij} = {}} 0 \qquad \text{for } i \neq j.$$

is the Kronecker delta. The energy is given by

$$\Sigma = \int \psi \hat{H} \psi \, d\tau \bigg/ \int \psi \psi \, d\tau, \tag{7.37}$$

where ψ is given by (7.35). Using the notation introduced in Chapter 6, Section 3A, we can write (7.37) as

$$E = \frac{C_1^2 H_{11} + 2C_1 C_2 H_{12} + C_2^2 H_{22}}{C_1^2 + C_2^2}. \tag{7.38}$$

Minimizing the energy given by (7.38) with respect to the mixing coefficients C_1 and C_2 leads to

$$\begin{aligned} \frac{\partial E}{\partial C_1} &= 0 = C_1(H_{11} - E) + C_2 H_{12}, \\ \frac{\partial E}{\partial C_2} &= 0 = C_1 H_{12} + C_2(H_{11} - E). \end{aligned} \tag{7.39}$$

Hence, the secular equation is

$$\begin{vmatrix} H_{11} - E & H_{12} \\ H_{12} & H_{11} - E \end{vmatrix} = 0. \tag{7.40}$$

This leads us to find two values for the energy, given by

$$E_{\pm} = H_{11} \pm |H_{12}|. \tag{7.41}$$

Substituting into either of Eqs. (7.39), the two values for E make it clear that

$$\begin{aligned} E_+ &= H_{11} + H_{12} \qquad \text{and} \qquad C_1 = C_2, \\ E_- &= H_{11} - H_{12} \qquad \text{and} \qquad C_1 = -C_2. \end{aligned} \tag{7.42}$$

The two wave functions that result, ψ_k and ψ_{-k}, are then

$$\psi_k = e^{2\pi ikx} + e^{-2\pi ikx}, \qquad \psi_{-k} = e^{2\pi ikx} - e^{-2\pi ikx}. \tag{7.43}$$

Converting these exponential functions to trigonometric ones gives

$$\psi_k = \cos 2\pi kx, \qquad \psi_{-k} = \sin 2\pi kx. \tag{7.44}$$

The spatial variation of electron density in response to the oscillating potential can be visualized with the aid of Fig. 7.25. Note that the cosine function $|\Psi_+|^2$ will always have lower energy since the electron density has its maxima at the ion core where the potential energy is a minimum, whereas the maxima of $V(x)$ and $|\Psi_-|^2$ coincide. Since both functions represent solutions at the boundary of the zone, the difference between their energies produces a gap.

There are three ways of representing electron waves in k-space: the so-called extended zone scheme that plots all values of $E(k)$ in a continuous and repetitive manner, the reduced zone scheme that makes use of the fact that all energies for a given band can be found in the first zone, and the repeated zone scheme, which plots all values and is the background for the extended and reduced schemes shown in Figs. 7.26 and 7.27. Plotting the energy in the extended zone with the free-electron parabola superimposed will allow us to discern the origin of the band gap energy. $E(k)$ follows the free-electron values except when near the zone boundary. The origin

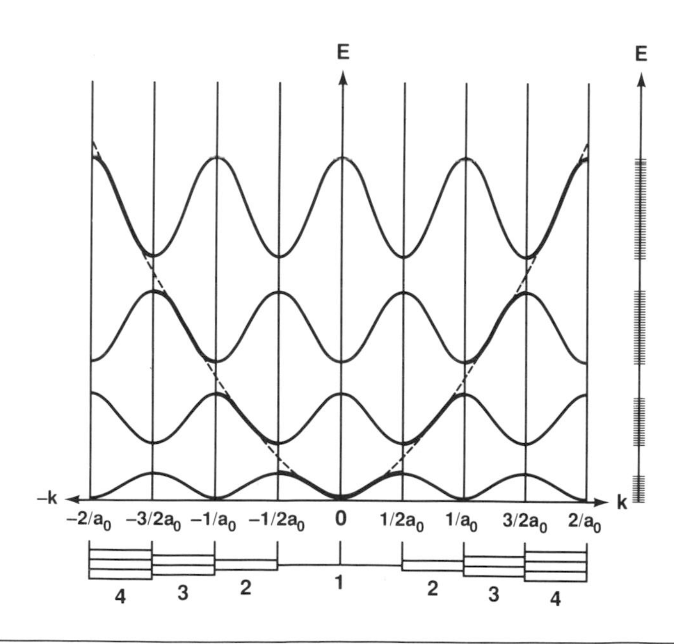

Figure 7.26.

The free-electron parabola (dashed line) is superimposed upon the extended zone plot. The sections belonging to the same Brillouin zone are emphasized by the heavy lines and are designated by the numerals 1, 2, 3, 4.

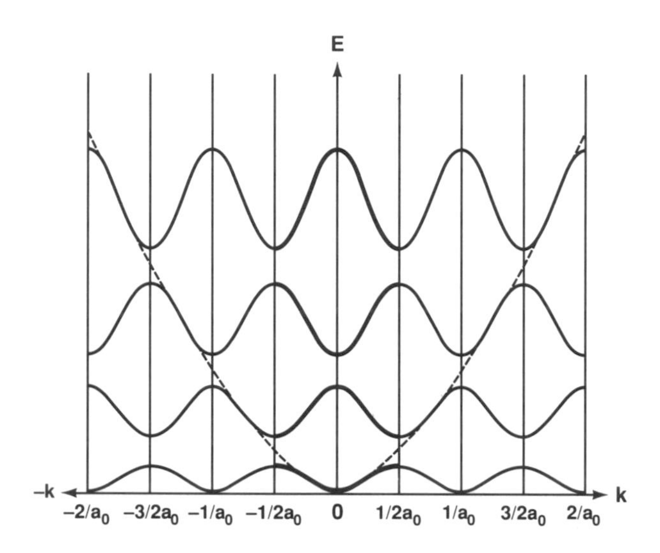

Figure 7.27.

The energy values as a function of k_n are contained within the range of $1/a_0$ in the reduced zone method.

of the k_n values in the zone is translated in a manner that takes account of the positive and negatives values of k. Instead of ranging over the values reported by (7.34), the origin is shifted so that

$$k_n = -\frac{1}{Na_0}, \ldots, -\frac{1}{2a_0}, 0, \frac{1}{2a_0}, \ldots, \frac{1}{Na_0}. \tag{7.45}$$

The zone for the first band runs from $-1/2a_0$ to $1/2a_0$ and contains N eigenfunctions. The zone for the second band extends from $-1/2a_0$ to $-1/a_0$ and from $1/2a_0$ to $1/a_0$ and contains the same number of eigenfunctions. Thus, all zones in R or reciprocal space are $1/a_0$ in length.

The reduced zone scheme is plotted in Fig. 7.27. This can be thought of as a simple translation of the successively higher band back into the first zone with all centered about the origin. It should be clear that $E(k)$ is, as it must be, the same in either form of representation. The important point that is made by both graphs is the existence of energy gaps that occur with a periodicity of $1/2a_0$ within which all energy states are forbidden.

There is inadequate space to derive here the expressions for Brillouin zones in three dimensions. Suffice it to say that the one-dimensional model is replaced by a three-dimensional vectorial representation of distance, which leads to polyhedra of various shapes that reflect the symmetry of the parent crystal. Other important omissions from our discussion include the tight-binding approximation and structure of Brillouin zones in compounds. However, the important features remain unaltered. That is, energy bands are created in all cases.

C. Band Filling and Brillouin Zones

Staying with the linear chain model, let us see how the Brillouin zones or energy bands fill as we increase the number of valence electrons per atom. The first two bands are plotted in Fig. 7.28. As we know, there is one energy state in the band for each participating atomic orbital, and, like the atomic orbitals, this band state can hold two electrons with opposing spins. Consequently, N monovalent atoms, such as the alkali metals, will fill half the first band, with the Fermi level being designated as E_1 In like manner, the alkaline earth metals, Be, Mg, Ca, etc., donate two electrons into the conduction band, filling it to the level E_2. At this point, all N levels of the first zone are filled. Moving on to the trivalent group, IIIB elements, e.g., Al, Ga, In (B is omitted because of its covalent bonding) fill the first zone and half fill the second zone.

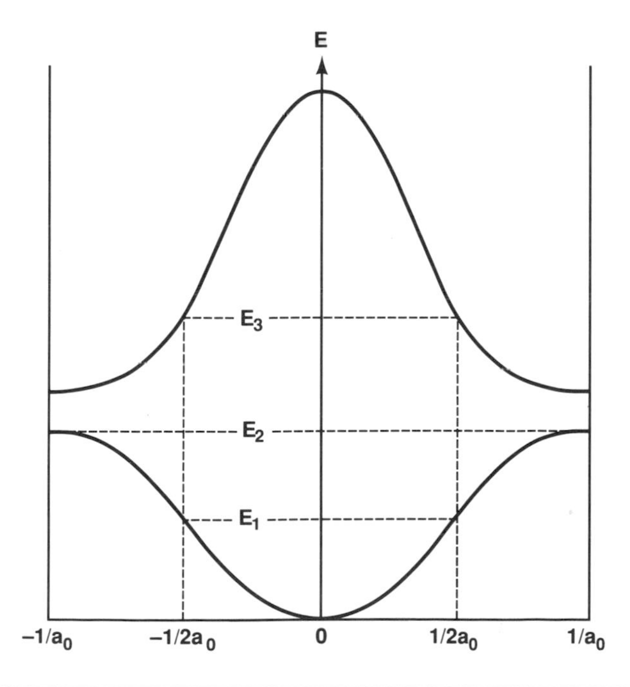

Figure 7.28.

Filling the first and second bands with electrons. (After S. L. Altman, *Band Theory of Metals*, reprinted with permission from Pergamon Press, PLC, p. 99, Copyright 1970).

The geometry becomes considerably more complicated in three dimensions, although no new physical principles need be introduced. An example of a Brillouin zone in three-dimensional k-space is that for the bcc structure as depicted by Fig. 7.29. The filling of the Brillouin zone in three dimensions can be imagined as follows: the coordinates of the electron gas in k-space (see Fig. 6.2) have their origin at the center of the zone. The Fermi sphere continues to expand with the addition of delocalized electrons as dictated by the number of valence electrons belonging to the element in question. Ultimately the Fermi level intersects the nearest Brillouin surfaces and encounters an energy gap at that point. Further expansion is limited to the residual volume available in the first zone, which restricts the expansion of the sphere. This is illustrated for two dimensions by Fig. 7.30. We can see from Fig. 7.30 that in two or three dimensions the band gaps depend upon the direction. Also, because of the increase in the density of states and because the electrons have no component of momentum normal to the zone face, the Fermi sphere flares out and flattens in the immediate vicinity of the Brillouin surface. Another difference between the one-dimensional case and those of higher dimensionality is that in the latter adjacent zones can overlap in certain directions. As a consequence, elements such as Be, Mg, and Ca that had appeared to be insulators in the one-dimensional description are in fact perfectly good metals. For example, in Fig. 7.30, if the band gap energy E_g across the six points of tangency of E_3 is such that $E_3 + E_g < E_a$, where E_a is the maximum energy at the six apical points, such as the one labeled a, then the next

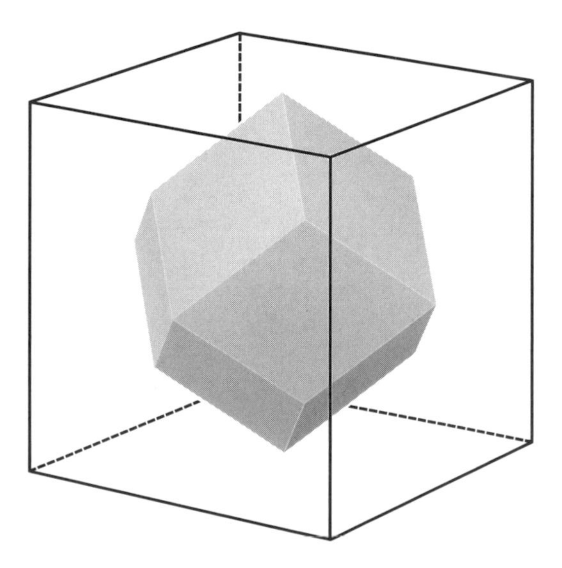

Figure 7.29.

The first Brillouin zone for the body-centered cubic structure.

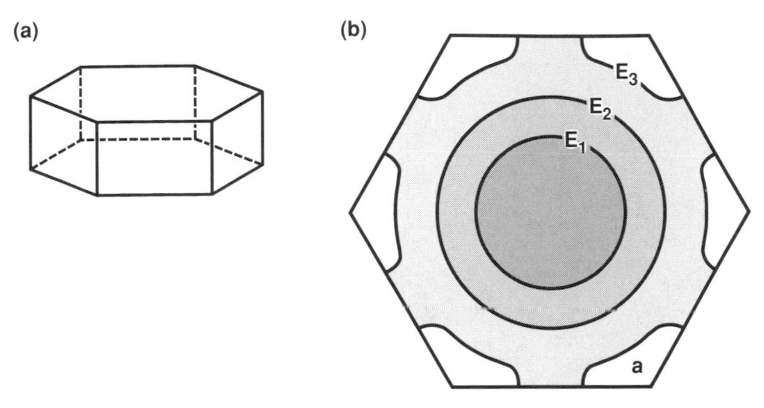

Figure 7.30.

(a) Three-dimensional first zone of graphite; (b) three distinct Fermi spheres of free electrons marked E_1, E_2, E_3 lie within the hexagonal Brillouin zone; the outermost, E_3, intersects the zone wall.

higher zone will begin to fill before the lower one has been completely filled. This has important consequences for the electrical properties of the solid.

The density of available states for electron occupancy is described by the sequence of graphs in Fig. 7.31. These drawings do not represent any particular solids, but contain the essential features that characterize insulators, semiconductors, and metals. The crosshatched areas indicate the electron distribution at $T > 0$ K. Looking at Fig. 7.31, we see that the expanding sea of nearly free electrons first intersects a zone surface at the maximum value of $N(E)$. The density of states starts out following the parabolic relation for a free-electron gas (a), but increases in the proximity of the band gap. This reflects the reverse in curvature that is clearly seen in the E versus $(a_0)^{-1}$ curves that occur near the half-integral values of $(na_0)^{-1}$ (see Figs. 7.26 or 7.27). Case (b) shows a partially filled zone, which in itself satisifies the conditions necessary for metallic conduction, since an electric current can flow through the continuum of unfilled states immediately above the filled ones. The applied potential is sufficient to unpair the electrons in the topmost levels and transport them via the unfilled states. However, even if this zone were completely filled up to E_1, this would still be a conductor because of the available empty states in the overlapping second zone. On the other hand, an insulator such as shown in case (c) has a substantial band gap between the completely filled zone and the closest-lying available empty states. A semiconductor, portrayed in (d), has a band gap of about 1 eV or less. Consequently, electrons can be thermally excited from the filled band into the empty conduction band at room temperature, where they are then free to move under the influence of an applied potential.

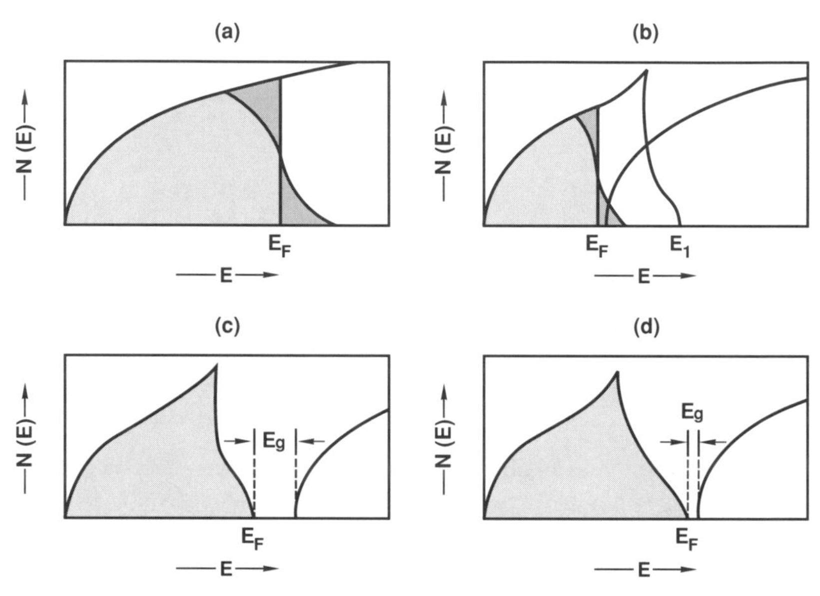

Figure 7.31.

The four fundamental configurations for the density of electron states, $N(E)$, in solids. The shaded areas designate occupied regions: (a) the purely hypothetical free electron gas at 0 K; (b) a metallic conductor; (c) an insulator with large E_g; (d) a semiconductor with small E_g.

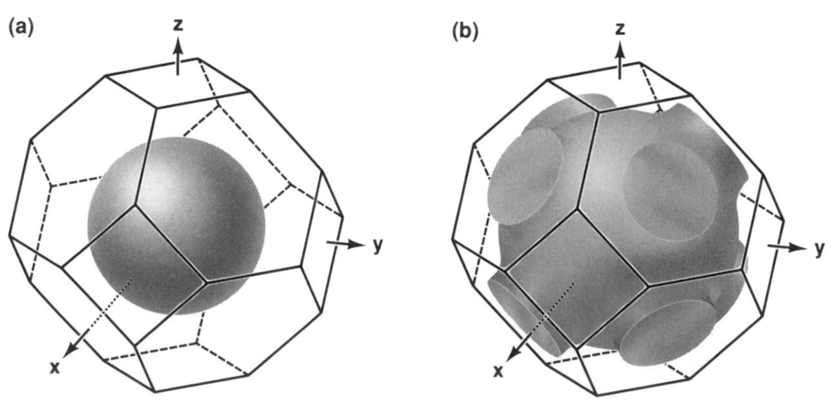

Figure 7.32

The first Brillouin zone of the face-centered cubic unit cell. (a) The number of occupied states is small and the Fermi surface has not intersected the zone surface. (b) The Fermi surface intersects the nearest zone faces with a resulting deformation of the free-electron sphere. (After W. Hume-Rothery, *The Structure of Metals and Alloys*, The Institute of Metals, p. 34, 1956.)

Finally, to extend our visualization of the expanding sea of electrons within the first Brillouin zone, we depict the situation for the fcc structure in Fig. 7.32.

6. Elemental Intrinsic Semiconductors

This discussion is limited to semiconductors having sp^3 tetrahedral bonding. There are many others that do not have this structure, but elemental Si and Ge along with the so-called III-V compounds will adequately demonstrate the fundamental concepts.

It is a necessary requirement for semiconduction that there be no overlapping or partially filled energy bands; otherwise metallic conduction would result, as explained in the last section. This is nicely illustrated by the group IVB elements, all of which, with the exception of Pb, are diamond cubic semiconductors. The atomization or sublimation enthalpy is a direct measure of the bond strength, and its values show the same trend as the respective band gap energies, as shown in Table 7.6.

The first four elements of group IVB have structures that are isotypic with diamond forming as they do strong covalent bonds. However, as the atomic number increases and the outermost electrons become better screened from the direct nuclear charge, the element gains in metallic characteristics. It is believed that the exaggerated difference in the band gap energy of diamond as compared to Si and Ge is caused by the screening provided by the $2p^6$ electrons of the completed $n = 2$ shells of Si and Ge. Because of the required orthonormality between the $2p$, $3p$, and $4p$ wave functions, the p electrons of these elements are forced much farther from their nuclei, thus encouraging their delocalization with accompanying reduction of the band gap. Tin is the borderline case, and is diamond cubic and a semiconductor below 286 K while being a body-centered tetragonal metal at higher temperatures. The group IVB elements terminate with Pb, which has only the face-centered cubic structure, and is a true metal.

Table 7.6. Group IVB elements: band gap and atomization energies.

Element	Band Gap (eV)	Atomization Energy (eV)
C	5.5	7.35
Si	1.1	4.21
Ge	0.7	3.50
Sn	0.1	3.1
Pb	0.0	2.0

Two major factors that influence the transition of these elements from insulator → semiconductor → metal are the energy differences between the atomic *s* and *p* orbitals and the interatomic distances. When the energy difference between orbitals is large, as in the case of Pb, hybridization is not favored, and this element is octahedrally coordinated by its purely *p* electron bonds. At the other extreme, the sp^3 bonds formed by diamond are extremely strong, and the antibonding conduction band is so much higher in energy that it is a true insulator.

7. Compound Semiconductors

Mooser and Pearson[7] have proposed a set of rules that we quote here that govern the bonding traits of semiconductors:

(1) The bonds are predominantly covalent.
(2) The bonds are formed by a process of electron sharing which leads to completely filled *s* and *p* subshells on all atoms in elemental semiconductors, although in compound semi-conductors it is only necessary that one or two atoms bonded together should contain a filled subshell.
(3) The presence of vacant "metallic" orbitals on some atoms in a compound does not prevent semiconductivity provided that these atoms are not bonded together; it may, however, give rise to pivotal resonance of the bonds in the molecule, in which case the coordination numbers of the atoms exceed their normal valencies.
(4) The bonds form a continuous array in one, two, or three dimensions throughout the whole crystal.

It is possible to describe the structures of most compound semiconductors within the context of the filling of the octahedral or tetrahedral interstitial positions, as is diagrammed in Fig. 7.33a,b (see also Figs. 1.18 and 1.19), displaying an fcc anionic lattice with metallic cations.[8]. The structures that result from the filling process show either tetrahedral coordination of the cation as a result of sp^3 hybrid bonds or else octahedrally coordinated cations reflecting the symmetry of pure *p* electron bonds. Derivative or extended structures are formed from the same basic unit, and this is shown in Fig. 7.34.

The so-called III-V semiconductors were among the earliest developed and, consequently, are the best-studied of all the compound semiconductors. They are com-

---

[7] E. Mooser and W. B. Pearson, *J. Chem. Phys.* **26**, 893 (1957).

[8] The following discussion is based upon and closely follows that of E. Mooser, *Science*, **132**, 1285 (1960).

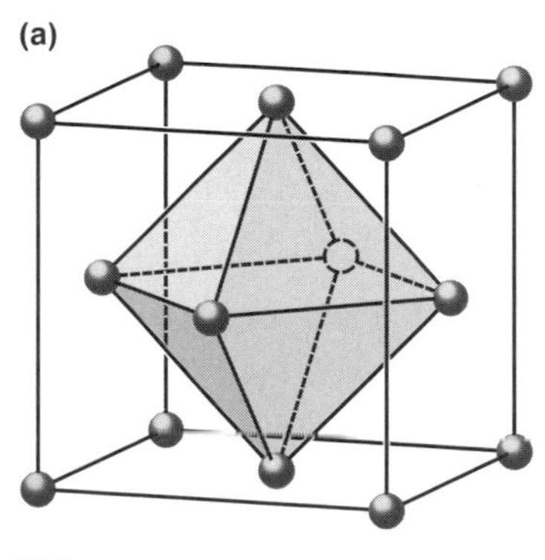

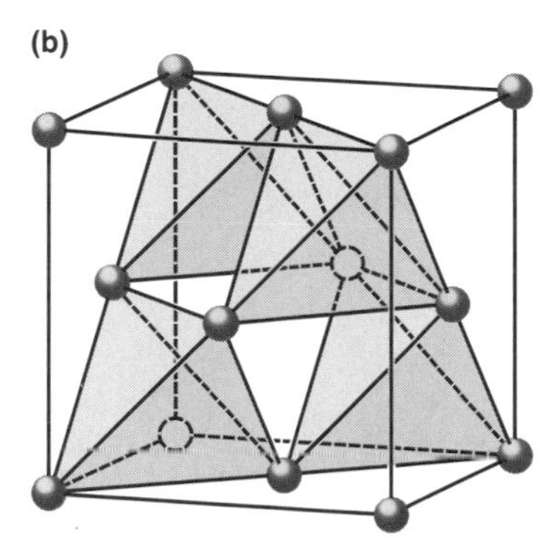

Figure 7.33.

(a) The octahedral and (b) the tetrahedral holes in the close-packed cubic structure. (After E. Mooser, *Science*, **132**, 1285, Copyright 1960 by the AAAS).

posed of equiatomic proportions of group IIIB and VB elements, and a representative sampling is listed in Table 7.7. These elements are tetrahedrally bonded again via sp^3 hybrid orbitals, and the compounds are isoelectronic with Si and Ge. They crystallize with the zinc blende structure (see Fig. 7.1), which perforce has the diamond cubic lattice. In their basic electronic and crystal structures, they are completely analogous to the group IV semiconducting elements.

The completed octets of electrons necessary for extended tetrahedrally bonded structures are also obtained from combinations of groups I and VII, II and VI, and IV and IV. These constituents also give rise to compositions that are semiconductors having the wurtzite (Fig. 7.2) as well as the zinc blende structure.

Finally, sp^3 bonded semiconductors can be obtained by varying the stoichiometry. An example of an $A_2^{II}X^{IV}$ compound is Mg_2Sn. Because Sn is more electronegative than Mg, its $5s$ and $5p$ orbital energies are lower than the corresponding $3s$ and $3p$ Mg orbital energies. This leads to the schematic band structure pictured in Fig. 7.22.

The reader should not conclude from the foregoing discussion that all semiconductors are sp^3 bonded elements or compounds. However, they do constitute the most important class.

Table 7.7. The semiconducting $A^{III}X^{V}$ compounds.

AlP	GaP	InP
AlAs	GaAs	InAs
AlSb	GaSb	InSb

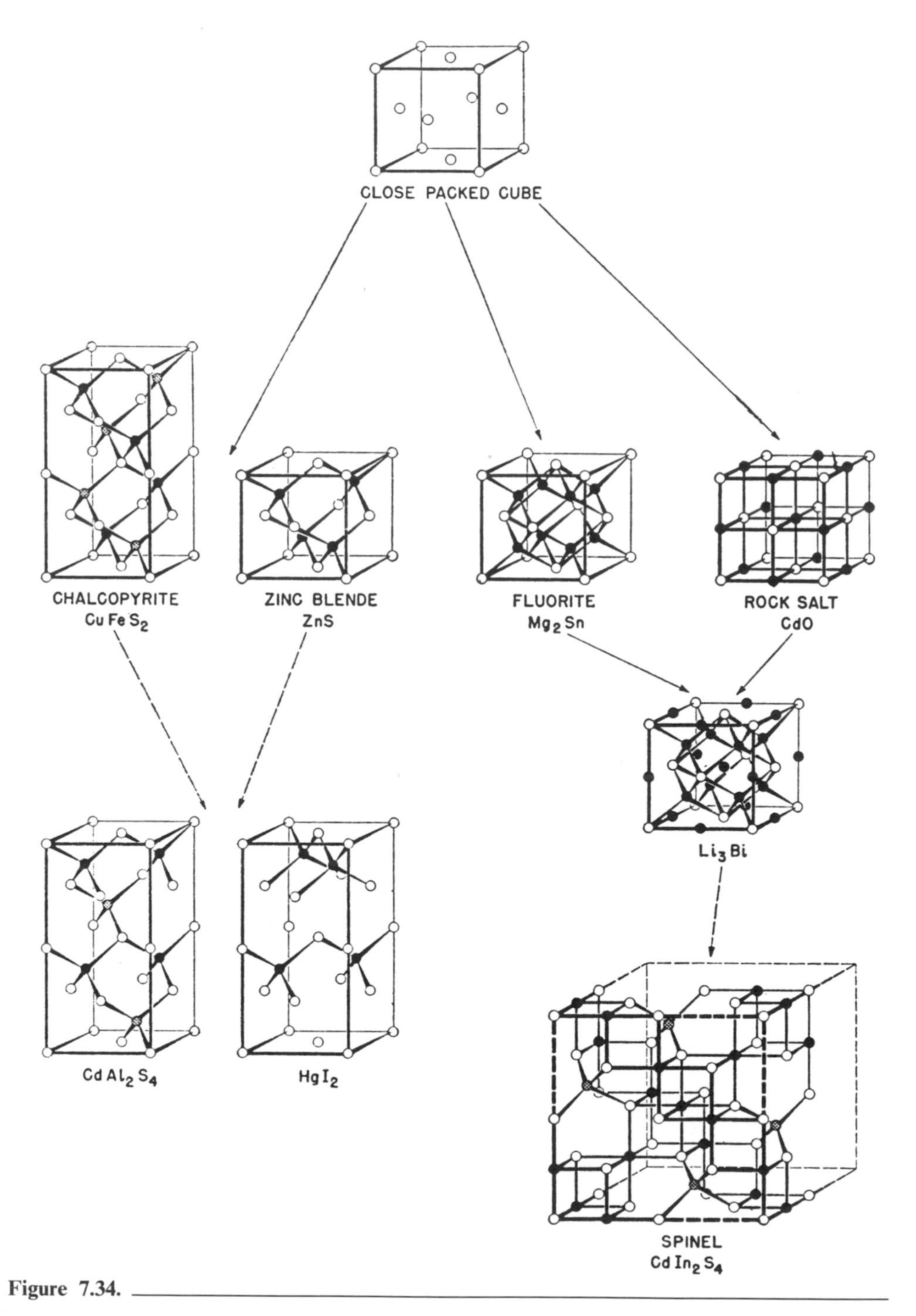

Figure 7.34.

The evolution of several examples of semiconducting crystal structures showing their relation to the elementary fcc lattice. (After Mooser, loc. cit.)

8. Extrinsic Semiconductors

Impurities can have a profound effect upon the concentration of conduction electrons, and because defects in semiconductors are frequently charged, they can influence such quantities as the vacancy concentration. The energy levels of the impurity can lie within the band gap and may be close to either the valence or conduction band, thus effectively lowering the band gap energy. The three possibilities are shown in Fig. 7.35. The actual distribution of ionized to neutral impurities will follow either Boltzmann or Fermi–Dirac statistics, which were discussed earlier in this Chapter (Section 4).

The effect of impurities whose number of bonding electrons differs from the host material can be illustrated by the substitution of pentavalent phosphorus or trivalent boron into tetravalent Si. The four outer electrons of Si form sp^3 hybridized covalent bonds. The substitution of phosphorus, with five outer electrons, leaves this impurity with one extra loosely bound electron, which is easily promoted into the conduction band. On the other hand, substitution of boron, with three outer electrons, leaves a deficit of a single electron and, hence, a hole or effective positive charge in the valence band. These electron reactions are illustrated in Fig. 7.36.

In order for the donors or acceptors to be of practical importance, they must lower the effective band gap energies at ambient room temperatures. Thus their ionization energies should be about kT, which at 298 K is 0.026 eV. The values for several popular dopants of Si and Ge are listed in Table 7.8.

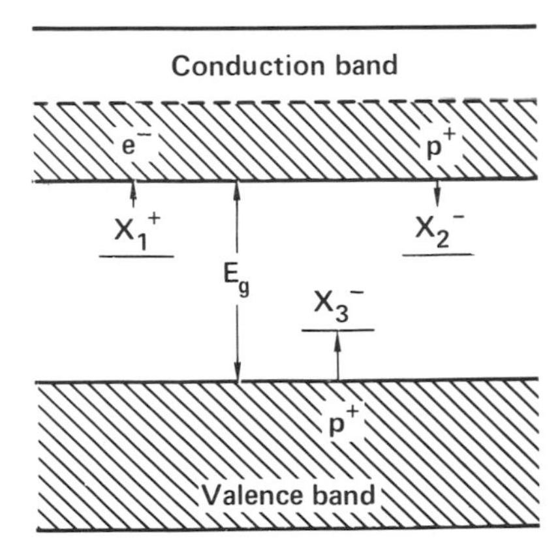

Figure 7.35.

Impurities, such as X_1, are donors becuase they will donate their electrons to the conduction band, whereas X_2 and X_3 are acceptors receiving electrons from the conduction or valence band, respectively.

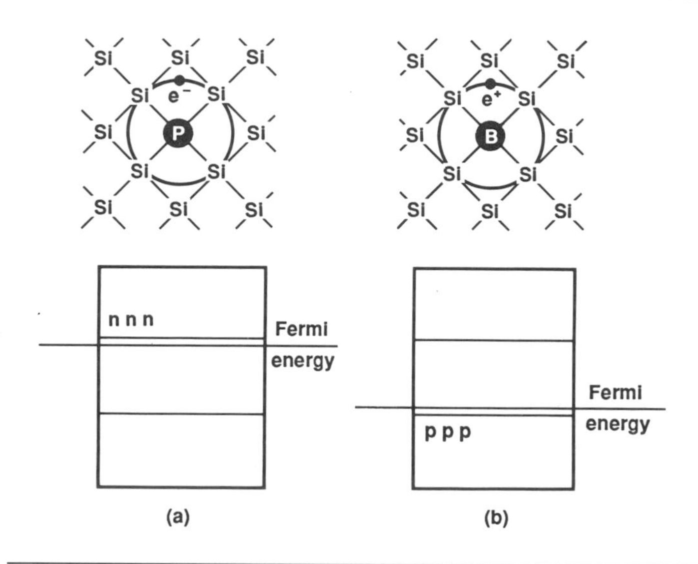

Figure 7.36.

The substitution of pentavalent P and trivalent B for Si atoms, showing how impurties may serve as donors or acceptors, thus producing the indicated shifts in the Fermi energy. (After W. Moore, *Seven Solid States*, Benjamin, New York, 1967.)

Table 7.8. Acceptor and Donor Ionization Energies.

	Acceptor Ionization Energies in eV of Trivalent Impurities				Donor Ionization Energies in eV of Pentavalent Impurities		
	B	Al	Ga	In	P	As	Sb
Si	0.045	0.057	0.065	0.16	0.045	0.049	0.039
Ge	0.0104	0.0102	0.0108	0.0112	0.0120	0.0127	0.0096

9. Graphite

Although mechanically fragile in comparison to diamond, nevertheless graphite is thermodynamically more stable except at very high pressures. This stability is reflected by the shorter interatomic distance between carbon atoms lying within the basal plane (see Fig. 1.22 for the crystal structure of graphite). This distance is 1.415 Å, which is very close to the value of benzene (1.39 Å) and is significantly smaller than the 1.54-Å value of the C—C separation in diamond. Thus, the intraplanar spacing is far smaller than for a carbon–carbon single bond, viz. $\sim$1.55 Å (re: Table 7.1), yet is appreciably larger than the 1.33-Å value of a double bond. The

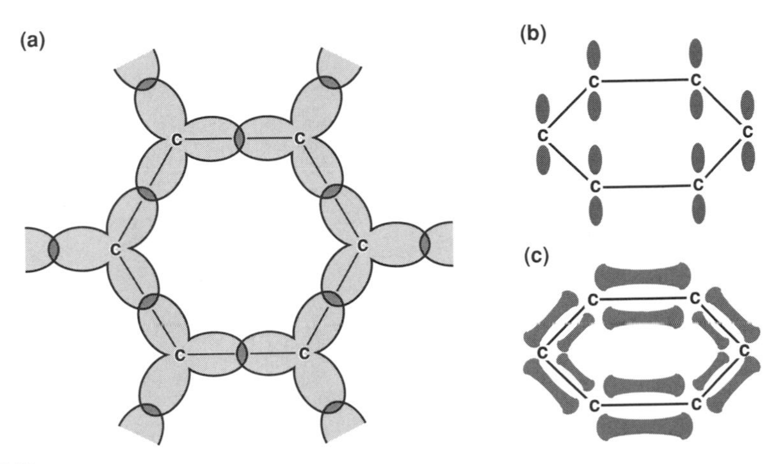

Figure 7.37.

(a) The planar σ-sp^2 trigonal hybrids bond covalently with their three nearest neighbors. (b) The fourth electron is imagined to be in fictitious p_z atomic orbitals. (c) These coalesce forming the π MO.

large interplanar space along the c-axis is 3.35 Å and is an expression of the relative weakness of the van der Waals bonding between layer planes.

Graphite finds numerous industrial uses, such as for electrical contacts, electrodes, refractory crucibles, and heat shields as well as providing fibers for reinforcing composite materials. It is nearly as good a thermal and electrical conductor as most metals within the basal plane and is an excellent insulator normal to them, and thus finds applications in both capacities. The strong directional dependence of its physical properties is clearly explainable in terms of the bonding. The $2s^2 2p^2$ electrons of carbon hybridize, forming three trigonal σ-sp^2 planar orbitals. The remaining electron half fills a π molecular orbital. These configurations are shown schematically in Fig. 7.37. Almost all of the cohesion is supplied by the σ-sp^2 bonds, and the electrons in the π orbital can be considered to occupy a two-dimensional network of interconnected Bloch states. Tight-binding calculations based upon this model reveal the absence of a significant band gap at the corners of the first Brillouin zone between the π bonding and π^* antibonding orbitals. The density of states is shown graphically in Fig. 7.38, and attention is called to the approximate contact between π and π^*.

Because the π orbital is completely filled at 0 K and the upper π^* is completely empty, graphite behaves as a true intrinsic semiconductor. Because the band gap is vanishingly small, [9] the slightest increase in temperature promotes electrons into the

[9] Graphite is perhaps a semimetal without any band gap whatsoever.

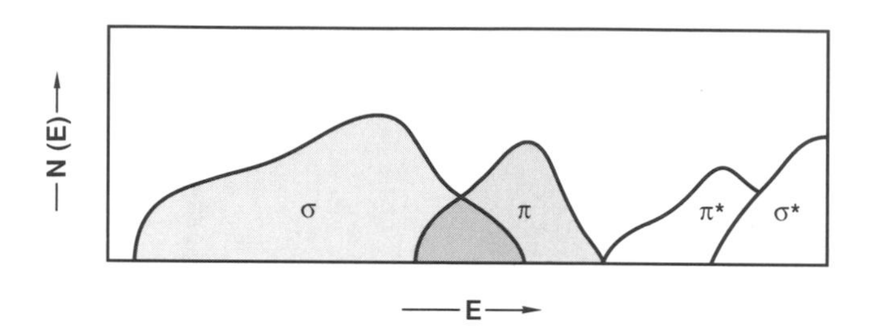

Figure 7.38.

The density of states of graphite as a function of energy.

π^* MO, transforming it into a conductor. The extreme interplanar separation precludes electronic conduction normal to the basal planes. Finally, because heat is transported primarily by the delocalized electrons so that the ratio of thermal to electrical conductivity is directly proportional to temperature (i.e., the Wiedemann–Franz law), graphite is an excellent thermal insulator in the *c*-direction and, conversely, an excellent conductor in the *a*-direction. The conducting and insulating properties, as would be expected, become more extreme as the graphite crystal becomes more perfectly ordered and the number of accidental cross-linkages between planes decreases.

10. Intercalate Compounds

> The term intercalation literally refers to the act of inserting into the calendar some extra interval of time, such as February 29 in a leap year. In chemistry, it describes the reversible insertion of guest species into a lamellar host structure with maintenance of the structural features of the host."[10]

Intercalation offers an extremely versatile method of altering the structural, electronic, optical and chemical properties of a solid.[11] Intercalation reactions can occur with a wide variety of layered structures, including simple metal oxides and sulfides, e.g., MoO_3, TiS_2, clays, intermetallic compounds, and others. Intercalates offer yet another interesting example of mixed bonding. The host compound is essentially

[10] M. S. Whittingham, *Intercalation Chemistry* (M. S. Whittingham and A. J. Jacobsen, eds.), Chap. 1, p. 1, Academic Press, 1982.

[11] Intercalation reactions are also possible at the molecular level but are outside the subject matter of this treatise.

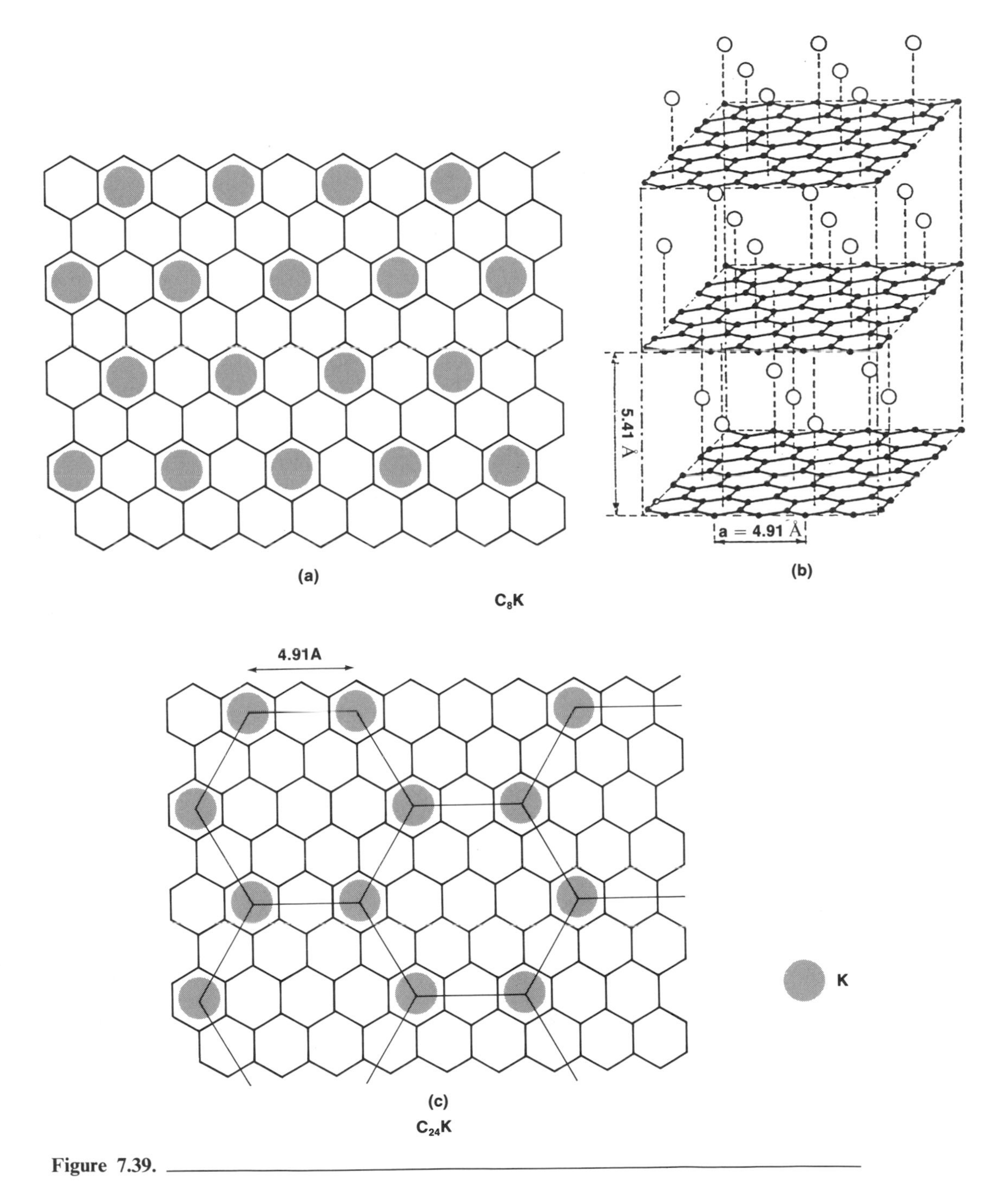

Figure 7.39.

(a) Two-dimensional portrayal of the K position of C_8K; (b) a three-dimensional exploded view of C_8K; (c) two-dimensional representation of $C_{24}K$. (Adapted from *Nouveau Traité de Chimie Minérale*, vol. VIII, Figs. 8, 9 pp. 387–388, Masson, Éditeurs, 1968.)

covalent, and the intercalating component can bond to it ionically or, in the case of graphite, with metallic-type bonds. Here, we will examine intercalation reactions only with graphite, which has been the most studied of all.

Graphite can accommodate both oxidizing and reducing agents, e.g., certain halogens or alkali metal atoms, in its interlayer spacing. The elements K, Rb, and Cs react readily at 200°C or less, whereas Li and Na react only at higher temperature and pressure, forming impurity compounds as well as poorly characterized intercalates. This difference in chemical behavior between these two groups of elements

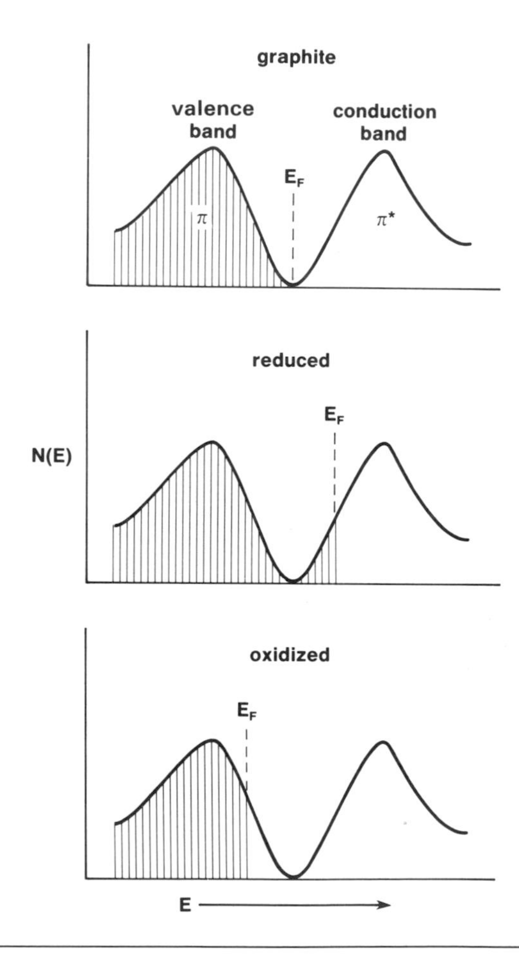

Figure 7.40.

The Fermi energy of pure graphite is shifted upward by accepting electrons, e.g., from intercalated alkali metals, and downward by the donation of electrons, e.g., to intercalating halogens.

is generally attributed to the ratio of the first ionization potential to the electron affinity of graphite, the latter being 4.6 eV. The ionization potentials of the three heaviest elements are all less than the electron affinity of graphite, whereas the values for Li and Na lie above it. The crystallographic positions of the K in C_8K and $C_{24}K$ are shown in Fig. 7.39. Of all the halogens, Br is the only one that reacts spontaneously at room temperature. Fluorine has not been intercalated, and iodine cannot be introduced by direct reaction but will intercalate when graphite is reacted with ICl or IBr. One can regard the alkali metals as reducing agents and the halogens as oxidizing agents. This is illustrated schematically in Fig. 7.40.

All intercalating reactants expand the inner layer spacing regardless of their size or charge. Unlike ordinary solid solutions, particularly interstitial solid solutions, the case of intercalation does not directly correlate with the size of the intercalating reactant, but rather with the energetics of the charge transfer. For example, elemental Ni with a metallic radius of 1.154 Å is essentially insoluble in graphite, whereas the much larger $NiCl_2$ molecule can be intercalated, yielding the well-defined $C_{13}NiCl_{2.04}$.

The chloride salts of most metals as well as many bromides can be intercalated. Table 7.9 contains a brief representative list of such intercalates.

The "stages" listed in Table 7.9 refer to the repetitive period of the intercalate configuration. Figure 7.41 shows schematically how this is done, where the offset patterns of the second- and third-stage structures arise from the resulting minimization of the strain energy.

Intercalation compounds provide an experimental means for studying two-dimensional magnetism and superconductivity. Furthermore, it is expected that such compounds will find useful commercial application in view of the ease with which their properties can be tailored to meet a variety of specifications.

Table 7.9. Intercalated metal chlorides.

Salt	Intercalate	Stages
$AuCl_3$	$C_{12.6}AuCl_3$, $C_{25.2}AuCl_3$	1, 2
	$C_{37.8}AuCl_3$, $C_{50.4}AuCl_3$	3, 4
$SbCl_5$	$C_{12}SbCl_5$, $C_{24}SbCl_5$	1, 2
	$C_{36}SbCl_5$, $C_{48}SbCl_5$	3, 4
	$C_{12}SbCl_{3.9}$, $C_{25}SbCl_{4.3}$	1, 2
$AlBr_3$	$C_9AlBr_3 \cdot Br_2$, $C_{18}AlBr_3$	1, 2
	$C_{24}AlBr_{3.3}$, $C_{33}AlBr_3$	2, 3
$PtCl_4$	$C_{42-51}PtCl_{4.5}$	3
$EuCl_3$	$C_{37.1}EuCl_3$	3

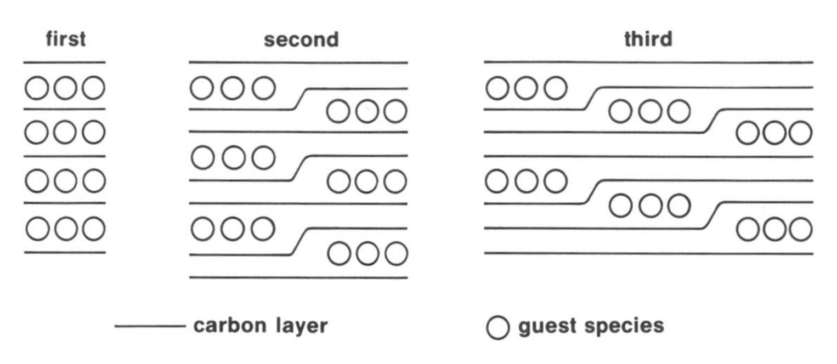

Figure 7.41.

The domain model for graphite staging. (Reprinted from *Intercalation Chemistry*, N. Bartlett and B. W. McQuillan, (Fig. 4, p. 27), Academic Press, 1982.)

11. Electron-Deficient Elements and Compounds

Three-center bonding is the result of electron deficiency and occurs in molecules formed by the lighter elements in groups IA, IIA, and IIIA when there are insufficient electrons to form the customary shared electron-pair bonds.

Probably the most widely studied electron-deficient solids are α-boron and its derivative compounds. The basic building block unit is the icosahedron illustrated in Fig. 7.42. The 12 B atoms that compose the icosahedral unit are interconnected

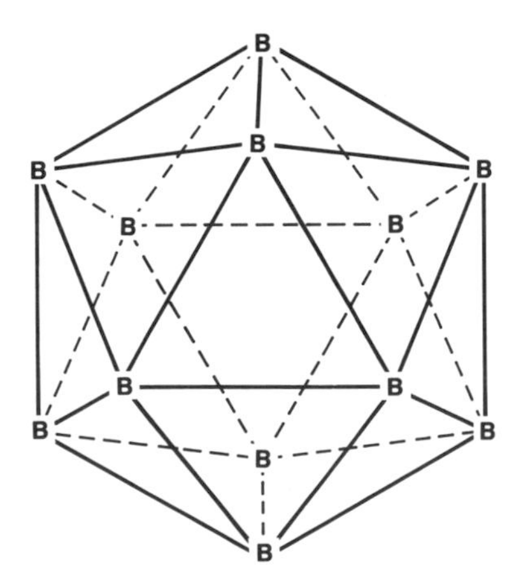

Figure 7.42.

The fundamental unit of α-rhombohedral boron is the icosahedron. It has 20 triangular faces created by the bonds between the 12 constituent B atoms.

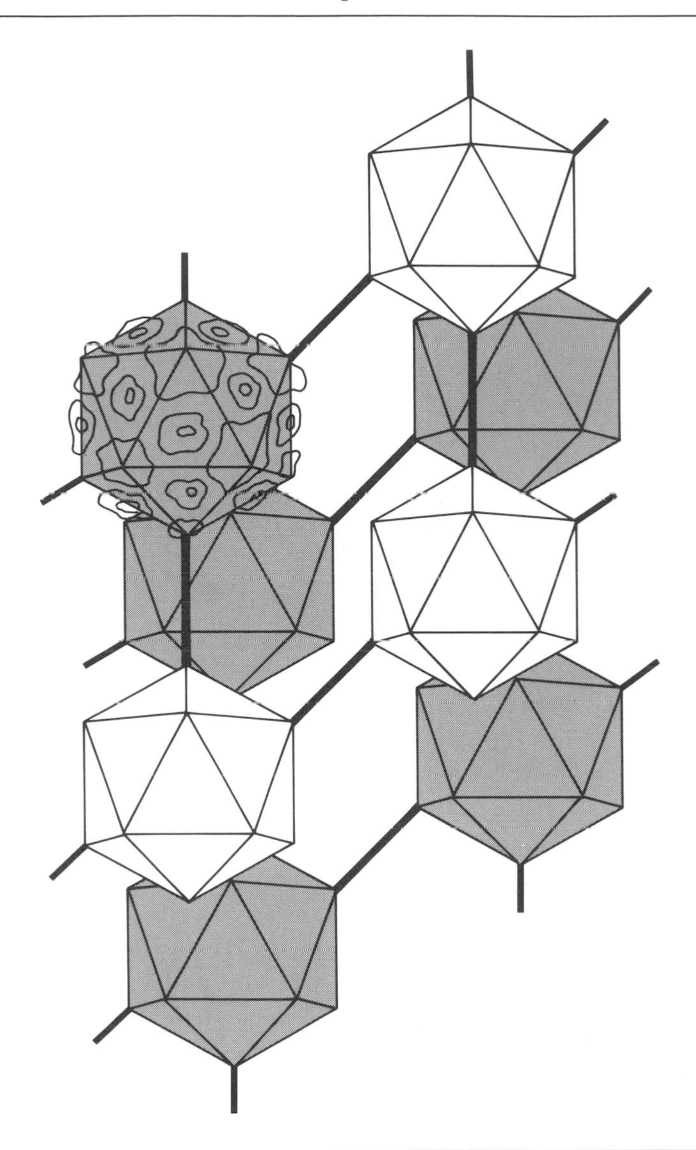

Figure 7.43.

Each icosahedron has six two-center bonds with each of six nearest-neighboring icosahedra, thus forming the structure of α-rhombohedral boron. (After D. Emin, *Phys. Today*, Jan. 1987, p. 2.)

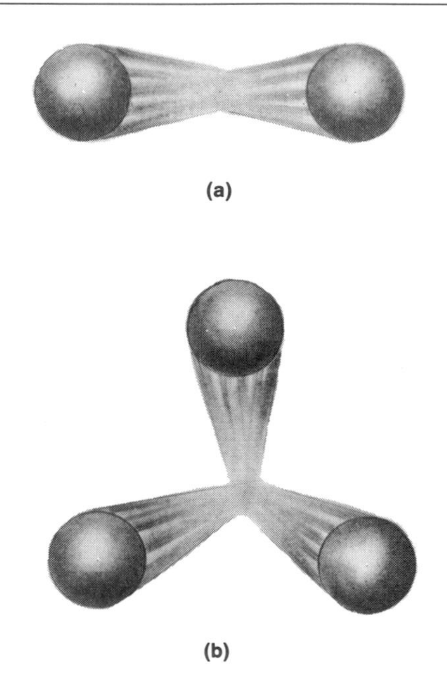

Figure 7.44.

A schematic drawing of (a) conventional two-center bond, (b) a three-center covalent bond.

by 30 bonds. There are six additional bonds connecting each icosahedron to its six nearest-neighboring similar units. This structure is shown in Fig. 7.43.

Each boron atom is bonded to five others within its own icosahedron, and, further, the 12 atoms collectively must somehow provide an additional six electrons for intericosahedral bonding. Normal electron-pair bonding would require each B atom to contribute 5.5 valence electrons whereas only three are available. It is this short fall that gives rise to the name "electron defficient." The actual bonding is as follows: based upon symmetry arguments it can be shown[12] that each icosahedron contains 13 bonding molecular orbitals. These are occupied by 26 of the 36 valence electrons, giving $26/20 = 1.3$ electrons for each of the 20 triangular faces. Of the remaining 10 electrons, 6 are used for conventional two-center bonding between neighboring units. The remaining four electrons are distributed among the remaining six atoms not involved in the intericosahedral bonding and give rise to the three-center bonding.

[12] H. C. Longuet-Higgins and M. de V. Roberts, *Proc. Roy. Soc. Lon. Ser. A*, **230**, 110 (1955).

The charge of the valence electrons is concentrated about the center of the triangular faces rather than between pairs of bonded atoms, as sketched in Fig. 7.44. Because the bonding orbitals are exactly filled, α-B is an insulator.

We have dwelt somewhat at length upon this unusual bonding scheme for two reasons. First, because it illustrates that chemical bonds need not be formed by the simple addition or modification thereof of atomic orbitals. Second, because compounds such as the boron carbides, which are structurally similar to α-B (see Fig. 7.45), are extremely refractory ($T_m = 2350°C$) and have as well potentially useful electronic properties.

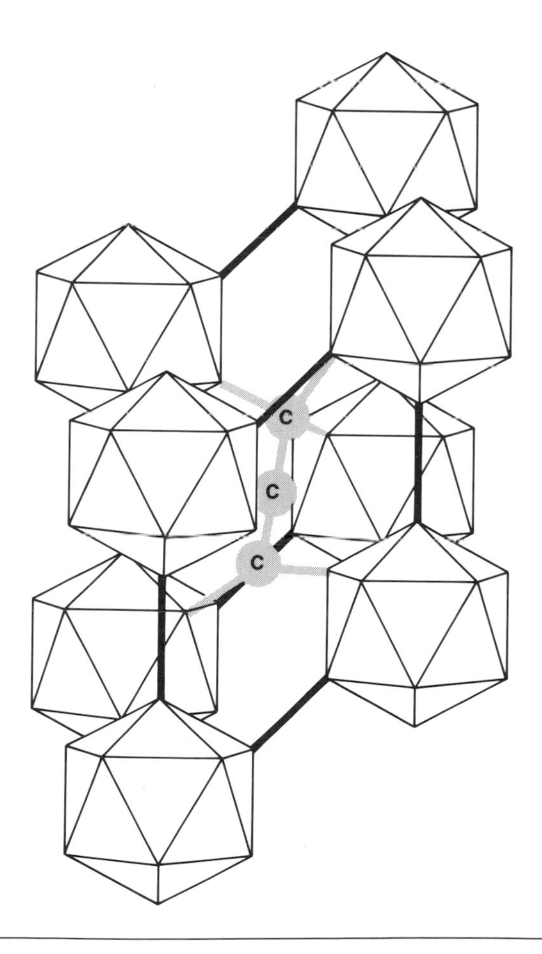

Figure 7.45.

In boron carbide a three-carbon chain replaces the direct linkage between B atoms. Carbon atoms can occupy only one or both end sites as well as one intericosahedral site.

Chapter 7 Exercises

1. Compare the fraction of electrons that can be excited thermally into the conduction band at room temperature for Ge, Si, and diamond ($E_g = 0.72$, 1.10, and 5.6 eV, respectively).

2. Suppose that 10^{12} (electron donors)/cm^3 are then injected into the semiconductor of problem 1. Calculate the conductivity at 300 K and 600 K for donor levels 0.1 eV, 0.2 eV, and 0.5 eV below the conduction band. (Hint: Use Fermi–Dirac statistics and ignore the intrinsic conductivity.)

3. Calculate the electronic contribution to the heat capacity of an intrinsic semiconductor. [Hint: Combine Eqs. (7.15) and (3.63).]

4. A semiconductor with a band gap of 1.5 eV has conduction electrons with a mobility of 1000 cm sec^{-1} dyne^{-1} and holes with 300 cm sec^{-1} dyne^{-1}. At 300 K the conductivity is 10^{-6} Ω^{-1} cm^{-1}. What is the conductivity at 600 K?

5. Prove that the existence of energy gaps at the Brillouin zone boundary of a one-dimensional lattice is equivalent to the condition for Bragg reflection of electron waves. (Remember that in one dimension the Bragg equation becomes $2d = n\lambda$. Also $\lambda = 2\pi/k$, where $k = n\pi/d$, n being the order of the diffraction and d the interplanar spacing.)

6. Sketch two-dimensional square, rectangular, and hexagonal lattices and their corresponding reciprocal lattices.

7. Show that the number of different k-states in the first Brillouin zone of a simple cubic lattice is equal to the number of lattice sites.

8. MoS_2, molybdenite, has the same crystal structure as graphite. Devise a molecular orbital scheme to describe the bonding.

Additional Reading

C. Klein and C. S. Hurlbut, Jr., *Manual of Mineralogy*, Wiley, New York (1985), bearing the same title and based upon the classic by J. D. Dana. It goes far beyond the original and is both a textbook as well as an exceptional reference work.

F. D. Bloss, *Crystallography and Crystal Chemistry*, Holt, Rinehart and Winston (1971). An exceptionally well-written textbook; deals more with structure than with bonding.

L. Pauling, *Nature of the Chemical Bond*, Cornell University Press (1960), especially Chaps. 3, 4, and 7.

P. A. Cox, *The Electronic Structure and Chemistry of Solids*, Oxford University Press (1987). A clearly written, nonmathematical resume of a great number of solid state phenomena. Assumes the reader to be familiar with the relevant quantum mechanics. A good place to start before resorting to more specialized texts.

S. L. Altman, *Band Theory of Metals—The Elements*, Pergamon Press (1970). A well-written, intermediate-level textbook, written by an expert in the field.

A. A. Levin, *Solid State Quantum Chemistry*, McGraw-Hill (1977). An extremely mathematical treatment of bonding in tetrahedrally coordinated semiconductors.

Chapter 8

Metallic Crystals

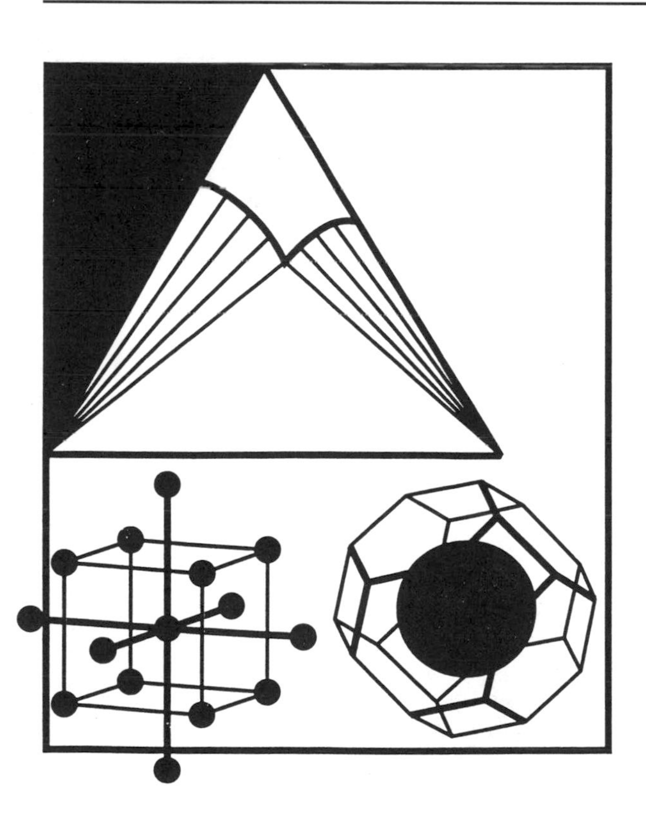

Chapter 8

Contents

Merely corrobative detail, intended to give artistic versimilitude
To an otherwise bald and unconvincing narrative.

—Gilbert & Sullivan, *The Mikado*

Metallic Crystals

Ionic, covalent, and dipolar bonds exist in isolated molecules as well as condensed phases. To the contrary, the metallic bond forms only in solid or liquid metals. It is by far the most difficult to explain theoretically, and quantitative *ab initio* predictions are the exception rather than the rule. Nevertheless, much of solid state chemistry is derived from the pioneering efforts of metallurgists, and the phenomenological behavior of metallic systems is in general well understood. Although the economic and technological importance of metals can scarcely be over emphasized, the metallic state is generally slighted by chemists.

There are two more or less distinct theoretical approaches to dealing with the metallic bond. One is a clear extension of molecular orbital theory and might be labeled the chemical method. The other attempts to calculate the energy differences between the nearly free valence electrons of the condensed and those of the free gaseous atoms: this could be called the physical method, and it relies upon the extension and refinement of free-electron theory.

Before taking up the various approaches to bonding in metals, a few generalities are in order. First, let it be noted that the enthalpies of melting are quite small for metals, usually between 5–10 kcal/mol. Thus, as loss of periodicity does not greatly weaken the metallic bond, it is probably safe to infer that periodicity does not play an important role in metallic bonding. Second, the presence of *d* orbitals significantly increases the bond strength of both liquid and solid metals.

We might also comment here on metallic ductility. Because of the nondirectionality of metallic bonding, there is no strong preference for one atomic position over another; the crystalline arrangement is primarily a geometrical convenience. Thus, the atoms can be displaced from their ideal position without greatly impairing the structure. Consequently, metals are ductile and yet of high strength.

Here we will tend to emphasize the intuitive empirical approach but fully acknowledge that lengthy computer calculations can provide more accurate answers on a case-by-case basis.

1. The Band Model

The band model unfortunately provides no satisfactory explanation of the cohesive forces between atoms in a metal, although it is a firm basis for understanding the magnetic and electron transport properties. What follows is an extension of the band theory concepts, already presented in Chapter 7, to include metallic substances.

It is clear that any description of bonding must also be compatible with the general properties that more or less define metallic character, viz. high electrical and thermal conductivity, metallic luster, ductility, etc. The most elementary model consistent with these facts is to consider the ion core embedded in a sea of valence electrons. The electrons provide the negative cement to hold the positive ion cores together while being delocalized and not tightly bound to any particular ion core. Quantum mechanics clearly shows that when two atoms are brought to within bonding distance of one another the atomic orbitals split to form two molecular orbitals—one of slightly higher, the other of slightly lower, energy than the original atomic orbitals. The addition of a third atom adds yet another molecular orbital to the now triatomic molecule and slightly alters the previously existing levels. This process proceeds to form a band of closely spaced energy levels, each level capable of containing two electrons in conformance to the Pauli exclusion principle. The levels are sufficiently close so that the energy states can be treated as a continuum, and the electrons occupying these states are called the conduction electrons. This delocalization of electrons and formation of conduction bands applies only to the outermost valence electrons. The inner electrons are somewhat perturbed from their free-atom levels, as shown schematically for Na in Fig. 8.1.

The Fermi energy E_F is defined as the maximum electron energy of the occupied states at 0K. It can be obtained from (7.17), which in three dimensions becomes

$$E_F = \frac{\hbar^2}{2m}\left(\frac{3\pi^2 N}{V}\right)^{2/3}, \tag{8.1}$$

where N is the number of electrons in a volume V. For an ordinary metal, E_F corresponds to a temperature of about 5×10^4 K. Thus, the curve for $kT = E_F/5$ shown in Fig. 8.2 is far above the melting temperature of any known substance, and is displayed only to illustrate the functional form. Figure 8.2 should be compared

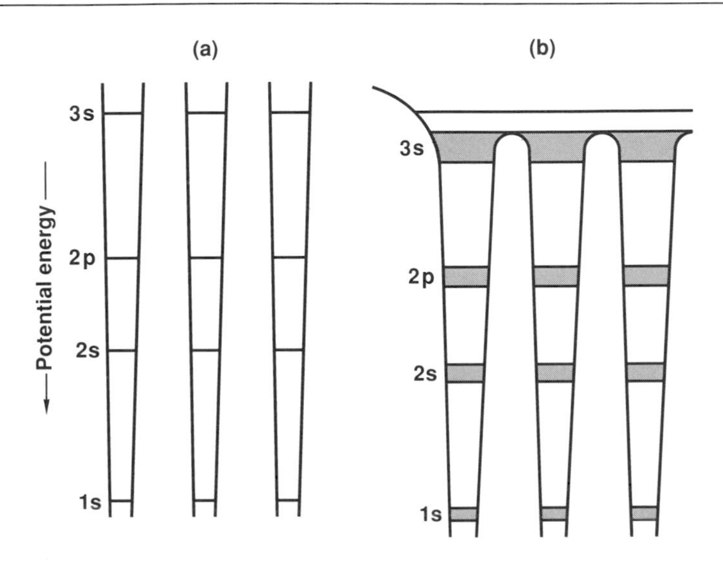

Figure 8.1.

The energy levels for Na in (a) the free atom and (b) in the crystalline state, where the formerly sharp inner levels have broadened and the 3*s* electrons have become delocalized. The delocalized 3*s* electrons can now be described by Bloch functions as explained in Chapter 7.

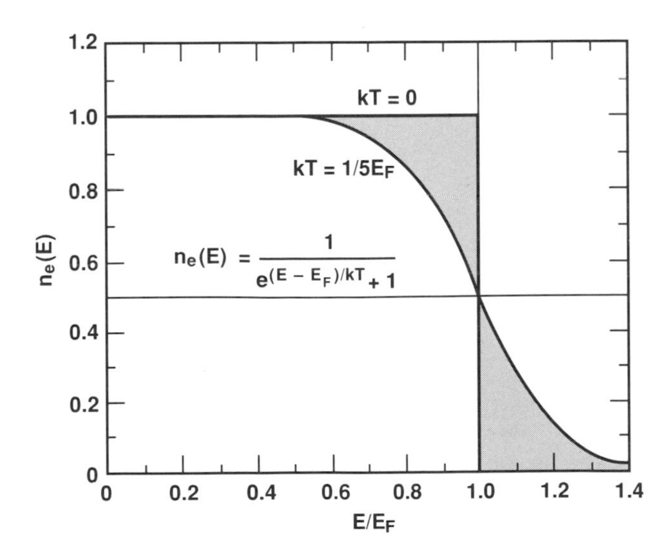

Figure 8.2.

The Fermi–Dirac distribution function $n_e(E)$ as an implicit function of temperature. As the temperature is raised above $T = 0$ K, electrons are promoted to energies $E > E_F$. (Adapted from C. Kittel, *Introduction to Solid State Physics*, 3rd ed., Wiley, New York, p. 204, 1967.)

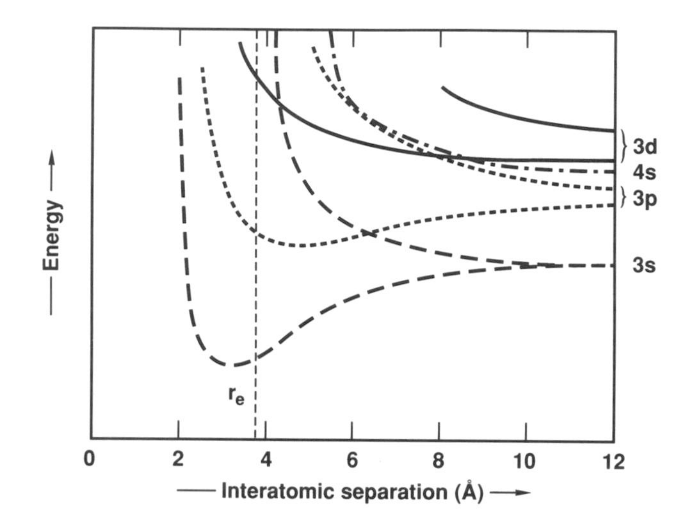

Figure 8.3.

The calculated band structure of Na, showing the overlapping bands as a function of interatomic separation.

with Fig. 7.20 in order to see how $n_e(E)$ is altered by a finite band gap such as is found in semiconductors.

When the conduction band is completely filled, as, for example, in the 3*s* band of Mg, it must overlap an unfilled band in order for the electrons to have access to the unfilled states necessary to maintain electrical conductivity. Such overlapping is illustrated, as a representative example, by the band structure of Na shown in Fig. 8.3, although in this case the 3*s* band is not filled and the overlapping is not necessary for conductivity.

2. The Metallic Bond

Because of its complexity, there is no elementary comprehensive theory of the metallic bond. There are several modern theoretical methods, in particular the pseudopotential approach, that are capable of quite accurately calculating the cohesive energies of the metallic elements and their alloys. All involve a sophistication beyond the level of this text, as well as considerable computational effort. We will review some of the earlier simpler methods that are now superseded, and then discuss some semiempirical procedures that still serve as the basis for an intuitive comprehension of the chemistry of alloys and intermetallic compounds.

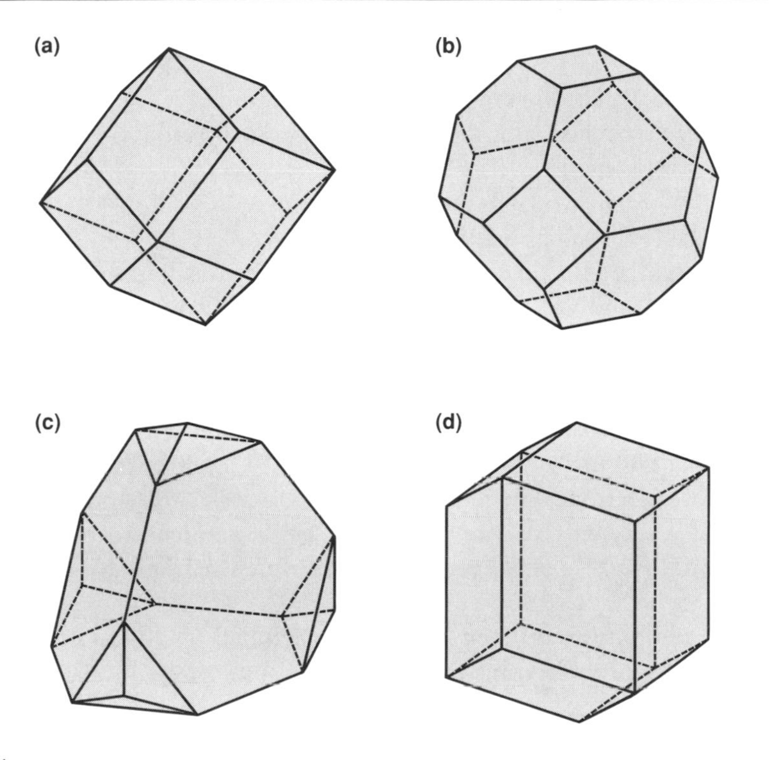

Figure 8.4.

Wigner–Seitz cells for the (a) fcc, (b) bcc, (c) diamond cubic and (d) hcp lattices.

A. The Wigner–Seitz[1] Approximation or Cellular Method

The Wigner–Seitz method confines the entire calculation to within a volume called the Wigner–Seitz cell. The cell is constructed by first selecting a central atom and then bisecting the distances between it and its near neighbors by a plane normal to the line of centers. Wigner–Seitz cells of four simple crystal structures are shown in Fig. 8.4. The cohesive energy is equated to the difference in the total energy of the valence electrons in the free atoms and the solid state, the calculation of the latter being confined, as previously stated, to the volume of a single cell.

The major contributions to the total electronic energy that affect cohesion, i.e.

[1] E. P. Wigner and F. Seitz, *Phys. Rev.* **43**, 804 (1933); **46**, 509 (1934); F. Seitz, *ibid.*, **47**, 400 (1935).

bond energy, arise from

1. The coulomb repulsion between positive ion cores.
2. The energy associated with the valence electrons arising from
   a. Their kinetic energy
   b. Their potential energy in the field of the ion cores
   c. Their coulomb self-potential
   d. The exchange interaction energy between electrons of parallel spin
   e. The correction to (d) for electrons of antiparallel spin
3. The exchange and correlation interactions between valence and core electrons, the latter deriving from their mutual repulsion.

The Wigner–Seitz approximation simplifies the calculation of the binding energy by simply neglecting all but the following interactions:

1. The valence electrons with the positive ion core within the same polyhedron. For example, the single 3*s* valence electron of Na is assumed to be confined to a single polyhedron (see Fig. 8.4a) and to interact with the single Na^+ located at its centre.

2. The mutual repulsion between ion cores is neglected. The effect of this omission is greatly reduced by the electrons that have been more or less localized about each ion core.

3. The Wigner–Seitz polyhedra are replaced by spheres for the purpose of the actual calculations. This, too, is not a bad approximation, since the polyhedra, particularly for the fcc lattice, are nearly spherical. In any case, the geometry of the actual Wigner–Seitz cells imposes geometric constraints upon the subsequent calculations that make them nearly intractable without significantly improving the final results.

4. $(\partial \Psi / \partial r)_{r_0} = 0$, where Ψ is the valence electron wave function and r_0 is the radius of the spheres that have replaced the polyhedra.

The energy at the bottom of the valence band can be obtained by solving the Schrodinger equation in three dimensions with $\Psi_k = \exp(i\mathbf{k} \cdot \mathbf{r}) U_k(\mathbf{r})$, the three-dimensional Bloch function. Ψ_r is substituted for the radial part of the complete wave function, which is then solved for $\mathbf{k} = 0$. This gives the energy, E_0, the lowest level in the band. Thus from

$$\frac{d^2\Psi_0}{dr^2} + \frac{2}{r} + \frac{d\Psi_0}{dr} + \frac{2m}{\hbar^2}[E_0 - V(r)] = 0, \tag{8.2}$$

one can obtain a solution for Ψ_0, having first assumed a semiempirical form of $V(r)$ based upon the potential experienced by the valence electron of the free atom and an initial value of E_0 just slightly lower than that in the free atom. Successive values of Ψ_0 are then calculated with E_0 as an adjustable parameter. E_0 may also be taken

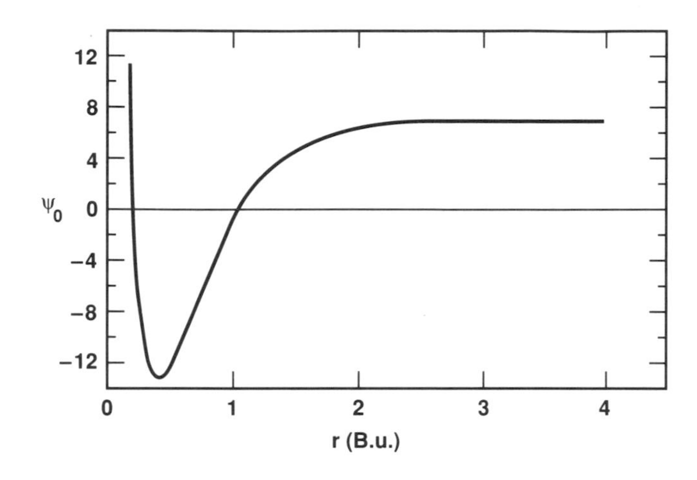

Figure 8.5.

The wave function Ψ_0 for the lowest 3*s* level in Na metal calculated by the Wigner–Seitz method. The radius r is given in Bohr (atomic) units (1 B.u. = 0.529 Å).

directly from the observed spectral levels without resorting to (8.2). It is also required that an acceptable solution, Ψ_0, place both nodes of the 3*s* radial wave function within the sphere of radius r_0, with $(d\Psi_0/dr)_{r_0} = 0$. This value is found by successive approximations, and the resulting form of the wave function is shown in Fig. 8.5.

The energy corresponding to the wave function in Fig. 8.5 is $E_0 - -8.2$ eV. To this is added the average kinetic energy of the valence electrons, which is obtained from the free-electron approximation. It is $(3/5)E_F$, calculated using (8.1). The value of E_F for Na is 3.24 eV, and therefore the electronic energy, E_e, is given by

$$E_e = E_0 + (3/5)E_F = -8.2 + 1.95 = -6.2 \text{ eV}.$$

This value is 1.2 eV lower than that for the free atom, and this cohesive energy is, perhaps fortuitously, in good agreement with the measured values. The results of similar calculations for Li and K along with those for Na are listed in Table 8.1. The relatively good agreement between the calculated and experimental values does not

Table 8.1. Cohesive energies.

	Calculated (eV)	Experimental (eV)
Li	1.6	1.69
Na	1.2	1.13
K	0.72	1.00

extend beyond the simple *s*-electron metals and even then the method is best suited for open structures such as the alkali metals, where Ψ_0 remains nearly constant over most of its range. The Wigner–Seitz method gives poor results for the cohesive energies of the transition elements, alerting us to the additional complexities introduced by the *d* electrons.

A method[2,3] employing renormalization atoms gives much better results for transition elements. This scheme lops off the free-atom wave functions at r_0 and renormalizes them to the volume of the Wigner–Seitz sphere. This procedure increases the *d*-electron charge density by less than 5%, but increases the *s* charge by factors of 2–3. The calculated cohesive energies for the 3*d* and 4*d* transition elements generally agree with the measured values to within ~20%.

B. The Miedema Rules for Alloy Formation

A scheme first proposed by Miedema[4–9] for calculating the heats of formation of alloys also uses a cellular approach. Unlike the Wigner–Seitz method, it is not grounded in quantum mechanics but is intrinsically empirical.[10] The concepts upon which the calculations are based evolved with many significant modifications over a period of several years, and, because of their intuitive and empirical origins, they cannot be justified in detail since they lack the terseness of mathematical rigor. Although several objections can be raised against this model, it can also be justified on the grounds that it mostly works and has the advantage of being simple. The cohesive energy is defined here as the enthalpy of mixing of the reference states, i.e., standard states, which are the pure solid elements, rather than the free atoms at infinite separation. Consequently, one calculates the differences between the heats of sublimation of the solid solution or compound and its pure solid constituents, which are quantities of similar magnitude. In contrast, using the electron energies of the free atoms as the standard state forces one to calculate the cohesive energy, a small quantity, as the difference between two large numbers.

---

[2] R. E. Watson, H. Ehrenreich, and L. Hodges, *Phys. Rev. Lett*, **24**, 829 (1970).

[3] C. D. Gelatt, Jr., H. Ehrenreich, and R. E. Watson, *Phys. Rev. B.* **15**, 1613 (1977).

[4] A. R. Miedema, F. R. deBoer, and P. F. deChatel, *J. Phys. F*, **3**, 1558 (1973).

[5] A. R. Miedema, *J. Less-Common Met*, **32**, 117 (1973).

[6] A. R. Miedema, R. Boom, and F. R. deBoer, *J. Less-Common Met*, **41**, 283 (1975)

[7] A. R. Miedema, *J. Less-Common Met*, **46**, 67 (1976).

[8] A. R. Miedema and P. F. deChatel, *Metall. Soc. AIME Proc.* (1980).

[9] A. R. Miedema, P. F. deChatel, and F. R. deBoer, *Physica (Amsterdam)*, **100**, 1 (1980).

[10] D. G. Pettifor, *Solid State Phys.*, vol. 40, Academic Press, 1987.

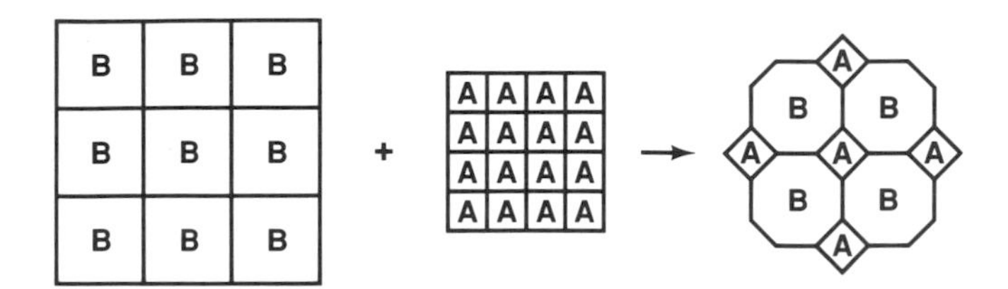

Figure 8.6.

The "macroscopic atom" model for alloy formation.

The formation energy of binary alloys is expressed in terms of two parameters plus their atomic volumes. The first parameter is simply the difference in the work functions of the constituents, $\Delta\phi$. The work function ϕ is, in this case, given by

$$\phi = V - E_F, \tag{8.3}$$

where ϕ is the energy required to remove an electron from the Fermi level to infinity, or the so-called vacuum level. Clearly ϕ depends upon the energy of the electron and the band from which it is taken. For the Miedema scheme E_F, which is also μ_e, the chemical potential of the electron, seems most suitable; V is the depth of the potential well in which the electrons reside referenced to the vacuum level. The second parameter, $\Delta\rho_e$, is the change in the charge density between the alloying species, and its values are estimated in an indirect manner.

The Miedema method reassembles the Wigner–Seitz cells in the fashion illustrated in Fig. 8.6. The $\Delta\phi$ term represents an attractive potential—the work functions being

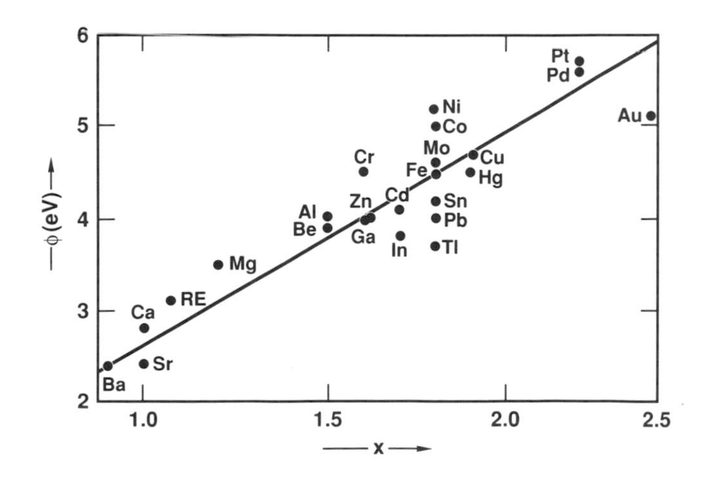

Figure 8.7.

Correlation between the work function ϕ and the Pauling electronegativities, x. [After Miedema *et al.*, *J. Phys. F*, **3**, 1560 (1973).]

in some ways analogous to the difference in electronegativities, and thus they reflect the degree of ionicity in the bonding. On the other hand, $\Delta\rho_e$ is essentially repulsive since it is the flow of charge across the cell boundaries that neutralizes the ionic character.

Several objections can be raised to this model, and most of them have been examined by its originator. However, it is noteworthy that indeed ϕ does correlate quite well in a simple manner with the Pauling electronegativities as shown in Fig. 8.7.

The charge density is obtained from various calculations and measurements. The Wigner–Seitz method as described in Section 2A suffices for the alkali metals, and band structure calculations provide an avenue for more complex elements. Perhaps, rather surprisingly, the compressibilities of the elements correlate in a linear manner with ρ_e as seen in Fig. 8.8.

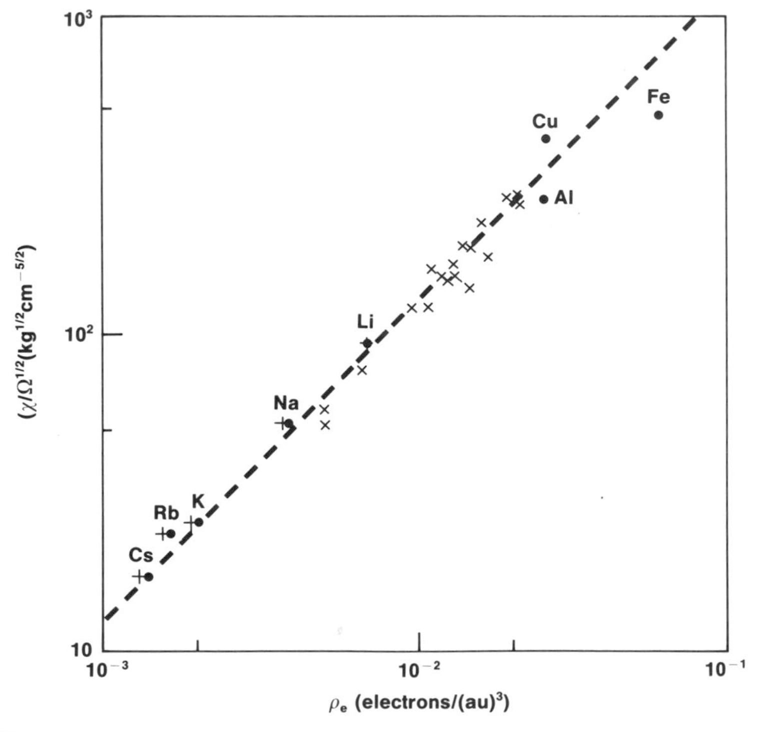

Figure 8.8.

The relationship between the charge density ρ_e at the boundary of the Wigner–Seitz cell and the bulk modulus χ^{-1} divided by the atomic volume. The x's designate nontransition elements for which ρ_e is derived from atomic wave functions. The labeled points are obtained from a variety of calculations listed in the text. (After Miedema *et al.*, p.1562, loc. cit.)

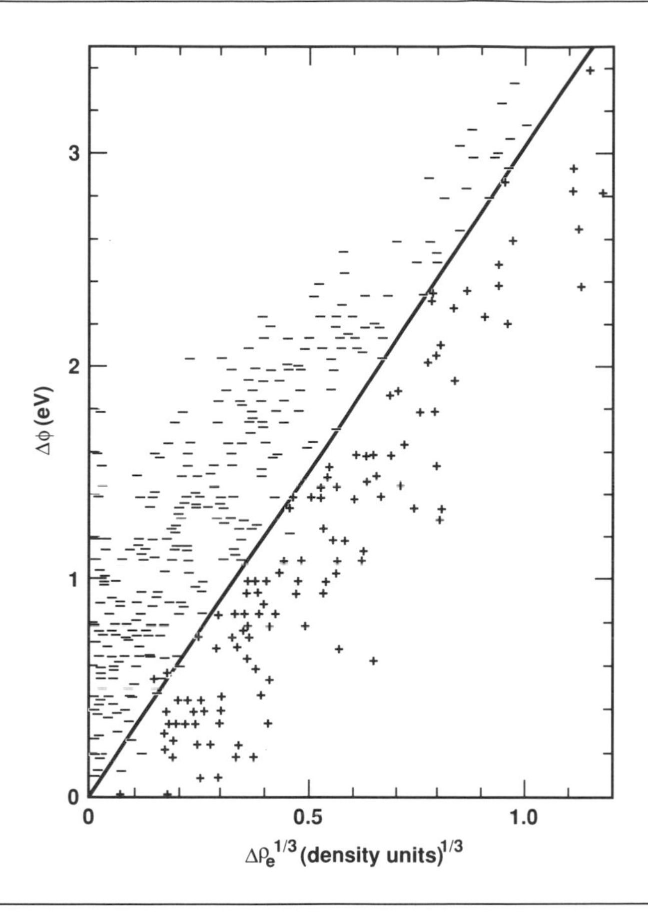

Figure 8.9.

The diagonal straight line separates the binary alloys whose measured enthalpies of formation are positive (+) from those that are negative (−). These solid solutions combine transition elements with other transition elements, as well as noble alkali, and alkaline earth metals. (After Miedema as presented by Pettifor in Ref. 10).

A graphical demonstration of the relevance of $\Delta\phi$ and $\Delta\rho_e$ to the enthalpies of alloy formation is given by Fig. 8.9, which shows a sharp division between positive and negative enthalpies. More or less by trial and error it was found that $\Delta\phi$ and $\Delta\rho_e^{1/3}$ could be used to calculate ΔH_f, the enthalpy of formation.[11] The formula ultimately evolved is

$$\Delta H_f = 2f(C)(C_A \Omega_A^{2/3} + C_B \Omega_B^{2/3})[(\rho_e^B)^{-1/3}]^{-1}[Q(\Delta\rho_e^{1/3}) - P(\Delta\phi)^2]^2, \tag{8.4}$$

[11] The reader is referred to Refs. 4–10 for a fuller semiquantitative justification of this choice.

Table 8.2. Enthalpies of formation of selected silicides.[a]

Compound	ΔH_{exp} (kcal/g-atom)	ΔH_{calc}[b] (kcal/g-atom)
TiSi	−15.5; −19	−23.5
CrSi	−6.4	−11.2
MnSi	−7.2	−14.0
FeSi	−9.4	−10.4
CoSi	−11	−11.4
NiSi	−10.7	−12.0
HfSi	−17	−26.1

[a] After Miedema *et al.*, *J. Phys. F.*, **3**, 1558 (1973).
[b] From (8.4).

where Ω_{A} and Ω_{B} are the atomic volumes and the term containing them accounts for the total interfacial area between cells.

Because the area of the interfaces between cells is the crucial factor in this model rather than the volume concentrations, the terms $f(C^S)$ is introduced as

$$f(C^s) = f(C^s_{\mathrm{A}}, C^s_{\mathrm{B}}) = C^s_{\mathrm{A}} C^s_{\mathrm{B}} \qquad \text{(disordered alloys)},$$

$$f(C^s) = f(C^s_{\mathrm{A}}, C^s_{\mathrm{B}}) = C^s_{\mathrm{A}} C^s_{\mathrm{B}}[1 - 8(C^s_{\mathrm{A}} C^s_{\mathrm{B}})^2] \qquad \text{(ordered alloys)}, \tag{8.5}$$

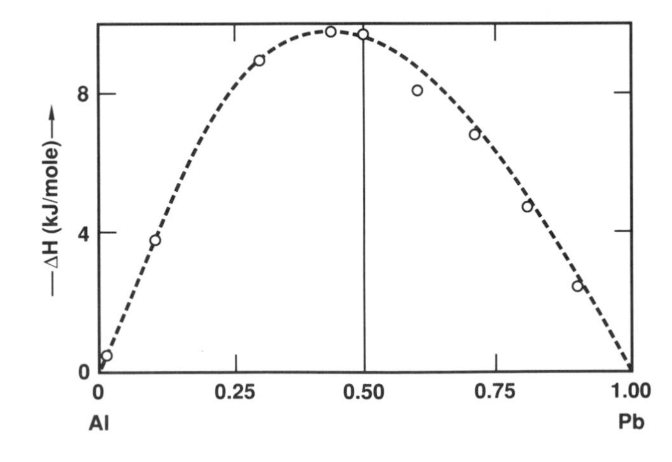

Figure 8.10.

The enthalpy of mixing, ΔH_{m}, corrected to 1200 K. The dashed line is derived from measured values and the circles are calculated from (8.4).

where

$$C_A^s = C_A V_A^{2/3}/(C_A V_A^{2/3} + C_B V_B^{2/3}),$$
$$C_B^s = C_B V_B^{2/3}/(C_A V_A^{2/3} + C_B V_B^{2/3}), \quad (8.6)$$

and where V_A and V_B are the molar volumes and P and Q are approximately constant for large groups of metals.

For some special cases, such as for the silicides listed in Table 8.2, Eq. (8.4) works quite well, but this is among the better agreements, which also includes borides, nitrides, and carbides.

The Miedema equation is also applicable to metallic solutions, as exemplified in Fig. 8.10.

3. The Pseudopotential

The pseudopotential method in general requires a level of mathematics as well as specialized wave mechanical concepts beyond the scope of a general textbook. Consequently, our presentation is limited, elementary, and nonrigorous. Nevertheless, this technique is widely used and cannot be skipped over without mention. What follows is simply stated without being derived.

The basic idea of the pseudopotential approximation is the replacement of the complicated true wave function Ψ by a simpler function Φ that, when substituted into the Schrödinger equation, reproduces the same eigenvalues as Ψ. The function Φ for the valence electrons is identical to Ψ outside the volume of the core electrons, but is a smooth function with all nodes eliminated within the core volume. Schematically the pseudopotential, V_{ps}, for an ion core might appear as sketched in Fig. 8.11.[12] As shown, the pseudopotential for the bare ion is $-A_0$ for $r < r_c$ and $-z/r$ for $r > r_c$, where z is the ionic charge.

To the ionic pseudopotential as shown in Fig. 8.11 is added the potential of a uniform electron gas. The gas is confined to the volume of one atomic (Wigner–Seitz) cell that has been replaced by a sphere of equal volume. Schematically, the pseudopotential of the ion core plus the electron gas might appear as shown in Fig. 8.12.

The total energy per electron is now written as follows:

$$U_0 = (3/5)E_F + U_X - U_S + E_H(k = 0) + U_C. \quad (8.7)$$

[12] Most of the following exposition follows directly from V. Heine, *Solid State Physics*, vol. 24, Academic Press, 1989; *The Physics of Metals*, vol. 1, *Electrons*, Chap. 1, Cambridge University Press, 1969.

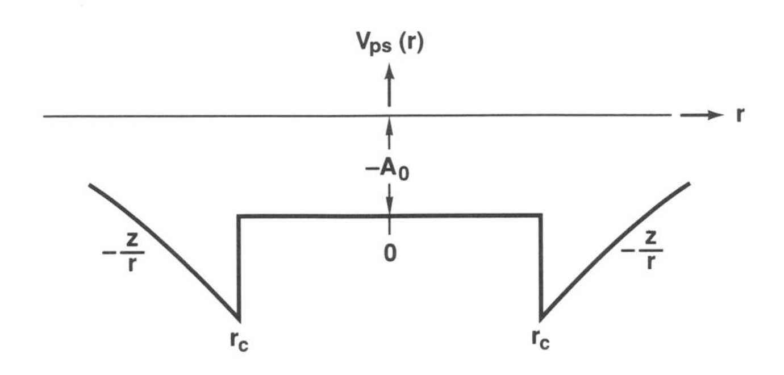

Figure 8.11.

The model of the pseudopotential of a bare ion. (Adapted from Ref. 12).

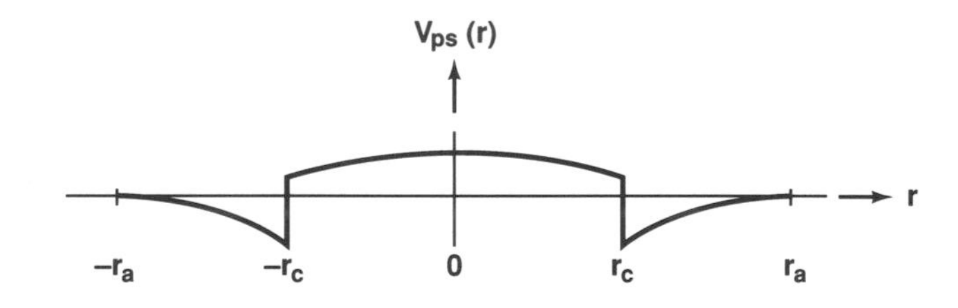

Figure 8.12.

The combined pseudopotentials of an ion core and a uniform electron gas of equal and opposite charge within a single atomic sphere of radius r_a. r_c is the effective core radius.

The first term on the right-hand side, $(3/5)E_F$, is the mean energy of the uniform electron gas. The second term is the exchange energy, U_X, which is obtained by the Hartree–Fock method.[13] For any free-electron gas,

$$\frac{3}{5}E_F + U_X = \frac{2.21}{r_s^2} - \frac{0.916}{r_s} \quad (\text{ryd})^{14} \tag{8.8}$$

where $r_s \equiv z^{-1/3}r_a$.

The next term, U_S, is the classical electrostatic self-energy, which has already been counted twice automatically by the Hartree–Fock approximation and must be subtracted. It is equal to $-1.2z^2/r_a$. $E_H(k = 0)$ is the mean value of the pseudopo-

[13] The Hartree–Fock approximation is dealt with by most standard texts on quantum mechanics. A particularly clear explanation of its application to a free-electron gas is given by S. Raimes, *The Wave Mechanics of Electrons in Metals*, North-Holland, 1961.

[14] 1 ryd = 13.60 eV.

tential within the sphere and is the sum of the pseudopotentials of the ion core and the electron gas. The former is shown in Fig. 8.11 and, for the electrons, is

$$\frac{ze^2}{r_a}\left[\frac{3}{2}-\frac{1}{2}\left(\frac{r}{r_a}\right)^2\right].$$

These terms combine to give

$$E_H(k=0)=\frac{-0.6z}{r_a}+\frac{3z}{r_a}\left(\frac{r_c}{r_a}\right)^2-A_0\left(\frac{r_c}{r_a}\right)\quad \text{(ryd)}. \tag{8.9}$$

The final term, U_C, accounts for the correlation energies of the free-electron gas and depends only on the local total electron density. Values for this quantity based

Table 8.3. Cohesive energies calculated by (8.10).[a]

	$-U_0$ (ryd)	
Element	Obs.	Calc.
Li	0.512	0.513
Na	0.460	0.457
K	0.388	0.369
Be	1.13	0.995
Mg	0.892	0.852
Zn	1.05	0.991
Cd	0.993	0.937
Hg	1.10	1.05
Ca	0.733	0.722
Ba	0.617	0.613
Al	1.38	1.32
Ga	1.47	1.38
In	1.36	1.28
Tl	1.43	1.38
Si	1.96	1.82
Ge	1.97	1.83
Sn	1.77	1.69
Pb	1.81	1.73
As	2.55	2.29
Sb	2.24	2.12
Bi	2.21	2.10
Se	3.23	2.84
Te	2.73	2.58

[a] Calculated by Weaire and reported by Heine, *Solid State Physics*, loc. cit.

upon a screened model have been calculated and reported by Animalu and Heine[15] for 25 different elements.

Collecting all the pertinent terms and substituting now transforms (8.7) to

$$U_0 = \frac{2.21}{r_s^2} - \frac{0.916}{r_s} - 1.8\frac{z}{r_a} + \frac{3z}{r_a}\left(\frac{r_c}{r_a}\right)^2 - A_0\left(\frac{r_c}{r_a}\right)^3 + U_C \quad \text{(ryd)}. \qquad (8.10)$$

Using this relatively simple formula, Weaire[16] has calculated the cohesive energy for 23 "simple" metals (in this sense "simple" means that *d*- and *f*-electron interactions are absent). These values are reproduced here in Table 8.3.

It is clear that (8.10) furnishes quite acceptable values for U_0 in all but a few cases.

In conclusion, we should remind ourselves that the foregoing is a structure-independent calculation. Taking account of the crystal structure by this approach occurs at the next step and lies beyond the scope of this presentation.

4. Metallic Solid Solutions

Metals have a greater proclivity for forming solid solutions than do either ionic or covalent compounds. The metallic bond is unrestricted by the requirements of charge balance or by the need to saturate an integral number of shared bonding electrons.

Although there is no rigorous theoretical basis for predicting the mutual solubilities of metals, there are empirical rules that can assist one's intuition. Such rules have been proposed in various forms by several researchers and they generally take into consideration the respective atomic sizes, electronegativities, and the number and kind of valence electrons. Most frequently, these rules are associated with the name of Hume-Rothery,[17,18] who can be paraphrased as follows:[18]

1. Difference between the electronegativities, ΔX, of the two metals. This is the most important factor, and increasing ΔX decreases the tendency for two atoms to unite in either liquid or solid solutions.

---

[15] A. O. E. Animalu and V. Heine, *Phil. Mag*, **12**, 1249 (1965).

[16] Private communication to V. Heine, reported in *Solid State Physics*, vol. 24, Academic Press, 1989.

[17] Wm. Hume-Rothery and G. V. Raynor, *The Structure of Metals and Alloys*, Part IV, The Institute of Metals, 1956.

[18] Wm. Hume-Rothery, *Phase Stability in Metals and Alloys*, Chap. 1 (P. S. Rudman, J. Stringer, and R. I. Jaffee, eds.), McGraw-Hill, 1967.

2. Size-factor effects related to the atomic diameters of the elements, which may be defined as the closest distances of approach of the atoms in the crystals of the elements. This crude definition correlates better with the facts than do more sophisticated atomic diameters or atomic volumes. Atoms with similar sizes tend to be mutually soluble.

To this should be added that when the electronegativities are greatly different, compound formation is favored over the formation of solid solutions, given that all other factors are favorable. Thermodynamically speaking, this asserts that when the enthalpic contribution to the free energy of formation significantly outweighs the entropy of mixing, compounds are favored.

A. The Size Effect

It is reasonable to assume that different elements with nearly the same atomic size would show greater mutual solubility than ones that are significantly disparate. When of markedly different sizes, local strains are induced that must raise the enthalpy of solution and ultimately restrict the solubility as well. Nevertheless, similar atomic sizes of the constituents do not guarantee appreciable solubility, although greatly unequal sizes generally preclude it. It should be remembered that the effective size of an atom in solution is naturally different from that in its elemental state and depends upon whether the valence electrons are shifted away from their ion core, or conversely, the atom in question attracts electrons from the surrounding matrix. There is more than one way to estimate the effective size, but, as noted by Hume-Rothery, the distance of closest approach in crystals of the pure elements, though a crude approximation, provides the best correlation with the facts. Table 8.4 lists the nearest-neighbor radii for the room temperature crystal structure and the radii for the 12-coordinated structure (i.e., Goldschmidt radii). These values are the same for the close-packed structures.

By simple trial and error it was found that elements whose radii differed by more than ~14% were nearly always insoluble in one another; needless to say, there are exceptions. However, it is noteworthy that the alkali metals, which are chemically very similar, reveal distinctly the effect of size upon solubility as shown in Table 8.5

When metallic elements are chemically similar and, furthermore, of nearly equal atomic size, mutual solubility is virtually ensured. Consider the transition elements as listed in Fig. 8.13. If using Fig. 8.13, one examines the mutual solubilities of pairs of elements related to one another, as depicted by the six arrows, it turns out that of the 57 alloys for which there are data, 39 have complete miscibility at some temperature, 12 have significant partial miscibility (20 at .% or more) of at least one

Table 8.4. Radii (Å) of the metallic elements.[a]

IA	IIA	IVB	VB	VIB	VIIB		VIII			IB	IIA	IIIA	IVA	VA	VIA
Li	Be											Al			
1.52	1.12											1.42			
1.56	1.12											1.42			
Na	Mg	Sc	Ti	V	Cr	Mn	Fe	Co	Ni	Cu	Zn	Ga	Ge	As	Se
1.85	1.60	1.60	1.46	1.31	1.25	1.12	1.23	1.25	1.24	1.28	1.33	1.21	1.22	1.25	1.16
1.91	1.60	1.60	1.46	1.35	1.28	1.36	1.27	1.25	1.24	1.28	1.37	1.35	1.39	—	—
K	Ca	Y	Zr	Nb	Mo	Tc	Ru	Rh	Pd	Ag	Cd	In	Sn	Sb	Te
2.31	1.96	1.81	1.60	1.43	1.36	1.35	1.33	1.34	1.37	1.44	1.48	1.62	1.40	1.45	1.43
2.38	1.96	1.81	1.60	1.47	1.40	1.35	1.33	1.34	1.37	1.44	1.52	1.67	1.58	1.61	—
Rb	Sr	La	Hf	Ta	W	Re	Os	Ir	Pt	Au	Hg	Tl	Pb	Bi	Po
2.46	2.15	1.87	1.58	1.43	1.37	1.37	1.35	1.35	1.38	1.44	1.50	1.71	1.74	1.55	1.68
2.53	2.15	1.87	1.58	1.47	1.41	1.37	1.35	1.35	1.38	1.44	1.55	1.71	1.74	1.82	—
Cs	Ba														
2.62	2.17														
2.70	2.24														

The lanthanide elements

Ce	Pr	Nd	Pm	Sm	Eu	Gd	Tb	Dy	Ho	Er	Tm	Yb	Lu
1.82	1.82	1.81	—	—	1.98	1.78	1.77	1.75	1.76	1.73	1.74	1.93	1.73
1.82	1.82	1.81	—	—	2.04	1.78	1.77	1.75	1.76	1.73	1.74	1.93	1.73

The actinide elements

Th	Pa	U	Np	Pu	Am	Cm	Bk	Cf	E	Fm	Mv	No
1.80	1.60	1.38	1.30	1.64	—	—	—	—		—	—	—
1.80	1.63	1.54	1.50	1.64	—	—	—	—		—	—	—

[a] The upper value is one-half the distance of closest approach. The lower is the Goldschmidt radius. (Taken from R. C. Evans, *Crystal Chemistry*, Cambridge University Press, p. 87, 1964.)

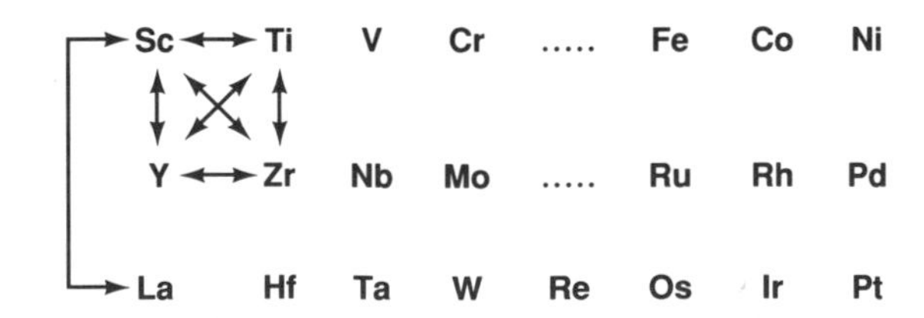

Figure 8.13.

The transition elements. Arrows define the ways all elements were combined to form binary alloys. Mn is omitted because of its generally anomalous behavior, and Tc for lack of data.

Table 8.5. The mutual solubilities of the solid alkali metals as a function of atomic size.

Elements	$\Delta r/r(\%)$[a]	Solubility[b]
Li–Na	20	IM
Li–K	42	IM
Li–Rb	47	IM
Li–Cs	54	IM
Na–K	22	IM
Na–Rb	28	IM
Na–Cs	34	IM
K–Rb	6.1	CM
K–Cs	17	CM
Rb–Cs	6.5	CM

[a] The 12-coordinated, Goldschmidt radii are used for the computation.

[b] IM = immiscible, CM = completely miscible.

constituent, and only 6 or ~8% of all these combinations show complete immiscibility. Further, all of these six have radii that differ by more than 11%, and all the completely miscible solid solutions have radii that differ by less. These observations were based upon the Goldschmidt radii, and the percentages based upon the smallest radius of the two alloying elements. Although we emphasize that exceptions are not uncommon, the size factor more often than not plays the decisive role in the ability of metallic elements to form solid solutions.

5. Intermediate Phases

Metallic elements frequently react to produce intermetallic compounds whose compositions, perforce, lie between those of the primary solid solutions. Because such compounds frequently embrace a range of composition, they are categorized by the general term *intermediate phase*. As previously remarked, such compounds are most likely to appear when the electronegativities of the components are significantly different. Factors other than the relative electronegativities and sizes that affect compound formation are:

1. A tendency for atoms near the ends of the short periods and B subgroups to complete their octets of electrons, and a similar but less marked tendency to fill the *d* shell in the later transition elements.
2. A tendency for definite crystal structures to occur at characteristic numbers of electrons per unit cell, which if all atomic sites are occupied, is equivalent to

saying that certain structures occur at characteristic electron–atom ratios or electron concentrations.

3. Orbital-type restrictions. Structures whose hybrid bonding orbitals involve a very high proportion of *d* wave functions may not mix with class I or class IV elements. Similarly, atoms giving rise to almost pure *p* bonding may not enter structures involving other types. A high value of the difference in electronegativities may overcome these restrictions.

The valence of elements in the metallic state is quite capricious, as pointed out by Pauling,[19] using Hg as his example. The amazing number of compounds that Hg forms with a single alkali metal is confusing because of the irregular stoichiometry, and becomes even more so when the stoichiometric ratios vary in the extreme among the different alkali metal elements, as shown by the following enumeration:

$$NaHg_4, NaHg_2, NaHg_8, Na_3Hg_2, Na_5Hg_2, Na_3Hg$$

$$KHg_{11}, KHg_8, KHg_4, KHg_3, KHg_{2.7}, KHg_2, KHg$$

$$LiHg_3, LiHg_2, LiHg, Li_2Hg, Li_3Hg, Li_6Hg.$$

It is evident from the foregoing list that the elementary rule of definite proportions as it applies to daltonides does not apply to these or, in fact, to any series of intermetallic compounds. Nevertheless, some systematics are apparent that permit classification and a qualitative rationalization of the stability of certain compositions in relation to their crystal structures.

A. Saltlike Structures

Three groups of intermetallics have the same crystal structures as three simple ionic salts, though not necessarily the same properties; they are the NaCl, fluorite, and antifluorite structures. The fluorite crystal structure is shown in Fig. 8.14, and the antifluorite structure reverses the positions of the anions and cations. For intermetallic compounds, one considers the more electronegative of the two constituents analogous to the anion. A partial list of compounds having these structures is given in Table 8.6. Although all the compounds in Table 8.6 have a nonmetallic character, their degree of metallicity varies over a considerable range. The NaCl compounds, being composed of elements of greatly differing electronegativities, possess the most saltlike (i.e., insulator) properties. Members of the antifluorite group vary among

---

[19] L. Pauling, *Nature of the Chemical Bond*, 3rd ed., p. 355, Cornell University Press, Ithaca, NY, 1960.

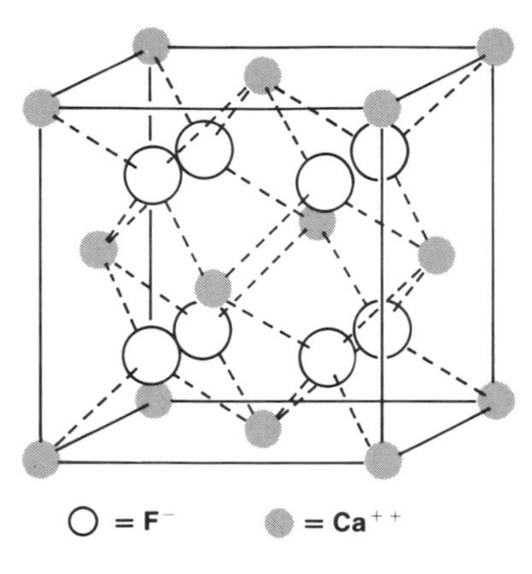

Figure 8.14.

The fluorite crystal structure.

Table 8.6. Intermetallic compounds with salt-like crystal structures.[a]

Sodium Chloride Structure	Antifluorite Structure	Fluorite Structure
MgSe	Mg_2Si	$PtSn_2$
CaSe	Mg_2Sn	Pt_2P
MnSe	Mg_2Pb	$PtIn_2$
PbSe	Cu_2Se	$AuAl_2$
CaTe	Ir_2P	
SrTe	LiMgN	
SnTe	LiZnN	
PbTe	LiMgAs	
	LiMgBi	
	AgMgAs	
	CuMgSb	
	CuCdSb	
	Li_3AlN_3	
	Li_3GaN_2	
	Li_5SiN_3	
	Li_5TiN_3	

[a] After W. Hume-Rothery and G. Raynor, p. 182, loc. cit.

themselves; e.g., Mg_2Sn is quite nonmetallic, having high electrical resistance, and is semiconducting in its behavior. On the other hand, the electrical resistance of Mg_2Pb is nearly 200 times less than Mg_2Sn, and it increases with temperature in the manner of a true metal. The electronegativities of the fluorite group are more nearly equal, and their properties, correspondingly, are more metallic.

B. Tetrahederally Coordinated Intermetallics

The zinc blende and wurtzite structures have been shown in Figs. 7.1 and 7.2. The compounds listed in Table 8.7 have these structures and are isoelectronic, with elemental Si and Ge having an average value of four valence electrons per atom and sp^3 hybrid bonding. It is therefore not surprising that many, if not most, members of this group have semiconductor characteristics. It is clear that the diamond cubic structures cited in Table 8.7 are not true intermetallics but are combinations of metals with metalloids or with groups VB and VIB.

C. Electron Compounds

The stability of electron compounds depends upon the size factor and the number of electrons outside of a full d shell. Electron compounds are characterized by having a definite electron per atom ratio associated with a specific crystal structure. The number of valence electrons per atom in metals or intermetallic compounds does not

Table 8.7.[a]

Zinc Blende Structure			Wurzite Structure
BeS	BeSe	BeTe	ZnS
ZnS	ZnSe	ZnTe	CdS
CdS	CdSe	CdTe	MnS
HgS	HgSe	HgTe	MgTe
MnS	MnSe	AlSb	CdSe
AlP	AlAs	GaSb	MnSe
GaP	GaAs	InSb	AlN
InP	InAs	SiC	GaN
			InN

[a] From W. Home-Rothery and G. Rayner, p. 184, loc. cit.

have universally accepted values; hence, there is a definite degree of arbitrariness right from the start. However, by assigning the total number of valence electrons according to the following scheme:

Metal	Cu	Ag	Au	Zn	Cd	Hg	Al	In	Ga	Sn	Si	Ge	Ni
Valence electrons	1	1	1	2	2	2	3	3	3	4	4	4	0

Hume-Rothery noted that specific crystal structures with narrow ranges of composition occur at definite electron–atom ratios. It has been shown that there is a good correlation between this ratio and the theoretical value, wherein the number of electrons is calculated to be just slightly in excess of the number contained within the Fermi sphere of free electrons, which just intersects the wall of the first Brillouin zone, as shown schematically in Fig. 8.15.

The energy to add additional electrons increases rapidly beyond this point because there are a decreasing number of states per unit energy. The alternative to filling the remaining higher energy states within the first zone is to commence filling states in the second zone that lie above the energy gap. At this juncture it is often advantageous for the solid solution or compound to transform its crystal structure to one whose corresponding first Brillouin zone can contain a greater number of electrons. Such alloys are conveniently characterized by their electron–atom ratio as in Table 8.8

The theoretical values in Table 8.9 correspond to the electron–atom ratio at the point of initial intersection of the Fermi sphere with the zone wall, i.e., point E_i in Fig. 8.15.

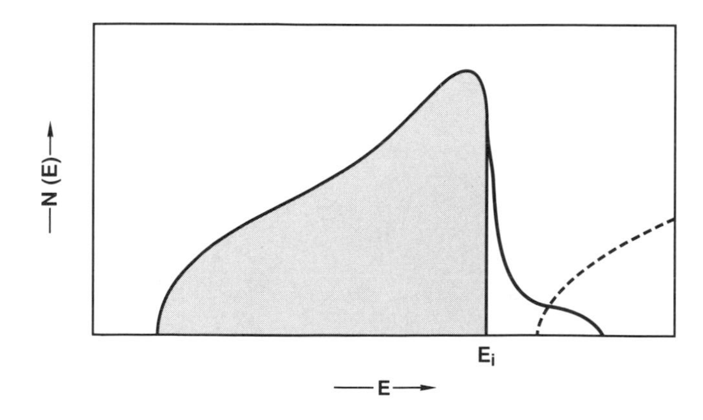

Figure 8.15.

A schematic graph showing the intersection of the Fermi electron gas (shaded area) with the face of the first Brillouin zone at E_i. The remaining unoccupied states decrease dramatically for $E > E_i$.

Table 8.8. Examples of electron compounds.[a]

Electron–Atom Ratio = 3:2			Electron–Atom Ratio = 21:13	Electron–Atom Ratio = 7:4
Body-Centered Cubic Structure	Complex Cubic ("β-manganese") Structure	Close-Packed Hexagonal Structure	Complex Cubic Structure ("γ-brass")	Close-Packed Hexagonal Structure
CuBe	Cu_5Si	Cu_3Ga	Cu_5Zn_8	$CuZn_3$
CuZn	AgHg	Cu_5Ge	Cu_5Cd_8	$CuCd_3$
Cu_3Al	Ag_3Al	AgZn	Cu_5Hg_8	Cu_3Sn
Cu_5Sn	Au_3Al	AgCd	Cu_9Al_4	Cu_3Ge
AgMg	$CoZn_3$	Ag_3Al	Cu_9Ga_4	Cu_3Si
AuMg		Ag_3Ga	Cu_9In_4	$AgZn_3$
AuZn		Ag_3In	$Cu_{31}Si_8$	$AgCd_3$
AuCd		Ag_5Sn	$Cu_{31}Sn_8$	Ag_3Sn
FeAl		Ag_7Sb	Ag_5Zn_8	$AuZn_3$
CoAl		Au_3In	Ag_5Cd_8	$AuCd_3$
NiAl		Au_3Sn	Ag_5Hg_8	Au_3Sn
NiIn			Au_5Zn_8	Au_5Al_3
PdIn			Au_5Cd_8	
			Au_9In_4	
			Fe_5Zn_{21}	
			Co_5Zn_{21}	
			Ni_5Zn_{21}	

[a] From W. Home-Rothery and G. Raynor, p. 197, loc. cit.

Table 8.9. Electron–atom ratios in electron compounds.[a]

Alloy	Fcc Phase Boundary	Minimum bcc Phase Boundary	β-Phase[b] Boundaries	Hcp Phase Boundaries
Cu–Zn	1.38	1.48	1.58–1.66	1.78–1.87
Cu–Al	1.41	1.48	1.63–1.77	
Cu–Ga	1.41			
Cu–Si	1.41	1.49		
Cu–Ge	1.36			
Cu–Sn	1.27	1.49	1.60–1.63	1.73–1.75
Ag–Zn	1.38		1.58–1.63	1.67–1.90
Ag–Cd	1.42	1.50	1.59–1.63	1.65–1.82
Ag–Al	1.41			1.55–1.80
Theoretical	1.36	1.48	1.53	

[a] Adapted from N. F. Mott and H. Jones, *The Theory of the Properties of Metals and Alloys*, Dover, New York, 1958, as presented by C. Kittel, *Introduction to Solid State Physics*, Wiley, New York, p. 579, 1967.

[b] The β-brass structure is complex cubic with 52 atoms per unit cell

D. Laves Phases

The so-called Laves phases are named after Fritz Laves and his co-workers, who were the first to recognize them. In previous sections we have discussed saltlike structures whose bonding depends upon differences in electronegativities, covalent-like structures with sp^3 bonding, and electron compounds. The Laves phases are much more numerous. They are formed by elements whose electronegativities need not be greatly different and from a broad assortment of constituents. The two features that all have in common are the radius ratio of the constituents, which is nearly always very close to 1.2, and the general formula for Laves phases, which is AB_2, although ternary phases, ABC, are also known. The component A is always larger than B, and a specific element may occupy either an A or B position in the crystal structure and, hence, also in the formula, depending upon the size of its partner. For example, in $CuBe_2$, copper is in the A position, but in $MgCu_2$ it is the B component. Both compounds have the same crystal structure.

All Laves phases conform to one of three crystal structures, for which $MgCu_2$, $MgZn_2$, and $MgNi_2$ are the prototypes. All three structures are quite closely related, and only $MgCu_2$ is illustrated here in Fig. 8.16.

Even a partial list of representative compounds would be extremely long, so only the briefest sampling of virtually hundreds of compounds is given in Table 8.10.

The stability of the Laves phases no doubt is connected with the close packing of the atoms allowed by these structures when $r_A/r_B \cong 1.2$. The large coordination number of the A sites is 16, where 4 of the 16 sites also contain A atoms. This relatively high degree of coordination must contribute significantly to the stability of these phases.

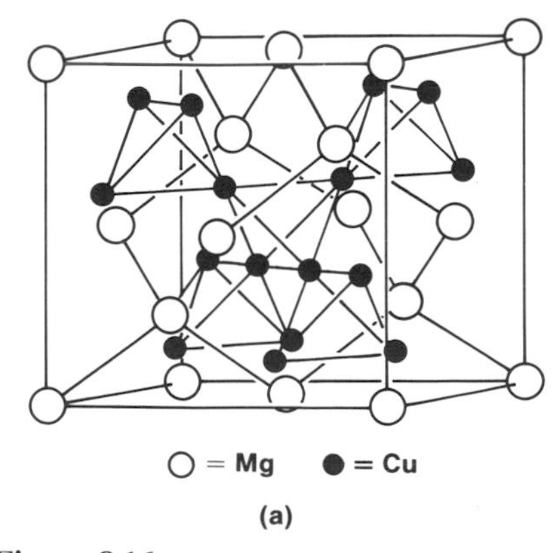

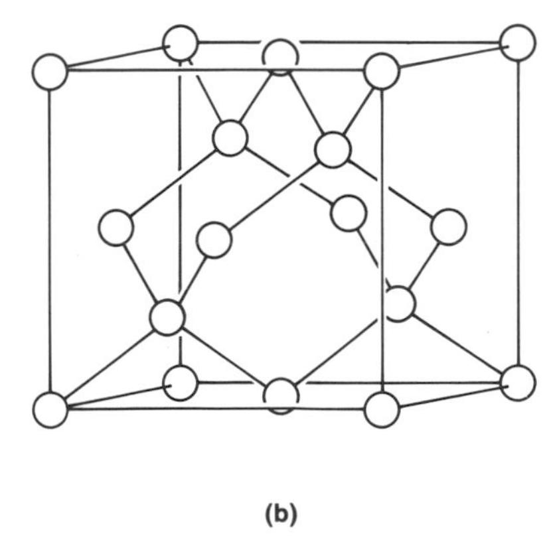

Figure 8.16.

(a) The $MgCu_2$ crystal structure; (b) the arrangement of the larger Mg atoms.

Table 8.10. Laves phases prototypic structure.

$MgZn_2$	$MgCu_2$	MgNi
$CaAl_2$	$BaMg_2$	$HfCr_2$
$CePt_2$	KNa_2	Mg_2CuZn_3
$DyMn_2$	$PuRe_2$	$NbZn_2$
$NbBe_2$	TaCrNi	$ReBe_2$
UFeNi	YRu_2	$ThMg_2$
$ZrV_{0.5}Ni_{1.5}$	$ZrTc_2$	UPt_2

E. The Engel–Brewer Rules

It is axiomatic that the strength of a chemical bond increases with the number of participating electrons. An example of this is the variation of the enthalpy of atomization ΔH_a, of the third row elements with their number of bonding electrons as shown in Table 8.11. The enthalpies for the solid elements are equivalent to the heats of vaporization, but are the heats of dissociation for S_2 and Cl_2. The valence states are chosen, bearing in mind that only electrons with unpaired intra-atomic spins are bonding. Thus, promotion of one of two 3*s* electrons into an empty 3*p* orbital is required before it can become a bonding electron. The bonding electrons of P, S, and Cl are simply the maximum number of *p* electrons obtainable with unpaired spins.

In the Engel–Brewer (E–B) scheme only the *s* and *p* electrons are effective in determining the crystal structure, although *d* electrons, when present, do contribute to the bond strength. The E–B rules are

1.0–1.5 bonding $s + p$ electrons $\Rightarrow$ bcc
1.7–2.1 bonding $s + p$ electrons $\Rightarrow$ hcp
2.5–3.0 bonding $s + p$ electrons $\Rightarrow$ fcc
4 bonding $s + p$ electrons $\Rightarrow$ diamond cubic

These figures for the electrons are in reasonable agreement with the Hume-Rothery estimates and, with some reasonable assumptions, can provide guidelines for the prediction of solid solution solubilities.

Table 8.11. Atomization enthalpy–number of bonding electrons.

	Na	Mg	Al	Si	P	S	Cl
ΔH_a (kcal)	26	35	79	109	77	50	29
Valence state	s	sp	sp^2	sp^3	s^2p^3	s^2p^4	s^2p^5
Bonding electrons	1	2	3	4	3	2	1

Table 8.12. Bonding energies (kcal) per electron per element at 0 K.[a]

No. of Electrons	(1)	(2)	(3)	(4)	(5)	(6)	(7)
	H						
	52						
	Li	Be	B	C	Ta	W	Re
	38	38	44	42	38	34	26
	Cl	S	N	Hf	Nb	Mo	Tc
	32	34	38	38	36	26	22
	Br	O	Ce, La, Lu, Y	Zr, Th	Pa	U	
	14	30	34	36	(30)	22	
	Na	Se	Ac, Gd, Tb	Ti	V		
	26	28	32	28	24		
		Te	Cm, Sc	Si			
		26	30	26			
		Ba					
		22					

[a] L. Brewer, *Structure and Bonding in Crystals*, Chap. 7 (M. O'Keefe and A. Navrotsky, eds.), Academic Press, 1981.

A quantity called the *bonding energy* (not to be confused with the more conventional bond energy) is defined as the sum of the sublimation and promotion energies. The latter is the energy necessary to promote the electrons of an isolated atom from their ground state to the same configuration postulated for the valence electrons in the solid. For example, the promotion energies for Mg, Al, and Si are those required to promote a $3s$ electron into a $3p$ orbital, thus maximizing the number of unpaired electrons, at 2, 3, and 4 for these elements, respectively. Thus, the energies of the valence electrons are explicit references to their energies in the free-atom state. These energy values are obtained from their atomic spectra. The bonding energies for some of the elements are shown in Table 8.12.

The effect of temperature requires additional considerations. Many metallic elements have more than one crystal structure, depending upon the temperature. Most often the transformation is from the close-packed to the bcc phase. Because the body-centered cubic structure is more open, it has a higher vibrational entropy. This increase in entropy stabilizes the bcc structure at high temperatures.

The application of the foregoing postulates and calculations requires a certain amount of arithmatic. Following the example given by Gokcen[20], we do an example calculation of the maximum solubility of Rh in Cr, Mo, or W. These solvent metals are all bcc and hence are postulated to have a maximum of 1.5 $(s + p)$ electrons.

[20] N. A. Gokcen, *The Statistical Thermodynamics of Alloys*, pp. 246-247, Plenum Press, 1986.

From their position in the periodic table, it is seen that these solvent elements have a total of six $(s + d)$ electrons. The element Rh has a total of nine $(s + d + p)$ electrons. Thus, the electron contribution from the group VIB elements is $6(1 - X_{Rh})$ and the contribution from each atom of Rh is $9X_{Rh}$. The sum must correspond to $d^5sp^{0.5}$ since 1.5 $(s + p)$ electrons are required to maintain the bcc structure. Hence, the total number of outer electrons is $6.5 = 6(1 - X_{Rh})$, and thus $X_{Rh} = 0.167$. This value, along with the results of similar calculations, is compared with measured values in Table 8.13.

A very simple picture of the bonding in the transition metals would continue the Engel–Brewer maxim of maximizing the number of unpaired valence electrons with the expenditure of the minimum amount of promotion energy. The simplest scheme might be the following sequence, where the numbers in parentheses are the total number of bonding electrons.

Ca	$4sp$	(2)	Fe	$4sp^3d^6$	(6)
Sc	$4sp^3d$	(3)	Co	$4sp^3d^7$	(5)
Ti	$4sp^3d^2$	(4)	Ni	$4sp^3d^8$	(4)
V	$4sp^3d^3$	(5)	Cu	$4sp^3d^9$	(3)
Cr	$4sp^3d^4$	(6)	Zn	$4sp^3d^{10}$	(2)
Mn	$4sp^3d^5$	(7)			

The fourth and fifth long periods would have the same starting sequence starting with Sr and Ba. Cd and Hg establish the strengths of the s-p bonds in the absence of any d electrons for the third and fourth transition series. Although the cohesive energy does track fairly well with the number of bonding electrons, as shown in Fig. 8.17, there is an obvious departure from the general parabolic behavior at the center of

Table 8.13. Maximum bcc solid solubility limits of selected elements in Cr, Mo, and W.[a]

		Experimental		
Solute	Calculated	Cr	Mo	W
Tc	50	—	50	—
Te	50	48.5	42	37
Ru	25	34	30	23
Os	25	31	19	18.5
Rh	16.7	~15	~20	~6
Ir	16.7	~12	16	~10
Pd	12.5	~5	6.5	~5
Pt	12.5	~10	15	4

[a] Values in at.% solute.

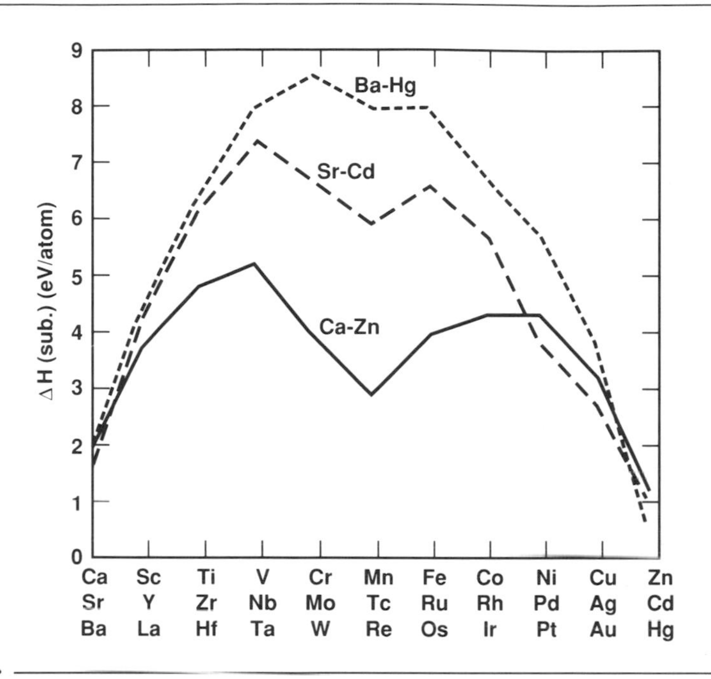

Figure 8.17.

Sublimation energies for transition metals of the three long periods. (After P. A. Cox, *The Electronic Structure and Chemistry of Solids*, Oxford University Press, p. 69, 1987).

each series. This deviation is explainable, at least qualitatively as follows. As the number of electrons per atom increases, their mutual electrostatic repulsion is minimized by maximizing the total spin; i.e., the optimum number of parallel spins maximizes the distance among intra-atomic electrons with the same angular momentum (this is the so-called Hund's rule in the case of free atoms). However, this leads to a lowering of the number of spin-coupled bonding pairs in solids. Consequently, for specific elements in which the bonding energy does not overcome the unpairing energy, the total cohesive energy declines. This argument is reinforced by the appearance of the magnetic elements, Fe, Co, and Ni, after the minimum at Mn, which gives irrefutable evidence for the existence of unpaired electrons.

F. Altmann, Coulson, Hume-Rothery Scheme

The perceptive reader might find it contradictory to exclude *d* electrons from playing any role in the determination of the crystal structure of metals in view of their evident effect upon the bond strength (see, e.g., Fig. 8.17) and obvious directionality. An approach that takes account of *d*-electron participation has been proposed by

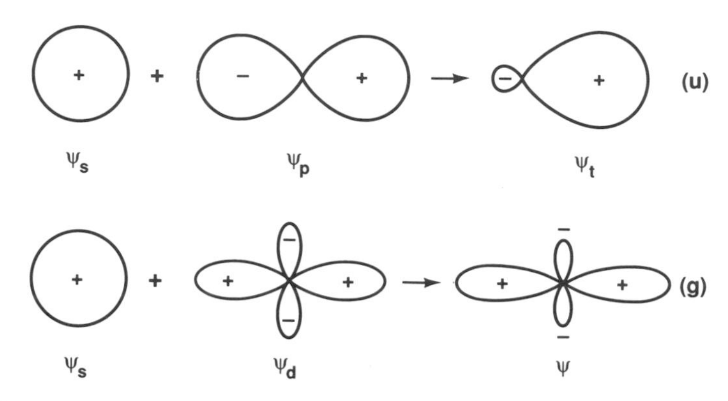

Figure 8.18.

Hybrids formed from *s*, *p*, and *d* atomic orbitals. ψ_s wave functions produce only ψ_g hybrids, whereas $\psi_g + \psi_u$ make only ψ_u hybrids.

Altmann, Coulson, and Hume-Rothery[21,22] in contradistinction to that of Engel and Brewer, which does not.

Hybrid orbitals that are made up of symmetrical wave functions, *gerade* (g), have inversion symmetry and are also *gerade* wave functions.[22] The combination of antisymmetric, *ungerade* (u), wave functions leads to an *ungerade* hybrid. An example of this is given by Fig. 8.18, showing the combining of ψ_s(g) with ψ_p(u) and ψ_s(g) with ψ_d(g).

With this in mind, the basic tenents of the Altmann–Coulson–Hume-Rothery scheme are as follows:

1. The (*u*) hybrid develops a highly directional charge distribution, well-adapted to the formation of strong localized bond. The (*g*) hybrid orbitals retain the charge symmetry about the nucleus and are more suited to the creation of delocalized electrons. The (*g*) hybrids form weaker bonds than do the (*u*)-type hybrids.

2. The total energy of the crystal referenced to the free atom includes the energy necessary for hybridization, which we said, in Chapter 7, consisted of a positive promotion energy plus the negative bonding energy. Using the band theory approach, we set the formation energy of the crystal at 0K equal to

$$\Delta E_f = E_A + E_F + E_0,$$

where E_A is the electron energy of the free atom, E_F is the mean Fermi energy, and E_0 is the potential energy of the electron in the crystal. Figure 8.19 shows the energies

[21] S. L. Altmann, C. A. Coulson, and W. Hume-Rothery, *Proc. Roy. Soc*, **A240**, 145 (1957).
[22] This presentation follows D. M. Adams, *Inorganic Solids*, pp. 268-275, Wiley, New York, (1974).

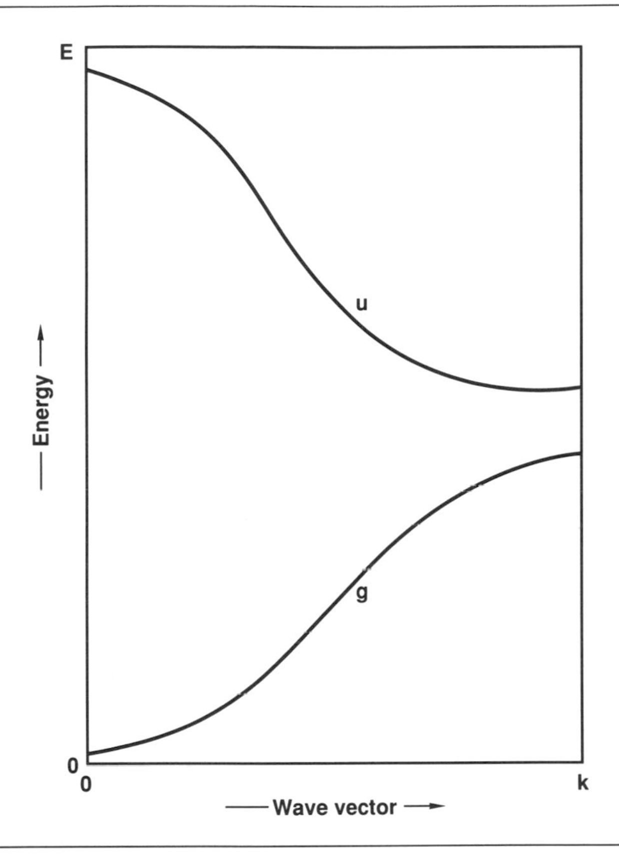

Figure 8.19.

The energy levels for the (u) and (g) hybrids as a function of **k**. Note that the energy of the (g) states is lowest. (After S. L. Altmann *et al.*, p. 150, loc. cit.)

of the two types of hybrid orbitals as a function of the wave vector **k** ($k = 2\pi/\lambda$; see Chapter 7, Section 5D). Because the density of states is greater at lower energy for the (g) orbitals, E_F can be assumed to be lower for the (g) state than for the (u). Hence, bonding favors the use of (g) hybrid orbitals over (u).

3. A lattice of N atoms bonded covalently requires $2N$ electrons. However, if only one-half the orbitals participate in the bonding, viz. the (g) orbitals, only N electrons are required. Thus, instead of being an insulator, the crystal has a half-filled first Brillouin zone and therefore is metallic. This situation is analogous to that of benzene, where the half-filled π orbitals contain the delocalized electrons, whereas the sp^2 hybrids provide normal covalent single bonds between adjacent carbon atoms.

The appropriate hybrid orbitals for the bcc, fcc, and hcp structures are obtained from matching the symmetry of the orbitals with that of the appropriate lattice. They are shown diagrammatically in Fig. 8.20 and are listed in Table 8.14.

Table 8.14. Hybrid orbitals compatible with metal crystal structures.[a]

Crystal Structure		Hybrids
Body-centered cubic	sd^3	[nearest neighbors (pure g)]
	d^4	
	d^3	[Next-nearest neighbors (pure g)]
Cubic close-packed	p^3d^3	(mixed g and u)
	sd^5	(pure g)
Hexagonal close-packed	sd^2	Neighbors in c.p. plane (pure g)
	pd^5	[Neighbors in planes above and
	spd^4	below (mixed g and u)]

[a] After S. L. Altmann *et al.*, p. 270, loc. cit.

The total bonding in the bcc structure is assumed to be the result of resonance among the sd^3, d^3, and d^4 hybrids.

The fcc p^3d^3 hybrids have lobes directed to the corners of a trigonal antiprism. The resonance of these four hybrid orbitals about the threefold axis of symmetry accounts for the bonding to the 12 nearest neighbors. Contrary to the Engel–Brewer rules, the assumption that the p^3d^3 determines the structure implies that the s states are without influence. Nevertheless sd^5 hybrids also have the symmetry appropriate for bonding in the fcc lattice.

It is postulated that bonding to the six in-plane neighbors occurs via the sd^2 hybrid in the hcp system. The arrangement of the three atoms above and below the central plane is not centrosymmetric, and the hybrids spd^4 and pd^5 have the appropriate symmetry.

This scheme as outlined merely tells us which hybrids have symmetries compatible with that of the nearest neighbors about a central atom in each of the three Bravais lattices. It says nothing about bond strength or the degree to which each hybrid is populated by the s, p, and d electrons. In the original proposal[23], the proportional contribution of the d orbital, f_d, was defined by

$$\text{bcc} \quad (sd^3)^a(d^4)^b(d^3)^c \qquad f_d = \frac{3a + 4b + 3c}{4a + 4b + 3c},$$

$$\text{fcc} \quad (p^3d^3)^a(sd^5)^b \qquad f_d = \frac{3a + 5b}{6a + 6b},$$

$$\text{hcp} \quad (spd^4)^a(pd^5)^b(sd^2)^c \qquad f_d = \frac{4a + 5b + 2c}{6a + 6b + 3c}.$$

[23] S. L. Altmann *et al.*, loc. cit.

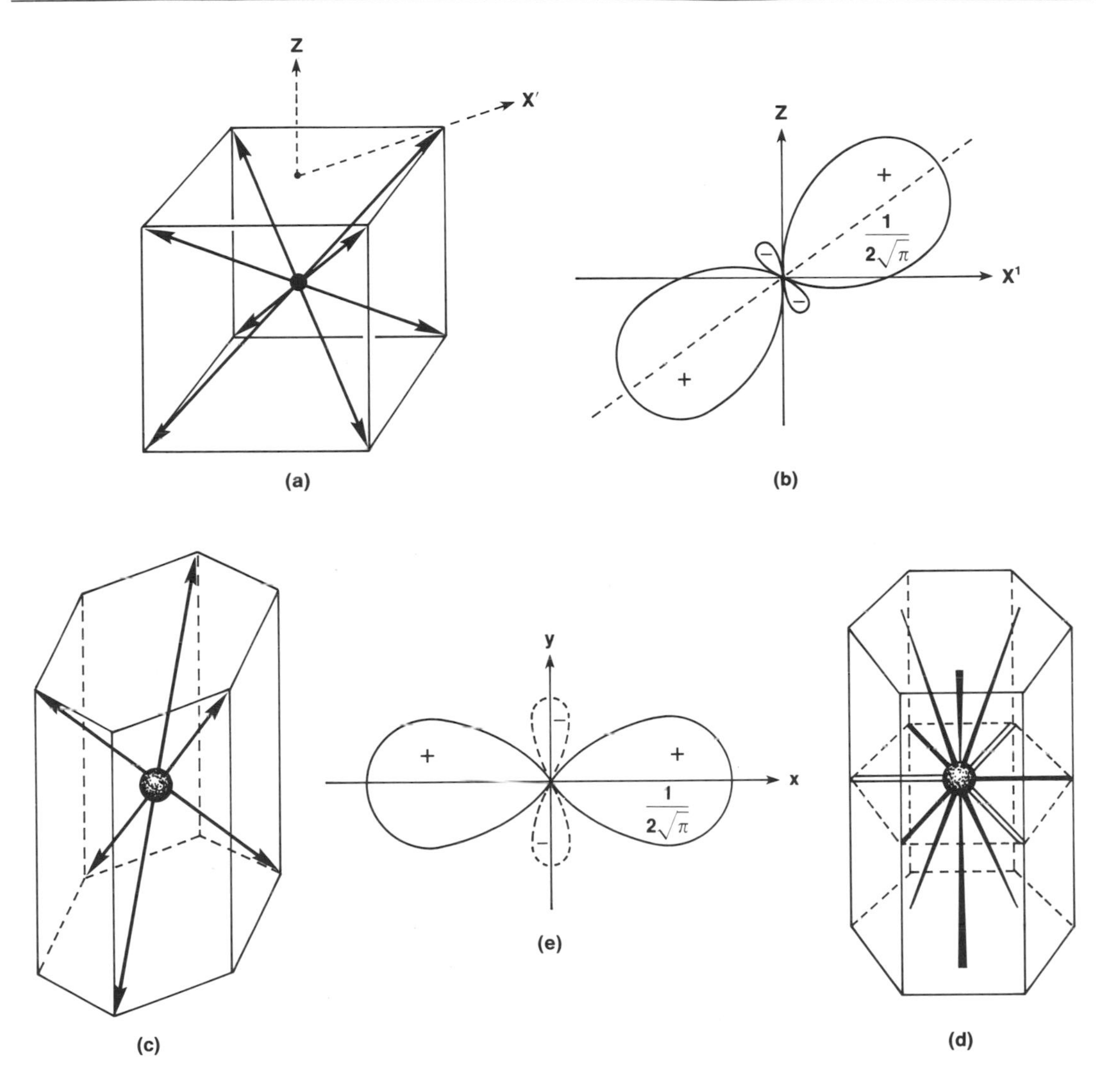

Figure 8.20.

The lobes of sd^3 and d^4 hybrids point towards the vertices of a cube and are represented in (a) by arrows. An sd^3 hybrid is shown to scale in (b), the d orbitals being from the t_{2g} set. (c) The basic unit for the fcc structure, a trigonal prism with p^3d^3 or sd^5 hybrids. (d) The hcp structure showing the disposition of mixed pd^5, spd^4 hybrids (above and below central plane), and sd^2 hybrids (from s, $d_{x_2-y_2}$, and d_{xy}) in the plane. (e) Detail of an sd^2 hybrid, drawn to same scale as Figure 8.17b. (Adapted from D. M. Adams, loc. cit. as taken from Altmann *et al.*, loc. cit.)

The values of a, b, and c are unknown and extremely difficult to estimate. However, for the sake of comparison let us assume that $a = b = c$. Then $f_d(\text{bcc}) = 0.91$, $f_d(\text{Hcp}) = 0.73$, and $f_d(\text{fcc}) = 0.66$. Hence, the participation of d electrons is in the order bcc > hcp > fcc. This trend is in reasonable agreement with the sequence of structures observed in the 3*d*, 4*d*, and 5*d* transition elements. The elements at the beginning and ends of the three rows have the hexagonal and fcc structures, respectively. Those at and about the middle of each row, where the maximum number of unpaired d electrons occurs, tend to be body-centered cubic.

6. Metallic Compounds

Metals need not be composed solely of metallic constituents. Many ionically bonded compounds show a strong metallic character. This is particularly true for transition metal oxides, sulfides, phosphides, and tellurides, but there are many others as well. Such compounds have the high electrical conductance and reflectivity that is customarily associated with conventional metals, but because of the predominantly ionic bonding do not share their ductility. The high conductivity in this case is not the result of electron hopping, as sometimes occurs with mixed valence compounds, nor is the current carried by the diffusion of ions. Just as for ordinary metals, it originates with partially filled bands of delocalized electrons. However, unlike ordinary metals, these conduction bands only involve the metal cations. Regardless of the theoretical approach, this metallic behavior is ascribed to the overlapping of the cationic wave functions. Unfortunately, once again the mathematical requirements of the relevant theory lie well beyond the general level of this text. The reader wishing to pursue this topic at an advanced level should consult the following reviews: D. Adler, *Solid State Physics*, vol. 21, p. 1, Academic Press, 1968; J. B. Goodenough, *Progress in Solid State Chemistry*, vol. 5, p. 149, Pergamon Press, 1971; N. F. Mott, *Metal-Insulator Transitions*, Barnes and Noble, 1974.

The bonding between nearest-neighboring cations is primarily of the form *cation–anion–cation*. The conduction band is formed from the interactions between unpaired *d* electrons or *s-d* hybrids. The bandwidth increases with decreasing separation. Current theory ascribes to intra-atomic *d*-electron repulsion the underlying driving force for band formation rather than the molecular orbital splitting because of the Pauli principle.[11]

Transition metal oxides are the largest and most thoroughly studied category of metallic compounds. The degree of metallicity is frequently a strong function of stoichiometry. An example is TiO, which behaves as a metal when stoichiometric but is otherwise a semiconductor. TiO has the rutile structure. Ti_2O_3 is semiconducting and has the corundum structure. However, the substitution of as little as 2% V turns pure Ti_2O_3 into a metal at all temperatures. ReO_3 has a distorted perovskite structure, and at room temperature possesses an electrical conductivity only about an order of magnitude less than that of copper. From this, it is seen that metallic behavior is not dependent upon a restrictive class of crystal structures. Other examples of metallic salts are LaI_2 and LaB_6, showing that this phenomenon is not restricted to transition metal cations or to oxides alone. The role of nonstoichiometric vacancies can also be important in determining the degree of metallicity, as is believed to be the case with TiO. Most compounds must still be examined on a case-by-case basis, and *ab initio* predictions of their properties frequently fail, although general trends can usually be rationalized.

Chapter 8 Exercises

1. Derive the expression for the Fermi energy of a free-electron metal (i.e., all electrons belonging to a free-electron gas) in terms of its density and atomic number.

2. Find the number of electron states per atom inside a sphere that just touches the faces of the first Brillouin zone of the simple cubic, body-centered cubic, and face-centered cubic lattices. (The Brillouin zones of the primitive cubic are bounded by $\{100\}$, the bcc by $\{110\}$, and the fcc by $\{111\}$, $\{200\}$.)

3. The lowest state of the valence electron is given by (8.2) by the Wigner–Seitz method subject to the boundary condition $(\delta\psi_0/\delta r)_{r_0} = 0$. Find the approximate value of E_0, assuming $\psi_0 = e^{-r} + e^{r-r_0}$.

4. Suppose that a single valence electron, as in the alkali metals, is distributed uniformly throughout a sphere of radius r_0 about each ion. Show that the coulombic energy per electron is 29.49 a_0/r_0 eV/electron, where a_0 is the Bohr radius.

Additional Reading

N. F. Mott and H. Jones, *The Theory of the Properties of Metals and Alloys*, Dover (1958). Although first published in 1936, this classic continues to provide an excellent, easily understood introduction to theories of metallic properties.

Wm. Hume-Rothery, *The Structure of Metals and Alloys*, The Institute of Metals (1956). The first edition of this volume, like the preceding, appeared in 1936. However, because it deals with the observed relations between atomic constitution and crystal structure, it remains a current, highly readable, factual account.

S. Raimes, *The Wave Mechanics of Electrons in Metals*, Interscience Publishers (1961). An excellent pedagogical text from which to learn the mathematics needed to describe electrons in metals.

D. G. Pettifor, *Solid State Physics*, vol. 40, Academic Press (1987). Gives lucid and comprehensive review of the Miedema method and an extensive bibliography.

V. Heine, *Solid State Physics*, vol.24, Academic Press (1970). Presents the fundamental approximation involved in the pseudopotential approach in a reasonably clear fashion. (Unfortunately, we have not discovered a well-written, reasonably elementary, short tutorial article on pseudopotential theory.)

P. S. Rudman, J. Stringer, and R. I. Jaffee, *Phase Stability in Metals and Alloys*, Part 1–3, McGraw-Hill (1966). L. Brewer presents a chemist's approach to phase stability. There are also several other articles by various authors that are generally nonmathematical and will be useful for the student wishing to learn more about metallic behavior.

Chapter 9

Polyphase Equilibrium

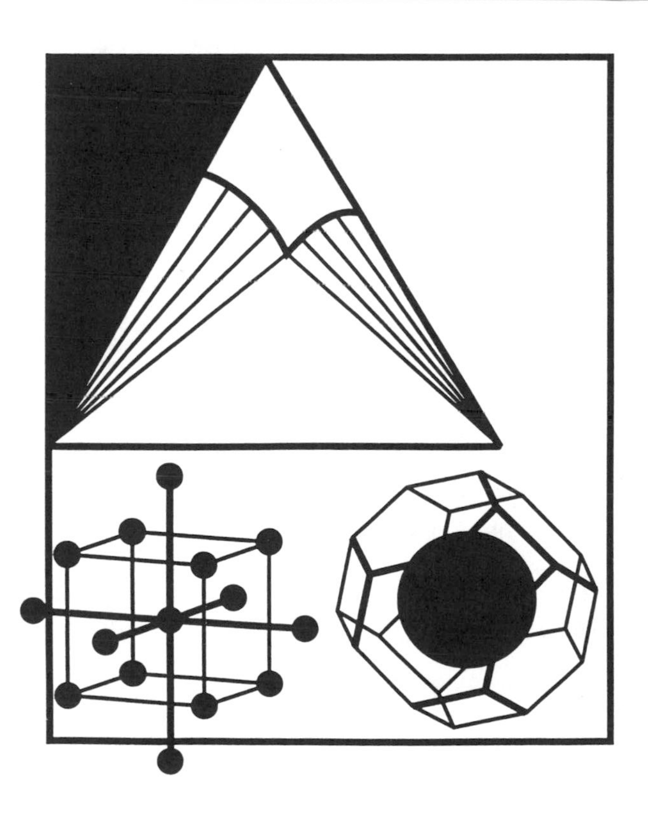

Chapter 9

Contents

The fascination of a growing science lies in the work of the pioneers at the very borderline of the unknown, but to reach this frontier, one must pass over well travelled roads; of these one of the safest and surest is the broad highway of thermodynamics.

—G. N. Lewis and M. Randall, *Thermodynamics*, McGraw–Hill (1923)

Polyphase Equilibrium

When solids or reaction products exceed their saturation limits in a solid solution, precipitation occurs, creating one or more additional phases, sometimes with surprising morphologies, such as the examples shown in Fig. 9.1. Then, the equilibrium state of the system becomes a mechanical mixture of the distinguishable phases, the amount and composition of each being determined by the temperature, pressure, and the total concentrations of the chemical constituents. Understanding the thermodynamic relations that govern such polyphase systems is a subject of considerable practical importance. Most commercial alloys, such as the many types of steels, contain several different phases, as do rocks and industrially useful ceramic materials, and, in fact, solid single-phase systems are infrequently encountered outside of the laboratory.

Because solid solutions are so frequently encountered, before we begin our examination of polyphase or heterogeneous equilibria, we will review the fundamental thermodynamics of single-phase solutions.

1. Ideal Solutions

The formation of an ideal solution does not generate heat when the constituents are combined; i.e., the enthalpy of mixing, ΔH_{m}, is zero. This is because they combine to form a random configuration, all the chemical bonds being of equal strength between all the atoms, ions, or molecules regardless of chemical species. As might be expected, constituents that have nearly the same physical and chemical properties in

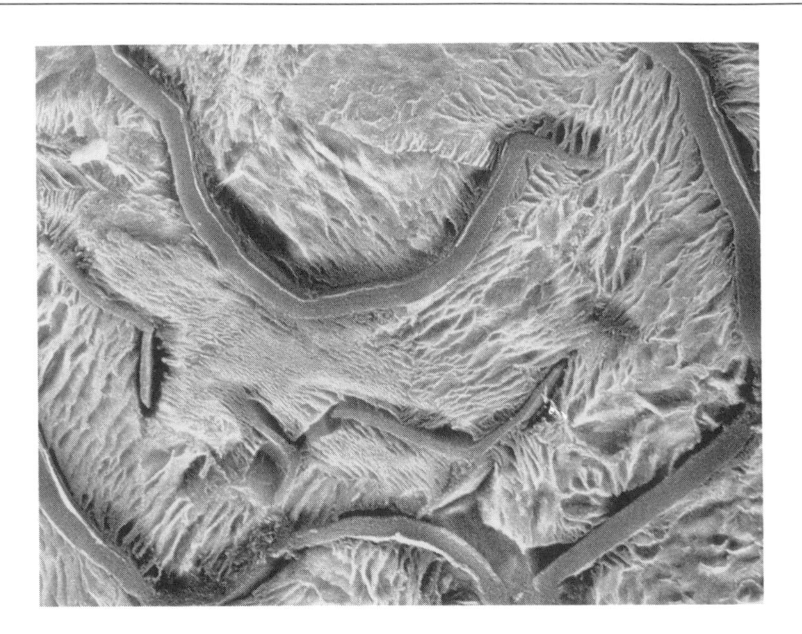

Figure 9.1.

Graphite flakes with pearlite. Scanning electron micrograph. (Courtesy of Rosemarie Koch, Center for Materials Research, Stanford University.)

the pure state; e.g., ionic or atomic radius, charge, crystal structure, etc., tend to form solutions that are more ideal than those of dissimilar components. The change in the free energy of the system that results from mixing the constituents to form an ideal solution derives solely from the increase in entropy. This result is easily shown, starting with (9.1),

$$\Delta G_{\mathrm{m}} = \sum_i X_i \, \Delta\mu_i, \tag{9.1}$$

which states that the Gibbs free energy of mixing, ΔG_{m}, for 1 mole of solution is simply the sum of the changes in the individual chemical potentials or partial molar free energies, $\Delta\mu_i$, for each of the i components multiplied by their respective mole fractions. The chemical potentials in turn are given by

$$\mu_i = \mu_i^\circ + RT \ln(P_i/P_i^\circ), \tag{9.2}$$

which assumes that the solution is ideal and that the perfect gas law is obeyed, μ_i° is the chemical potential of an arbitrarily defined standard state, and P_i and P_i° are the vapor pressure of the ith component in solution and in the pure state, respectively.

Thus, rearranging (9.2) gives

$$\Delta\mu_i = RT \ln(P_i/P_i^\circ). \tag{9.3}$$

Raoult's law not only is valid for, but also can serve as the definition of ideal solutions, and is

$$P_i = X_i P_i^\circ. \tag{9.4}$$

Now substituting for P_i/P_i° from (9.4), we obtain, as $\Delta H_m = 0$, for (9.1),

$$\Delta G_m = RT \sum_i X_i \ln X_i, \tag{9.5}$$

thus equating the free energy to the ideal entropy of mixing times the temperature. Because X_i is always a fraction ΔG_m is always negative or zero (at $T = 0$).

The same result also is quickly derived by statistical methods. We define a crystal to have N lattice sites upon which we randomly place the (i) constituents so that

$$\sum_i N_i = N. \tag{9.6}$$

The configurational entropy, ΔS_c, of this ensemble of localized atoms is given by

$$\Delta S_c = k \ln \Omega, \tag{9.7}$$

where the total number of distinguishable configurations is given by

$$\Omega = \frac{N!}{N_1! N_2! \cdots N_i!}. \tag{9.8}$$

Substituting (9.8) into (9.7) yields

$$\Delta S_c = k \ln N! - k \sum_i \ln N_i!. \tag{9.9}$$

Applying Sterling's approximation and gathering terms leads to

$$\Delta S_c = -kN\left[\sum_i \frac{N_i}{N} \ln N_i - \ln N\right]. \tag{9.10}$$

Because the sum of the mole fractions must add up to unity, i.e.,

$$\frac{N_1}{N} + \frac{N_2}{N} + \cdots + \frac{N_i}{N} = \sum X_i = 1, \tag{9.11}$$

we can combine terms in (9.11) to obtain

$$\Delta S_c = -kN\left[\sum \frac{N_i}{N} \ln \frac{N_i}{N}\right] = -R \sum X_i \ln X_i, \tag{9.12}$$

which, when substituted into the equation for ΔG_m, viz. $\Delta G_m = \Delta H_m - T\Delta S_m$, yields (9.5) since $\Delta H_m = 0$ by definition. ΔS_c, the configurational entropy, is clearly identical to the entropy of mixing.

The entropy of mixing for an ideal binary solution has its maximum at the equimolar composition about which it is symmetrically disposed according to (9.12), as is easily shown by direct differentiation. Because of (9.5), there is a corresponding minimum in the free energy of mixing at the 50–50 composition, which decreases with increasing temperature as shown by Fig. 9.2

The enthalpy of mixing for nonideal solutions is, of course, nonzero, and the complete equation for the free energy of formation, or mixing, is

$$\Delta G_m = \Delta H_m - T\,\Delta S_m.$$

The contribution of the $T\,\Delta S_m$ term generally outweighs that from ΔH_m. For example, for the 50–50 CuNi alloy, $\Delta H_m = 425$ cal, whereas $T\,\Delta S_m = 1.33 \times 10^3$ cal, thus being over three times larger at 973 K. At 423 K, the solid solution $Bi_{.6}Tl_{.4}$ is formed, with $\Delta H_m = -532$ cal and $T\,\Delta S_m = 691$ cal. Although $T\,\Delta S_m$ obviously goes to zero with the temperature doing the same, ΔH_m has a weak temperature dependence. This is because $C_p(T)$(reactants) $- C_p(T)$(product) is generally quite small. At absolute zero, ΔH_m is simply the difference between the bond energies of the reactants and the resultant solution.

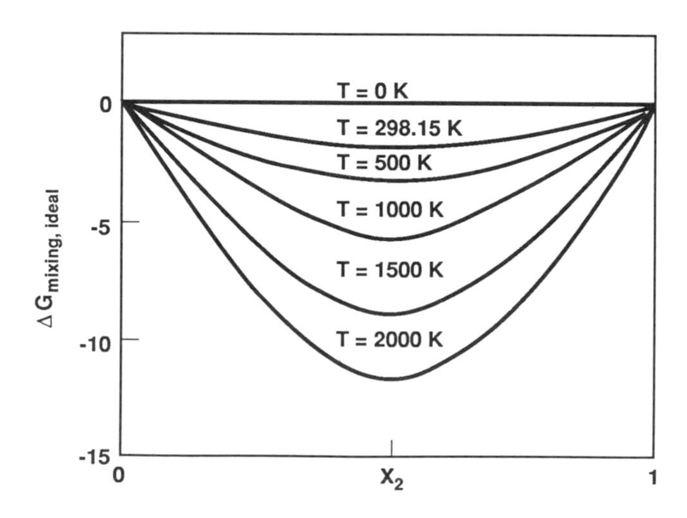

Figure 9.2.

Free energy of ideal mixing in binary system as a function of temperature.

2. The Liquid–Solid Phase Diagram for Ideal Binary Solutions

Our discussion of the representation and interpretation of phase equilibria begins with the detailed examination of the binary phase diagram for an ideal solid solution that melts to form an ideal liquid. Such phase or, as they are sometimes called, equilibrium diagrams (we will use both terms interchangeably) are closely approximated by systems in which both constituents are closely alike in atomic[1] size and chemical properties. This conforms to a general qualitative rule that governs miscibility, viz. that species with similar characteristics tend to be mutually soluble. Thus, ionic compounds dissolve in polar solvents, hydrocarbons, on the other hand, in nonpolar solvents, and metals in liquid or molten metals. This idea is intuitively appealing. For example, consider what happens when one tries to dissolve, say, NaCl in benzene, C_6H_6. This solvent has no permanent dipole moment to screen the ionic charges of Na^+ and Cl^-, which continue to exert a coulombic attraction, and, hence, the salt remains insoluble. Water, on the other hand, creates solvent cages about the ions with the positive end of the dipole, which is H, screening the Cl^- ions and the negative O doing the obverse for the Na^+. Compounds such as CH_3OH and C_2H_5OH have both polar, OH, and nonpolar, CH_3, C_2H_5, groups, but because the hydroxyl group is dominant they both are completely miscible with H_2O. However, as the nonpolar portion of the molecule, C_nH_{2n+}, increases, the compound becomes less and less soluble in polar solvents until ultimately it will only dissolve in nonpolar solvents.

An extension of this reasoning has been applied with fair success to metallic solid solutions. These predictions are frequently referred to as the Hume-Rothery rules, named after the metallurgist first to formalize them. There is appreciable solubility when the constituents are less than about 15% different in size, have the same crystal structure, exhibit no appreciable difference in electronegativity, and are of the same valence. Elements having the most valence electrons will dissolve to a greater extent in those having fewer, rather than the reverse. When these conditions are not satisfied, the solubility is generally severely limited. Frequently, one or more intermediate phases (compounds) will form in such cases.

The Cu–Ni system shown in Fig. 9.3 closely approximates the ideal symmetrical cigar-shaped phase diagram for such systems. The upper boundary of the two-phase area is called the liquidus, the lower one the solidus.

The compositions of the liquids and solids in equilibrium with one another within the two-phase region are given at each temperature by the intersection of the

[1] Although for brevity we will continue to use the word "atom" or "atomic", the reader should infer the terms "ionic" or "molecular", when applicable.

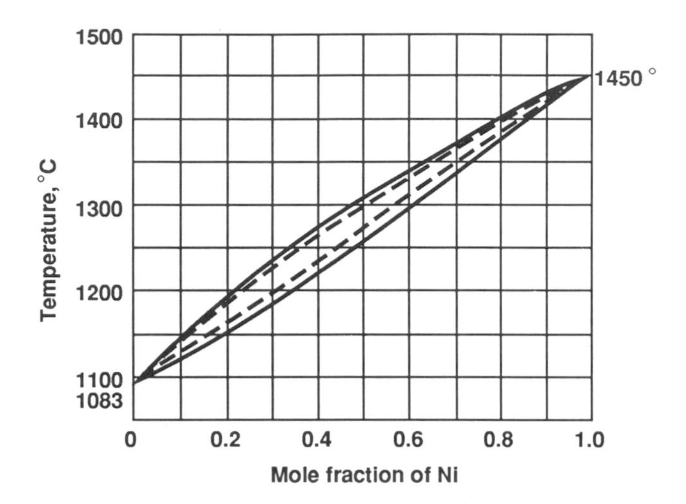

Figure 9.3.

The phase diagram for the Cu–Ni system is described by the solid lines, and the one calculated from the ideal solution model is outlined by the dashed lines. The latter was calculated by means of (9.25).

horizontal isotherm with the liquidus and solidus. For example, at 1300°C a liquid having ~50 at.% Ni is in equilibrium with a solid containing ~60% Ni as given by the intersection of a tie line with the phase boundaries.

The composition of the two equilibrium phases is necessarily different at all temperatures. This comes about because of the change in standard state in going from solid to liquid, and vice versa, depending upon the melting temperature of each pure component. This is made explicit by deriving the thermodynamic equation relating the compositions of the equilibrium solid and liquid phases.

Consider Fig. 9.4, which represents the phase diagram at a fixed pressure for an ideal solid solution in equilibrium with its, also ideal, melt phase. Equilibrium is premised upon the condition that the free energy of the entire system is at a minimum, and, in addition, the chemical potential of each component has the same value in all coexisting phases. These are the constraints for the derivation of a general expression for the shape and extent of the two-phase region shown in Fig. 9.4.

We begin by writing the equation for the total free energy of formation of the liquid phase,

$$\Delta G_{\mathrm{l}} = X_{\mathrm{B}}^{\mathrm{l}} \Delta G_{\mathrm{B}}^{\mathrm{m}} + RT(X_{\mathrm{A}}^{\mathrm{l}} \ln X_{\mathrm{A}}^{\mathrm{l}} + X_{\mathrm{B}}^{\mathrm{l}} \ln X_{\mathrm{B}}^{\mathrm{l}}), \tag{9.13}$$

where $X_{\mathrm{B}}^{\mathrm{l}}$ is the mole fraction of component B in the liquid, and $\Delta G_{\mathrm{B}}^{\mathrm{m}}$ is the free energy of melting of pure B at the temperature specified for the calculation of ΔG_l.

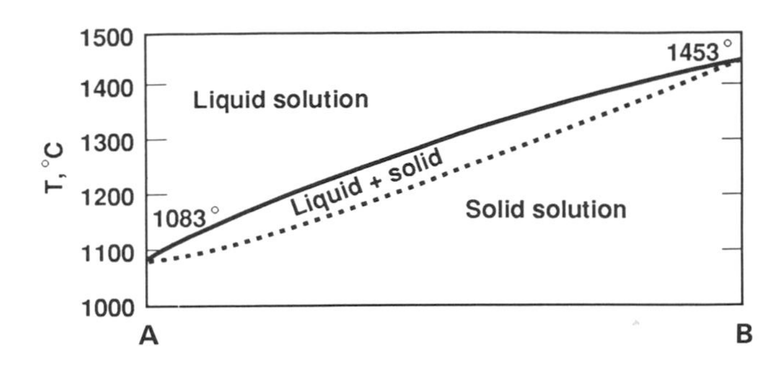

Figure 9.4.

Phase diagram for a binary system with ideal solid and ideal liquid solutions.

The second term on the right-hand side is the ideal entropy of mixing, which has been already discussed. At equilibrium the chemical potential of each component is the same in all coexisting phases. Hence, the chemical potentials for each component are the same in the solid as in the liquid. Hence,

$$\mu_{\mathrm{B}}^{\mathrm{l}} = \mu_{\mathrm{B}}^{\mathrm{s}}.$$

But, by definition, for ideal solutions

$$\mu_{\mathrm{B}}^{\mathrm{l}} = \mu_{\mathrm{B}}^{\circ,\mathrm{l}} + RT \ln X_{\mathrm{B}}^{\mathrm{l}}$$

and

$$\mu_{\mathrm{B}}^{\mathrm{s}} = \mu_{\mathrm{B}}^{\circ,\mathrm{s}} + RT \ln X_{\mathrm{B}}^{\mathrm{s}}.$$

Equating the above gives

$$\mu_{\mathrm{B}}^{\circ,\mathrm{l}} - \mu_{\mathrm{B}}^{\circ,\mathrm{s}} = RT \ln(X_{\mathrm{B}}^{\mathrm{s}}/X_{\mathrm{B}}^{\mathrm{l}}).$$

Now at any temperature below $T_{\mathrm{B}}^{\mathrm{m}}$, the equilibrium melting temperature of B at the appropriate specified pressure, $\mu_{\mathrm{B}}^{\circ,\mathrm{l}} - \mu_{\mathrm{B}}^{\circ,\mathrm{s}} > 0$, because of the free energy of melting at a nonequilibrium temperature. Hence, $X_{\mathrm{B}}^{\mathrm{s}}/X_{\mathrm{B}}^{\mathrm{l}}$ is finite and greater than unity. An analogous relation with the obverse final result holds for component A. However, the standard state for each of the components must be the same state (i.e., solid, liquid, or gas) as it is in the system for which the chemical potentials are specified.

Returning now to $\Delta G_{\mathrm{B}}^{\mathrm{m}}$, which is the free energy of melting of pure B at temperatures other than $T_{\mathrm{B}}^{\mathrm{m}}$, the equilibrium melting point, we calculate this quantity in terms of the enthalpy of fusion (latent heat of melting), using the Clausius–Clapeyron relation

$$\left(\frac{\partial(\Delta G_{\mathrm{B}}^{\mathrm{m}}/T)}{\partial T^{-1}}\right)_p = \Delta H_{\mathrm{B}}^{\mathrm{m}}. \tag{9.14}$$

Integrating (9.14) gives

$$\int_{T_{\rm B}^{\rm m}}^{T} d\left(\frac{\Delta G_{\rm B}^{\rm m}}{T}\right) = \int_{T_{\rm B}^{\rm m}}^{T} \Delta H_{\rm B}^{\rm m}\, d(T^{-1}), \tag{9.15}$$

and recalling that $\Delta G_{\rm B}^{\rm m} = 0$ at $T_{\rm B}^{\rm m}$ gives

$$\Delta G_{\rm B}^{\rm m}(T) = \Delta H_{\rm B}^{\rm m}(1/T - 1/T_{\rm B}^{\rm m}). \tag{9.16}$$

In an entirely analogous manner, we can derive an expression for the free energy of formation of the solid phase. This is given by

$$\Delta G_{\rm s} = X_{\rm A}^{\rm s}\,\Delta G_{\rm A}^{\rm f} + RT(X_{\rm A}^{\rm s} \ln X_{\rm A}^{\rm s} + X_{\rm B}^{\rm s} \ln X_{\rm B}^{\rm s}), \tag{9.17}$$

where, $X_{\rm A}^{\rm s}\,\Delta G_{\rm A}^{\rm f}$ is the mole fraction of A in the solid times the free energy of freezing and accounts for the fact that component A in the pure state is a liquid across the entire two-phase region. Therefore, we must again use a fictitious standard state, viz. solid A above its equilibrium melting point. The free energy of freezing is the negative value of the free energy of melting, so $\Delta G_{\rm A}^{\rm f}$ is now given by

$$\Delta G_{\rm A}^{\rm f} = -\Delta H_{\rm A}^{\rm m}(1/T - 1/T_{\rm A}^{\rm m}). \tag{9.18}$$

Substituting (9.16) and (9.18) into (9.13) and (9.17), respectively, gives

$$\Delta G_{\rm l} = -X_{\rm B}^{\rm l}\,\Delta H_{\rm B}^{\rm m}(1/T - 1/T_{\rm B}^{\rm m}) + RT(X_{\rm A}^{\rm l} \ln X_{\rm A}^{\rm l} + X_{\rm B}^{\rm l} \ln X_{\rm B}^{\rm l}), \tag{9.19}$$

$$\Delta G_{\rm s} = X_{\rm A}^{\rm s}\,\Delta H_{\rm A}^{\rm m}(1/T - 1/T_{\rm A}^{\rm m}) + RT(X_{A}^{\rm s} \ln X_{\rm A}^{\rm s} + X_{\rm B}^{\rm s} \ln X_{\rm B}^{\rm s}). \tag{9.20}$$

By definition,

$$\frac{\partial\,\Delta G_{\rm l}}{\partial X_{\rm B}^{\rm l}} \equiv \mu_{\rm B}^{\rm l} \tag{9.21}$$

and

$$\frac{\partial\,\Delta G_{\rm s}}{\partial X_{\rm B}^{\rm s}} \equiv \mu_{\rm B}^{\rm s}, \tag{9.22}$$

and at equilibrium $\mu_{\rm B}^{\rm l} = \mu_{\rm B}^{\rm s}$.

Performing the differentiations stated in (9.21) and (9.22) and equating the chemical potentials results in

$$\ln \frac{X_{\rm B}^{\rm s}}{X_{\rm B}^{\rm l}} = \frac{\Delta H_{\rm B}^{\rm m}}{R}\left(\frac{1}{T} - \frac{1}{T_{\rm B}^{\rm m}}\right). \tag{9.23}$$

By an identical procedure we also obtain

$$\ln \frac{X_{\rm A}^{\rm s}}{X_{\rm A}^{\rm l}} = \frac{\Delta H_{\rm A}^{\rm m}}{R}\left(\frac{1}{T} - \frac{1}{T_{\rm A}^{\rm m}}\right). \tag{9.24}$$

There are additional constraints on the system, namely material balance in each phase. Thus,

$$X_B^s + X_A^s = 1, \qquad X_B^l + X_A^l = 1.$$

These relations in combination with (9.23) and (9.24) give, by algebraic manipulation,

$$X_B^l = \frac{1 - \exp[(\Delta H_A^m/R)(1/T - 1/T_A^m)]}{\exp[(\Delta H_B^m/R)(1/T - 1/T_B^m)] - \exp[(\Delta H_A^m/R)(1/T - 1/T_A^m)]}, \tag{9.25}$$

and

$$X_B^s = X_B^l \exp\left[\frac{\Delta H_B^m}{R}\left(\frac{1}{T} - \frac{1}{T_B^m}\right)\right].$$

Thus, knowing the melting points and the heats of fusion, we may calculate the complete solidus and liquidus curves. The calculated curves of Fig. 9.3 were obtained by this procedure.

A system that behaves in nearly ideal fashion will conform, but not exactly follow, these equations over the entire range of composition. The Cu–Ni shown in Fig. 9.3 is such an example.

3. Nonideal Solutions

Those solutions whose departure from ideality is the easiest to deal with have their chemical potentials or partial molar free energies given by

$$\begin{aligned} \bar{G}_A - G_A^\circ &= RT \ln a_A = RT \ln \gamma_A X_A, \\ \bar{G}_B - G_B^\circ &= RT \ln a_B = RT \ln \gamma_B X_B. \end{aligned} \tag{9.26}$$

So for this simple class of solutions the thermodynamic activity differs from ideality by a temperature and composition-dependent multiplicative term γ, called the activity coefficient. The excess free energy of formation, ΔG_f^x is defined by subtracting the ideal free energy of formation from Eqs. (9.26):

$$\Delta G_f^x = (X_A \Delta\bar{G}_A + X_B \Delta\bar{G}_B) - (X_A \Delta\bar{G}_A^i + X_B \Delta\bar{G}_B^i), \tag{9.27}$$

where the ideal partial molar free energies are given by

$$\begin{aligned} \Delta\bar{G}_A^i &= RT \ln X_A, \\ \Delta\bar{G}_B^i &= RT \ln X_B. \end{aligned} \tag{9.28}$$

Consequently, the excess free energy of formation is given by

$$\Delta G_f^x = X_A\, RT \ln \gamma_A + X_B RT \ln \gamma_B. \tag{9.29}$$

We should here recall that the ideal solution model requires zero heat of mixing. The Bragg–Williams model, which we here introduce, attributes the entire deviation from ideality to the heat of mixing while preserving the concept of randomness for the entropy of mixing. This is expressed by

$$\Delta G_f^x = \Delta H_f^x. \tag{9.30}$$

Solutions described by this model are called *regular solutions*. A more restricted model, characterized by the relation

$$\Delta H_f^x = cX_A X_B \qquad \text{with c a constant,}$$

which is equivalent to the Bragg–Williams model described next, is also usually referred to as a *regular solution* in the literature.

A straightforward model-dependent expression can be developed for ΔH_f^x, the excess enthalpy of formation. The simplest of the models restricts the total bond energy of the atom or ion to only nearest-neighbor interactions. It also neglects the effect of distributions of bond types about a single atom upon its overall binding energy to the matrix. However, these limitations do not impose a serious error on our result, since about 90% of the total bond energy in most real solids is accounted for by the bonding between nearest neighbors. A more serious approximation is the assumption that allows us to calculate the configurational entropy on the basis of random mixing. This is, of course, in direct contradiction to our assumption of preferred types of bonds. We commence our derivation for a solid at 0 K which postpones the problem of accounting for the vibrational or thermal energy. In our usual notation, we let X_A and X_B be the mole fractions of components A and B, N the total number of atoms, and z the number of nearest neighbors. The probability of selecting an A atom at random from the ensemble is simply X_A. The average number of A atoms that are nearest neighbor to the A atom initially designated is zX_A. Now the total number of A – A pairs is given by

$$NzX_A^2/2 = N_{AA}, \tag{9.31}$$

and a similar expression describes the number of B – B pairs. The factor of 2 in the denominator corrects for otherwise counting each pair twice. The number of A – B pairs is given by

$$NzX_A(1 - X_A) = N_{AB}. \tag{9.32}$$

To each of the three types of bonds we assign different energies of formation, which in an obvious notation are ε_{AA}, ε_{BB}, and ε_{AB}. The internal energy of the system is now given by

$$E = \frac{NzX_A^2}{2}\varepsilon_{AA} + \frac{Nz(1 - X_A)^2}{2}\varepsilon_{BB} + NzX_A(1 - X_A)\varepsilon_{AB}$$

$$= \frac{Nz}{2}\left[X_A\varepsilon_{AA} + X_B\varepsilon_{BB} + 2X_AX_B\left(\varepsilon_{AB} - \frac{\varepsilon_{AA} + \varepsilon_{BB}}{2}\right)\right]. \tag{9.33}$$

The term

$$\Delta E_f = NzX_AX_B\left(\varepsilon_{AB} - \frac{\varepsilon_{AA} + \varepsilon_{BB}}{2}\right) \tag{9.34}$$

arises from the mutual interactions of A and B.

To extend (9.34) to temperatures above 0 K, we must add a heat capacity term, so that

$$\Delta E_f = NzX_AX_B\left(\varepsilon_{AB} - \frac{\varepsilon_{AA} + \varepsilon_{BB}}{2}\right) + \int_0^T C_V\, dT. \tag{9.35}$$

The ideal entropy of mixing is given by (9.12), and the thermal or vibrational contribution to the entropy is given by $\int_0^T (C_V/T)\, dT$. We have now arrived at our final expression, which is

$$\Delta F_f \simeq \Delta G_f = NzX_AX_B\left(\varepsilon_{AB} - \frac{\varepsilon_{AA} + \varepsilon_{BB}}{2}\right) + RT(X_A \ln X_A + X_B \ln X_B)$$

$$+ \int_0^T \Delta C_P\, dT + \int_0^T \frac{\Delta C_P}{T}\, dT. \tag{9.36}$$

Here we equated ΔF to ΔG and made the requisite corresponding change of C_V to C_P. The error involved in doing this at pressures of about 1 atm is negligible. Also the ΔV of mixing is generally small. We will often resort to this procedure throughout the text when the goal is to link formulas derived from models or statistical thermodynamics with experimental data that generally are obtained at constant pressure rather than constant volume.

At constant temperature, according to (9.36), the excess free energy becomes

$$\Delta G_f^x = NzX_AX_B\left(\varepsilon_{AB} - \frac{\varepsilon_{AA} + \varepsilon_{BB}}{2}\right) + RT(X_A \ln X_A + X_B \ln X_B). \tag{9.37}$$

$$\varepsilon \equiv \varepsilon_{AB} - \frac{\varepsilon_{AA} + \varepsilon_{BB}}{2}; \tag{9.38}$$

ΔG_f^x has a minimum symmetrical about $X_A = X_B = 0.5$ when

$$\varepsilon < \frac{2RT}{Nz}.$$

The minimum about $X_A = 0.5$ grows deeper as ε_{AB} becomes more negative. Positive ε's are discussed in the next section.

The ideal entropy of mixing also has a minimum symmetrical about $X_A = 0.5$, so that, neglecting the last two temperature-dependent terms in (9.36), ΔG_f^x is always symmetrical as shown in Fig. 9.5. In this figure we have identified $Nz\varepsilon$ with ΔH^x. More realistic theories requiring computers take account of the energy differences among locally different configurations and extend the interactions beyond just the nearest neighbors.

We now proceed to examine what happens when ΔH^x becomes increasingly positive rather than negative.

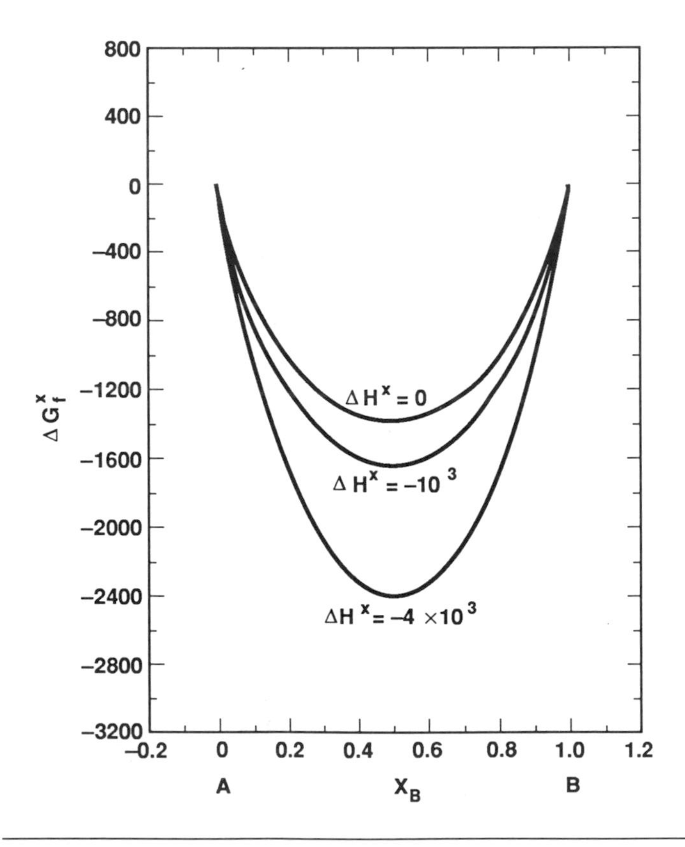

Figure 9.5.

The free energy of formation at $T = 1000$ K decreases from the ideal solution value as, ΔH^x is decreased from 0 to -1000 cal/mole to -4000 cal/mole.

4. Phase Separation

More often than otherwise, complete miscibility does not exist between the components, but only a limited range of solubility at each of the extrema (i.e., X_A, $X_B \simeq 0.1$), with two separate A-rich and B-rich solid solutions accounting for the intermediate range of concentrations. This separation into two equilibrium phases results from the rather sensitive balance between the enthalpic and entropic terms of (9.36). We have already remarked that the contribution of the entropy of mixing, $T\,\Delta S_m$, to the free energy of formation, ΔG_f, is always positive. However, the enthalpic terms ΔH_f, which stems from the energy of chemical bonding, may have either sign. A negative enthalpy signifies a preference for bonds between unlike atoms and, hence, an increased stability of the solution referenced to the pure components; a positive enthalpy, of course, reflects the opposite situation, which leads ultimately to phase separation. This can be shown by displaying a series of numerical solutions to (9.36), as shown in Fig. 9.6, employing the dimensionless parameter ε/RT.

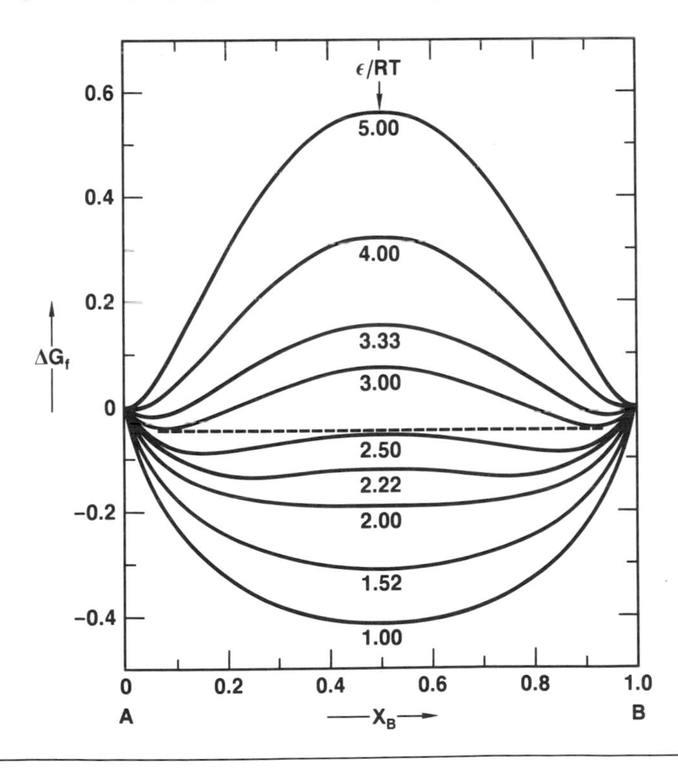

Figure 9.6.

The free energy of formation of binary solid solutions are given by (9.36) as a function of bond energies and temperature. (Adapted from J. M. Honig, *Studies in Modern Thermodynamics*, vol. 4, Elsevier, p. 266, 1982.)

The entropic contribution is a strong function of temperature; the enthalpy, to the contrary is not. The quantity ε, which is the binding energy at 0K, changes with increasing temperature only in response to the thermal expansion that increases the time-averaged bond length. As a crude approximate example, consider that the linear expansion of NaCl is $\sim$0.1% in going from 23 K to 273 K. If the bond energy scales as the reciprocal bond length (i.e., r^{-1}), as would be the case for purely coulombic interactions, a single bond between Na^+ and Cl^- then would decrease in energy by only 0.1%, or for the 12 nearest neighbors 1.2% over this range of 250 K. Also the temperature effects due to thermal vibrations that are incorporated into the ΔC_p term are small because the difference between the heat capacities of the reactants and products is generally small. Thus, the results of the isothermal calculations presented in Fig. 9.6 appear to be at least partially justified.

It is clear that when ε/RT is very large the system tends toward separation into its pure components. As the temperature increases, and ε/RT perforce decreases, two localized minima in the free energy commence to appear near the initial and terminal compositions. These correspond to two coexisting equilibrium solid solutions. Continuing downward as the temperature rises, the entire system tends toward a single-phase solid solution as the two minima are dragged toward the center of the diagram. A dashed line tangential to the localized minima for $\varepsilon/RT = 3.00$ designates the two solid solutions, with $X_B \cong 0.1$ and 0.9, respectively, which are here in equilibrium with each other.

We will return for a more thorough investigation of free-energy diagrams after discussing the phase rule and the practical business of reading phase diagrams. For now, it is sufficient to point out that the Bragg–Williams model is competent to establish the chemical conditions for the unmixing of a single solid solution into equilibrium separate phases, which will be dealt with again in the next chapter.

5. Equilibrium Among Phases

In the preceding sections, we sought to give the reader some familiarity with the subject of heterogeneous equilibria by introducing and discussing a few elementary examples and their relevant energetics. We now turn to the rigorous thermodynamic basis that governs the equilibrium among phases, before proceeding further to more complex cases.

A rigorous definition of the term "phase" must include the concept of a phase boundary. This phase boundary, by definition, is a surface that separates substances that are not congruent in all of their structural elements. Such elements include the crystal structure, concentrations of the component elements, bond lengths, and the more obvious discontinuities between differing states of matter (i.e., solids, liquids, and gases). Tersely put, different phases have different symmetries. The equilibrium state of a polyphasic system requires the total free energy of the isolated system to be at a minimum with respect to spontaneous change in any of the pertinent intensive state variables. The state variables most commonly encountered are temperature, pressure, and chemical concentrations. We can express the Gibbs free energy as the complete differential in terms of these variables as follows:

$$\begin{aligned} dG &= \left(\frac{\partial G}{\partial T}\right)_{P,n_i} dT + \left(\frac{\partial G}{\partial P}\right)_{T,n_i} dP + \sum_i \left(\frac{\partial G}{\partial n_i}\right)_{T,P} dn_i \\ &= -S\,dT + V\,dP + \sum_i \mu_i\,dn_i, \end{aligned} \tag{9.39}$$

where n_i represents the number of moles of the ith component, and the other symbols have their usual meanings. Although the condition for equilibrium is that $dG = 0$ for the entire system, let us consider its implication for each term in (9.39). For the entire system, the entropy must be at a maximum if the internal energy is at a minimum. This allows us to write

$$\delta S = \sum_i \delta S_i = \frac{dq_1}{T_1} + \frac{dq_2}{T_2} + \cdots + \frac{dq_i}{T_i} = 0, \tag{9.40}$$

where i designates the different phases that can irreversibly exchange quantities of heat dq. Let us suppose that heat is exchanged between hotter and colder phases so that

$$\left(\frac{dq_1}{T_1} + \frac{dq_2}{T_2} + \cdots + \frac{dq_j}{T_j}\right) - \left(\frac{dq_{j+1}}{T_{j+1}} + \frac{dq_{j+2}}{T_{j+2}} + \cdots + \frac{dq_i}{T_i}\right) = 0. \tag{9.41}$$

Now the amount of heat lost by $(dq_{j+1} \cdots dq_i)$ must exactly equal that gained by $(dq_1 \cdots dq_j)$, since by definition the system is isolated.

Consider an element of mass m_i at an initial temperature of T_i. Let an amount of heat dq_i be transferred from m_i to a contiguous element m_j, the latter being at an initial temperature T_j. Because in an isolated system heat flows only from a hotter to a colder reservoir (the so-called zeroth law of thermodynamics), T_j is always less then T_i. If we decompose (9.41) into a series of paired terms requiring that $dq_i = dq_j$, then for all pairs

$$dq_i/T_i - dq_j/T_j > 0. \tag{9.42}$$

However, this contradicts (9.41), since the sum of such pairwise terms must also be greater than zero. Consequently, (9.41) and, hence, (9.40) are valid only when all T_i and T_j are equal and the system is truly isothermal. Thus, our initial requirement of equilibrium for the entire system demands that there be no thermal gradients, and, hence, all phases are at the same temperature.

The next term expresses the requirement for mechanical stability. If, in fact, the pressure exerted by or on the several phases in equilibrium varied, then $P\,dV$ work could be done in an isolated isothermal system at constant volume. Let us designate the various phases by Greek letter superscripts. Then

$$P^{\alpha}\delta V^{\alpha} + P^{\beta}\delta V^{\beta} + P^{\gamma}\delta V^{\gamma} + \cdots = 0, \tag{9.43}$$

since the total free energy remains constant. Because of the constraint of constant volume on the total system,

$$\delta V^{\alpha} + \delta V^{\beta} + \delta V^{\gamma} + \cdots = 0, \tag{9.44}$$

from which it is both necessary and sufficient that

$$P^{\alpha} = P^{\beta} = P^{\gamma} = \cdots. \tag{9.45}$$

The validity of (9.45) is easily shown by first rearranging (9.43) so that

$$P^{\alpha}\delta V^{\alpha} = -(P^{\beta}\delta V^{\beta} + P^{\gamma}\delta V^{\gamma} + \cdots). \tag{9.46}$$

Substituting for δV^{a} from (9.44) gives

$$P^{\alpha}(\delta V^{\beta} + \delta V^{\gamma} - \cdots) = P^{\beta}\delta V^{\beta} + P^{\gamma}\delta V^{\gamma} + \cdots, \tag{9.47}$$

from which (9.45) results. Thus, at equilibrium there can be no pressure gradients within the system. Turning at last to the chemical potentials of the various components, we have

$$\begin{aligned} dG = 0 = {} & \mu_1^{\alpha}\delta n_1^{\alpha} + \mu_1^{\beta}\delta n_1^{\beta} + \mu_1^{\gamma}\delta n_1^{\gamma} + \cdots, \\ & \mu_2^{\alpha}\delta n_2^{\alpha} + \mu_2^{\beta}\delta n_2^{\beta} + \mu_2^{\gamma}\delta n_2^{\gamma} + \cdots, \\ & \mu_n^{\alpha}\delta n_n^{\alpha} + \mu_n^{\beta}\delta n_n^{\beta} + \mu_n^{\gamma}\delta n_n^{\gamma} + \cdots. \end{aligned} \tag{9.48}$$

Because matter is conserved in an isolated system

$$\sum \delta n_1 = 0, \quad \sum \delta n_2 = 0, \quad \text{etc.}, \tag{9.49}$$

and, hence, for (9.48) to be valid it is necessary and sufficient that

$$\begin{aligned} \mu_1^{\alpha} &= \mu_1^{\beta} = \mu_1^{\gamma} = \cdots, \\ \mu_2^{\alpha} &= \mu_2^{\beta} = \mu_2^{\gamma} = \cdots, \\ \mu_n^{\alpha} &= \mu_n^{\beta} = \mu_n^{\gamma} = \cdots, \end{aligned} \tag{9.50}$$

the proof of this being the same as the preceding one for establishing that a system at equilibrium must be homogeneously isobaric. Thus, according to (9.50) the chemical potential of each component is the same in all the phases in equilibrium with one another. Were this not so, material could be exchanged spontaneously between phases with a concomitant lowering of the free energy of the total system, which violates our initial assumption, namely, that the entire system is at its free-energy minimum.

One frequently sees simple multicomponent equilibrium diagrams that show no solubility of any component in the pure end member. Were this the case, it would clearly violate (9.50). However, thermodynamics must prevail over mere graphics. Consequently, when a tie line connects a solution or compound with a pure component, it can be assumed that the solubility of the missing component is too small to present graphically.

6. The Phase Rule

The *phase rule* defines the number of *degrees of freedom* which in turn is defined as the number of independent intensive variables that can be varied without changing the number or composition of the coexisting phases. The Gibbs phase rule is derived from a consideration of the results of our preceding discussion of phase equilibria. Its derivation, like the one preceding, is limited to three variables: temperature, pressure, and bulk composition. If the number of phases is called p and the number of components c, there are then pc composition variables. To this, one adds temperature and pressure making $(pc + 2)$ variables in all. However, if we express the composition of each phase in terms of mole fractions, the number of composition variables is reduced by p. Letting the Greek letter superscripts represent the various phases we have

$$\begin{aligned} X_1^\alpha + X_2^\alpha + \cdots + X_n^\alpha &= 1 \\ X_1^\beta + X_2^\beta + \cdots + X_n^\beta &= 1 \\ X_1^\omega + X_2^\omega + \cdots + X_n^\omega &= 1. \end{aligned} \tag{9.51}$$

Thus, the number of degrees of freedom, f, is

$$f = p(c - 1) + 2 = pc - p + 2 \tag{9.52}$$

since

$$\alpha + \beta + \cdots + \omega = p.$$

However, at equilibrium we have additional constraints imposed by the equality of the chemical potentials

$$\begin{aligned} \mu_1^\alpha &= \mu_1^\beta = \cdots \mu_1^\omega, \\ \mu_2^\alpha &= \mu_2^\beta = \cdots \mu_2^\omega, \\ \mu_n^\alpha &= \mu_n^\beta \cdots \mu_n^\omega. \end{aligned} \tag{9.53}$$

Each equation or equals sign represents a condition of constraint that in total number is $c(p - 1)$. This must be subtracted from (9.52) to finally arrive at the *phase rule*, which is

$$f = pc - p + 2 - cp + c = c - p + 2. \tag{9.54}$$

Thus, the phase rule states that the number of degrees of freedom equals the number of components minus the number of phases plus 2. It is worth emphasizing that it in no way predicts the number of phases in equilibrium, although it places an upper limit since f can never be negative.

Let us see how the phase rule is applied to a simple phase diagram such as the one shown in Fig. 9.4. Because this exact diagram is valid for only a particular pressure, that value is implicitly fixed and pressure can no longer be regarded as a variable. As a result, within the isobaric plane of the diagram $f = c - p + 1$. Confining ourselves to the region below the solidus, we have $c = 2$, $p = 1$, and, hence, $f = 2$. This says that we can vary two variables independently, say temperature and composition, as long as our substance remains a single-phase solid solution. The same result applies to the region above the liquidus. However, within the two-phase region, $f = 2 - 2 + 1 = 1$, and we have but a single degree of freedom. One can easily convince oneself that this is the case by noting that by selecting a temperature within the two-phase area, one fixes the composition of the equilibrium phases to be those joined by the isothermal tie line. If the temperature is varied along a vertical isocomposition line, then the composition of the equilibrium phases differs continuously with temperature as given by the continuous succession of tie lines. Thus, either the bulk composition or temperature within the two-phase region can be the single independent variable.

7. Phase Diagrams for Single Component Systems

The intensive variables temperature and pressure along with a value of unity for the composition set a maximum of 2 degrees of freedom according to (9.54). The phase diagram for water is universally used as the example to introduce and explain one-component systems, and we will continue to honor this tradition. The equilibrium diagram for water is shown by Fig. 9.7, and we immediately see that the two phase regions in parametric P-T space are represented by single lines, unlike the areas proven necessary for two component systems in Section 9.2.

Now if we exactly follow the pressure–temperature path given by the curve AO, the system will consist of ice and vapor. If the pressure is increased above the values given by AO, the water vapor is converted entirely to ice. Let us follow the locus of arrows a-b-c-d invoking a piston and cylinder concept. At the initial point, a, the cylinder is entirely filled with vapor. This single-phase region has 2 degrees of freedom according to (9.45) since T and P can be independently varied as long as one remains below AOB, and hence, point a is arbitrary. We now isothermally compress the system until T and P have the values given by the intersection of ab with AO. At this precise point we have a two-phase, ice–vapor, system and but a single degree of freedom, which is to say that by selecting our temperature as T_a, and requiring two phases, specifies the value of P. Continuing to compress the water leads to the single phase, ice, and within the region bounded by AOC P and T can be varied independently. If we halt the compression at b and isobarically heat the cylinder, we will ultimately attain a two-phase, ice–water, system at the exact intersection of bc and CO, which,

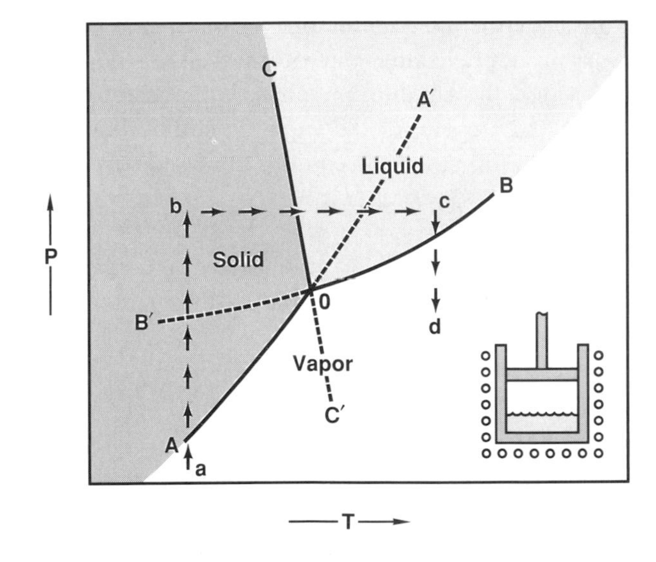

Figure 9.7.

The phase diagram of a single-component system such as H_2O. The open areas are single phases, solid, liquid, or vapor as labeled. We can imagine the creation of this diagram by enclosing a sufficient amount of water within a cylinder capped with a tight-fitting piston as shown in the insert and this entire assembly then made capable of being uniformly heated or refrigerated.

of course, has but 1 degree of freedom, P_b being the independent variable. Continuing to c brings us again into the single-phase region, and isothermal expansion along cd gives results analogous to those just described—that is, a two-phase system at the intersection of cd and OB and a single vapor phase at lower pressures. Here again, there is a single degree of freedom at the intersection, since T_c can be chosen arbitrarily.

The intersection of AO, CO, and BO defines the invariant triple point. Returning to our construct of the water-filled cylinder with piston, we can visualize the interior as ice floating on water with a gap above it filled with water vapor at the critical triple point pressure.

The dashed lines OA', OB' and OC' represent the hypothetical extrapolations of the loci of the two-phase systems into regions of metastability. The curve OB' shows schematically the metastable undercooling of water in equilibrium with its vapor. Such a state is not difficult to achieve experimentally, not only with H_2O but with other liquids as well. Such metals, as for instance Ni, have been retained in the liquid state at several hundred degrees below the equilibrium freezing temperature. On the other hand, superheating into the liquid or vapor regions as portrayed by OA' and OC' has been observed only under very special circumstances. However, in both cases, one would not expect the thermodynamic properties to change abruptly but rather to continue to vary in a predictable monotonic fashion beyond the equilibrium transition point. Because the equilibrium phase, or phases, is by definition, at a free-energy minimum, the pressure of the vapor in equilibrium with the metastable phase must be greater than that for the equilibrium condition. This follows from

$$[G_{\mathrm{e}} = G^\circ + RT \ln P_{\mathrm{e}}] < [G_{\mathrm{m}} = G^\circ + RT \ln P_{\mathrm{m}}], \tag{9.55}$$

where the subscripts e and m stand for equilibrium and metastable, respectively. Thus, the alternation of stable with metastable pathways about an invariant point is shown to be a thermodynamic necessity.

A. Solid State Transformations

Needless to say, an invariant point need not only result from the intersection of solid, liquid, and vapor phases but from any three phases coexisting in equilibrium. The pressure–temperature diagram for elemental sulfur exhibits three triple points. In Fig. 9.8, AB gives the equilibrium vapor pressure of rhombic sulfur, BE does the same for the monoclinic form, and EF records the vapor pressure over the liquid. As the rhombic phase is heated isobarically at modest pressures, it expands until at last it becomes unstable with respect to the less dense monoclinic structure, to which it then transforms. The transformation at point B on our graph shows two solids that in

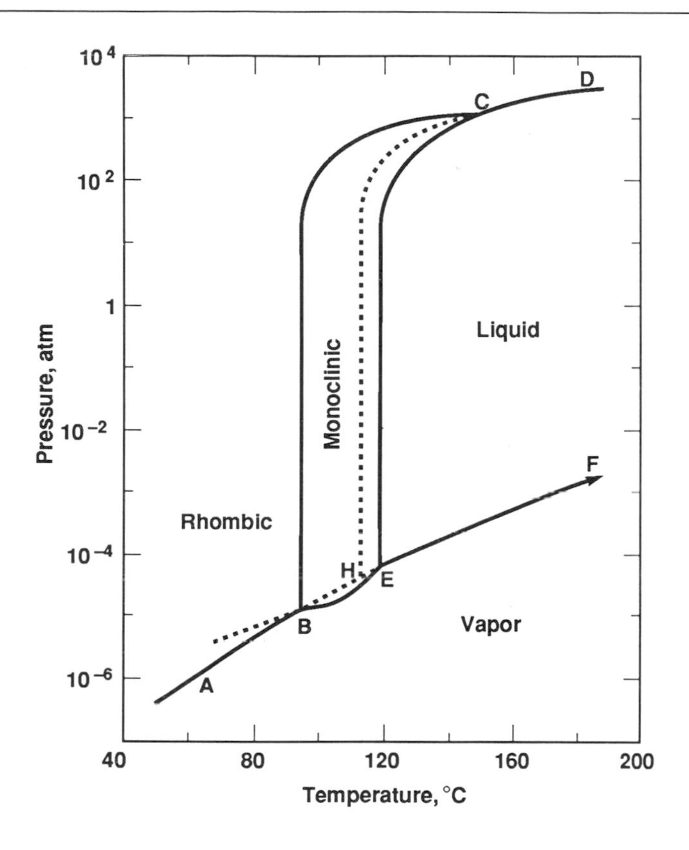

Figure 9.8.

The sulfur system on a *PT* diagram. Note the logarithmic pressure scale.

combination with the vapor provides us with our first triple point at 0.010 Torr and 95.5°C. With continued heating and a corresponding increase in *P*, the monoclinic phase ultimately melts at 0.025 Torr and 120.0°C, giving rise to a second triple point at *E*. Because the monoclinic is the least dense of the solid phases, it reverts back to rhombic, with increasing pressure, and the third invariant point, *C*, is where the two solids are in equilibrium with the liquid at 1290 atm and 154.8°C. At higher temperatures and pressures, the rhombic phase remains in equilibrium with the melt along *CD*.

As stated before, it is essentially impossible to superheat a solid past its equilibrium melting temperature. However, it is possible, in some cases, to raise the temperature of a solid above its equilibrium transition to a different crystal structure. The rate at which such a transformation proceeds is very much a function of the crystallographic complexity of the participating structures. Simple melting or vaporization consists mainly in the transition from an ordered to a disordered state, although liquids just slightly above their melting point generally possess a certain degree of short-range order. Such transitions can proceed much more quickly than to a final substance of geometric intricacy. When orthorhombic sulfur is rapidly heated at $P = 6 \times 10^{-5}$ atm, it will pass into the monoclinic region melting at 113.8°C as given by point H in Fig. 9.8. The dashed line from H to C is the metastable melting curve for the rhombic phase, and EH is the vapor pressure curve for metastable undercooled liquid. We will take up the kinetics of phase transformations in detail in a later chapter.

B. Thermodynamics of Single-Component Phase Transitions

The thermodynamic description for a single-component phase transition is especially simple. We begin, as usual, by writing the complete differential for the change in the Gibbs free energy, ΔG, in going from one phase to another.

$$d\,\Delta G = \left(\frac{\partial\,\Delta G}{\partial P}\right)_T dP + \left(\frac{\partial \Delta G}{\partial T}\right)_P dT = \Delta V\, dP - \Delta S\, dT. \tag{9.56}$$

At equilibrium both phases coexist and $\Delta G = 0$. Hence,

$$\frac{dP}{dT} = \frac{\Delta S}{\Delta V}, \tag{9.57}$$

or

$$\frac{dP}{dT_{tr}} = \frac{\Delta H_{tr}}{T_{tr}\,\Delta V}, \tag{9.58}$$

which is another form of the Clausius–Clapeyron equation (9.14). Here, ΔV is the difference in the molar volume of the phases, T_{tr} is the transition temperature, and ΔH_{tr} is the associated enthalpy.

Integrating (9.58) gives

$$\int_{T^\circ_{tr}}^{T_{tr}} d\ln T_{tr} = \int_1^P \frac{\Delta H_{tr}}{\Delta V}\, dP, \tag{9.59}$$

which becomes

$$\ln\left(\frac{T_{tr}}{T^\circ_{tr}}\right) = \frac{\Delta H_{tr}}{\Delta V}(P - 1). \tag{9.60}$$

We have arbitrarily selected 1 atm and the corresponding transition temperature T°_{tr} as the lower limits of the integrals as the standard state for any substance is conventionally taken to be its customary structural form at 1 atm pressure. It should be remembered that ΔH_{tr} and ΔV are both implicitly dependent upon pressure and temperature, and if $P \gg 1$ atm this dependence must be explicitly accounted for by adding the pressure-dependent terms. Consequently, to remind ourselves of this, we conclude with

$$\ln\left(\frac{T_{tr}}{T_{tr}^{\circ}}\right) = \frac{\Delta H_{tr}(P, T)}{\Delta V(P, T)}(P - 1). \tag{9.61}$$

8. Binary Phase Diagrams—Construction and Nomenclature

Fortunately, most all the salient features portrayed in all phase diagrams can be adequately illustrated by binary systems. It is obvious that the number of variables to be represented cannot exceed the dimensionality of the representational space. Thus, a planar sheet of paper is suitable for graphing but two variables. By expressing concentrations in the form of mole fractions, we can reduce the composition variables of the binary system to 1 and we are then free to choose another variable, allowing us a total of three. Most frequently, this is temperature, though pressure is sometimes the parameter of greater importance. To display four variables requires a three-dimensional model, and such constructs are rare but by no means unknown.

A. Temperature–Composition Diagrams

We have described in Fig. 9.2 the phase diagram for an ideal solid and liquid, which is closely approximated by the Cu–Ni system shown in Fig. 9.3. Perhaps the next most complicated behavior is exhibited by a system that has a minimum point resulting from the coalescence of the solidus and liquidus, and for the sake of brevity we also introduce the concept of a *miscibility gap* in the same diagram, as shown in Fig. 9.9.

Although the solid solution varies continuously above the miscibility gap from end member to end member (i.e., there is no distinction between α and α' above the miscibility gap—that is, from pure A to pure B—the gap itself defines a region where two solid solutions having the same crystal structure but different compositions coexist in equilibrium. A tie line is drawn at $T = T'$, and its intersections with the phase boundaries delineated by the downward concave miscibility gap fix the compositions of α and α'.

It can be shown that when the solidus and liquidus coalesce they must both individually be at their maxima or minima. To prove this, we again commence with

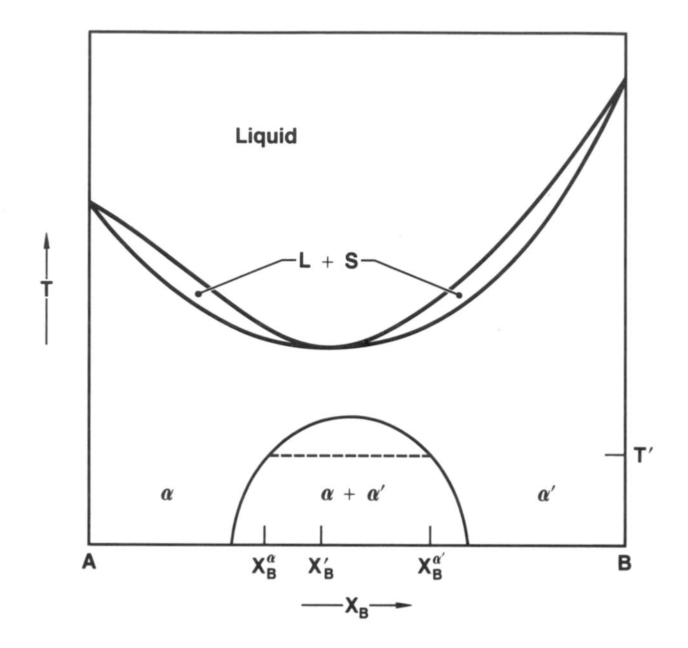

Figure 9.9.

The liquid solution has a negative departure from ideality, and the solid solution has a positive one. The latter is demonstrated by the miscibility gap, which is a two-phase region containing a mixture of α and α' phases.

the complete differentials of the free energy of the liquid and solid phases.

$$
\begin{aligned}
dG^l &= -S^l\,dT + X_A^l\,d\mu_A + X_B^l\,d\mu_B, \\
dG^s &= -S^s\,dT + X_A^s\,d\mu_A + X_B^s\,d\mu_B.
\end{aligned}
\tag{9.62}
$$

Because of equilibrium, we can equate dG^l to dG^s and, upon rearranging, obtain

$$(S^l - S^s)\,dT = (X_A^l - X_A^s)\,d\mu_A + (X_B^l - X_B^s)\,d\mu_B. \tag{9.63}$$

Dividing (9.63) by dX_B results in

$$(S^l - S^s)\frac{dT}{dX_B} = (X_A^l - X_A^s)\frac{d\mu_A}{dX_A} + (X_B^l - X_B^s)\frac{d\mu_B}{dX_B}. \tag{9.64}$$

The right-hand side of (9.64) equals zero because each component has its respective composition the same in both the liquid and solid phases at the point of tangency. However, $S^l - S^s > 0$, being equal to $\Delta H_{tr}/T_{tr}$. Thus, dT/dX_B must equal zero, corresponding to a maximum or a minimum. Because this maximum or minimum applies to both the solid and liquid phases because of their identical composition,

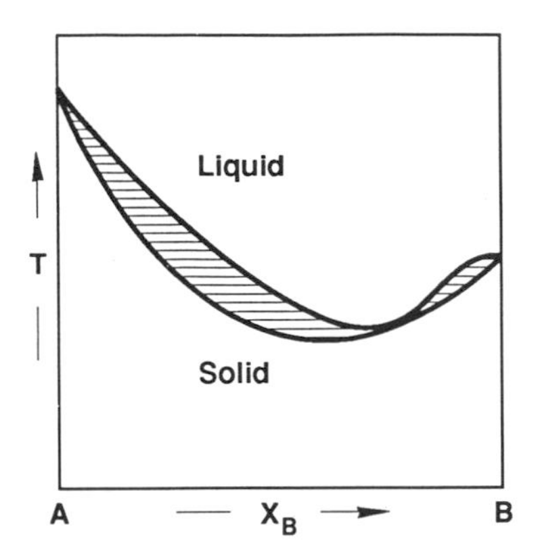

Figure 9.10.

This figure shows the liquidus and the solidus passing through minima at different compositions, which has been shown to be incorrect.

the curves of $X_B(T)$ for both must pass identically through the same maximum or minimum.

It has been pointed out[2] that this conclusion is invalid when $d\mu_B/dX_B$ and/or $d\mu_A/dX_B$ approaches infinity, which occurs when X_B approaches zero, which is to say infinite dilution as X_A approaches unity. The other instance of invalidity is in the vicinity of a composition corresponding to the formation of an extremely stable compound. As a consequence of this analysis, constructions such as shown by Fig. 9.10 are thermodynamically impossible.

B. Metastability Criteria

Another rather simple criterion for recognizing thermodynamically impossible constructions is based upon the conditions required for metastability. It has been consistently observed that all thermodynamic properties, such as enthalpies and entropies, extrapolate smoothly into their metastable states without noticeable discontinuities. This fact is essentially established by the definition of "metastable", for if a transition did occur at the equilibrium point the resulting state of the system, whether stable or metastable, could not be the metastable successor of its equilibrium predecessor. Thus, we can justifiably extrapolate in a monotonic manner the equilibrium phase boundaries as shown by the dashed lines in Fig. 9.11.

---

[2] L. S. Darken and R. W. Gurry, *Physical Chemistry of Metals*, Chap. 12, p. 318, McGraw-Hill, 1953.

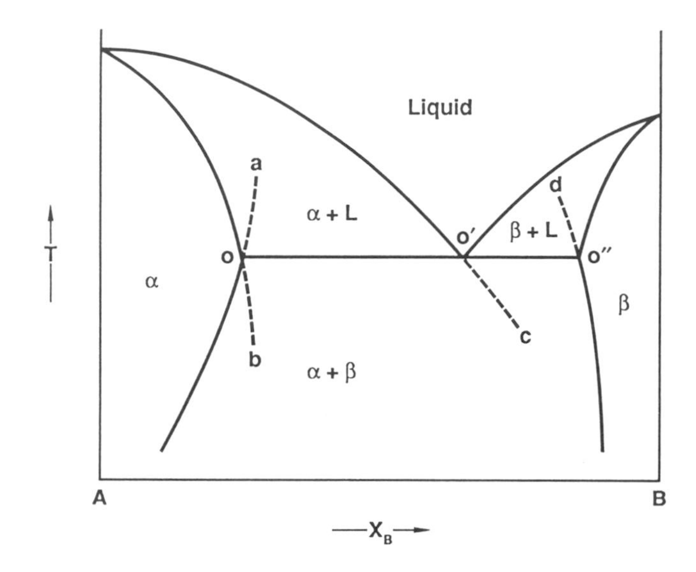

Figure 9.11.

Hypothetical phase diagram with dashed lines indicating permissible metastable phase boundaries.

Let us examine the dashed lines *ob* and *o′c* in Fig. 9.11, which are the extrapolations of the α-liquid phase boundaries into the metastable region. The thing to notice is that both metastable boundaries project into two-phase regions. In a similar fashion, the metastable α given by *oa* is in equilibrium with metastable β, whose composition is described by *o″d*. Again the metastable boundaries extrapolate into the equilibrium two-phase region. The general form of this phase diagram is thermodynamically correct.

Now consider Fig. 9.12, which includes the phases β and γ, which have a considerable range of compositions. Note that this diagram has been constructed so that the phase boundaries β-L and α-β intersecting at point *o′* allow extrapolation of *o′d* and *o′c* to extend into the β phase, which will prove to be thermodynamically incorrect.

In like manner the extrapolations of the liquidi of β-L and γ-L, *o‴g* and *o‴f*, respectively, extend into the liquid, single-phase field. A moment's reflection will show that all such extrapolations are impossible. For if metastable *o′c* truly existed, it would have to be in equilibrium with metastable α, having compositions given by *oa*, and similarly *o′d* would have to be in equilibrium with *ob′*. Yet in both cases all phases are enriched in component *b*, which is clearly impossible. The same is true for *o″e* in equilibrium with *o‴f*. On the other hand, suppose *ob* is proposed as the

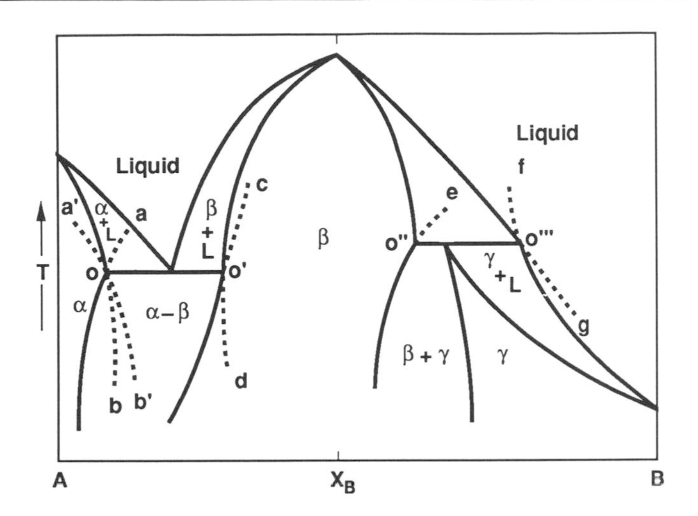

Figure 9.12.

Illustrating correct and incorrect constructs in the presence of compound β. See text for complete explanation.

true equilibrium boundary, and oa is the locus of the metastable compositions. Now it would appear that a mass balance could be maintained as metastable β gains in component B according to $o'c$ and metastable α loses B in agreement with oa. However, the construct necessarily constricts the equilibrium α–β phase field, with consequent increasing solubility of B in α and A in β with decreasing temperature. This again violates thermodynamic reasoning because solute concentrations must, perforce, in simple nonreactive systems, decrease with decreasing temperature or else the entropy of mixing would increase with decreasing temperature. Thus, we may conclude that the lines denoting the single, and two-phase regions must intersect, so that an angle of less then 180° is formed embracing the single-phase field. If this is so, these boundaries will extrapolate as required into two-phase regions only.

9. The Lever Law

The lever law is a straightforward calculation of the relative amounts of two phases in equilibrium connected by an isothermal tie line. The necessary condition that matter is conserved is stated by

$$X_{\mathrm{B}}(n_\alpha + n_{\alpha'}) = X_{\mathrm{B}}^{\alpha} n_\alpha + X_{\mathrm{B}}^{\alpha'} n_{\alpha'}, \tag{9.65}$$

where X_{B}, X_{B}^{α}, and $X_{\mathrm{B}}^{\alpha'}$ are defined in Fig. 9.9 and n_α and $n_{\alpha'}$ are the number of moles of each of the two phases. Rearranging (9.65) gives

$$(X_B - X_B^{\alpha})n_{\alpha} = (X_B^{\alpha'} - X_B)n_{\alpha'}.$$

Hence,

$$\frac{n_{\alpha}}{n_{\alpha'}} = \frac{X_B^{\alpha'} - X_B}{X_B - X_B^{\alpha}}. \tag{9.66}$$

Thus, the number of moles of α' phase is proportional to the line length $X_B X_B^{\alpha'}$ and, conversely, n_{α} is proportional to $X_B^{\alpha} X_B$.

Another useful way of expressing the lever law is

$$\frac{n_{\alpha}}{n_{\alpha'} + n_{\alpha}} = \frac{X_B^{\alpha'} - X_B}{X_B^{\alpha'} - X_B^{\alpha}}. \tag{9.67}$$

10. Eutectics

The next complicating features arise when the miscibility gap as shown in Fig. 9.8 has risen to intersect the minimum point in the solidus–liquidus curve. Such an equilibrium diagram is shown by Fig. 9.13.

The various phases illustrated by a polyphase equilibrium diagram are generally denoted by Greek letters unless they are well-defined stoichiometric compounds. Primary solid solutions are presumed to be continuous with the pure end member and, hence, must possess the same crystal structure. In Fig. 9.13 we can recognize three two-phase regions, two consisting of liquid in equilibrium with a solid solution and the third containing a mixture of the two primary solid solutions.

An invariant point occurs at T_e, the eutectic temperature, for all compositions lying between X_e^{α} and X_e^{β} that designate the compositions of the eutectic solid phases. Precisely at T_e three phases coexist in equilibrium: the solids X_e^{α}, X_e^{β}, and the liquid X_e^{L}. These compositions at this temperature must be invariant since $f = 2 - 3 + 1 = 0$.

Let us now examine the solidification sequence that results from the cooling of three different liquids having initial compositions designated by L_1, L_2, and L_3. Starting with L_1, the liquid cools without change in composition until the temperature intersects the liquidus for the α phase at T_{α}^{i}. At this temperature, precipitation commences, and the composition X_{α}^{1} is given by the intersection of T_{α}^{i} tie line with the solidus. If the cooling could proceed in a continuous manner while maintaining equilibrium, it would follow the path delineated by the dotted line extension of L_1. This requires the continuous redisolving of the precipitates to reestablish equilibrium with the liquid at each successively lower temperature.

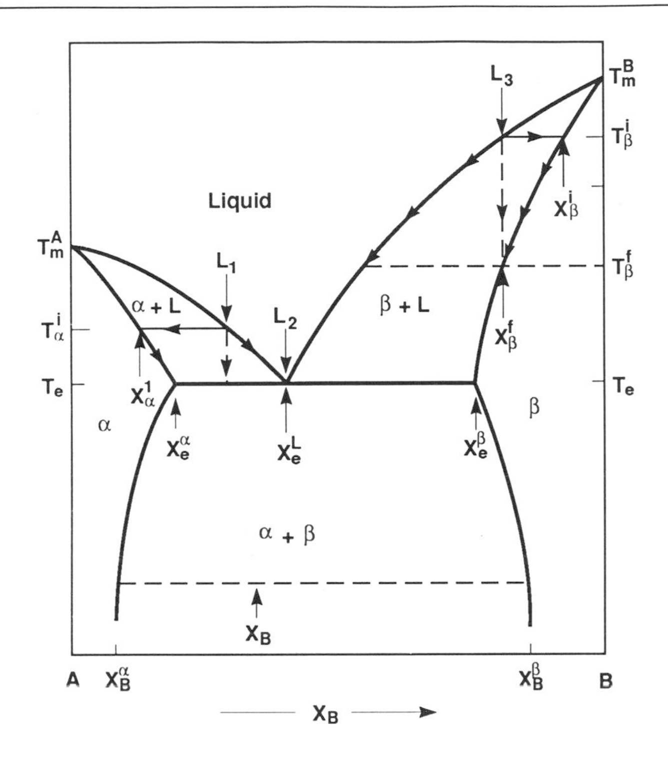

Figure 9.13.

The phase diagram for a binary system with two primary solid solutions, α and β, and a eutectic.

In essentially all laboratory measurements or commercial processes, solidification proceeds too rapidly for resolution and, hence, is a nonequilibrium process. As the freezing continues, both the liquid and the solid, with which it is in transient equilibrium, become progressively enriched in constituent B. The composition of the liquid moves smoothly down the liquidus while that of the solid does likewise down the solidus. When cooled infinitesimally below T_e, the remainder of the liquid solidifies as two solids having the compositions X_e^α and X_e^β. This last process generally occurs very quickly with a minimum of undercooling. The primary α and β precipitates are generally large grains of nonuniform concentration, with the primary α crystallites becoming more enriched in B proceeding outward from the center, and conversely, primary β precipitates are enriched in component A. Such structures are called "cored" grains or crystallites. Solidification of L_2, the eutectic composition, leads to a mixture having only the compositions X_e^α and X_e^β. This mixture is intimately mixed, uncored, and frequently shows a lathlike structure.

Liquids whose compositions lie either to the left of X_e^α or to the right of X_e^β freeze, yielding pure primary α or β, respectively, although, as before, they usually will present a cored structure. The freezing of L_3 illustrates this sort of pathway. The initial solid to appear upon cooling has the composition X_β and the final precipitate X_β^f. The equilibrium path, as before, is the direct vertical line connecting L_3 and X_β^f, with the compositions of the liquid and solid being given by the continuum of isothermal tie lines.

The area enclosed by the isotherm T_e, and the solubility limit of B in α and β, respectively, consists of a mixture of the solid solutions α and β. Again we should remind ourselves that the equilibrium proportions of α and β with their appropriate

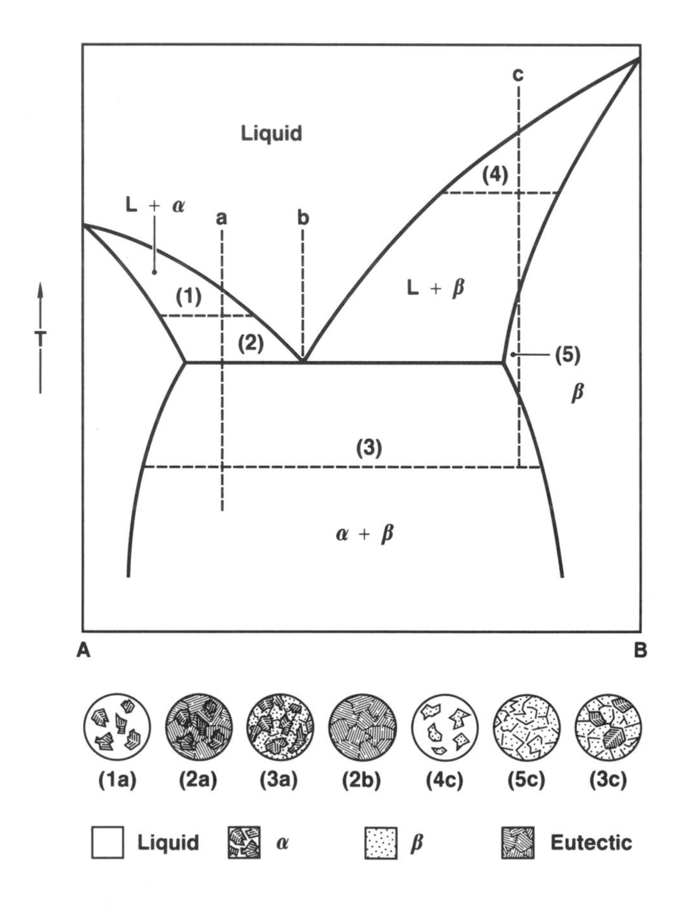

Figure 9.14.

Sketches (1a)–(3c) show schematically the equilibrium structures corresponding to the intersections of compositions a,b,c, with tie lines (1), (2), (3), and (4). Except for (2b), significant annealing must be done to attain equilibrium. Shown are the following: (1a) primary α + liquid, (2a) primary α + eutectic $\alpha + \beta$, (3a) primary α and β, (2b) eutectic $\alpha + \beta$, (4c) primary β + liquid, (5c) primary β, (3c) primary α and β.

compositions are often difficult, if not impossible, to achieve. As we have previously stated, the freezing is complete for all compositions occurring at T_e or above. Consequently, equilibrium between the solids already formed can only be achieved via diffusion, which is commonly a very slow process.

The pathway taken by a particular solidification process is frequently revealed by the sequence of resulting microstructures obtained along the way. Highly polished specimens representing the freezing and heat treatment of selected compositions are examined under a microscope and subjected to electron beam microprobe analysis. The results of such analysis can often be interpreted in the light of the relevant phase diagram, or, conversely, elements of an unknown phase diagram can be deduced from such findings. Figure 9.14 offers five examples of different microstructures and two liquid–solid mixtures that result from the cooling of three different compositions.

11. Complex Phase Reactions

Thus far, the discussion of heterogeneous equilibria has been limited to simple solutions and eutectic decompositions. Although binary systems cannot demonstrate a large variety of phase reactions, there nevertheless remain for our examination a few reactions that occur with great frequency. For the sake of completeness, we will list and label them all, but the nomenclature, though useful, is not the aim of this section. It is rather to provide the groundwork for the correct interpretation of phase diagrams.

All invariant reactions that are observed are schematically defined by Fig. 9.15. Some, such as the eutectic and peritectic, are very common; others, such as the

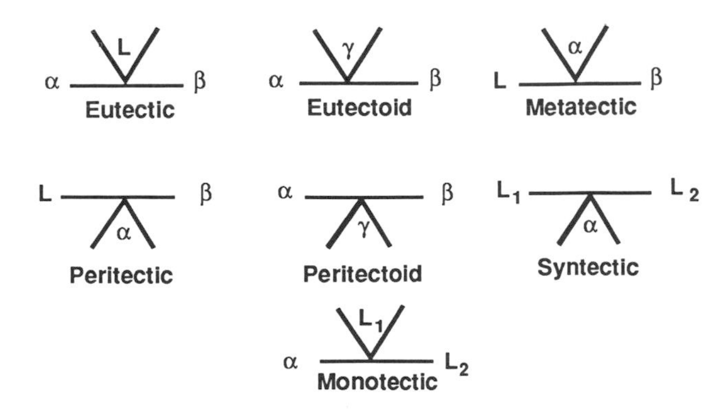

Figure 9.15.

Schematic illustration of various observed invariant points in binary systems. α, β, γ stand for solid phases and L for liquid. The implied sense of the temperature has it rising toward the top of the figures.

eutectoid and peritectoid, are not rare, and the three remaining, viz. the morotectic, syntectic, and metatectic, are seldom encountered. Notice that the suffix -tectic implies one or more liquid phase reactants, whereas -tectoid applies to cases where all are solids. The eutectoid has the same general form as the eutectic, except that all three products at the invariant temperature and pressure are solids.

Peritectic, or incongruently melting compounds, as they are often called, occur very frequently. This class includes such important examples as Nb_3Sn, the most widely used superconducting compound. Incongruently melting compounds, by definition, cannot be in equilibrium with liquids of the same composition. We have already derived (see Section 8.A, this chapter) the necessary constraint for congruent

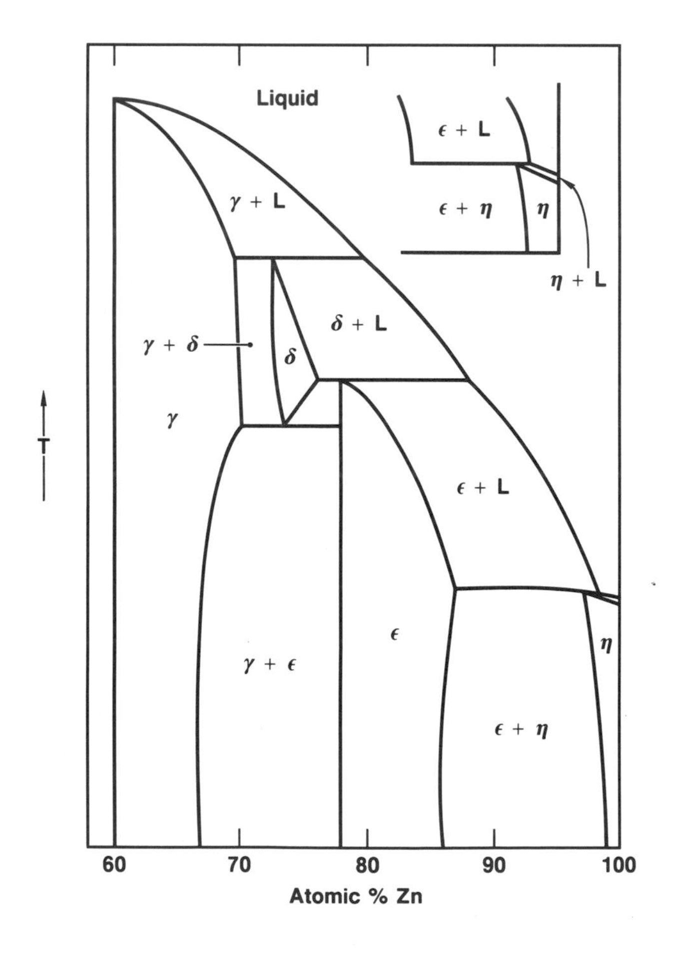

Figure 9.16.

A section of the Cu–Zn equilibrium diagram.

melting, which requires either a maximum or minimum in the melting curve, and which, conversely, peritectic melting does not.

Peritectoid and eutectoid decompositions are defined by the same general features as the peritectic and eutectic constructs. The Zn-rich half of the Cu–Zn phase diagram is shown in Fig. 9.16. It illustrates several of the features we have been discussing; of note, are the following:

γ-phase	Only partially shown, melts congruently
δ-phase	melts incongruently (peritectic) and also has a eutectoid decomposition
ε-phase	peritectic melting point
η-phase	peritectic melting primary solid solution (see insert)

The Cu–Zn phases also illustrate the phenomenon of nonstoichiometry, which gives rise to compounds having a finite range of composition, or, in other words, the compound is also capable of being a solid solution. Extensive solubility of one or more constituents is characteristic of intermetallic and silicate compounds. In nature, Dalton's law of definite multiple proportions is more the exception than the rule and is valid most often only for simple ionic compounds. However, even within this category there are several common compounds possessing a significant range of composition, such as FeS_{1+x} and FeO_{1+x}. In the case of iron sulfide $x \geq 0$, but for ferrous oxide (wüstite), x is always greater than zero, being $0.049 \leq x \leq 0.174$. Hence, FeO per se does not exist.

12. Free-Energy Diagrams

Graphs of the free energy of formation of polyphase systems as a function of temperature and composition can provide additional insight into heterogeneous equilibria. Although the creation of such graphs introduces no new concepts, they provide considerable assistance to the visualization of the thermodynamic variation in the phase stabilities.

Earlier in this chapter it was shown that the free energy of a single binary phase at constant temperature varies roughly parabolically with composition. (See Figs. 9.2 and 9.5.) We preserve this form for the schematic illustration of the free energy in the diagrams that follow. The phase diagram for a binary system without compounds and a single melting point minimum is shown by Fig. 9.17(1).

At $T \geq T_A$, the liquid has a lower free energy at all compositions. At $T \geq T_B$, the liquid has a lower free energy at all compositions, except in the range $X_B \leq X_b$, where it is in equilibrium with the solid of composition $X_B = X_a$. There are two phases at T_1: one exists when $X_c \leq X_B \leq X_d$, the second when $X_e \leq X_B \leq X_f$. Figure 9.17(5)

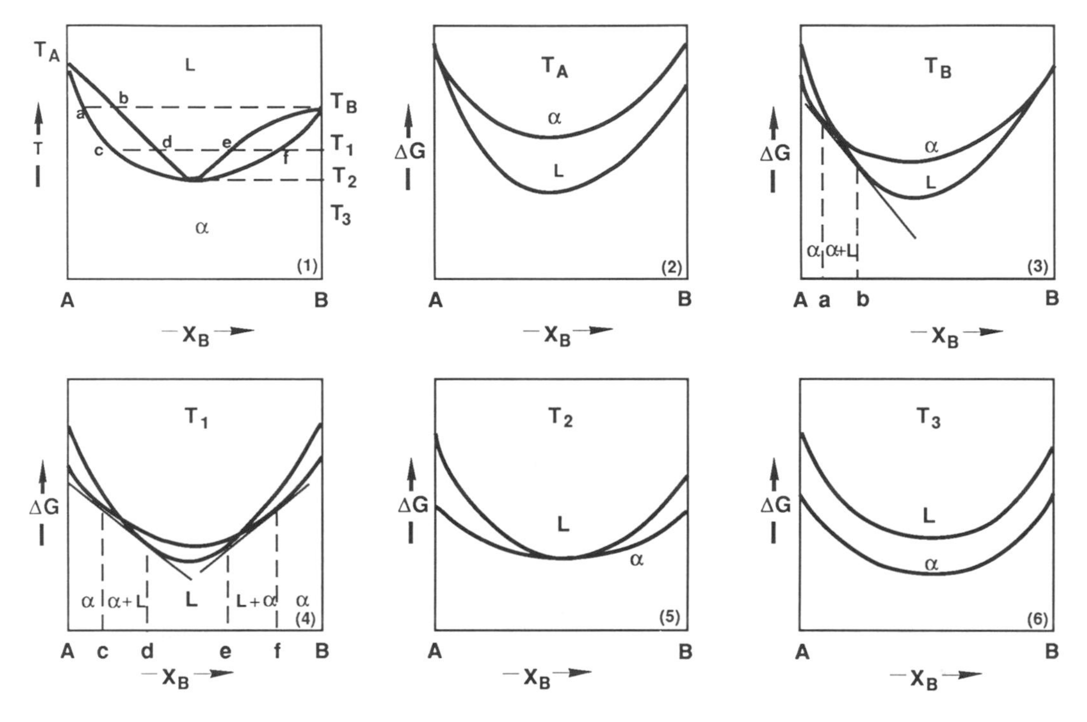

Figure 9.17.

(1) The phase diagram, (2, 3, 4, 5, 6) are the free-energy diagrams corresponding to isotherms at T_A, T_B, T_1, T_2, T_3 in (1), respectively, as a function of composition.

shows the free energies of the liquid and α phase equal at the melting point minimum, but that of the α phase lower everywhere else. At the lowest temperature only the α phase is stable.

The lines of tangency between phases in equilibrium with each other correctly expresses the condition that

$$\left(\frac{\partial \Delta G}{\partial X_B}\right)^{\alpha} = \left(\frac{\partial \Delta G}{\partial X_B}\right)^{\beta}, \tag{9.68}$$

or that

$$\mu_B^{\alpha} = \mu_B^{\beta}, \tag{9.69}$$

as is required for equilibrium.

It can also be seen that the minimum for the overall free energy of the system is given by the tangent held in common; i.e., given the individual free-energy curves for α, β, and the liquid, then all values of ΔG lie above their tangents. Further, the free

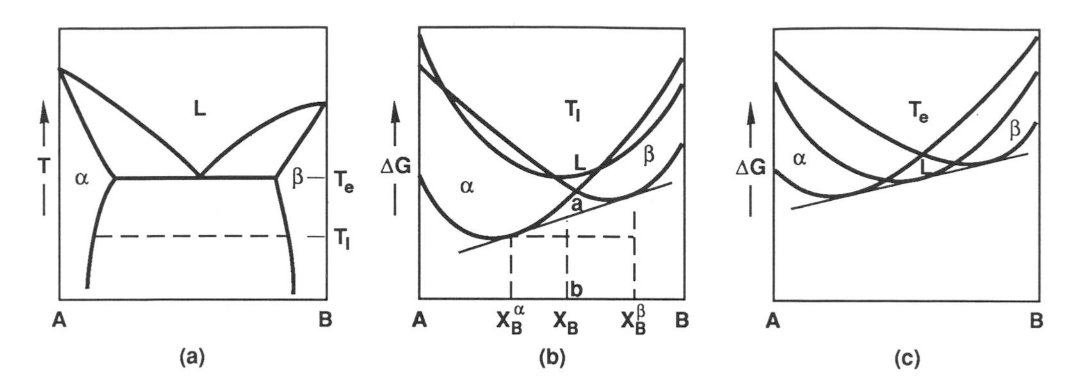

Figure 9.18.

(1) The equilibrium diagram for a binary system with a eutectic; (2) the free-energy diagram corresponding to T_1 (3) the free-energy diagram corresponding to T_e, the eutectic temperature.

energy of formation of the system as a whole is given by the intersection of the tangent by a vertical perpendicular line extending from the value of the overall composition. This can be shown with the aid of Fig. 9.18.

An analytical expression for ΔG at X_B is

$$\Delta G(X_B) = \Delta G^\alpha + \left(\frac{\Delta G^\beta - \Delta G^\alpha}{X_B^\beta - X_B^\alpha}\right)(X_B - X_B^\alpha). \tag{9.70}$$

The simple geometric relations between similar triangles allow us to write

$$\frac{ab - \Delta G^\alpha}{X_B - X_B^\alpha} = \frac{\Delta G^\beta - \Delta G^\alpha}{X_B^\beta - X_B^\alpha}, \tag{9.71}$$

which, after rearranging, leads to

$$ab = \Delta G^\alpha + \left(\frac{\Delta G^\beta - \Delta G^\alpha}{X_B^\beta - X_B^\alpha}\right)(X_B - X_B^\alpha), \tag{9.72}$$

and thus $ab = \Delta G$ as was to be shown.

13. Ternary Systems

Three-component systems can be represented by planar isothermal diagrams by simply substituting for temperature the third component as a variable. However, this different method of representation required to graph ternary systems, like so many other innovations, was invented by Josiah Willard Gibbs.

A. Construction of Ternary Phase Diagrams

The so-called Gibbs triangle is shown in Fig. 9.19. This method of graphing is based on the geometric principle that states that, within an equilateral triangle, the sum of the distances drawn parallel to the three sides from any point is always the same and is equal to the side of the triangle. Hence, by setting the latter to unity and expressing the composition in mol fractions, the composition of any system is given by the relative length of the three parallels. It is convenient to indicate the counting procedure by the three arrows shown in Fig. 9.19. At point x, for example, the length of the line parallel to AC and pointing to line AB represents $0.3A$, the one parallel to AB and pointing to line BC represents $0.5B$, and the one parallel to BC and pointing to line AC represents $0.2C$. Note that because of symmetry, the system is invariant to rotation by 60° if the arrow directions are altered appropriately, i.e., by the same rules as given before. As two other illustrations, the compositions at y and z are given by

$$y = 0.1\text{A}, \quad 0.35\text{B}, \quad 0.55\text{C},$$

$$z = 0.6\text{A}, \ 0.1\text{B}, \quad 0.3\text{C}.$$

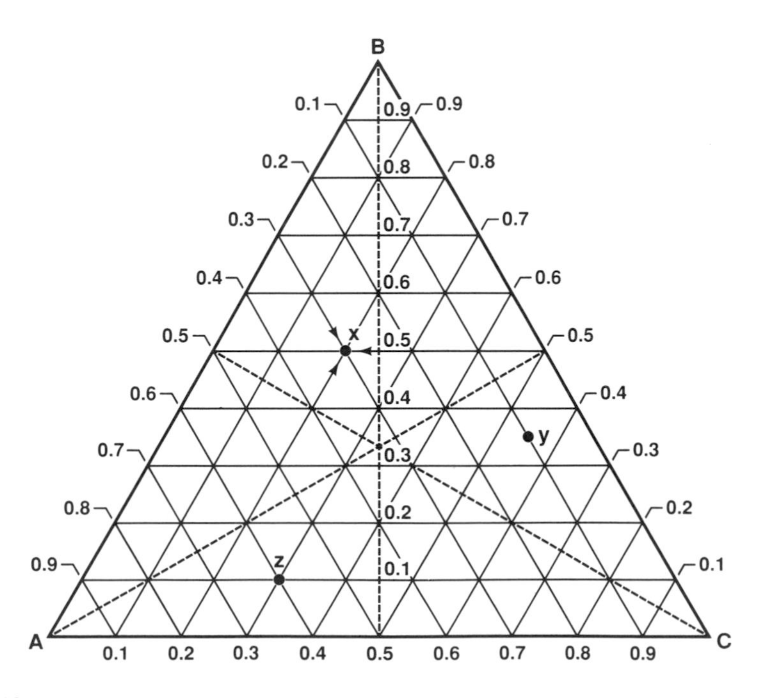

Figure 9.19.

The Gibbs triangle for the graphing of ternary systems.

A point situated on one of the sides of the triangle represents, of course, two components only.

An entirely equivalent procedure relies on the fact that, in an equilateral triangle, the sum of the perpendiculars from any point is equal to the length of the altitude of the triangle. Hence, by setting the latter equal to unity and expressing the compositions in terms of mol fractions, the composition of any system is given by the relative lengths of the three perpendiculars. The results are clearly the same as already illustrated, based on Fig. 9.19.

Some other useful properties of the equilateral triangle are the following, as illustrated in Fig. 9.20. If a line is drawn through any corner, for example *B*, to any point *D* on the opposite side, as also illustrated in Fig. 9.20, then all points on the line *BD* represent a constant ratio of *A* to *C* with variable amounts of *B*. Any line parallel to a side, for example, *EF* in Fig. 9.20, represents a constant proportion of one component, in this case *B*, with variable amount of the others. The equivalent of the lever law is illustrated by line *PQ*. Any point on this line, such as *R*, represents a mixture of *P* and *Q*, the amounts being in the proportion of *RQ* to *RP*, respectively.

The interpretation of ternary diagrams in their simplest forms can be visualized

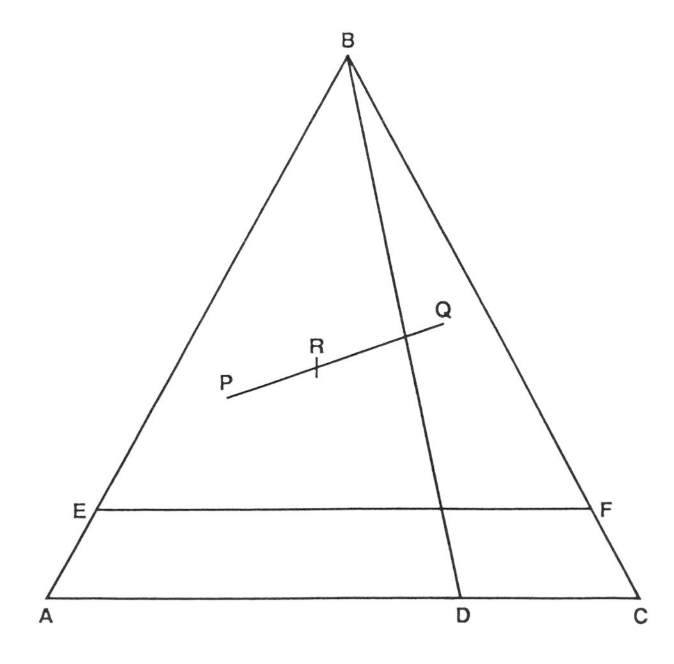

Figure 9.20.

Illustration of some useful properties of the equilateral triangle.

as the superposition of three binary systems. We consider first three binaries, AB, AC, and BC in which there is complete solubility from end member to end member in both the liquid and solid states as shown in Fig. 9.21. The temperature T for which this figure is drawn is marked on each of the three binary diagrams. From where this isotherm intersects phase boundaries, perpendiculars are dropped to the sides of the triangle. These intersections are connected with their counterparts on opposite sides, preserving the proper sequence of solids and liquids. The precise pathway taken across the interior of the Gibbs triangle, lacking further data, is quite simply a matter of guesswork. It has been assumed in this example that the ternary solid solution forms with the complete solubility of all components. The tie lines connecting liquid and solid phase gradually change slope from a congruence with AB to one with BC. The lever law can be applied as in the case of binary systems to points that lie within the two-phase region.

All the phase reactions that were cited for binary systems (e.g., eutectic, peritectic, etc) find expression in the ternary ones as well. It is beyond the scope of this treatment to discuss the entire catalog, and, in fact, we only will examine in detail a single case, viz. that of a ternary eutectic in which all three binaries possess a single eutectic. Let us imagine temperature to be recorded on an imaginary axis perpendicular to the

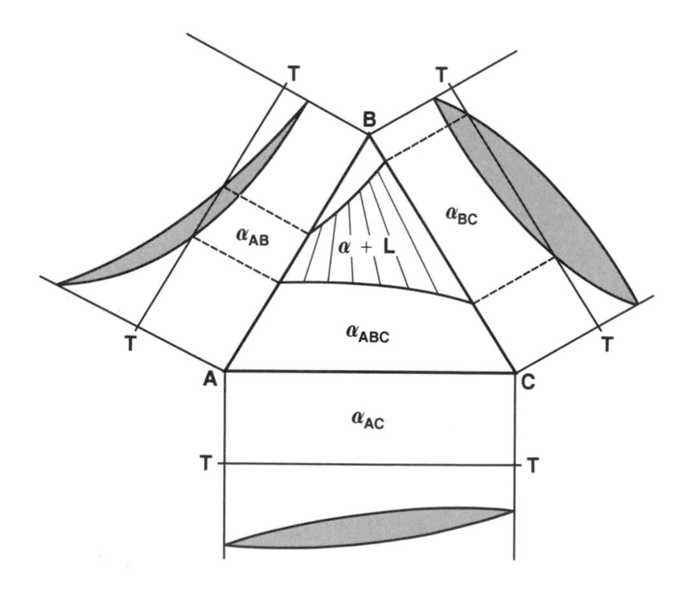

Figure 9.21.

A simple two-phase ternary system at a temperature T decomposed into its three constituent binaries.

plane of a continuum of Gibbs triangles. Then a sequence of phase diagrams taken at periodic intervals while descending in temperature might appear as shown in Fig. 9.22.

The subdivision of the ternary system into its binary members is shown in Fig. 9.22a. A temperature T is selected such that $T_A > T > T_B$. The binaries are

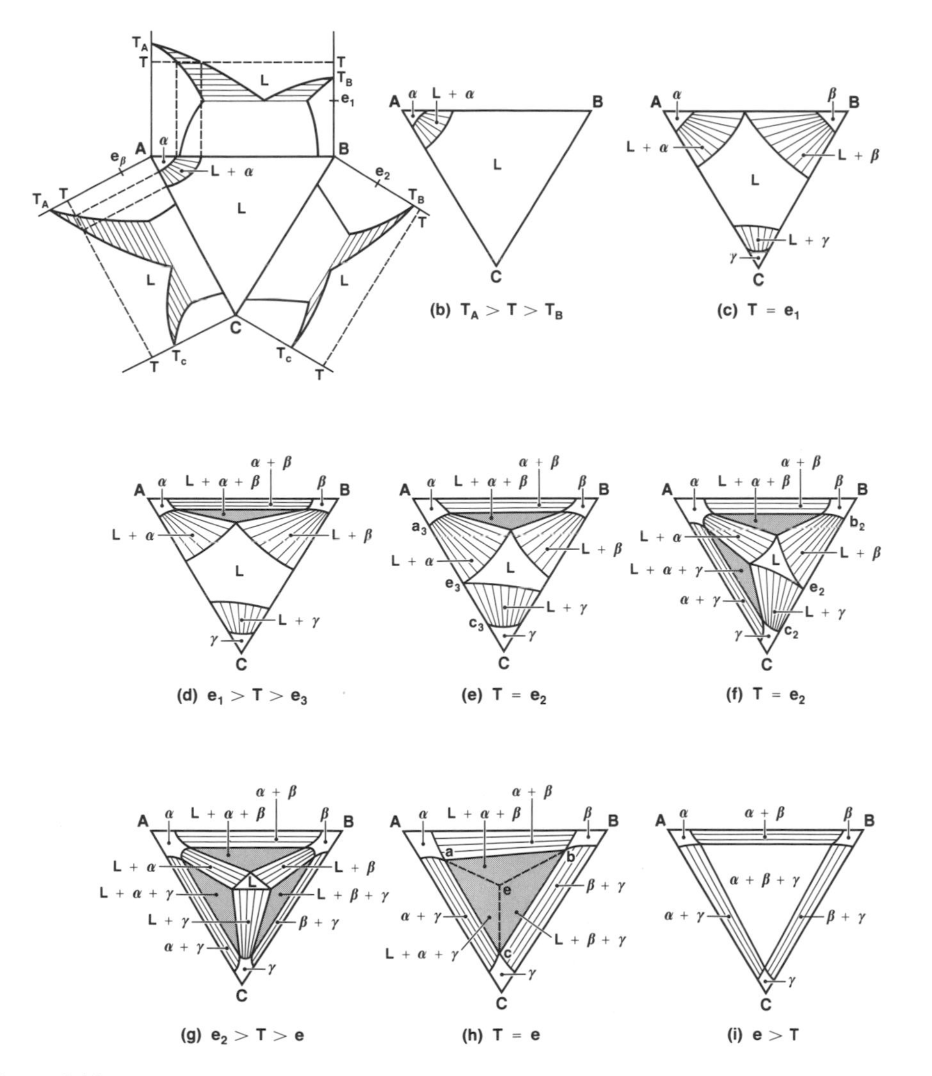

Figure 9.22.

The freezing sequence of a ternary system composed of binary eutectics. See text for explanation. (After A. Prince, *Alloy Phase Equilibria*, p. 176, Elsevier, 1966.)

constructed so that $T_A > T_B > T_C$, and their eutectic temperatures are in the order $e_1 > e_2 > e_3$. By dropping perpendiculars from the intersections of the dotted lines, T, with the phase boundaries in systems A–B and A–C, the boundaries between α and the liquid are defined on the sides of the Gibbs triangle. The extension of these boundaries to within the triangle is a matter of interpolation, since no specific information is given about the ternary solutions. Nevertheless, the general form of the phase diagram corresponding to each stage in the progressive cooling is easily intuited, and they are shown schematically in Figs. 9.22b–i. A ternary eutectic is expected and is shown in Fig. 9.22h for $T = E$. The systems depicted in the sequence of the figure appear, and indeed are, quite complicated. They are rather simple, however, compared with many practical metallic alloy, ceramic, and glass systems, where the formation of polymorphs and compounds must often be considered. Such complicated systems are outside our present scope.

Instead of the more complicated systems, it is worthwhile to consider briefly a simpler but interesting type of three-component system, namely water and two salts

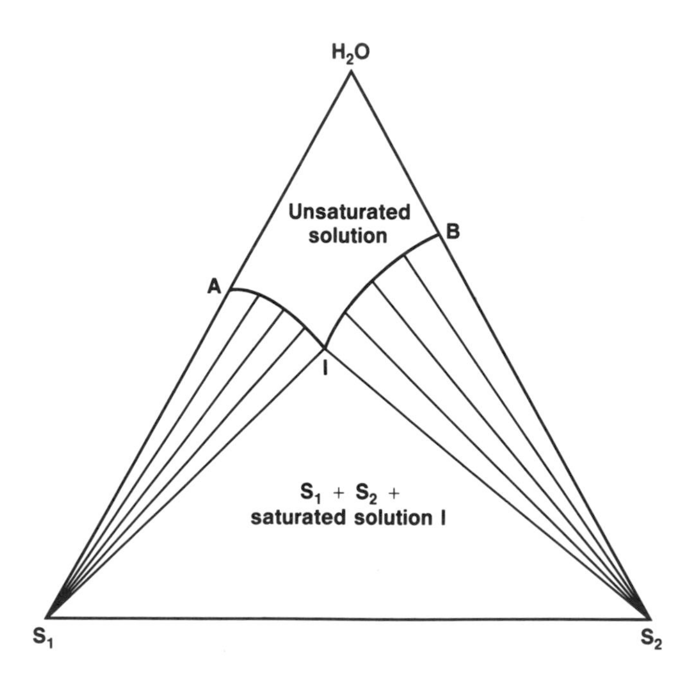

Figure 9.23.

Water and two salts that do not form solid solutions with one another.

with an ion in common (so as to remain a ternary). When the two salts, S_1 and S_2, do not form a compound or a solid solution, the isothermal solubility curves look like Fig. 9.23, if the temperature is kept above the freezing point of water. Along *AI* the solid phase in equilibrium with the solution is S_1, along *BI* it is S_2. The point *I* is referred to as an isothermal invariant point. At *I* there are three phases (i.e., two solid and one liquid) and, hence, there are at this point 2 degrees of freedom that are used up by specifying the pressure and temperature. An example of such a system is NH_4Cl–$(NH_4)_2SO_4$–H_2O. When the two salts form a continuous series of solid solutions, there is no isothermal invariant point because there are only two phases present. The composition of the solid phase varies with the composition of the solution from which it separates, as shown by the tie lines in Fig. 9.24. An example of such a system is $Ba(NO_3)_2$–$Pb(NO_3)_2$–H_2O.

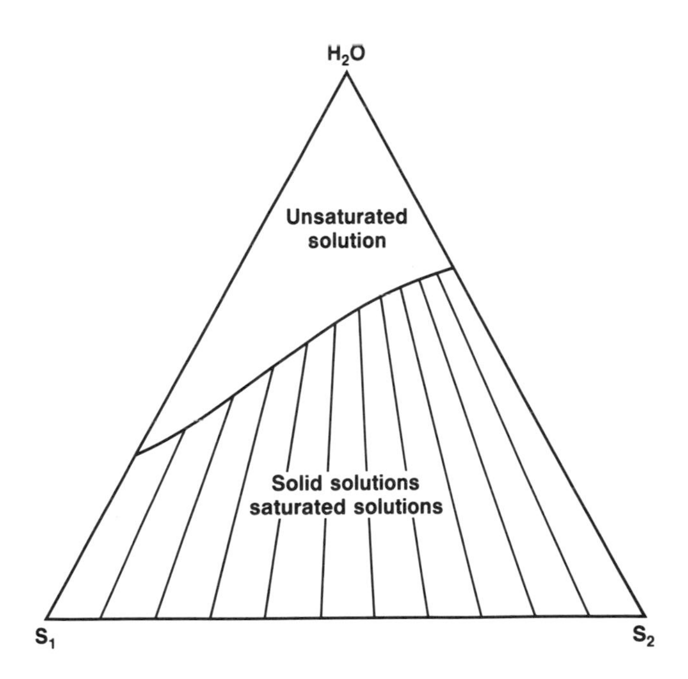

Figure 9.24.

A hypothetical phase diagram showing a continuum of anhydrous solid solutions extending from one pure end member, S_1, to the other, S_2. Each composition of the mixed salt (i.e., solid solution between S_1 and S_2) is in equilibrium with a saturated aqueous solution of specific composition, as shown by the partial representation of the continuum of tie lines.

Chapter 9 Exercises

1. Making use of the following data and the regular solution concept, reconstruct a phase diagram for Cu–Ni and compare it with the calculated one shown in Figure 9.2.

X_{Cu}	ΔG_f° (cal) $T = 973$ K
0.1	−415
0.2	−578
0.3	−654
0.4	−679
0.5	−667
0.6	−625
0.7	−559
0.8	−469
0.9	−334

2. Mo and Ti form nearly ideal solid and liquid solutions. Calculate an approximate phase diagram, knowing only that the heats of melting are $\Delta H_m = 7777$ cal/mol at 2890 K for Mo and $\Delta H_m = 3692$ cal/mol at 1943 K for Ti.

3. Extremely pure solids are produced by zone refining. This process repetitively sweeps the solid with a thin molten zone generated by a traveling furnace or by moving the solid through a focused hot zone. What thermodynamic conditions optimize the purification and determine whether an impurity is concentrated at the right or left-hand side with respect to the motion of the melt phase?

4. Suppose you were given a close-ended tube furnace that could be evacuated, and the temperature at each end could be varied independently. Given a binary solid solution AB, where only A is appreciably volatile, and given the vapor pressure of A as a function of temperature, tell explicitly how you could determine and calculate μ_A and μ_B as a function of temperature and composition. You can assume $f_A = P_A$, but the activities $a_A \neq X_A$ and $a_B \neq X_B$.

5. A binary solid solution, AB, which behaves as a regular solution has an enthalpy of mixing $\Delta H_m = 5X_A X_B$ kcal/g-atom. At 1000 K calculate the compositions of α and α' in equilibrium and the composition at the inflection points.

6. An equation giving the transformation enthalpy (latent heat) along the equilibrium path is

$$\frac{d\,\Delta H_{tr}}{dT} = \Delta C_p + \frac{\Delta H_{tr}}{T} - \Delta H_{tr}\left(\frac{\partial \ln \Delta V_{tr}}{\partial T}\right)_p.$$

Derive this equation, starting with

$$d\,\Delta H_{tr} = \left(\frac{\partial\,\Delta H_{tr}}{\partial T}\right)_P dT + \left(\frac{\partial\,\Delta H_{tr}}{\partial P}\right)_T dP.$$

7. Let the three liquids designated L_1, L_2, and L_3 in Fig. 9.13 cool while the system exchanges heat with an isothermal reservoir of lower temperature. Sketch the cooling curves (i.e., time vs. temperature) for each so as to qualitatively illustrate the salient differences between them.

8. Derive (9.67)

9. Calculate maximum (critical) temperature and composition of a binary system with a miscibility gap when the free energy of formation of the single-phase system for $T > T_c$ is given by

$$\Delta G_f = X_A G_A^\circ + X_B G_B^\circ + RT(X_A \ln X_A + X_B \ln X_B$$
$$+ X_A X_B(3150X_A) + (2300X_B)(1 - T/4000) \text{ cal.}$$

10. Draw the phase diagram for a binary system at $P = 1$ atm from the following data: $T_A^m = 1063°C$, $T_B^m = 451°C$, compound AB_2 melts at 464°C. At 447°C the compound is in equilibrium with essentially pure A, and the ratio (moles A)/(moles AB_2) = 0.20. At 416°C the compound is in equilibrium with nearly pure B, and the ratio (moles B)/(moles AB_2) = 0.33.

11. At 600 K in an equiatomic solution Cd and Hg have their respective activity coefficients $\gamma_{Cd} = 0.576$ and $\gamma_{Hg} = 0.692$.
a. Calculate the vapor pressure of each component.
b. Calculate the free energy of mixing, ΔG_m, of 1 mole of solution.
c. Assuming the entropy of mixing to be ideal, what is ΔH_m?

12. Derive rigorously the lever law for ternary systems (re. Fig. 9.20).

13. The Au–Ni phase diagram is given by Fig. 10.6. For the pure Au, $T_m = 1336$ K, $\Delta H_m^\circ = 2995$ cal/mol, $C_p(s) - C_p(l) = 1.34 - 1.24 \times 10^{-3}T$ cal/K-mol, and for Ni, $T_m = 1726$ K, $\Delta H_m^\circ = 4210$ cal/mol, $C_p(s) - C_p(l) = 3.20 - 1.80 \times 10^{-3}T$ cal/K-mol. The excess free energy of the solid phase is given by

$$\Delta G_f^x(s) = X_{Au} X_{Ni}(5770X_{Au} + 9150X_{Ni} - 3400X_{Au}X_{Ni})(1 - T/2600) \text{ cal,}$$

as derived from activity data and the location of the critical point. It has been assumed that an analogous equation for the liquid is

$$\Delta G_f^x(l) = X_{Au} X_{Ni}(C_1 X_{Au} + C_2 X_{Nt}).$$

a. Sketch the free-energy curves of the solid and liquid phases at 1000 K and 1300 K. Sketch also the activity of Ni across the composition range at the same temperatures.

b. Estimate the parameters C_1 and C_2 from the position of the minimum in the solidus and liquidus lines ($X_{Ni} = 0.42$, $T = 1223$ K).

Additional Reading

N. F. Rhines, *Phase Diagrams in Metallurgy*, McGraw-Hill (1956).

A. Prince, *Alloy Phase Equilibria*, Elsevier (1966). A very thorough presentation of the construction and interpretation of binary and ternary phase diagrams with a brief review of the pertinent thermodynamics.

A. M. Alper, ed., *Phase Diagrams*, Academic Press (1970). A complete coverage of heterogeneous equilibria, including thermodynamics, phase diagrams, solidification, computer calculations, and experimental methods.

N. L. Bowen, *The Evolution of Igneous Rocks*, Princeton University Press (1956). The first systematic study of geological systems, now regarded as a classic introduction to silicate systems.

L. S. Darken and R. W. Gurry, *Physical Chemistry of Metals*, Chaps. 11–13, McGraw-Hill (1953). A very clearly written introductory level treatment. Covers phase relations, heterogeneous equilibria and free-energy diagrams. Does not discuss ternary systems.

C. H. P. Lupis, *Chemical Thermodynamics of Materials*, North-Holland (1983). An elegant, comprehensive text; more advanced than others in this list, though couched in terms of classical thermodynamics; lots of examples and applications as well as theory. Highly recommended.

Chapter 10

Thermodynamics of Heterogeneous Equilibria

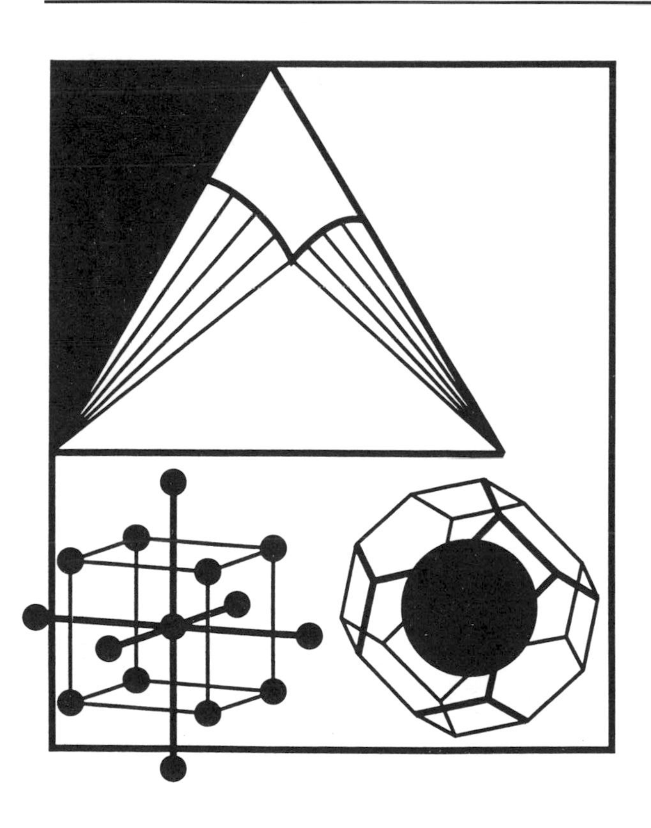

Chapter 10

Contents

There's many a good tune played on an old fiddle.

—Samuel Butler, *The Way of All Flesh*

Thermodynamics of Heterogeneous Equilibria

The preceding chapter was devoted primarily to the construction and interpretation of phase diagrams. Although extensive use was made of thermodynamic principles and relations, they mainly were applied toward achieving a better understanding of the factors governing phase equilibria. We here turn the tables to ask, given the phase diagram, what thermodynamic parameters and values can we extract from it. In addition, we will examine in greater depth some of the concepts touched on in Chapter 9, such as regular solutions and phase separation, while also taking up the subject of gas–solid reactions.

1. Partial Pressure, Fugacity, and Thermodynamic Activity

Let us start by establishing the relations between the fugacity, activity, and the partial pressure, of which three only the last one is a directly measurable chemical property. First, writing the total differential of the Gibbs free energy with respect to temperature, pressure, and composition leads to

$$dG = \left(\frac{\partial G}{\partial T}\right)_{P,n_i} dT + \left(\frac{\partial G}{\partial P}\right)_{T,n_i} dP + \sum \left(\frac{\partial G}{\partial n_i}\right)_{T,P} dn_i. \tag{10.1}$$

Substituting well-known identities for the partial derivatives transforms (10.1) to

$$dG = -S\,dT + V\,dP + \sum_i \mu_i\,dn_i. \tag{10.2}$$

At constant temperature and composition two terms vanish, leaving

$$dG = V\,dP. \tag{10.3}$$

Integration of (10.3) requires an equation of state relating pressure to the volume. The most elementary of these is given by the perfect gas law, viz. $PV = RT$. Substituting this into (10.3) leads to

$$\int_{G_1}^{G_2} dG = RT \int_{P_1}^{P_2} \frac{dP}{P},$$

or

$$\Delta G = RT \ln (P_2/P_1). \tag{10.4}$$

The ideal gas law is quite accurate up to pressures of a few atmospheres, especially when the gas is monatomic or is a homonuclear diatomic molecule rather than a polar one, such as CO, or NH_3, or H_2O.

The conditions are nearly ideal for applying (10.4) to the equilibrium vapor pressures of many solids because they are generally monatomic and have pressures of only a few Torr. However, it is those cases in which the gas does not behave ideally that brings us to the concept of fugacity.

The fugacity was invented by G. N. Lewis[1] as a device for preserving the convenient logarithmic form of (10.4). It is defined by $\Delta\bar{G}$, which is the actual isothermal, reversible pressure–volume work in going from the initial state f^0 to a final state f. In keeping with this,

$$\Delta\bar{G} = RT \ln(f/f^0). \tag{10.5}$$

In turn, $\Delta\bar{G}$ is given by the actual area under a P-V isotherm as shown in Fig. 10.1. Thus, fugacity can be thought of as a fictitious pressure that makes (10.5) valid.

As we have stated previously, the vapor pressures of most solids, even at elevated temperatures, are so small as to allow the universal application of the ideal gas law, and, hence, the substitution of pressure for fugacity.

We now take up our discussion of the thermodynamic activity, a, which is defined in terms of the fugacity:

$$a_i \equiv \frac{f_i}{f_i^0} \simeq \frac{P_i}{P_i^\circ}. \tag{10.6}$$

Since the activity always refers to a component in solution, the subscript i in (10.6) denotes the fugacities and partial pressures of the ith component. The quantities f_i^0

---

[1] G. N. Lewis, *Z. Phys. Chem*, **38**, 205 (1901).

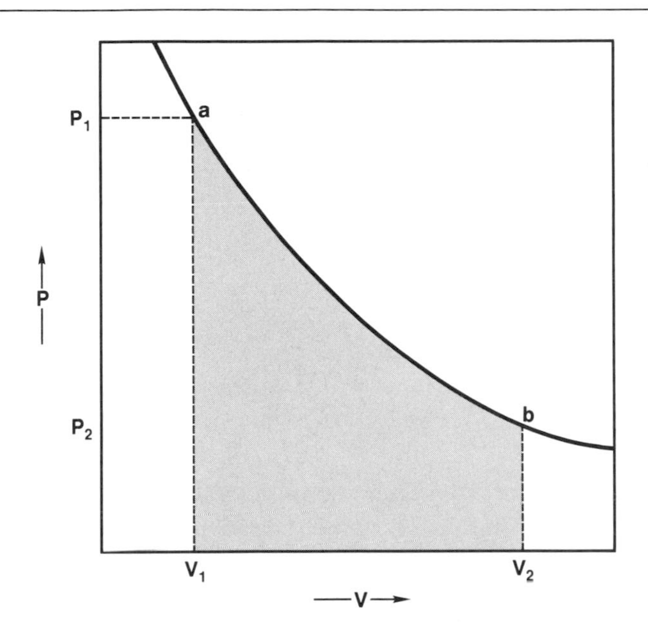

Figure 10.1.

The line *ab* gives the measured equilibrium values of P and V at constant T. The shaded area is proportional to the reversible work done by the system in going from P_1V_1 to P_2V_2. This line can be closely approximated by the ideal gas law when the pressures are low.

andP_i° are the fugacity and vapor pressure of pure i and, as a consequence of (10.6), the activity of a pure substance is unity. Combining (10.5) and (10.6) gives

$$\Delta\bar{G}_i = RT \ln a_i \simeq RT \ln(P_i/P_i^\circ). \tag{10.7}$$

Thus, a_i is ultimately defined by the difference in partial molal free energy, $\Delta\bar{G}_i$ between the substance in its standard state and in solution. The partial molal free energy is also referred to as the chemical potential, μ_i, and we will use, from time to time, both terms interchangeably.

Although gases generally do behave ideally in the pressure range of interest to us, solutions almost never do, and this is particularly true of solid solutions. Consequently, a knowledge of the vapor pressure as a function of temperature for a substance in the pure and dissolved state provides a convenient means of determining the partial molal free energy, enthalpy, and entropy. Of course, these quantities can also be obtained from calorimetry and electromotive force measurements, but generally this requires a somewhat greater effort. We illustrate the use of partial pressures with the following reaction, which mixes n_A moles of pure component A with n_B moles of B to form a solution:

$$n_A A + n_B B \rightarrow (A_{n_A} B_{n_B})_{soln}, \tag{10.8}$$

and for the present example we do not care whether the reactants and products are all solids or liquids as long as they are all in the same physical state. The equations for the free energy of formation are the sum of

$$\begin{aligned} n_A \bar{G}_A - n_A \bar{G}_A^\circ &= n_A RT \ln a_A = n_A RT \ln(P_A/P_A^\circ), \\ n_B \bar{G}_B - n_B \bar{G}_B^\circ &= n_B RT \ln a_B = n_B RT \ln(P_B/P_B^\circ), \end{aligned} \tag{10.9}$$

to give

$$\Delta G_f = n_A \Delta\bar{G}_A + n_B \Delta\bar{G}_B = RT \ln(P_A/P_A^\circ)^{n_A}(P_B/P_B^\circ)^{n_B}. \tag{10.10}$$

Now if the vapor pressures are known as a function of the temperature, the values of ΔH_f and ΔS_f can be calculated by taking the derivatives of ΔG_f with respect to the appropriate temperature function. It is of some importance to note in passing that the vapor pressure of only one component is needed to calculate ΔG_f for a binary solution if it is also known as a function of composition as well as temperature. By employing the Gibbs–Duhem[2] equation, we calculate the activity of the other component, starting with

$$X_A \, d\bar{G}_A + X_B \, d\bar{G}_B = X_A \, d \ln a_A + X_B \, d \ln n_B = 0, \tag{10.11}$$

which leads to

$$\ln a_A = -\int_{X_A=1}^{X_A} \frac{X_B}{X_A} \, d \ln a_B. \tag{10.12}$$

2. Activities Derived from Phase Diagrams

If the solution is saturated with respect to one of the components, which itself dissolves a negligible amount of the other reactant, then its activity in the solution is essentially unity, since it is in equilibrium with the nearly pure substance; this is by no means an uncommon situation. For example, we see in Fig. 10.2 that Au dissolves about 52.5 at.% Fe at 850°C, whereas Fe dissolves less than 2%Au. (Incidentally, this illustrates the last of the Hume-Rothery rules cited in the previous chapter, viz. that the element with the most outer or valence electrons is the solute with the greatest solubility.) Consequently, the tie line connecting these two phases shows

---

[2] See Appendix H for derivation.

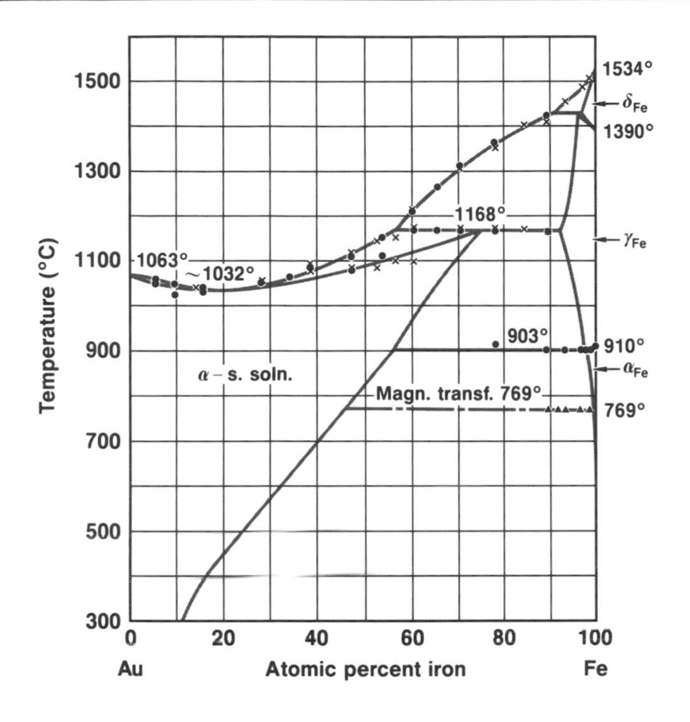

Figure 10.2.

The Au–Fe phase diagram illustrating the significant solubility of Fe in Au and the negligible solubility of Au in Fe. (After M. Hansen and K. Anderko, *The Constitution of Binary Alloys*, McGraw-Hill, p. 204, 1958.)

nearly pure α-Fe to be in equilibrium with an Fe-rich solid solution. If this solid solution follows the regular solution model, we can write

$$\Delta \bar{G}_{Fe} = RT \ln a^{\alpha}_{Fe} = RT \ln \gamma^{\alpha}_{Fe} X^{\alpha}_{Fe}. \tag{10.13}$$

Because of the equilibrium with nearly pure bcc α-Fe, we write for the activity of the dissolved Fe at the boundary of the Au-rich α-solid solution,

$$a^{\alpha}_{Fe} = \gamma^{\alpha}_{Fe} X^{\alpha}_{Fe} \simeq 1,$$

and then calculate the activity coefficient γ^{α}_{Fe} as

$$\gamma^{\alpha}_{Fe} = 1/X^{\alpha}_{Fe} = 1/0.525 = 1.91. \tag{10.14}$$

Let us compare this with the measured values given in Table 10.1. Considering the high concentration of Fe for which the activity coefficient γ^{α}_{Fe} was calculated, ($X_{Fe} > 0.5$), the agreement is surprisingly good except for the lowest concentrations.

Table 10.1. Activity and activity coefficients for Au–Fe solid solutions ($T = 850°C$).

X_{Fe}	a_{Fe}	γ_{Fe}	X_{Au}	a_{Au}	γ_{Au}
0.0	0.000	1.118	1.0	1.000	1.000
0.1	0.158	1.575	0.9	0.885	0.983
0.2	0.384	1.918	0.8	0.761	0.951
0.3	0.622	2.073	0.7	0.650	0.929
0.4	0.820	2.050	0.6	0.561	0.935
0.5	0.955	1.909	0.5	0.496	0.992
	(± 0.045)	(± 0.09)		(± 0.02)	(± 0.04)

From Hultgren *et al.*, *Selected Values of the Thermodynamic Properties of Binary Alloys*, American Society for Metals, p. 273, 1973.

The activity coefficients for γ_{Au}^{α} are also given and are very close to unity over the entire range of composition, illustrating that if Henry's law is obeyed even approximately, so is Raoult's law. This will be dealt with in more detail in the next section.

3. Henry's Law

Solutions that are ideal, and hence obey Raoult's law (9.4), must also obey Henry's law, which is

$$P_B = k_B X_B, \tag{10.15}$$

where component B is the solute in a solution and k_B is the Henry's law constant.

Henry's law can be rationalized by imagining a solution sufficiently dilute so that most of the solute atoms, because of their relatively small number, have only solvent atoms for nearest neighbors. As long as the concentration remains sufficiently low, the solute vapor pressure will scale as a linear function of composition, since all solute atoms are presumed to have and to preserve the same chemical environment. Obviously, in this dilute range, the solvent atoms are bonded mostly to one another, and their chemical environment is about the same as in the pure substance. It follows that the vapor pressure of the solvent is about the same as for the pure solvent after being corrected to the actual concentration, hence Raoult's law.

Not only is this qualitative explanation consistent with both Raoult's and Henry's laws, but it is easy to show that if either of these laws strictly applies it is then a thermodynamic necessity that the other be valid as well.

This can be shown by starting with the Gibbs–Duhem equation (see Appendix H for the derivation)

$$n_A\, d\mu_A + n_B\, d\mu_B = 0. \tag{10.16}$$

Integrating (10.16) leads to

$$\int_{\mu_B^\circ}^{\mu_B} d\mu_B = -\int \frac{n_A}{n_B}\, d\mu_B,$$

which becomes, after introducing the ideal solution approximation and, hence, Raoult's law,

$$\mu_B - \mu_B^\circ = -RT \int \frac{X_A}{1 - X_A}\, d \ln X_A = RT[\ln(1 - X_A) + \text{const.}], \tag{10.17}$$

and finally

$$\mu_B - \mu_B^\circ = RT \ln k' X_B \simeq RT \ln (P_B/P_B^\circ). \tag{10.18}$$

Since $P_B/P_B^\circ \simeq a_B$, we can equate

$$a_B = k' X_B, \tag{10.19}$$

and because P_B° is a temperature-dependent constant for each pure substance,

$$P_B = k'_B P_B^\circ X_B = k_B X_B, \tag{10.20}$$

which was the result to be shown. Clearly all the interactions between the solute and solvent atoms or ions are contained within k_B. We can make this more explicit by recalling that for *regular solutions*, as presented in the previous chapter,

$$a_i = \gamma_i X_i,$$

and so we can identify the activity coefficient γ_i with the Henry's law constant. Thus, solutions that obey Henry's law are regular solutions.

The analysis of Henry's law can be carried a step further by expressing the activity coefficient in terms of the Bragg–Williams model. Rewriting (9.34) leads to

$$\Delta E_f = zNX_A X_B\left(\varepsilon_{AB} - \frac{\varepsilon_{AA} + \varepsilon_{BB}}{2}\right) = \frac{zN_A N_B}{N_A + N_B}\, \varepsilon, \tag{10.21}$$

where N_A and N_B are the numbers of A and B atoms, z is the number of nearest neighbors, and $[\varepsilon_{AB} - (\varepsilon_{AA} + \varepsilon_{BB})/2] = \varepsilon$ is the energy required to form an A – B bond starting with pure A and pure B. Neglecting the heat capacity terms, we now rewrite (9.36) as

$$\Delta G_f = \frac{zN_A N_B \varepsilon}{N_A + N_B} + kT[N_A \ln N_A + N_B \ln N_B - (N_A + N_B) \ln(N_A + N_B)], \tag{10.22}$$

where the entropy of mixing is that for a random solution. The chemical potentials of A and B are obtained from the partial differentiation of ΔG_f in the usual manner and are then related to the activities. Thus,

$$\frac{\partial \Delta G_f}{\partial N_A} = \mu_A - \mu_A^\circ \cong RT \ln \frac{P_A}{P_A^\circ} \cong RT \ln a_A$$

$$= \frac{zN_B\varepsilon}{N_A + N_B} - \frac{zN_AN_B\varepsilon}{(N_A + N_B)^2} + RT \ln \frac{N_A}{N_A + N_B}, \quad (10.23)$$

which in terms of mole fractions (10.23) becomes

$$RT \ln a_A = z\varepsilon(1 - X_A)^2 + RT \ln X_A, \quad (10.24)$$

or

$$a_A = X_A \exp[z\varepsilon(1 - X_A)^2/RT]. \quad (10.25)$$

Comparing with $a_A = \gamma_A X_A$, we obtain

$$\gamma_A = \exp[z\varepsilon(1 - X_A)^2/RT]. \quad (10.26)$$

An analogous procedure leads to

$$\gamma_B = \exp[z\varepsilon X_B^2/RT]. \quad (10.27)$$

These are the γ_i of a *regular solution*. At high dilution the above γ_i values approach a constant that differs from 1, a result arising directly from a nonzero value of ε. For negative ε (i.e., attraction between unlike atoms), $\ln \gamma_i$ is negative and $\gamma_i < 1$, a negative deviation from Raoult' law, and the converse, of course, for a positive ε. Clearly Henry's law is an approximation because γ_i approaches a constant value only asymptotically as $X_i \to 0$.

The foregoing results may seem at first contradictory to the definition of ideality, which is so often cited as the absence of a heat of mixing. However, this requirement for ideality is both necessary and sufficient for the definition of Raoult's law. It is only sufficient but not necessary in order for Henry's law to be valid. Even the existence of a finite heat of mixing does not deter the solute vapor pressure from following Henry's law if the solution is suitably dilute. Figure 10.3 illustrates how both components of a binary Ag–Au solid solution approach the Raoult's and Henry's law approximations as their respective mole fractions approach unity and zero, respectively. The phase diagram for the Ag–Au system closely approximates that for ideal solid and liquid solutions (see Fig. 10.3) as would be expected, considering the similar chemical properties of these two elements. Nevertheless, the values of ΔG_f^x, the excess free energy of formation, are quite large for both elements, even at infinite dilution. Their values are $\Delta G_f^x(\mathrm{Ag}) = -2930$ cal-mol^{-1} and $\Delta G_f^x(\mathrm{Au}) = 3765$ cal-mol^{-1} for the solutes at infinite dilution where Raoult's and Henry's laws are obeyed.

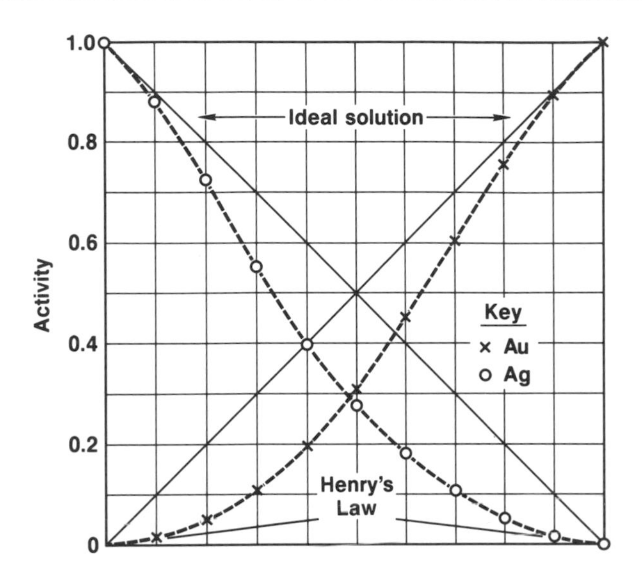

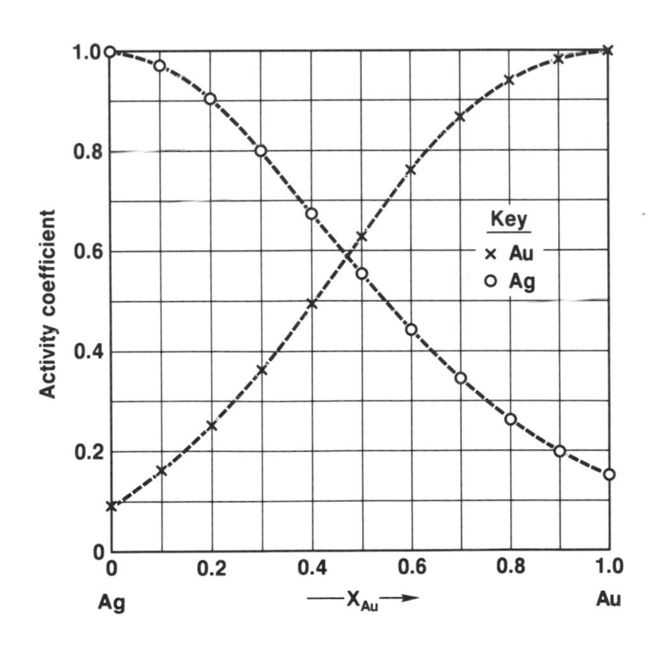

Figure 10.3.

The activities and activity coefficients for Ag–Au solid solutions at 800°C as a function of concentration. (Data taken from Hultgren *et al.*, pp. 28, 29, loc. cit.)

4. Phase Separation

Having pursued the Bragg–Williams model as far as (10.27), we can now analytically demonstrate the conditions required for the decomposition of a solution into two phases, a subject discussed in a more qualitative fashion in Chapter 9, Section 4. At equilibrium, the chemical potential of each component must be at its minimum regardless of the number of phases present. To find this minimum, we start with

$$\frac{\partial(\mu_A - \mu_A^\circ)}{\partial N_A} = RT\,\frac{\partial \ln a_A}{\partial N_A} = 0, \tag{10.28}$$

and substituting from (10.25) for the term on the right-hand side, we arrive at

$$\frac{1}{X_A} - \frac{2z\varepsilon}{RT}(1 - X_A) = 0, \tag{10.29}$$

or

$$X_A^2 - X_A + RT/2z\varepsilon = 0. \tag{10.30}$$

Now solving for X_A leads to

$$X_A = (1 \pm \sqrt{1 - 2RT/z\varepsilon})/2. \tag{10.31}$$

Consequently, X_A has two roots that are symmetrical about $X_A = 1/2$:

$$X_A = 1/2 \pm 1/2\sqrt{1 - 2RT/Z\varepsilon} \tag{10.32}$$

as long as $2RT/z\varepsilon < 1$. It should be noted that (10.29) is not generally valid and is not a requirement for equilibrium between phases. It is true here because the symmetry that is built into the Bragg–Williams model gives both phases in equilibrium identical free energies of formation, as shown in Fig. 9.6.

A further consequence of (10.31) is the definition of a critical temperature T_c, which is the temperature at which the homogeneous solution commences to separate, upon cooling, into two phases of differing composition:

$$T_c = z\varepsilon/2R, \tag{10.33}$$

since clearly (10.31) has no real solution for $T_c > z\varepsilon/2R$.

5. The Equilibrium Constant

Because of its extensive application to the treatment of solid–gas reactions, we review here the derivation of the equilibrium constant before proceeding to that subject.

For ease of visualization, we can relate the standard free energy of formation to the equilibrium constant starting with the following cycle:

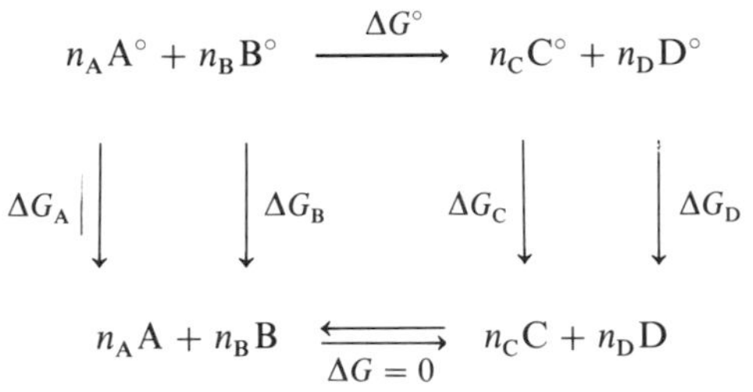

$$\begin{array}{ccc} n_A A^\circ + n_B B^\circ & \xrightarrow{\Delta G^\circ} & n_C C^\circ + n_D D^\circ \\ \Delta G_A \downarrow \quad \downarrow \Delta G_B & & \Delta G_C \downarrow \quad \downarrow \Delta G_D \\ n_A A + n_B B & \underset{\Delta G = 0}{\rightleftharpoons} & n_C C + n_D D \end{array}$$

Figure 10.4.

where the top line shows the reactants in their standard states, whereas the bottom line shows reactants and products in equilibrium with one another. It must be emphasized that n_A, n_B, n_C, and n_D are the stoichiometric coefficients, which are invariant and do not represent concentration. Consequently, they have the same values on both lines.

Let us consider two separate but equal pathways of going from the upper left-hand corner of the cycle to the lower right-hand side. Then, because G is a state function independent of path, we are allowed to write

$$\Delta G^\circ + n_C \Delta G_C + n_D \Delta G_D = n_A \Delta G_A + n_B \Delta G_B,$$

and upon rearranging, obtain

$$-\Delta G^\circ = n_C \Delta G_C + n_D \Delta G_D - (n_A \Delta G_A + n_B \Delta G_B),$$

or

$$-\Delta G^\circ = n_C(\bar{G}_C - G_C^\circ) + n_D(\bar{G}_D - G_D^\circ) - n_A(\bar{G}_A - G_A^\circ) - n_B(\bar{G}_B - G_B^\circ). \quad (10.34)$$

Substituting from (10.9) yields

$$-\Delta G^\circ = RT \ln\left[\frac{(a_C)^{n_C}(a_D)^{n_D}}{(a_A)^{n_A}(a_B)^{n_B}}\right]. \quad (10.35)$$

We substitute according to (10.4) for each of the partial molar free-energy terms to obtain

$$\begin{aligned} -\Delta G^\circ &= RT\left[\left(n_C \ln \frac{P_C}{P_C^\circ} + n_D \ln \frac{P_D}{P_D^\circ}\right) - \left(n_A \ln \frac{P_A}{P_A^\circ} + n_B \ln \frac{P_B}{P_B^\circ}\right)\right] \\ &= RT \ln\left[\left(\frac{P_C}{P_C^\circ}\right)^{n_C}\left(\frac{P_D}{P_D^\circ}\right)^{n_D} \Big/ \left(\frac{P_A}{P_A^\circ}\right)^{n_A}\left(\frac{P_B}{P_B^\circ}\right)^{n_B}\right]. \end{aligned} \quad (10.36)$$

If all components behave ideally, the activities, given by the ratio of the partial pressures, are equal to the mole fractions, so we arrive by substitution at

$$\Delta G^\circ = -RT \ln \frac{X_C^{n_C} X_D^{n_D}}{X_A^{n_A} X_B^{n_B}}. \tag{10.37}$$

The equilibrium constant is defined as

$$K \equiv \frac{X_C^{n_C} X_D^{n_D}}{X_A^{n_A} X_B^{n_B}},$$

and (10.33) can be written as

$$\Delta G^\circ = -RT \ln K \tag{10.38}$$

for an ideal system.

The closed cycle shown in Fig. 10.4 is meant to represent the reactions and free energy changes between condensed phases of A, B, C, and D. In actuality, (10.37) is derived from the implied complementary cycle involving only the equilibrium vapor pressure of each reactant and product. Because equilibrium between the gases and condensed phases is specified, the free energies of the components are equal in both phases. Basing our calculations upon the gaseous rather than the condensed phases permits us to use exceptionally simple equations of state, such as the perfect gas law, to calculate the change in free energy caused by the reaction, and is the major, if not the sole, reason for doing the calculation in this indirect manner.

6. Solid–Gas Reactions

The decomposition of compounds to form gases leaving a different compound as a residue constitutes a very large class of polyphasic reactions. Included are such very basic commercial processes as smelting, whereby oxide ores are reduced to their metal values along with the formation of CO and CO_2, or sulfide ores are roasted to oxidize the sulfide ions to SO_2. The reverse reactions, namely the oxidation and sulfiding of metals, encompasses most of the common corrosive reactions as well as such useful industrial processes as internal oxidation and internal nitriding of metals.

A. Basic Equations

The only terms contained in the equilibrium constants for solid–gas reactions are the pressures of the gases involved. As a result, if the temperature dependence of the vapor pressure is known, one can apply the Clausius–Clapeyron relation in the usual way to obtain the entropy and enthalpy from the temperature dependence of the free energy. Let us take as our example the dissociation of a metal oxide, MO_2, according to

$$MO_2 \rightarrow M + O_2, \tag{10.39}$$

in which MO_2 and M are solids and O_2 is a gas. At equilibrium the sum of the chemical potentials of the products must equal that of the reactants as expressed by

$$\mu_M + \mu_{O_2} = \mu_{MO_2}. \tag{10.40}$$

Recalling that

$$\mu_i = \mu_i^\circ + RT\ln(a_i) \simeq \mu_i^\circ + RT\ln(P_i/P_i^\circ), \tag{10.41}$$

we obtain by substitution

$$\begin{aligned}\Delta G^\circ &= \Delta\mu_M^\circ + \Delta\mu_{O_2}^\circ - \Delta\mu_{MO_2}^\circ \\ &= RT\left[\ln\frac{P_{MO_2}}{P_{MO_2}^\circ} - \ln\frac{P_M}{P_M^\circ} - \ln\frac{P_{O_2}}{P_{O_2}^\circ}\right],\end{aligned} \tag{10.42}$$

where the superscript $^\circ$ as usual denotes the standard state. However, by convention the standard state of any substance, be it solid, liquid, or gas, is taken to be its customary equilibrium form at an ambient pressure of 1 atm and at the temperature chosen for the calculation. Thus,

$$P_{MO_2}/P_{MO_2}^\circ = P_M/P_M^\circ = 1$$

as the free energies, and hence the vapor pressures of the solid phases, respond insignificantly to small changes in the ambient pressure, such as from 1 atm to near vacuum. The O_2 in its standard state is also at 1 atm pressure, but it is reduced to the pressure in equilibrium with MO_2, which represents a substantial change in its free energy. Consequently, only the last term in (10.42) is significant, and this equation reduces to

$$\Delta G^\circ = -RT\ln P_{O_2}, \tag{10.43}$$

where it also has been assumed that the partial pressure of O_2 obeys the perfect gas law. In passing, it should be mentioned that (10.43) is valid only if the composition of MO_2 is constant over the entire temperature range. This is frequently not the case, although the variation in the actual magnitudes of the relevant thermodynamic quantities is generally, but not always, negligible.

The standard enthalpy of formation is obtained from

$$\frac{\partial(\Delta G^\circ)}{\partial T^{-1}} = \Delta H^\circ = -R\frac{\partial\ln P_{O_2}}{\partial T^{-1}} \tag{10.44}$$

or

$$\frac{\partial \ln P_{O_2}}{\partial T^{-1}} = -\frac{\Delta H^\circ}{R}, \tag{10.45}$$

where ΔH° is the enthalpy per mole of O_2. Although the composition of the oxide may vary with temperature and partial pressure of oxygen, causing a variation in ΔH°, (10.45) remains valid at all temperatures.

Gas–solid equilibria can be very complicated, producing a wide spectrum of compounds if the cation has more than one valence state. This is illustrated by Fig. 10.5, which shows the many oxides that can be formed by Ti and V as a function of the partial pressure of oxygen.

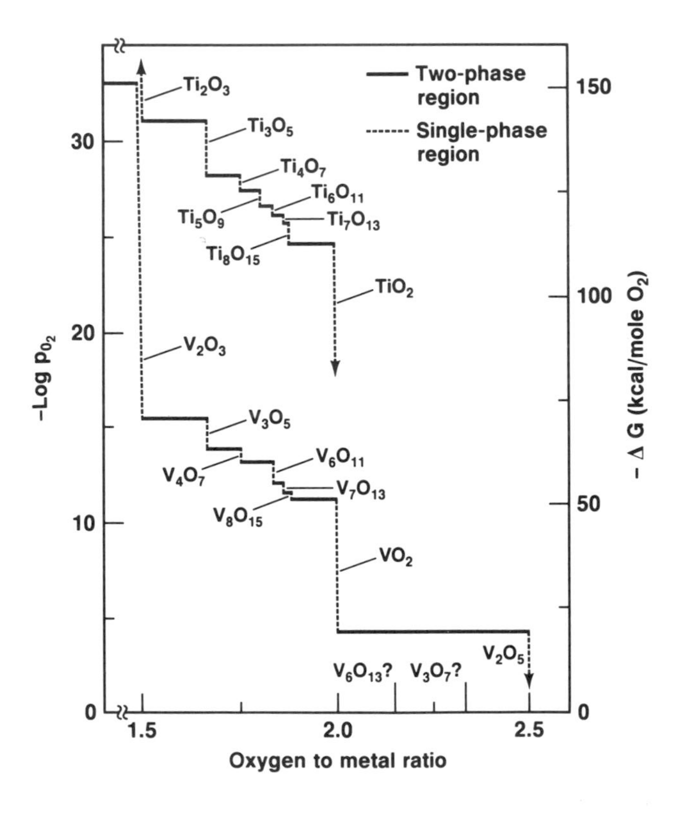

Figure 10.5.

The variation in oxygen partial pressure leads to the formation of several stable oxides of Ti and V at 1000 K. (After T. B. Reed *The Chemistry of Extended Defects in Non-Metallic Solids* (L. Eyring and M. O'Keefe, eds.), North-Holland, p. 27, 1970.)

7. Equilibrium between Compounds and Gases

A. Oxidation of Pure Elements

More often than otherwise, the equilibrium vapor pressure of metal oxides is too small to control. For example, at 1000 K the partial pressure of O_2 in equilibrium with NiO is about 10^{-13} atm, which is below the background of the very finest laboratory vacuums. However, the equilibrium constant for

$$\mathrm{Ni}(s) + \tfrac{1}{2}\mathrm{O}_2(g) \rightarrow \mathrm{NiO}(s), \qquad K_R = P_{\mathrm{O}_2}^{-1/2} \tag{10.46}$$

can be determined by using a mixture of gases with a smaller oxidative potential; generally, CO–CO_2 or H_2–H_2O. Then, for example, the actual reactions under investigation are

$$(1)\quad \mathrm{Ni}(s) + \mathrm{CO}_2 \rightarrow \mathrm{NiO} + \mathrm{CO}, \qquad K_1 = \frac{P_{\mathrm{CO}}}{P_{\mathrm{CO}_2}} \tag{10.47}$$

or

$$(2)\quad \mathrm{Ni}(s) + \mathrm{H}_2\mathrm{O} \rightarrow \mathrm{NiO} + (1/2)\mathrm{H}_2, \qquad K_2 = \frac{P_{\mathrm{H}_2}^{1/2}}{P_{\mathrm{H}_2\mathrm{O}}}. \tag{10.48}$$

All the relevant thermodynamic parameters associated with the oxidation of Ni can be had from combining the measured values of the temperature-dependent equilibrium constants for (10.46) with similar data for the oxidation of carbon monoxide:

$$(3)\quad \mathrm{CO} + 1/2\mathrm{O}_2 \rightarrow \mathrm{CO}_2, \qquad K_3 = \frac{P_{\mathrm{CO}_2}}{P_{\mathrm{CO}} P_{\mathrm{O}_2}^{1/2}}. \tag{10.49}$$

Clearly, adding (10.49) to (10.47) gives (10.46), the reaction of interest, and $K_1 K_3 = K_R$. The standard free energy of formation of NiO from elemental Ni and O_2, ΔG_f°, is obtained from

$$\begin{aligned}\Delta G_f^\circ &= \Delta G_1^\circ + \Delta G_3^\circ \\ &= -RT[\ln K_1 + \ln K_3] \\ &= -RT\left[\ln \frac{P_{\mathrm{CO}}}{P_{\mathrm{CO}_2}} + \ln \frac{P_{\mathrm{CO}_2}}{P_{\mathrm{CO}} P_{\mathrm{O}_2}^{1/2}}\right].\end{aligned} \tag{10.50}$$

It goes without saying that tabulated values of the free energies for all such common reactions as those just given have been in the literature for decades, and the foregoing is merely an illustrative example. Application of the Clausius–Clapeyron equation to (10.50) permits the calculation of ΔH_f° and ΔS_f° in the usual manner.

B. Carbides

The thermodynamics of carbides, many of which are the basis of industrially important compounds and solid solutions such as steel, can be determined from gas–solid equilibria. Let us suppose that we are interested in the direct reaction of a metallic element with carbon. The industrially useful transition element carbides (e.g., WC, W_2C, TiC, ZrC, MoC, etc.) are extremely refractory and are among the highest-melting compounds known. Consequently, the equilibria of reactions such as (10.51) lie far to the right,

$$xM(s) + C \rightarrow M_xC, \tag{10.51}$$

and the equilibrium concentrations of the pure metal and carbon are difficult, if not impossible, to measure. However, the carburizing reactions of the metal with carbon monoxide or methane afford a direct solution to this problem. Consider the following:

$$(1)\quad xM(s) + 2CO \rightarrow M_xC + CO_2, \qquad K_1 = P_{CO_2}/P_{CO}^2, \tag{10.52}$$

$$(2)\quad xM(s) + CH_4 \rightarrow M_xC + 2H_2, \qquad K_2 = P_{H_2}^2/P_{CH_4}. \tag{10.53}$$

Next we consider the reduction of CO_2 to CO,

$$(3)\quad CO_2 + C(s) \rightarrow 2CO, \qquad K_3 = P_{CO}^2/P_{CO_2}. \tag{10.54}$$

Adding reactions (10.54) to (10.52) returns us to (10.51), the reaction of interest. Thus, the free energy of formation of M_xC associated with (10.51) is given by

$$\Delta G_f^\circ = \Delta G_1^\circ + \Delta G_3^\circ = -RT[\ln K_1 + \ln K_3]. \tag{10.55}$$

We can employ an analogous process using (10.53), which is combined with the hydrogenation of carbon according to

$$(4)\quad 2H_2 + C \rightarrow CH_4, \qquad K_4 = P_{CH_4}/P_{H_2}^2, \tag{10.56}$$

which gives for the formation of M_xC,

$$\Delta G_f^\circ = -RT[\ln K_2 + \ln K_4]. \tag{10.57}$$

Similar reactions of metals with N_2, NH_3 will supply the same sort of information about metal nitrides. Halides, sulfides, and hydrides can be studied by the same technique.

C. Oxidation of Solid Solutions

There is less likelihood that measurements exist for the oxidation of solid solutions (i.e., alloys) than is the case for pure metals, and, in the absence of such data, estimating the relevant thermodynamic parameters can become a necessary exercise.

Suppose we begin by examining the two-phase alloy system, $\alpha + \alpha'$, of the Au–Ni solid solutions as shown in Fig. 10.6, and let us determine the critical oxygen pressure for the oxidation of Ni according to (10.58).

$$\mathrm{Au}_x\mathrm{Ni}_y + \mathrm{O}_2 \rightarrow \mathrm{Au}_x\mathrm{Ni}_y + \mathrm{NiO}. \tag{10.58}$$

At a given temperature, there will be a critical pressure below which NiO will not form or, if formed, it will dissociate, although the rate of this reaction could be imperceptably small, depending on the temperature. Because we are concerned only with the initial stage of the oxidation, only an infinitesimal amount of Ni need be converted to the oxide, and the composition of the alloy thus is taken as constant and the activities unchanged. NiO is in its standard state, and hence its activity is unity. The equilibrium constant for (10.58) then is given by

$$K = P_{\mathrm{O}_2}^{1/2}/a_{\mathrm{Ni}}, \tag{10.59}$$

where a_{Ni} is the activity of Ni dissolved in Au. The standard free energy for the oxidation of Ni is 34.9 kcal, which specifies the dissociation pressure, in this case, at 700°C. Therefore,

$$\ln K = -(34.9 \times 10^3)/RT,$$

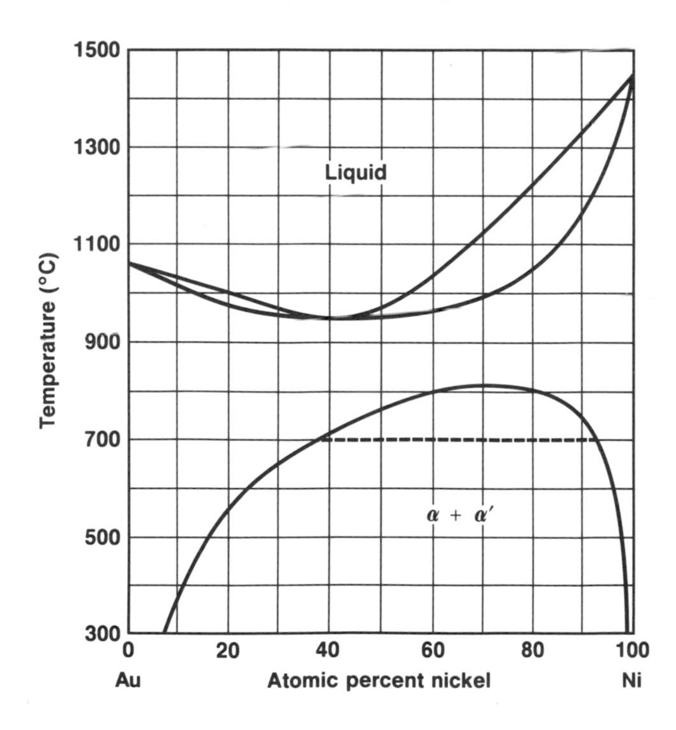

Figure 10.6.

The equilibrium diagram for Au–Ni. (After M. Hansen and K. Anderko, p. 220, loc. cit.)

or $K = 1.46 \times 10^{-8}$. The mole fractions of Ni in the α and α' phases, respectively, are 0.365 and 0.935 at 700°C, as approximately determined by reading Fig. 10.6. Because of the high concentration of Ni in the α' phase, we assume that Raoult's law is obeyed and set $a_{Ni}(\alpha') = X_{Ni}(\alpha') = 0.935$. Since $a_{Ni}(\alpha)$ must equal $a_{Ni}(\alpha')$, we can now also equate $a_{Ni}(\alpha) = 0.935$. Substituting this value for the activity of Ni into (10.59) along with the numerical value for K leads to

$$P_{O_2} = 1.86 \times 10^{-16} \text{ atm}.$$

The critical oxygen pressure for the oxidation of alloys with lower concentrations of Ni can be estimated by first calculating an activity coefficient such as

$$a_{Ni}(\alpha) = \gamma_{Ni}^{\alpha} X_{Ni}^{\alpha} = a_{Ni}(\alpha') = X_{Ni}^{\alpha'}, \tag{10.60}$$

leading to

$$\gamma_{Ni}^{\alpha} = X_{Ni}^{\alpha'}/X_{Ni}^{\alpha}. \tag{10.61}$$

This activity coefficient can be used, at the same temperature, for the calculation of the activity at other concentrations, but it is frequently a poor approximation.

For example, in the present case, at 700°C, $\gamma_{Ni}^{\alpha} = 2.56$, for $X_{Ni} = 0.356$, which appears to be in fair agreement with that obtained by interpolation of the measured values, which gives $\gamma_{Ni} \cong 2.25$. However, this fortuitous near agreement disappears quickly, as the measured value of γ_{Ni} is 2.0 at $X_{Ni} = 0.4$ and drops further to ~ 1.7 at $X_{Ni} = 0.5$. Thus, assuming γ_{Ni} to be constant over a significant range of composition can lead to substantial errors unless the substance conforms to the behavior prescribed by the regular solution model. This is in contrast to the Au–Fe system, which, on chemical grounds, one would expect to be quite similar and for which γ_{Fe} is very nearly constant over the range $0.2 \leq X_{Fe} \leq 0.5$ (see Table 10.1).

Chapter 10 Exercises

1. For an ideal gas $dG = RT \ln P$. Derive this expression, starting with $G = H - TS$.
2. Derive expressions for $\bar{H}_B - H_B^\circ$ and $\bar{V}_B - V_B^\circ$ for the solute in a dilute solution, based upon solubility as a function of temperature.
3. Derive a formula for the fugacity of a gas that has as its equation of state

$$PV = RT + P_0 P + P_1 P^2,$$

   where V is the molar volume and P_0 and P_1 are temperature-dependent numerical quantities.
4. Suppose a regular solution is formed by components A and B and $\varepsilon = 6$ kcal. At a temperature of 900 K, calculate the composition of the two phases that are in equilibrium, assuming both to be fcc.

5. In the transition of aragonite ($CaCO_3$) $\rightarrow$ calcite ($CaCO_3$) the volume change is $\Delta V = 2.75$ cm^3-mol^{-1}. At what pressure does aragonite become the stable crystal structure at 298 K.
6. Write the equilibrium constant expression for the reaction

$$CaCO_3(s) + 2H^+(aq) + SO_4^{2-}(aq) + H_2O(l) \Leftrightarrow CaSO_4 \cdot 2H_2O(s) + CO_2(g),$$

assuming that solids are pure crystalline phases and that the gas is ideal.
7. The oxide tarnish on silver dissociates when the Ag is heated. Calculate the dissociation energy, given that $\Delta G_f^\circ = -7000 + 15.25T$(cal-mol^{-1}) gives the free energy of formation of Ag_2O.
8. The oxygen vapor pressures corresponding to the coexistence of the two oxides CuO and Cu_2O are

T K	1173	1223	1273	1303	1350
P_{O_2} (atm)	0.0208	0.0498	0.1303	0.225	0.504

Estimate ΔH° and ΔS° of the reaction

$$2CuO = Cu_2O + (1/2)O_2,$$

and express its ΔG° as a function of temperature.
9. Consider the equilibrium reaction at 1000°C:

$$2CO \rightarrow C \text{ (in austinite)} + CO_2$$

(austinite is the fcc solid solution of C in Fe). The values of the partial pressures are

wt.% C	P_{CO}	P_{CO_2}
0.13	0.891	0.109
0.45	0.9660	0.0340
0.96	0.9862	0.0138
1.50 (saturated with graphite)	0.9928	0.0072

Find the activity of carbon in each case, taking graphite as the standard state.
10. Referring to Fig. 9.16, assume that the η phase is an ideal solution of Cu in Zn. The maximum solubility at the peritectic temperature, 424°C, is 2.7 at.% Cu, and at the lowest temperature shown in the figure, 100°C, it is 0.3 at.%. Show how you would estimate the activity coefficients γ_{Cu} in the ε phase at these temperatures. How might you then estimate γ_{Cu} at the composition $X_{Zn} = 0.8$?
11. Draw schematic free-energy vs. composition diagrams for the Au–Fe system at 1000 K, 1100 K, 1168 K, and ~1435 K (refer to Fig. 10.2).

Additional Reading

Most of the additional reading references in Chapter 9 apply equally to Chapter 10. Two exceptions are

N. A. Gokcen, *Statistical Thermodynamics of Alloys*, Plenum Press (1986). A very clear, well-organized presentation that relates statistical methods to classical thermodynamics with frequent examples based upon experimental results. Deals also with semiconductors and empirical methods of estimating thermodynamic parameters.

C. Wagner, *Thermodynamics of Alloys*, Addison-Wesley (1962). This little book is a true classic, very readable; written by one of the true pioneers in the field of solid state thermodynamics.

Chapter 11

The Chemistry of Interfaces

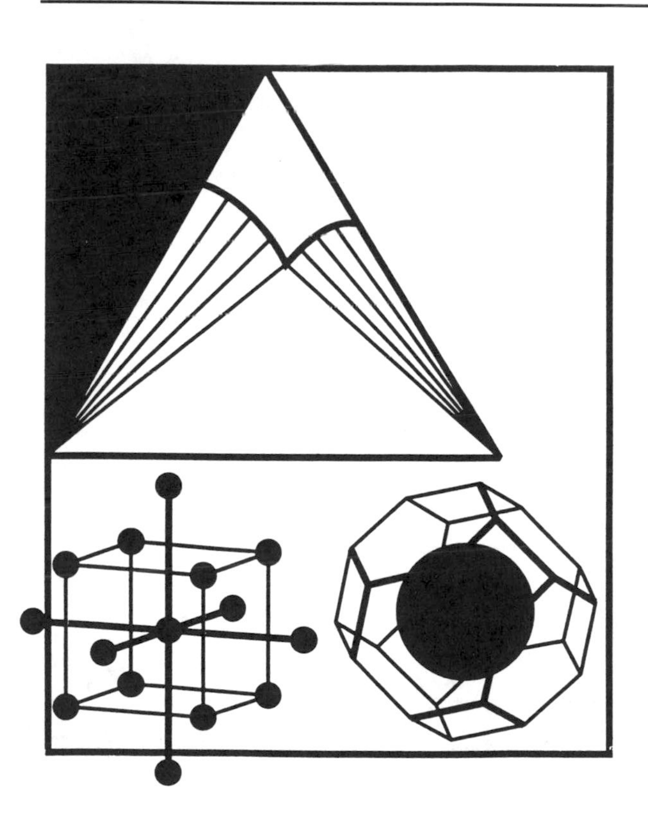

Chapter 11

Contents

Had I been present at the Creation, I would have given some useful hints for the better ordering of the universe.

—Alfonso the Wise, King of Castile (circa 1260)

The Chemistry of Interfaces

A surface exposed to only a vapor phase is called simply a surface. One that is adjoined by another condensed phase is referred to as an interface. In the first instance the atoms or molecules of the surface layer have different properties from those of the interior, having fewer nearest neighbors, and hence there are apt to be dangling bonds. An interfacial surface defines a change of character, compositional or topographical, in the bonding on either side, although bonds of some sort span the interface.

Because the surface molecules are subject to an inwardly directed force, it requires work to expand the surface in a lateral direction, since such a distortion incrementally alters the equilibrium bond distances between the surface and interior atoms. This is equivalent to the surface being under lateral tension and is the origin of the term "surface tension." For a plane surface the surface tension is defined as the force acting parallel to the surface and perpendicular to a line of unit length located arbitrarily within the surface. There is a great deal of chemistry embodied by the term "surface chemistry," and we can but touch on a small part of it here. Among the many processes and phenomena included in this field are catalysis, gas separations and analysis, sintering, control of the nucleation of solid phases, charge states in semiconductors, and yet others left uncited. An example of the effect of the excess surface free energy is shown in Fig. 11.1. It is the driving force for particle growth of UO_2 during sintering at elevated temperatures and stems directly from the reduction in total surface area and, hence, the surface free energy.

We begin our discussion of the thermodynamics of surfaces with liquids because they are more amenable to measurement and visualization than the corresponding quantities for solids. Nevertheless, the general conclusions are the same and equally valid for both states.

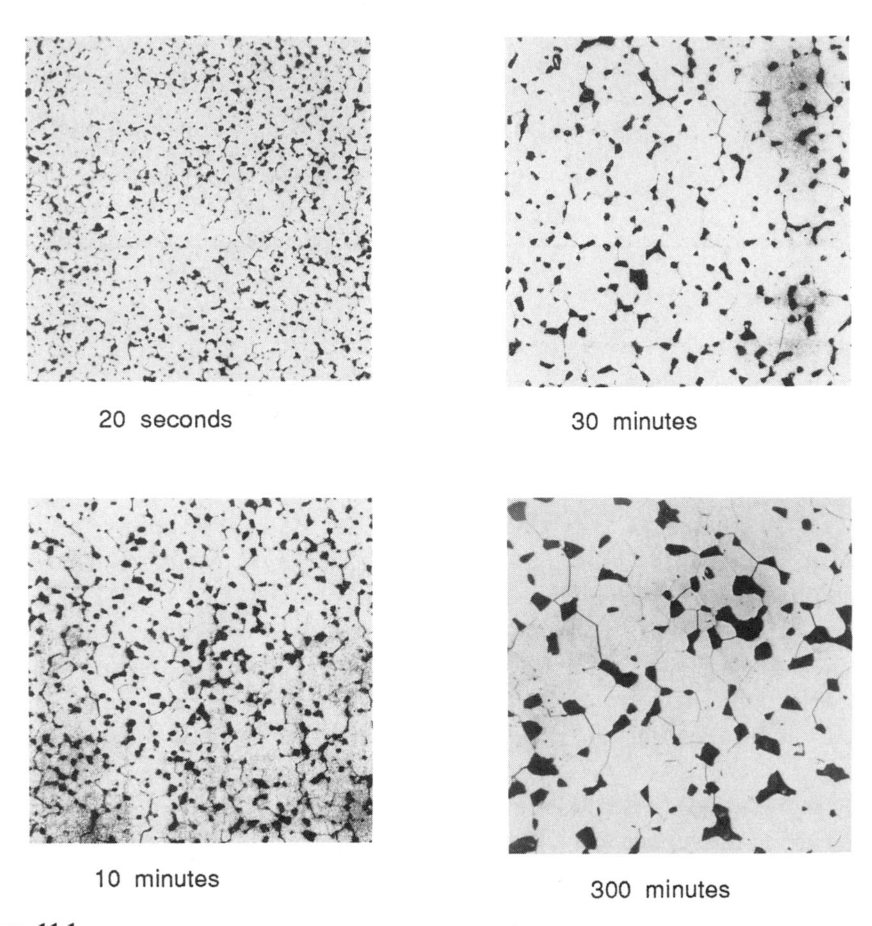

Figure 11.1.

Microstructure of UO_2 during sintering at 1600°C (400 ×). (After B. Francois and W. D. Kingery, *Proc. Intl. Conf. on Sintering and Related Phenomena*, Gordon and Breach, p. 512, 1965.)

1. Surface Free Energy

Commencing with the total differential of G, the Gibbs free energy, we have

$$dG = \left(\frac{\partial G}{\partial T}\right)_{P,\mu_i,A} dT + \left(\frac{\partial G}{\partial P}\right)_{T,\mu_i,A} dP + \left(\frac{\partial G}{\partial n_i}\right)_{T,P,A} dn_i + \left(\frac{\partial G}{\partial A}\right)_{T,P,\mu_i} dA$$

$$= -S\,dT + V\,dP + \sum_i \mu_i\,dn_i + \gamma\,dA, \tag{11.1}$$

where μ_i is the chemical potential of the ith constituent, A is the surface area, and

$$\partial G/\partial A \equiv \gamma.$$

Thus, γ is free energy per unit area of surface. Also, for an isothermal, isobaric system of unchanging composition, $\gamma\, dA = dw$ and is thus the reversible work necessary to incrementally create additional surface, and is the analog in two dimensions of $P\, dV$ work in three. This is convincingly illustrated by the direct measurement of pressure within a bubble. Imagine a capillary tube to be barely submerged below a liquid surface, as shown in Fig. 11.2, and an excess positive pressure, ΔP, exerted within the capillary. To satisfy the requirement that $P\, dV = \gamma\, dA$, that is, that the pressure–volume work be exactly equal and opposite to the contractile force per unit area times the increase in area, dA, one must have

$$Pd(\tfrac{4}{3}\pi r^3) = \gamma d(4\pi r^2).$$

Thus, $P = 2\gamma/r$ or

$$\gamma = Pr/2. \tag{11.2}$$

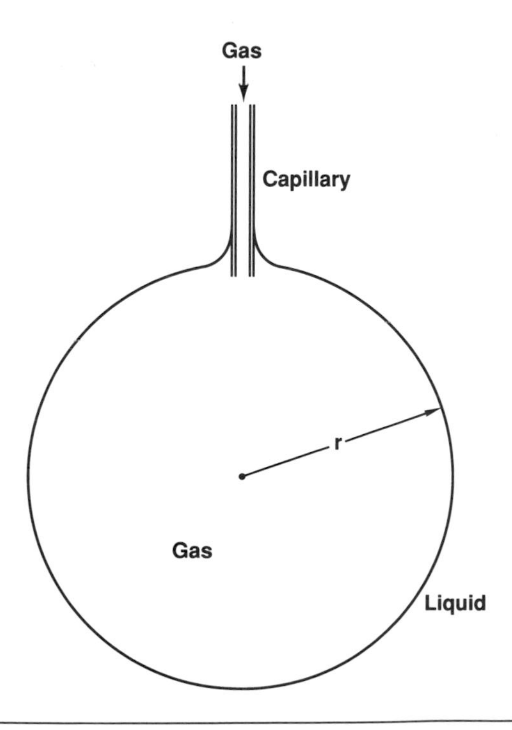

Figure 11.2.

The bubble method for determining γ.

Note that the surface free energy has the dimensions of force per unit length, which, of course, it must if $\gamma\, dA$ is to have the dimensions of work. There are several methods for determining the surface free energy of liquids; determining it for solids is considerably more difficult and will be dealt with in a later section of this chapter.

Strictly, one should speak of the *excess* surface free energy, since the total free energy of the surface layer also includes that derived from the vibrational, electronic, and, in the case of liquids, rotational and translational degrees of freedom. However, such niceties of nomenclature are seldom observed, and we will continue to refer to γ as the surface free energy.

2. The Chemical Potential as a Function of Surface Curvature

The following derivation introduces the important concept of the dependence of the chemical potential on the radius of curvature of the surface or interface. Because the radius of curvature r^{-1} increases with decreasing particle size, this effect can equally be called the dependence of the chemical potential upon particle size. In fact, the most frequent expressions of this phenomenon are generally encountered when dealing with very small particles, although it also is manifested by the tips of sharply pointed solids. Consider an ensemble of small, free-floating droplets at constant temperature and pressure and in which all the constituents are in equilibrium with their respective vapor pressures. Then

$$dG_{T,P} = \Sigma \mu_i\, dN_i + \Sigma \gamma\, dA. \tag{11.3}$$

We can simplify the arithmetic without altering the generality of the conclusion by first considering a single-component system. Then expanding (11.3), we have

$$dG_{T,p} = \mu_g\, dn_g + \mu_l\, dn_l + \gamma\, dA. \tag{11.4}$$

In a closed system $dn_g = -dn_l$. Hence, with $dG = 0$ at equilibrium,

$$\mu_g = \mu_l + \gamma \left(\frac{dA}{dn_l}\right)_{T,P}. \tag{11.5}$$

Now let the liquid be dispersed into M identical spherical particles of radius r. The total area of the liquid phase is

$$A = M 4\pi r^2, \tag{11.6}$$

and the corresponding total volume is

$$V = n_l \Omega = M(4/3)\pi r^3, \tag{11.7}$$

where Ω is the molecular volume and n_l is the total number of molecules in the liquid phase. As a result of (11.6) and (11.7),

$$A = 3n_l\Omega/r, \tag{11.8}$$

where n_l and r are independent variables, since the surface area can be altered by changing either the radius or the number of the spheres. Now, expressing the change in area as the complete differential leads to

$$dA = \left(\frac{\partial A}{\partial n_l}\right)_r dn_l + \left(\frac{\partial A}{\partial r}\right)_{n_l} dr,$$

so that

$$\frac{dA}{dn_l} = \left(\frac{\partial A}{\partial n_l}\right)_r + \left(\frac{\partial A}{\partial r}\right)\frac{dr}{dn_l}. \tag{11.9}$$

Using (11.6) and (11.7) and substituting into (11.9) gives

$$\begin{aligned} \frac{dA}{dn_l} &= \frac{3\Omega}{r} - \frac{3n_l\Omega}{r^2}\frac{dr}{dn_l} \\ &= \frac{3\Omega}{r} - \frac{3n_l\Omega}{r^2}\frac{r}{3n_l} \\ &= \frac{2\Omega}{r}. \end{aligned} \tag{11.10}$$

Substituting this result into (11.5) gives

$$\mu_g = \mu_l + \gamma 2\Omega/r. \tag{11.11}$$

Assuming the ideal gas law to be applicable allows one to write

$$\mu_g = \mu_l + RT\ln(P/P_\infty), \tag{11.12}$$

where P_∞ is the vapor pressure in equilibrium with a surface of infinite extension. Now substituting (11.11) into (11.12) results in

$$\begin{aligned} \ln(P/P_\infty) &= 2\gamma\Omega/RTr, \\ P/P_\infty &= \exp(2\gamma\Omega/RTr). \end{aligned} \tag{11.13}$$

If the radius of the droplets is 10^{-6} cm., the following examples reveal the magnitude of the increased vapor pressure due to the very small size.

Substance[a]	γ (dynes/cm)	$\Omega \times 10^{23}$ (cc/molecules)	P/P_{∞}
Hg	460.6	2.445	1.735
H_2O	69.85	2.980	1.108
CCl_4	25.02	16.013	1.216

[a] E. A. Moelwyn-Hughes, *Physical Chemistry*, Pergamon Press, p. 936, 1961.

The foregoing examples are estimates at best, since it is doubtful if the macroscopic value of γ remains constant down to particles of 200 Å in diameter. Nevertheless, the importance of this surface phenomenon in profound when one considers the nucleation of a different phase from its parent material. Clearly, embryos of the new phase, such as, for example, rain drops from water-saturated air, must begin with the coalescence of only two molecules and then continue to grow. Thus, in the initial stages, the vapor pressure of the water must exceed considerably the equilibrium value for a macroscopic surface, in part explaining the significant degree of undercooling required by some homogeneously nucleated phase transformations.

As our second example, let us consider the precipitation of a compound from a saturated solution. At equilibrium the chemical potential of the dissolved substance, μ_d, must equal the chemical potential of the solid, μ_s, plus the additional free energy necessary to enlarge the surface of a very small particle:

$$\mu_d = \mu_{s,\infty} + 2\gamma\Omega/r. \tag{11.14}$$

Also, when the ideal solution model pertains,

$$\mu_d = \mu^\circ + RT \ln X, \tag{11.15}$$

where X is the mole fraction of solute. However,

$$\mu_{s,\infty} = \mu^\circ + RT \ln X_\infty,$$

and, consequently,

$$RT \ln(X/X_\infty) = 2\gamma\Omega/r, \tag{11.16}$$

which is a positive number, thus showing the enhanced solubility resulting from the very small particle size of the solid. A feeling for the possible magnitude of this effect is obtainable from the following measured values:

Substance[a]	$\Omega \times 10^{23}$ (cc/cm^3)	r (cc $\times 10^5$)	C (mmol/L)
$CaSO_4$	7.56	20	15.3
		3	18.2
$BaSO_4$	7.72	18	0.00983
		1	0.01778

[a] E. A. Moelwyn-Hughes, p. 939, loc. cit.

3. The Gibbs Adsorption Isotherm

Many of the interesting and useful properties of interfaces relate to the adsorption of foreign atoms or molecules. The Gibbs equation[1] relates the surface concentration of the adsorbate to γ and the bulk chemical potentials. In passing, it is noted that the adsorption of foreign species lowers γ, and thus also the interfacial free energy.

We begin by considering the interface between two phases α and β. One system is real with an exchange of different species across the boundary, and the other is ideal with no mutual interadsorption, so that each phase maintains its bulk composition and properties right up to the discrete interface. The difference in the free energies of these two systems is the free energy of formation of the real interface. The thickness of the real interface, λ, may have any value starting with a monolayer of molecular dimensions as the minimum (see Fig. 11.3).

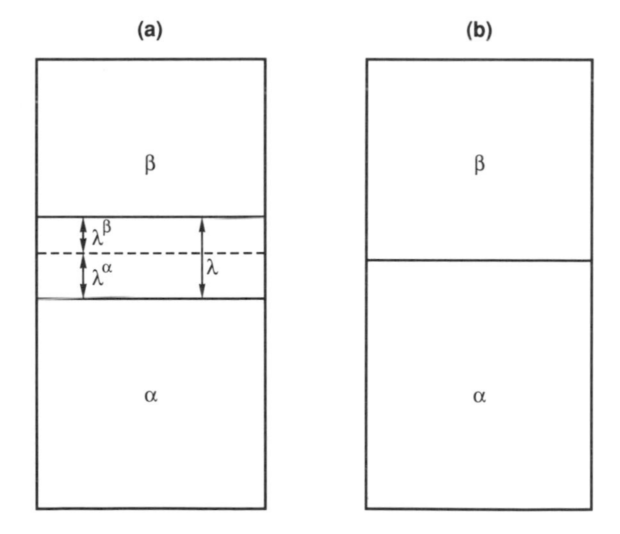

Figure 11.3.

(a) A real interface with varying composition of components 1 and 2 across the interface; (b) the ideal interface without adsorption.

The change in Helmholtz free energy of the ideal system at constant temperature and volume is given by

$$dF_{T,V} = \mu_1 \, dn_1 + \mu_2 \, dn_2 + \gamma \, dA, \tag{11.17}$$

[1] J. W. Gibbs, *Collected Works of J. Willard Gibbs*, vol. I, pp. 233–237, Yale University Press, 1957.

which in terms of the α and β phases becomes

$$\begin{aligned} dF^{\alpha}_{T,V} &= \mu^{\alpha}_1\, dn^{\alpha}_1 + \mu^{\alpha}_2\, dn^{\alpha}_2, \\ dF^{\beta}_{T,V} &= \mu^{\beta}_2\, dn^{\beta}_2 + \mu^{\beta}_1\, dn^{\beta}_1. \end{aligned} \tag{11.18}$$

At equilibrium, μ_1 and μ_2 are equal in both the α and β phases, so we may drop the superscripts. Then the free energy of formation of the interface is

$$F^{\sigma} = F - (F^{\alpha} + F^{\beta}), \tag{11.19}$$

where F^{σ} denotes the free energy of the real interface shown in Fig. 11.3. Then differentiating (11.19) gives

$$dF^{\sigma}_{T,V} = dF_{T,V} - (dF^{\alpha}_{T,V} - dF^{\beta}_{T,V}), \tag{11.20}$$

and substituting (11.17) and (11.18) into (11.20), we arrive at

$$dF^{\sigma} = \mu_1(dn_1 - dn^{\alpha}_1 - dn^{\beta}_1) + \mu_2(dn_2 - dn^{\alpha}_2 - dn^{\beta}_2) + \gamma\, dA. \tag{11.21}$$

This is generalized to an n-component system in an obvious manner, giving

$$dF^{\sigma}_{T,V} = \sum_{i=1}^{n} \mu_i\, dn^{\sigma}_i + \gamma\, dA, \tag{11.22}$$

where, as before, the superscript σ designates the difference between the real and ideal surface values of the terms so designated.

Integrating (11.22) gives

$$F^{\sigma}_{T,V} = \sum_{i=1}^{n} \mu_i n^{\sigma}_i + \gamma A. \tag{11.23}$$

Differentiating (11.23) and subtracting (11.22) (Euler's theorem: see Appendix H), leaves us with

$$\Sigma n_i\, d\mu_i + A\, d\gamma = 0. \tag{11.24}$$

This derivation and the relation (11.24) may be thought of as the analog of the derivation of the Gibbs–Duhem equation for an interface that requires the incorporation of the change in the interfacial free energy (i.e., $A\, d\gamma$). It is the reduction in γ by the excess superficial concentrations that furnishes the driving force for the adsorption. The excess superficial concentrations are defined by

$$\Gamma_i = \frac{n^{\sigma}_i}{A} = \frac{n_i - n^{\alpha}_i - n^{\beta}_i}{A}, \tag{11.25}$$

where n_i is the total number of molecules in the two-phase system. The Gibbs adsorption isotherm now can be written as

$$d\gamma = -\Sigma\Gamma_i\, d\mu_i. \tag{11.26}$$

If the surface in Fig. 11.3 is adjusted so that $\Gamma_1 = 0$, for example, one atomic distance below the surface of a solid upon which has been adsorbed a monolayer of an essentially insoluble component, such as a gas, the relative adsorption is given by

$$d\gamma = -\Gamma_2\, d\mu_2 = -\Gamma_2 RT\, d\ln f_2. \tag{11.27}$$

If the perfect gas model applies, the fugacity $f_2 = P_2$ and

$$d\gamma = -\Gamma_2 RT \ln P_2. \tag{11.28}$$

If the adsorbed layer consists of two components and behaves ideally, the two-dimensional analogs of Raoult's and Henry's laws are valid. Stated without derivation for the case where component A is at a much greater concentration than component B, and hence in three dimensions would correspond to the solvent, allows the approximation

$$d\gamma = -\Gamma_A\, d\mu_A - \Gamma_B\, d\mu_B \cong -\Gamma_B\, d\mu_B,$$

$$d\gamma = -\Gamma_B RT\, d\ln a_B \cong -\Gamma_B RT\, d\ln X_B$$

for ideal behavior and

$$\Gamma_B = \frac{X_B}{RT}\frac{d\gamma}{dX_B}. \tag{11.29}$$

It has been shown (Chapter 10, Section 3) that if Raoult's law pertains, then so does Henry's law. In two dimensions the latter is written in terms of the excess superficial concentration:

$$\Gamma_B = k_H P_B. \tag{11.30}$$

4. The Segregation of Impurities at Interfaces

Impurities, coming from the interior of the solid, tend to concentrate at the available interfaces as predicted by (11.25), and the decreases in γ with increasing superficial impurity concentration provides the driving force. However, it is easier to treat this situation as a partitioning of the impurities between the surface and the interior. The latter are considered to be two separate states in which the impurities reside with a concentration-independent difference in their partial molar enthalpies.

Because the congregation of impurities at interior interfaces, such as grain boundaries and dislocations, frequently causes embrittlement resulting in mechanical failure, this phenomenon is of more than just academic interest. The derivation of the equations necessary to calculate the relative concentrations begins with the following definitions:

- $\bar{H}_{\mathrm{I}}$ Partial molal enthalpy of the impurity I in the bulk solid
- $H_{\mathrm{I}}^{\mathrm{s}}$ Partial molal enthalpy of I in the surface
- N Number of available lattice sites for I in the solid
- N_{I} Number of I atoms in the solid
- n Number of available surface sites
- n_{I} Number of I atoms on the surface

The configurational partition function Ω_{c} is given by

$$\Omega_{\mathrm{c}} = \frac{N!}{n_{\mathrm{I}}!(N - N_{\mathrm{I}})!} \times \frac{n!}{n_{\mathrm{I}}!(n - n_{\mathrm{I}})!}. \tag{11.31}$$

Then the configurational entropy, or entropy of mixing, is

$$\Delta S_{\mathrm{c}} = R \ln \Omega_{\mathrm{c}}. \tag{11.32}$$

The free energy for adsorption is now given by

$$\Delta G_{\mathrm{a}} = -n_{\mathrm{I}}(\bar{H}_{\mathrm{I}} - \bar{H}_{\mathrm{I}}^{\circ}) + n_{\mathrm{I}}(\bar{H}_{I}^{s} - H_{\mathrm{I}}^{\circ}) - RT \ln \Omega_{\mathrm{c}}. \tag{11.33}$$

Here we have neglected ΔS_{v}, the change in the vibrational entropy. The standard state for the impurity H_{I}° is arbitrary, but, of course, must be the same for both bulk, and adsorbed atoms. Applying Stirling's approximation to (11.31) and making use of the conservation relation (viz. $dn_{\mathrm{I}} = -dN_{\mathrm{I}}$) as well as the approximations that $n_{\mathrm{I}} \ll n$ and $N_{\mathrm{I}} \ll N$, we can minimze the free energy, obtaining

$$\frac{\partial \Delta G_{\mathrm{a}}}{\partial n_{\mathrm{I}}} = 0 = -\Delta \bar{H}_{\mathrm{I}} + \Delta H_{\mathrm{I}}^{\mathrm{s}} - RT\left[\ln \frac{n_{\mathrm{I}}}{n - n_{\mathrm{I}}} - \ln \frac{N_{\mathrm{I}}}{N - N_{\mathrm{I}}}\right]. \tag{11.34}$$

Rearranging (11.34) and cancelling the implicit H_{I}° terms, we arrive at

$$\frac{n_{I}}{n - n_{I}} = \frac{N_{\mathrm{I}}}{N - N_{\mathrm{I}}} \exp\left[\frac{-(H_{\mathrm{I}}^{s} - H_{\mathrm{I}})}{RT}\right]. \tag{11.35}$$

Using the definitions

$$\frac{n_{\mathrm{I}}}{n} = X_{\mathrm{I}}^{\mathrm{s}} \qquad \frac{N_{\mathrm{I}}}{N} = X_{\mathrm{I}} \qquad Q = \bar{H}_{\mathrm{I}} - H_{\mathrm{I}}^{\mathrm{s}},$$

which is the binding energy of I to the surface relative to the interior bulk sites, (11.35) becomes

$$X_{\mathrm{I}}^{\mathrm{s}} = \frac{X_{\mathrm{I}} e^{Q/RT}}{(1 - X_{\mathrm{I}}) + X_{\mathrm{I}}^{Q/RT}} \cong \frac{X_{\mathrm{I}} e^{Q/RT}}{1 + X_{\mathrm{I}} e^{Q/RT}}, \tag{11.36}$$

when $X_{\mathrm{I}} < 1$.

The same equation can be derived by simply applying the law of mass action. Beginning with the reaction

$$N_{\mathrm{I}} + (n - n_{\mathrm{I}}) \rightleftarrows n_{\mathrm{I}}, \tag{11.37}$$

where all the symbols have their previous meanings, the equilibrium constant, as always, must be expressed in units of concentration (not numbers of molecules), which we will take to be mole fractions as previously defined. Thus, we obtain

$$K_{\mathrm{eq}} = \frac{X_{\mathrm{I}}^{\mathrm{s}}}{X_{\mathrm{I}}(1 - X_{\mathrm{I}}^{\mathrm{s}})} = e^{-\Delta G^{\circ}/RT}. \tag{11.38}$$

Then rearranging yields

$$X_{\mathrm{I}}^{\mathrm{s}} = X_{\mathrm{I}} e^{-\Delta G^{\circ}/RT} - X_{\mathrm{I}} X_{\mathrm{I}}^{\mathrm{s}} e^{-\Delta G^{\circ}/RT}$$

or

$$X_{\mathrm{I}}^{\mathrm{s}} = \frac{X_{\mathrm{I}} e^{-\Delta G^{\circ}/RT}}{1 + X_{\mathrm{I}} e^{-\Delta G^{\circ}/RT}}. \tag{11.39}$$

Omitting the vibrational entropy makes

$$-\Delta G^{\circ} = -(\Delta H_{\mathrm{I}}^{\mathrm{s}} - \Delta \bar{H}_{\mathrm{I}} = \bar{H}_{\mathrm{I}} - H_{\mathrm{I}}^{\mathrm{s}} \equiv Q. \tag{11.40}$$

Substituting Q for $-\Delta G^{\circ}$ in (11.39) reproduces (11.36).

5. The Langmuir Adsorption Isotherm

Using the law of mass action, Langmuir[2] derived adsorption isotherms widely applicable to the adsorption of gases on solid surfaces. Generally speaking, there are two kinds of adsorption: physical or chemical. The former involves a weak binding of the absorbate to the surface by van der Waals forces. The second mode, usually referred to as *chemisorption*, involves the formation of chemical bonds. Enthalpies of adsorption for the first case fall in the range of 0 to 5 kcal, whereas for chemisorption the range is 20 to 100 kcal/mole. Adsorption at liquid–liquid and liquid–vapor interfaces is most often of the physical sort, although it also occurs with some frequency in gas–solid interfaces,

Chemisorption does not occur indiscriminately over all portions of the interface but at specific surface sites that, because of the local composition or topography, are

---

[2] I. Langmuir, *J.A.C.S.*, **38**, 2221 (1916).

most suitable for binding the adsorbate; these are called *active surface sites*. The derivation starts by considering a gas of species M to be in equilibrium with the appropriate concentration of adsorbed molecules, as schematically shown in Fig. 11.4. It is furthermore supposed that the surface contains a specific concentration S_0 of active sites capable of capturing M-type molecules. The adsorbed molecules are represented by SM and, written as chemical reactions, are given by

$$(S_0 - SM) + \mathrm{M} \rightleftarrows SM. \tag{11.41}$$

If the fractional occupation of surface active sites is defined as

$$\theta \equiv SM/S_0,$$

we can write for the equilibrium constant for (11.41),

$$K = \frac{(SM)}{(S_0 - SM)(\mathrm{M})} = \frac{\theta S_0}{S_0(1-\theta)P_\mathrm{M}} = \frac{\theta}{P_\mathrm{M} - \theta P_\mathrm{M}}, \tag{11.42}$$

where $\mathrm{M} \propto P_\mathrm{M}$, the partial pressure of M. Continuing the algebraic rearrangement of (11.42) leads to

$$\theta = KP_\mathrm{M}/(1 + KP_\mathrm{M}), \tag{11.43}$$

which is the usual form of the Langmuir adsorption isotherm. For a surface having a spectrum of active adsorption sites (11.43) is generalized to

$$\theta = \sum_i \frac{X_i K_i P_\mathrm{M}}{1 + K_i P_\mathrm{M}}. \tag{11.44}$$

The chemical potential of the adsorbed species can be expressed in terms of the surface coverage by

$$\mu_{SM} = \mu_{SM}^\circ + RT \ln\left[S_0\left(\frac{\theta}{1-\theta}\right)\right], \tag{11.45}$$

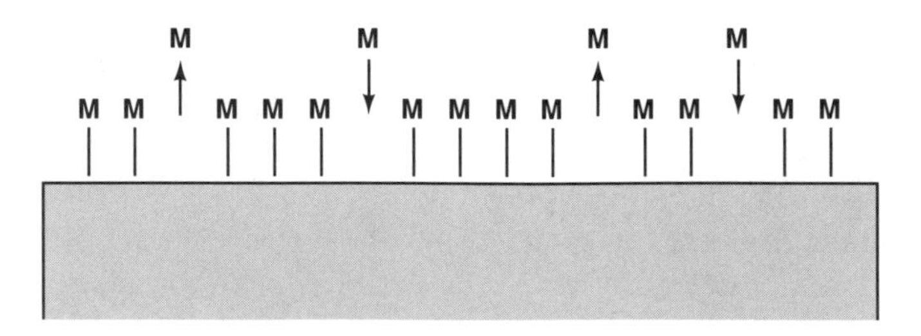

Figure 11.4.

A gas M in equilibrium with an adsorbed layer of itself.

where the standard state corresponding to μ_{SM}° is the chemical potential for a single molecule adsorbed on a surface of infinite dimensions. At equilibrium μ_M (gas) $= \mu_{SM}$, which allows us to write

$$\mu_M^{\circ}(\text{gas}) - \mu_{SM}^{\circ} = RT \ln \frac{S_0}{P}\left(\frac{\theta}{1-\theta}\right). \tag{11.46}$$

Plotting θ as a function of P at constant temperature produces a graph whose general features resemble Fig. 11.5.

If a diatomic molecule dissociates upon adsorption and the two atoms need to occupy adjacent sites, then, following Langmuir, the rate of adsorption is

$$\left(\frac{d\theta}{dt}\right)_a = c_1 P(1-\theta)^2, \tag{11.47}$$

where P is the pressure of the molecular gas and c_1 is a proportionality constant. The desorption is also quadratic in θ, since two atoms have to come together to form a molecule prior to desorption. Thus, for desorption

$$\left(\frac{d\theta}{dt}\right)_d = -c_2\theta^2. \tag{11.48}$$

At equilibrium, the net rate [sum of (11.47) and (11.48)] is zero, and the adsorption isotherm becomes

$$\theta^2/(1-\theta)^2 = cP$$

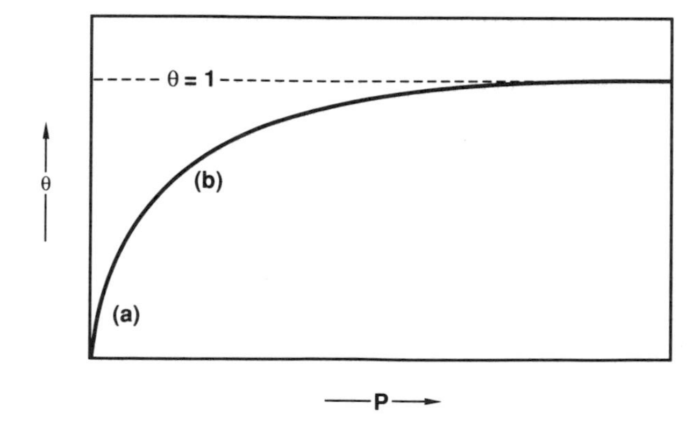

Figure 11.5.

General shape of the Langmuir adsorption isotherm in regions (a) $\theta \propto P$, (b) $\theta \propto P^{1/m}$, where m is an empirical constant.

or

$$\theta = (cP)^{1/2}/(1 + (cP)^{1/2}), \tag{11.49}$$

where $c = c_1/c_2$.

6. The BET Isotherm

More often than not, additional adsorbate molecules will condense on the initial monolayer before the latter has completely saturated all the available surface sites. These successive layers are bound less tightly, and their relative concentrations (i.e., the first, second, third, etc., layers) will follow the customary thermodynamic distribution functions. The Brunauer–Emmett–Teller (BET) isotherm takes account of these successive layers of adsorbate and thus is an extension of the Langmuir treatment.

The adsorption of gases as a function of pressure on nonporous surfaces generally (but not always) follows either of the two isotherms portrayed in Fig. 11.6a,b. The first, Fig. 11.6a, shows the adsorption of a monolayer in which the gas ultimately saturates all of the absorption sites without condensation of a second layer. Figure 11.6b shows the isotherm appropriate for the case where additional superficial layers develop before saturation by the initial monolayer.

As in the usual case for a thermodynamic description of chemical reactions, one can proceed via a classical or statistical approach; we will adhere to the former because it more closely parallels our previous derivation of the Langmuir isotherm. Defining $S_0, S_1, S_2, S_3, \ldots, S_i, \ldots$ as the areas covered by $0,1,2,3, \ldots, i, \ldots$ molecular

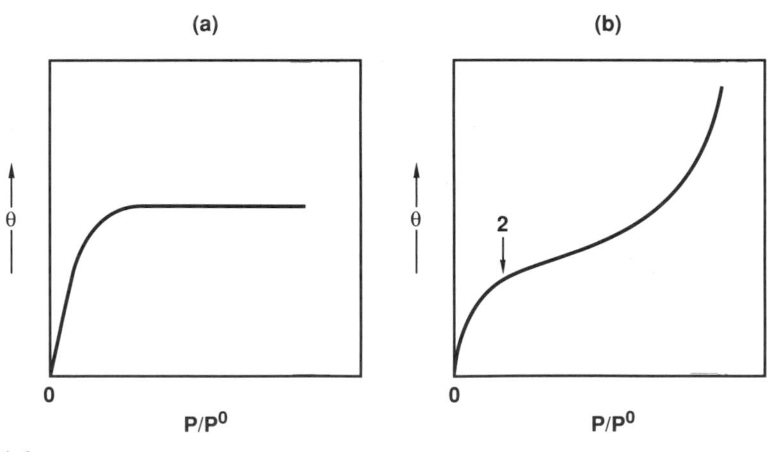

Figure 11.6.

Idealized adsorption isotherms: (a) Langmuir isotherm described by (11.43); (b) frequently encountered isotherm with second layer of adsorbate starting as noted at ~2.

layers of absorbate, we have at equilibrium for the first layer only:

$$a_1 P S_0 = d_1 S_1 e^{-E_1/RT} \tag{11.50}$$

or

$$S_1 = bPS_0 e^{E_1/RT}, \tag{11.51}$$

where a_1 and d_1 are rate constants for adsorption and desorption, E_1 is the heat of adsorption of the first layer, and $b = a_1/d_1$. Equation (11.51) is formally equivalent to the Langmuir equation. Similar relations exist in equilibrium, between the first and second layers, and so on. In the development of the BET isotherm, two major approximations are made. First, the proportionality constant b is taken to be the same for all layers. Second, the heat of adsorption for the second and for all layers beyond the second is taken to be the heat of liquifaction, E_l. Thus, for the second layer

$$S_2 = bPS_1 e^{E_l/RT} = (bP)^2 S_0 e^{E_l/RT} e^{E_1/RT}. \tag{11.52}$$

The general form for i layers is then

$$s_i = (bP)^i S_0 e^{E_1/RT} e^{(i-1)E_l/RT}. \tag{11.53}$$

It is convenient to let

$$bPe^{E_l/RT} = x, \tag{11.54}$$

to give

$$S_i = bPe^{E_1/RT} S_0 x^{i-1}. \tag{11.55}$$

Now the total active surface area of the solid is given by

$$A = \sum_{i=0}^{\infty} S_i, \tag{11.56}$$

and the total volume of adsorbed gas is

$$V^s = V_0^s \sum_{i=1}^{\infty} iS_i, \tag{11.57}$$

where V_0^s is the volume of gas adsorbed per unit area by a completed monolayer. Hence, for a surface of area A we can write

$$\frac{V^s}{AV_0^s} = \frac{V^s}{V_m^s} = \frac{\sum_{i=1}^{\infty} iS_i}{\sum_{i=0}^{\infty} S_i}, \tag{11.58}$$

where V_m^s is the volume corresponding to a monolayer over the entire surface A.

The summations in (11.58) may be expressed as

$$\sum_{0}^{\infty} S_i = S_0 + S_0 bPe^{E_1/RT}(1 + x + x^2 + \cdots)$$

$$= S_0 + \frac{S_0 bPe^{E_1/RT}}{1 - x}, \tag{11.59}$$

$$\sum_{1}^{\infty} iS_i = S_0 bPe^{E_1/RT}(1 + 2x + 3x^2 + \cdots)$$

$$= \frac{S_0 bPe^{E_1/RT}}{(1 - x)^2}, \tag{11.60}$$

and (11.58) becomes

$$\frac{V^s}{V^s_m} = \frac{bPe^{E_1/RT}}{(1 - x)(1 - x + bPe^{E_1/RT})}. \tag{11.61}$$

Let P_0 be the saturation pressure of the liquid layer and set $x = P/P_0$. Note that as $x \to 1$, $V^s \to \infty$ at $P = P_0$ because of the infinite number of layers. Thus, (11.61) may be written as

$$\frac{V^s}{V^s_m} = \frac{bPe^{E_1/RT}}{(1 - P/P_0)(1 - P/P_0 + bPe^{E_1})}, \tag{11.62}$$

which is the BET adsorption isotherm.

For matching to experiments, (11.62) may be rearranged by recalling that $bP_0 \exp(E_l/RT) = 1$, as

$$\frac{P}{V^s(P_0 - P)} = \frac{e^{\Delta E/RT}}{V^s_m} + \frac{P(1 - e^{\Delta E/RT})}{P_0 V^s_m}, \tag{11.63}$$

where $\Delta E = E_l - E_1$.

The BET method is most commonly used to determine the total surface area of solids. The most commonly used adsorbate is N_2. Figure 11.7 shows particularly good agreement between (11.63) and the experimental data.

In conclusion, it should be remarked that, although the BET approach is extemely useful for the determination of surface areas, it is not a rigorous theory, and, in fact, has been shown to violate the principle of microscopic reversibility.[3]

---

[3] T. L. Hill, *J.A.C.S.*, **72**, 5347 (1950).

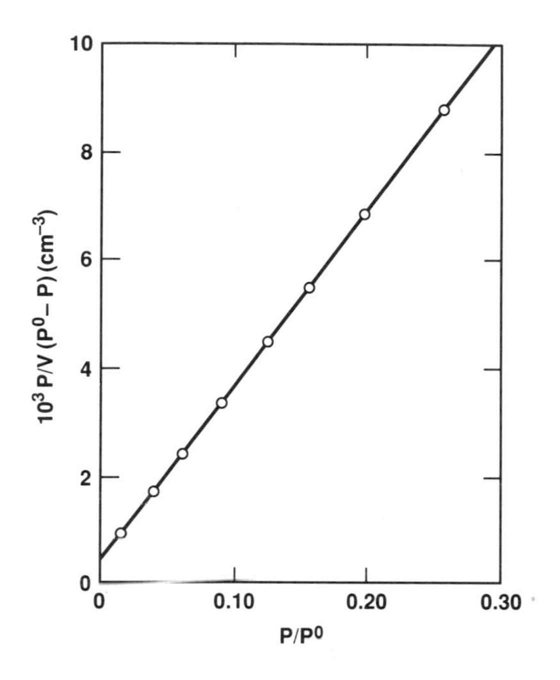

Figure 11.7.

Adsorption of N_2 on a fused Cu catalyst at 90.1 K plotted according to (11.63). (After Brunauer, 1944), as presented by R. Avegard and D. A. Haydon, *An Introduction to the Principles of Surface Chemistry*, Cambridge University Press, p. 165, 1973).

7. The Rate and Mechanism of Thermal Desorption

Thermal desorption is a relatively simple chemical process. Starting with the escape of an adsorbate from the surface into the gas phase, i.e.,

$$A_A \rightarrow A_G, \tag{11.64}$$

and assuming a first-order reaction allows us to write

$$\frac{-dA_A}{dt} = k_d A_A, \tag{11.65}$$

or in terms of surface coverage, θ,

$$\frac{-d\theta_A}{dt} = k_d \theta_A. \tag{11.66}$$

The temperature dependence of k_d is given in terms of the free energy of activation[4], $\Delta G_d^\ddagger$ by

$$k_d = \nu e^{-\Delta G_d^\ddagger/RT}. \tag{11.67}$$

The quantity ν has the dimensions of frequency and, as is the case for all molecular reactions, is about 10^{12}–10^{13} sec^{-1}, which is within the normal range of molecular vibrations.

If the adsorbate must first react to form a molecule before desorbing, and this is the rate-controlling step, the reaction is written

$$A_A + B_A \rightarrow AB_G, \tag{11.68}$$

and the corresponding rate equation is

$$\frac{-d\theta_A}{dt} = \frac{-d\theta_B}{dt} = k_d \theta_A \theta_B, \tag{11.69}$$

with the case $\theta_A = \theta_B$ already given by (11.48).

If the temperature is sufficiently high so that the adsorbed A and B behave as a two-dimensional gas, elementary kinetic theory leads to

$$k_d = d_{AB}(2RT/\mu)^{1/2} e^{-E/RT}, \tag{11.70}$$

where d_{AB} is the collision diameter and μ is the reduced mass of molecular AB, and E is the energy of adsorption.

On the other hand, if collisions between A and B are the result of diffusive motion, the activation energy for desorption must contain the activation energies for surface diffusion, viz. $\Delta G_A^\ddagger$ and $\Delta G_B^\ddagger$. For example, if $\Delta G_A^\ddagger < \Delta G_B^\ddagger$, which is to say that A is more mobile than B, and assuming that desorption is rapid, then at low values of θ,

$$k_d \cong \nu \exp(-\Delta G_A^\ddagger/RT). \tag{11.71}$$

The foregoing model is overly simplified in that it assumes $\Delta G_A^\ddagger$ and $\Delta G_B^\ddagger$ to be single-valued. They are, in fact, weighted averages that reflect the varying binding strengths of the active adsorption sites and, further, are a function of θ.

---

[4] A general discussion of activation energy is given in Section 3 of Chapter 13.

8. Coupled Adsorption and Bulk Diffusion

It has been assumed up to now that atoms adsorbed on the surface do not diffuse into the substrate, but this is often not the case. Adsorbed hydrogen atoms[5], in particular, often migrate into the bulk, and conversely, of course, atoms in the bulk may segregate to the surface. Well-known experimental examples are, for instance, hydrogen uptake by niobium and tantalum films.[6] We are dealing now with a more complicated equilibrium and kinetic problem, and make use of the concepts already treated and enlarge on them by coupling surface and bulk processes.

We have to consider, in the simplest case, four fluxes as indicated in Fig. 11.8. We use the results of Section 5 to express fluxes f_1 and f_2 as

$$f_1 = 2k_1(1-\theta)^2, \tag{11.72}$$

$$f_2 = -k\theta^2, \tag{11.73}$$

where θ is the coverage factor as defined for Langmuir adsorption and f_3 and f_4 describe the coupling to the bulk and are written

$$f_3 = -\nu\theta(1-X), \tag{11.74}$$

$$f_4 = \beta(1-\theta)X. \tag{11.75}$$

where ν and β are jump frequencies and X is the atomic fraction of H atoms in the bulk, f_3 is the flux from the surface into the bulk, and f_4 is the return flux. ν and β are taken to be simple activation terms given by

$$\nu = \nu_0 e^{-E_A/kT}, \tag{11.76}$$

$$\beta = \beta_0 e^{-E_B/kT}. \tag{11.77}$$

The energy balance of all the processes shown in Fig. 11.8 is

$$-E_A + E_B + E_D = E_S, \tag{11.78}$$

where E_S is the heat of solution of hydrogen in the metal.

The coupled rate equations for the surface and bulk processes are then for small X

$$\frac{d\theta}{dt} = 2k_1(1-\theta)^2 - k\theta^2 - \nu\theta + \beta(1-\theta)X, \tag{11.79}$$

$$N_1 \frac{dX}{dt} = \nu\theta - \beta(1-\theta)X, \tag{11.80}$$

where N_1 is the number of layers of the solid film in planar geometry.

---

[5] See, for example, G. Alefeld and J. Volkl, Eds., *Hydrogen in Metals*, Springer, New York, 1978.

[6] J. W. Davenport, G. J. Dienes, and R. A. Johnson, *Phys. Rev.*, **25B**, 2165 (1982).

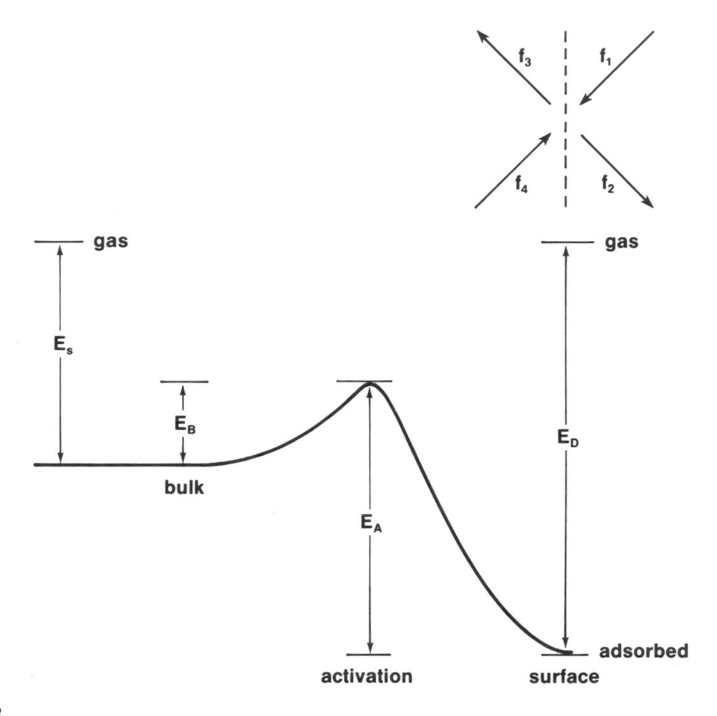

Figure 11.8.

Idealized potential surface for hydrogen in and on metal. The energy of the hydrogen atom at a surface site relative to H_2 in the gas phase is E_D. There is an activation barrier E_A to go into the bulk where the solution energy is E_S. f_1 is the flux of H_2 molecules incident from the gas phase, f_2 the flux of desorbing atoms, f_3 the flux from surface to bulk, and f_4 the "backflow" from bulk to surface. (After Ref. 6.)

The total uptake per unit surface, u, is given by

$$u = N_1 X + \theta, \tag{11.81}$$

and from (11.79) and (11.80) we obtain

$$\frac{du}{dt} = 2k_1(1-\theta)^2 - k\theta^2. \tag{11.82}$$

At equilibrium the time derivatives vanish, giving

$$\frac{\theta}{1-\theta} = \frac{\beta}{\nu} X \tag{11.83}$$

and

$$\theta/(1-\theta) = (2k_1/k)^{1/2}, \tag{11.84}$$

which lead to the well-known Sievert law of solubility, namely

$$X = C_s(P/P_0)^{1/2}e^{E_s/kT}, \tag{11.85}$$

where C_s is a constant, that incorporates the entropic term.

A general solution of the coupled master equations, (11.79) and (11.80), is not available. We shall consider first two limiting cases and discuss later the steady-state approximation. If the surface barrier is very small, then the bulk and surface concentrations may be approximated as equal, and we may take $\theta = X$. This limiting case is also equivalent to the assumption that the rate-limiting step is arrival from the gas phase. From (11.81) then,

$$(N_1 + 1)\frac{dX}{dt} = 2k_1(1 - X)^2 - kX^2, \tag{11.86}$$

which immediately integrates to

$$\frac{X}{1 - X} = \left[\frac{2k_1}{k}\right]^{1/2} \tanh\left[\frac{(2k_1k)^{1/2}}{N_1 + 1}\right]. \tag{11.87}$$

The second limiting case occurs when the rate-limiting step is the surface-to-bulk transfer. In this case the surface and gas come into equilibrium before there is any significant transfer to the bulk. In this case we solve (11.79) with $X = 0$ and $\nu = 0$ and then insert the equilibrium value of θ into (11.80) to find the uptake rate. The result is

$$N_1\frac{dX}{dt} = \frac{\beta}{1 + (\beta/\nu)X_m}(X_m - X), \tag{11.88}$$

where X_m is the equilibrium concentration.

Generally, a more useful approximation is obtained from the steady-state solution. In this case, X and θ are assumed to obey the steady-state relation

$$X = \frac{\nu}{\beta}\frac{\theta}{1 - \theta}, \tag{11.89}$$

even though X and θ separately have not attained their equilibrium values. With this approximation and with $d\theta/dt$ very small compared with $N_1(dX/dt)$, the integration can be performed to yield, for charging,

$$(1/2)(1 - b)^2 \ln(1 + y) - (1/2)(1 + b)^2 \ln(1 - y) - b^2y = at, \tag{11.90}$$

where $y = X/X_m$ and $a = 2k_1/(N_1X_m)$. For degassing, by setting $k_1 = 0$ in (11.79), one obtains

$$\frac{1}{y} - 1 + 2\frac{\beta}{\nu}X_0 \ln(y) + \frac{\beta^2}{\nu^2}X_0^2(1 - y) = \frac{1}{N_1}K\frac{\beta^2}{\nu^2}X_0 t, \tag{11.91}$$

where y is now X/X_0 and X_0 is the concentration at $t = 0$.

These rather complicated-looking equations can be neatly fitted to experimental data and describe the rates of hydrogen uptake by niobium and tantalum films with generally very good results.[7] The parameters of (11.90) and (11.91) are determined by fitting the experimental kinetic data. The energy terms have been evaluated for the hydrogen–niobium system, giving $E_s = 16.5$ kcal/mole of H_2, $E_D = 25$ kcal/mole of H_2 and $E_A - E_B = 8.5$ kcal/mole of H_2.

9. Chemisorption—Chemical Bonding to Surfaces

Bonds between a surface (substrate) and adsorbed molecules or atoms are influenced by the same factors as those that control bonding in ordinary molecules. Substrate–adsorbate bonds range from being predominantly ionic to predominantly covalent. Physisorption, in contrast to chemisorption, makes use of van der Waals forces as exemplified by the adsorption of rare gases on metal surfaces.

One can think of chemical bonds between solid surfaces and the adsorbed species in exactly the same terms as about bonding between ions or atoms in crystals or molecules.

Accurate images of surfaces with details at the scale of individual atoms are now obtainable from a variety of experimental methods.[8] What they reveal are patterns of extremely irregular structure that are reflected by the complexity and variability of the chemical bonding to surfaces. A list of generalities pertaining to chemisorption is quoted here.

1. Several binding sites, even on the more atomically homogeneous low-Miller-index surfaces, are distinguishable by their structure and binding strength for different adsorbates. As a result, a sequential filling of binding sites is frequently observed.
2. The structure of the binding sites and the strength of binding change from crystal face to crystal face.
3. The character of the surface bond may change markedly with temperature. This is perhaps the most novel feature of the surface chemical bond as compared with other types of chemical bonds.
4. The presence of surface irregularities (steps and kinks) and of other adsorbates can markedly influence the surface chemical bond.

---

[7] M. A. Pick, J. W. Davenport, Myron Strongin, and G. J. Dienes, *Phys. Rev. Letts*, **43**, 286 (1979).

[8] See, for example, G. A. Somorjai, *Chemistry in Two Dimensions: Surfaces*, Chap. 2, Cornell University Press, 1981.

5. The unique electronic and structural environment of the surface may cause the formation of distinct surface phases for multicomponent systems, which have no analogue in the bulk-phase diagrams.
6. A model that assumes that the surface bond is localized and is similar to that in multi–nuclear cluster compounds fairly well represents many of the surface bonds.
7. Chemisorption always weakens the bonds of the solid whether electrons are displaced toward the surface or toward the adsorbate.

In addition to the foregoing, chemisorption frequently causes the surface atoms of the solid itself to rearrange, and occasionally the adsorbate rearranges, creating a two-dimensional ordered phase.

A. Heats of Adsorption

The enthalpy of adsorption of small molecules on the planes of single crystals can vary by as much as 20 kcal from the strongest to the weakest binding sites. Also, the heats of adsorption vary with θ becoming lower as θ increases and adsorbate interactions increase. The measured values of the enthalpies most frequently reported are the so-called *isosteric heats* of adsorption, ΔH_{st}, derived from an application of the Clausius–Clapeyron principle, viz.

$$\left(\frac{\partial \ln P}{\partial T}\right)_\theta = \frac{\Delta H_{st}}{RT^2} \tag{11.92}$$

and evaluated from a series of isotherms at a constant value of the coverage θ. An example of such a series is shown in Fig. 11.9.

ΔH_{st} along with other energies involved in the process of adsorption from the gas phase are defined in Fig. 11.10, which represents a schematic potential energy diagram.

The heat of adsorption can change abruptly when the structure of the adsorbed layer itself changes. Such changes are reflected in the amounts of gas adsorbed as a function of pressure as in the vicinity of $P = 0.3$ in Fig. 11.9. A clearly demonstrated change in ΔH_{st} for the adsorption of carbon monoxide on Pd (111) planes at $\theta \cong 0.5$ is seen in Fig. 11.11.

It has been proposed that at low coverage the adsorbed CO occupies surface sites as shown in Fig. 11.12a, but with increasing coverage the adsorbate compacts itself to conform to Fig. 11.12b. The onset of this transition is revealed by the more or less abrupt decrease in ΔH_{st}.

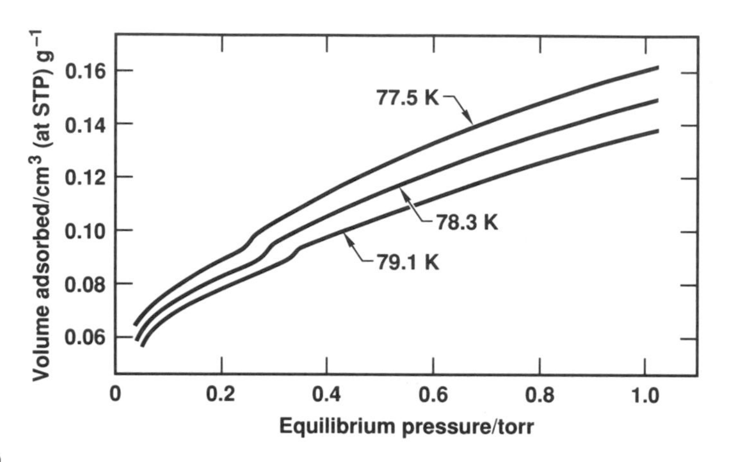

Figure 11.9.

Isotherms for Kr adsorbed on AgI. (From E. W. Sidebottom, W. A. House, and M. J. Jaycock, *J. Chem. Soc. Faraday Trans. I*, **72**, 2709 (1976).

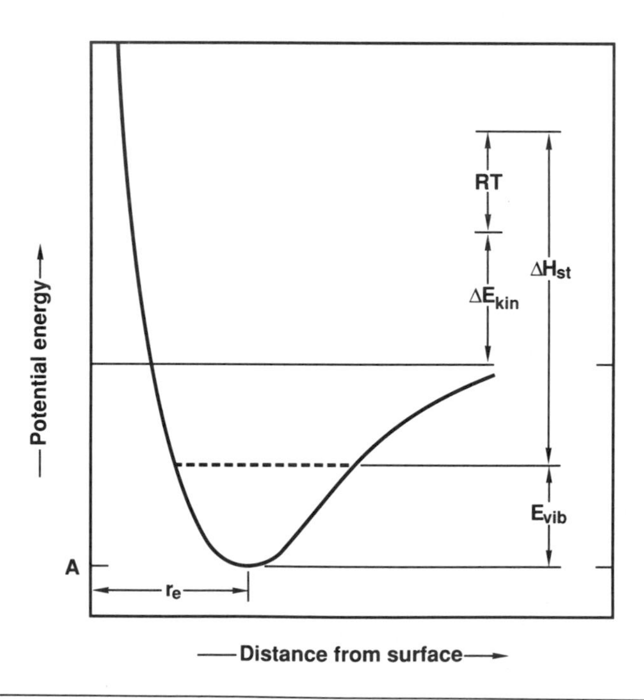

Figure 11.10.

The value of ΔH_{st} varies with temperature, which affects the energy of E_{vib}, the vibrational energy of the adsorbed species as well as that of the gas phase. It also includes ΔE_{kin}, the kinetic energy necessary to overcome the van der Waals attraction to the surface.

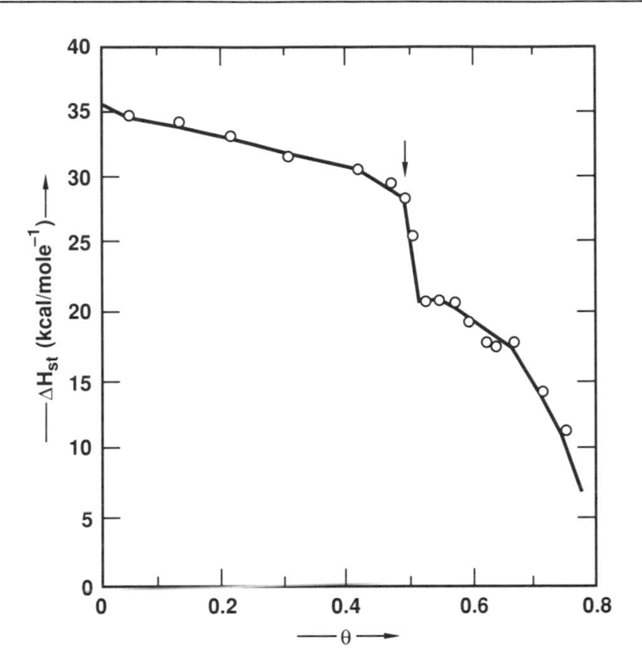

Figure 11.11.

The isosteric heat of adsorption of CO (From Alefeld and Volkl, loc. cit. as presented in Somorjai, loc. cit.)

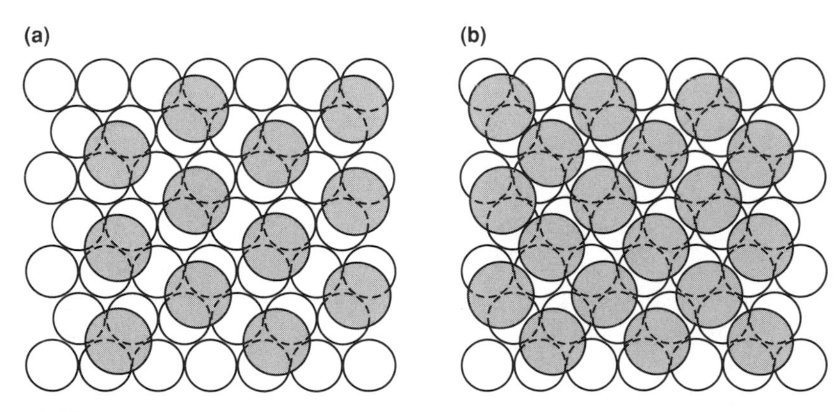

Figure 11.12.

Structural models for a CO adsorption on Pd(111); (a) $\theta = 1/3$; (b) compressed structure at saturation $\theta = 1/2$. [After H. Conrad, G. Ertl, J. Koch, and E. E. Latta, *Surf. Sci.* **43**, 463 (1974).]

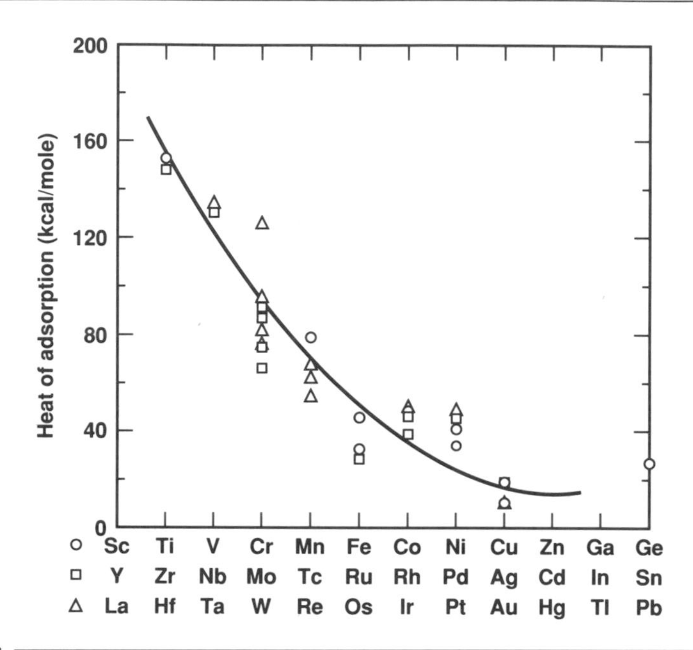

Figure 11.13.

Heats of adsorption of CO on polycrystalline transition metal surfaces. (Adapted from Gabor A. Somorjai: *Chemistry in Two Dimensions: Surfaces*, Copyright 1981 by Cornell University. Used by permission of the publisher, Cornell University Press).

There is seldom, if ever, a single definitive heat of adsorption because surface preparation and cleanliness are not reproducible between different laboratories and even between batches of the same substrate material. Nevertheless, there are general trends in ΔH_{st} reflecting variations in the electronic structure of the substrate as shown for the adsorption of CO in Fig. 11.13. The decreasing values of ΔH_{ad} seen in Fig. 11.13 are also observed for the adsorption of H_2, O_2, N_2, CO_2, C_2H_2, and NH_3, where the pattern for the heats of adsorption is Ti, Ta > Nb > W, Cr > Mo > Fe > Mn > Ni, Co > Rh > Pt, Pd > Cu, Au. Although this sequence clearly follows the decreasing number of empty d orbitals, no simple theory based upon this connection has yet been established.

B. Bonding to Metal Surfaces

One can consider the surface and adsorbate to be a single entity, and the bond between them can range from completely covalent to being entirely ionic. Following

Higuchi[9] and applying the formalism of molecular orbital theory, we write the total wave function as

$$\Psi = c_i\psi_i + c_c\psi_c, \tag{11.93}$$

where the subscripts i and c stand for ionic and covalent respectively. Following the variational procedure given in Chapter 6, Section 3A, we arrive at

$$1/c_i^2 = 1 + (E - H_{ii})/(E - H_{cc}). \tag{11.94}$$

Consistent with our previous notation $H_{ii} \equiv \int \psi_i H \psi_i \, d\tau$ and $H_{cc} \equiv \int \psi_c H \psi_c \, d\tau$. The weighting factors describe the fractional ionic character of the bond as $c_i^2 + c_c^2 = 1$. The problem is now to find ways of determining c_i, H_{ii}, and H_{cc}.

Initially let us assume the bond to be entirely ionic; then $H_{cc} = 0$. An empirical formula is substituted for H_{ii}, viz.

$$H_{ii} = A - I + (8/9)(e^2/r_e), \tag{11.95}$$

where A and I are the electron affinity of the metal and ionization potential of the adsorbate, and r_e is the equilibrium distance between the adsorbate atom and the substrate surface. The electron affinity and ionization terms are reversed when the direction of the electron transfer is reversed. The last term is derived from the potential function

$$V = -e^2/r + B/r^9, \tag{11.96}$$

which represents the interaction energy between a single negative and a single positive ion. The first term on the right gives the coulombic attraction, and the last term gives the repulsion at small values of r, and B is an empirical constant. Minimizing the potential energy V with respect to r allows B to be defined in terms of e and r_e and thus defines the last term in (11.95).

In order to calculate H_{cc} for a purely covalent bond, the approximation $\Delta H_{cc} = (1/2)(\Delta H_s^A + \Delta H_s^M)$ is used; i.e., one-half the enthalpies of sublimation of the adsorbate and the metal[10] (needless to say, better estimations are readily available). A partial degree of ionicity can be introduced into the otherwise purely covalent wave function using measurements of the dipole moment μ, taken from known values of the contact potentials and the expression

$$\mu = c_i^2(er_e). \tag{11.97}$$

---

[9] I. Higuchi, T. Ree, and H. Eyring, *J.A.C.S.*, **79**, 1330 (1957).

[10] I. Higuchi *et al.*, loc. cit.

With these approximate ad hoc formulas, a value of unity was calculated for c_i, in other words, 100% ionic bonding between Na or Cs and a W metal surface. However, c_i^2 was found to be 0.97 and 0.67 for Ba and Sr on W.

Essentially the same values for the heats of adsorption can be obtained without reference to quantum mechanics by a completely empirical method. Consider a neutral atom to be ionized at infinite separation from the surface, the energy involved being eI. The electron is now transferred to the Fermi level of the metal, with an associated release of an amount of energy $e\phi$, where ϕ is the work function. The adsorbate ion A^+ is now brought to the surface where it is attracted by the classical image force, the energy of this interaction being given by $e^2/4r_e$ (the image force is the collective result of the polarization induced by A^+ in the electron cloud of the metal). Combining these three terms, we arrive at

$$\Delta H_a = (eI - e\phi - e^2/4r_e). \tag{11.98}$$

However, neither (11.95) nor (11.98) yield very satisfactory results, as is evident from Table 11.1.

Modern theory based upon molecular orbital theory has gone far toward explanations of the qualitative differences between adsorption on metals, semiconductors, and insulators. Probable surface coverages and structures as well as catalytic activity are frequently understood on a case-by-case basis. Unfortunately, this material cannot

Table 11.1. Calculated and experimental initial heats of chemisorption.[a]

System	$N\phi$	N_i	$Ne^2/4r_e$	ΔH_a (calc.)	ΔH_a (exp.)	Reference[c]
Na on W	104	118	44.5 (1.83 Å)[b]	30.5	32.0	11
K on W	104	99.6	35.9 (2.27 Å)	40.3	—	
Cs on W	104	89.4	31.1 (2.62 Å)	45.7	64.0	12

[a] Values in kcal/g-atom.
[b] Values in parentheses, r_e.
[c11] M. Boudart, *J. Amer. Chem. Soc.*, **74**, 3556 (1952).
[c12] J. C. Mignolet, *Bull. Soc. Chim. Belg.*, **64**, 126 (1955).
(As presented by A. Clark, *The Chemisorptive Bond*, Academic Press, 1974).

[13] See, for example, R. Hoffman, *Solids and Surfaces* VCH, Angew. Chemie, **26**, 846 (1988).

be condensed and systematized to a length suitable for a general textbook and remains the province of the specialized reference work.[13]

10. Effects of Surface Structures

Although the general thermodynamic expressions for liquid interfaces apply equally to solids, the crystallography and morphology of the solid surface are by no means inconsequential. The actual values of the interfacial free energy are strongly anisotropic with respect to crystallographic orientation, and the degree of chemisorption and catalytic efficacy varies in the extreme in response to these parameters.

A. The Terrace–Ledge–Kink Model

The so-called Terrace–Ledge–Kink (TLK) model is not so much a model as it is a purely phenomenological description. Consider a low-index plane such as outlined by the dashed lines in Fig. 11.14a,b. According to this simple model the surface free energy as a function of θ is given by

$$\gamma_\theta = \gamma_0 \cos|\theta| + \alpha \sin|\theta|, \tag{11.99}$$

where γ_0 is the surface energy per unit area for $\theta = 0$ and, hence, corresponds to the terraces, and α is the increase per unit length of step added. The energy is independent of the sign of θ, hence the absolute values. The major conclusion from this analysis is that there is a cusp at $\theta = 0$, and the magnitude of this singularity is proportional to the free-energy increase per additional step. This is portrayed by Fig. 11.15

Planes with high Miller indices are frequently defined in terms of the indices of the component terraces and ledges. For example, a (755) plane can be decomposed into a (111) terrace five atoms wide, denoted by 5(111) and bounded by two (100) steps one atom high denoted by 2(100). The convention is to write this as (755) = 5(111) × 2(100)).

Surface irregularities such as steps and kinks have a profound effect upon chemisorption and catalytic efficacy. For example, ethylene is merely chemisorbed on Ni (111) planes at 200°C, but dissociates on a stepped surface of 6(111) × 100. Most hydrocarbons will respond in a similar way to change of orientation of Pt surfaces.

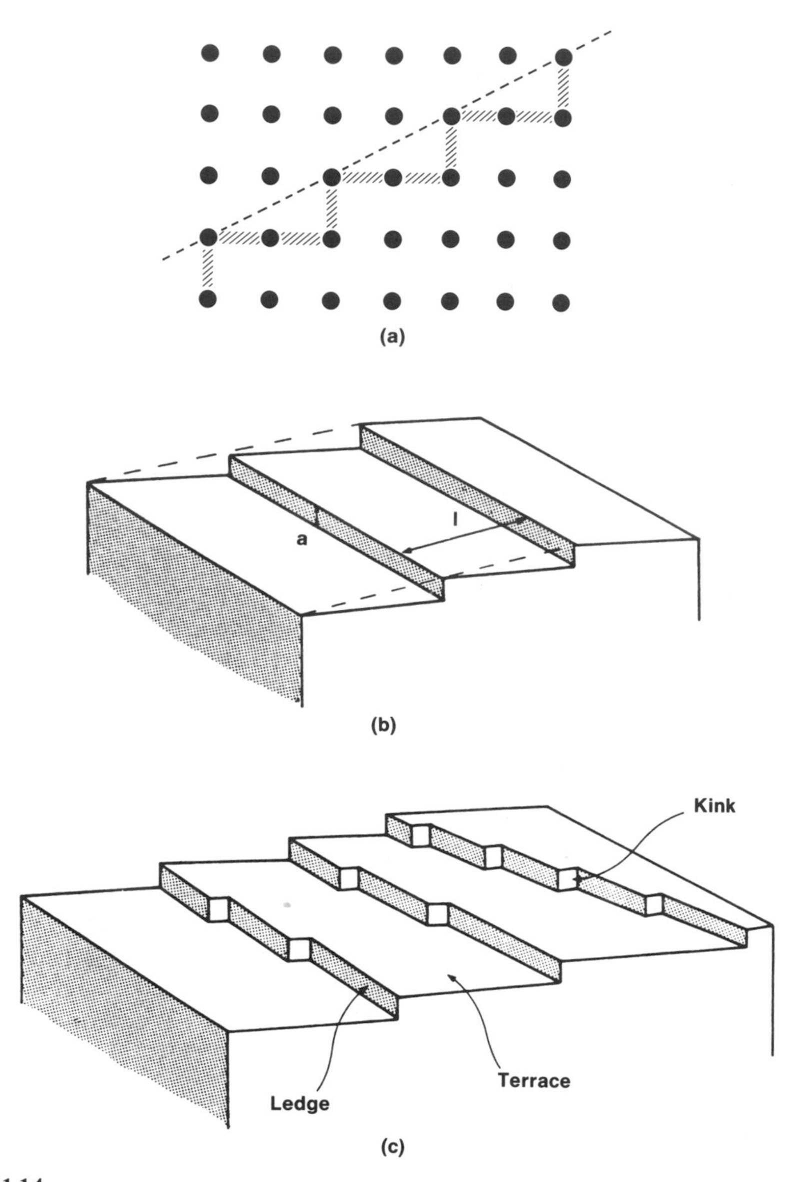

Figure 11.14.

An idealized representation of a low-index plane of an elementary crystal structure (a) in two dimensions showing ledge and terrace formation; (b) the same in three dimensions: (c) the same as (b) with the addition of kinks. (Taken in part from D. P. Woodruff, *The Solid–Liquid Interface*, Cambridge University Press, p. 4 1973).

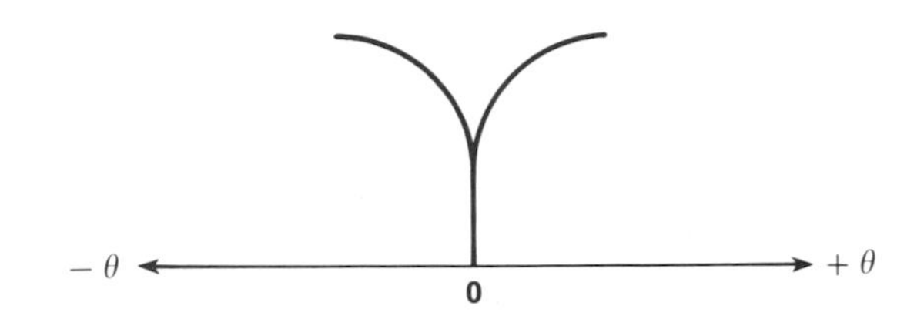

Figure 11.15.

The dependence of γ upon θ as given by (11.99). θ is the misorientation angle.

11. Surface Reactions—Carbon

The general subject of chemical reactions that take place on solid surfaces is too extensive to be examined here. In particular, the field of catalysis, driven by economic forces, contains hundreds of examples of surface reactions encompassing all the usual categories, such as oxidation, reduction, pyrolysis, hydration, hydrogenation, and so forth. Consequently, our discussion will be confined to the single category of epitaxial controlled surface reactions, with carbon serving as the single example.

A. Pyrolitic Graphite

Pyrolytic graphite, generally called pyrographite, is highly ordered within the basal plane (see Fig. 1.22), but the usual ...ababab... stacking sequence normal to the basal planes is erratic. It can be made by pyrolyzing almost any carbonaceous compound on a heated mandrel of polycrystalline graphite at temperatures in excess of $\sim 1350°C$. The continuation of this cracking process leads to the deposition of contiguous layers of graphite on the growing surface. The density of pyrographite approaches 2.25 g/cm_3, which is the theoretical value. By contrast, graphites produced by the usual coking methods are 20–30% less dense. The electrical conductivity along the basal plane exceeds that of very pure copper, whereas it is an excellent insulator parallel to the c-axis. Further, the tensile strength of pyrographite exceeds that of stainless steels at very high temperatures. This unusual combination of physical properties has made pyrographite a substance of major commercial importance. The morphology of this form of carbon appears as shown in Fig. 11.16.

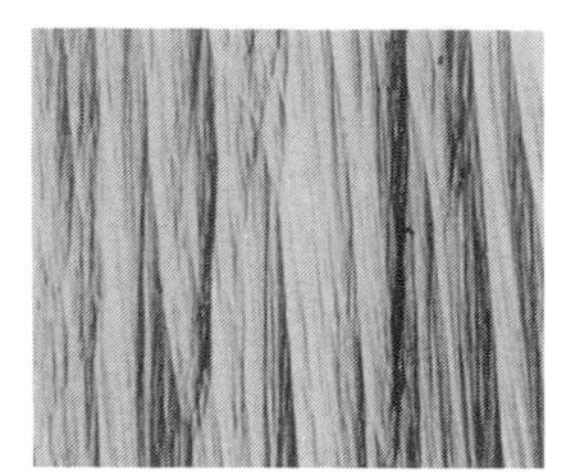

Figure 11.16.

(a) Low-magnification photomicrograph of pyrolitic graphite showing the well-developed cones (grains) with the (*c*) axis parallel to the plane of the page; (b) the cone structure looking perpendicular to the plane of the page.

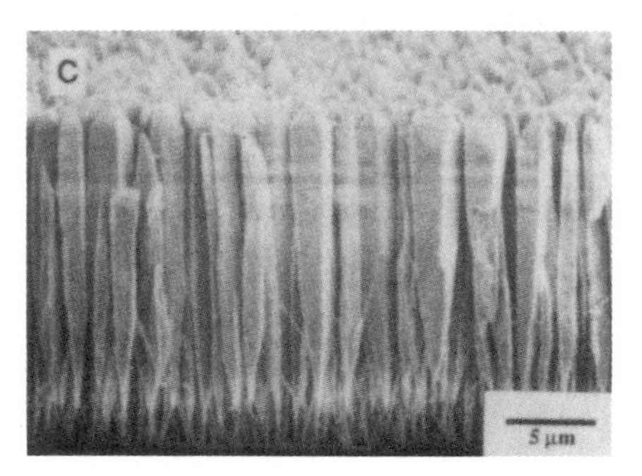

Figure 11.17.

Photomicrographs of CVD diamond thin-film structure illustrating growth morphology in polycrystalline coatings. (A) Top surface of as-grown CVD film showing dominant (100) morphology at high relative methane concentrations. (B) Surface of dominant (100) morphology after etching in oxygen-containing plasma. (C) Cross section of the film seen in (B) showing both the columnar morphology and the increasing size and perfection of the grains with continued growth. (Used through courtesy of Y. Sato as presented by W. A. Yarbrough and R. Messier, *Science*, **247**, 692, Copyright 1990 by the AAAS).

B. Synthesizing Diamond by Vapor Deposition

An unusual surface reaction is the production and growth of diamond films at low pressures, where diamond is the metastable form of carbon. It is well known that the reconstructive phase transformation diamond → graphite needs a high activation energy. Thus, if a diamond film can be prepared, it is expected to be quite stable. The techniques developed over the last 15 years or so rely on various ways of cracking hydrocarbons in the presence of a large excess of hydrogen and on the use of chemical vapor deposition (CVD) to form the diamond film. All such processes require additional energy to be supplied to the gas phase and surface species. A typical gas mixture may consist of 99% H_2 and 1% hydrocarbon, say CH_4. The crucial step in developing these techniques was the realization of the specific role of atomic hydrogen. Atomic hydrogen apparently stabilizes the sp^3 bonding configuration of diamond and selectively removes graphite from the surface, leaving behind the diamond deposit. The detailed mechanism is, however, not yet known.

The appropriate CVD environment for low-pressure diamond production may be achieved experimentally in a variety of ways. Three important ones are: (A) hot filament-assisted chemical vapor deposition (HFCVD), in which a hot filament is used to generate atomic hydrogen near the substrate; (B) a plasma-assisted deposition using a microwave cavity; and (C) a DC discharge method. In all these processes, atomic hydrogen is necessary.

The identification of vapor-grown diamond has been verified by diffraction and Raman spectroscopy, and the crystals show thermal conductivity and hardness characteristic of diamond.

Films of diamond have been prepared on many different substrates (e.g., Ni, Cu, BN, and fcc-Fe). Diamond films grown on silicon substrates by plasma-assisted CVD from hydrogen (1%)-methane produced highly faceted polycrystalline films. Photomicrographs reveal the morphology of the films, and examples are reproduced in Fig. 11.17.

The morphology varies because of the thermal gradient across the substrate.

Anticipated applications for diamond films include a wide variety of electronic and optical devices as well as for hard coatings.

Chapter 11 Exercises

1. Derive an expression for the dependence of the vapor pressure upon the size for a small single cubic crystal.

2. The surface tension of Hg as a function of temperature is given by the empirical equation

$$\gamma(\text{dynes-cm}^{-1}) = 463.3 + 8.32 \times 10^{-2}T - 3.125 \times 10^{-4}T^2,$$

where T is expressed in K. What is the total surface energy (enthalpy) at 50°C and 100°C?

3. Let N molecules be distributed among M sites. Assume the total number of configurations that a single molecule can assume with respect to its neighbors is z and the interaction energy between adsorbed molecules is zero. From statistical considerations show that the chemical potential of the adsorbed species is
   a. $\mu = kT \ln[\theta/z(1 - \theta)]$.
   b. derive an expression for μ when the interaction energy ε between nearest-neighbor adsorbate molecules is finite.

4. Explain why any solid is completely wetted by its own liquid at the melting point.

5. Consider a binary system of two phases, α and β, separated by an equilibrium interface. Derive an expression for the change in the interfacial free energy in response to applying hydrostatic pressure to the entire system.

6. Pure charcoal adsorbs 0.894 cm^3 of N_2 (at 25°C and $P = 1$ atm), at 4.60 atm and 195 K and also at 35.4 atm and 273 K. Calculate the isosteric enthalpy of adsorption as well as the entropy and free energy.

7. When diffusion is sufficiently rapid, the grain boundaries of a polycrystalline substance become grooved. This is because the crystallites attempt to spherodize in order to minimize their surface area. Based upon the geometry of such grooves, derive an expression for γ_b/γ_0, the ratio of the surface free energy of the interfacial boundary between adjacent grains and that of the free surface. (Hint: Begin by considering Fig. 11.15.)

8. It has been proposed that the bond energy E_{MA} of a molecule A_2 adsorbed on a metal surface M can be represented by the expression

$$E_{MA} = (1/2)(E_{MM} + E_{A_2}) + (x_M - x_A)^2,$$

where E_{MM} and E_{A_2} are the metal–metal and molecular bond energies and X_M and X_A are the electronegativities of M and A. Discuss the chemical basis for this empirical equation and its possible shortcomings. Can you think of ways in which it could be improved?

Additional Reading

G. A. Somorjai, *Chemistry in Two Dimensions: Surfaces*, Cornell University Press (1981). A largely descriptive text describing surface reactions and the instruments used for their investigation.

M. J. Jaycock and G. D. Parfitt, *Chemistry of Interfaces*, Wiley, (1981). A concisely written introduction to the subject containing the relevant mathematics and illustrations of pertinent apparatus.

R. Aveyard and D. A. Haydon, *An Introduction to the Principles of Surface Chemistry*, Cambridge University Press (1973). Contains the customary thermodynamical treatment of surfaces and emphasizes liquid–liquid and liquid–solid interfaces.

D. P. Woodruff, *The Solid-Liquid Interface*, Cambridge University Press (1973). Deals mainly with the interfacial aspects of nucleation and growth of solids from liquid (melt phases).

N. H. March, *Chemical Bonds outside Metal Surfaces*, Plenum Press (1986). Emphasizes theory, but is somewhat too terse to learn from. Major contribution is its broad scope combined with an extensive bibliography of primary references.

W. A. Yarborough and R. Messier, A review entitled "Current Issues and Problems in the Chemical Vapor Deposition of Diamond," *Science*, **247**, 688 (1990).

G. Alefeld and J. Volkl, Eds. *Hydrogen in Metals*, Springer (1978). A comprehensive treatment of this subject.

Chapter 12

Crystal Defects

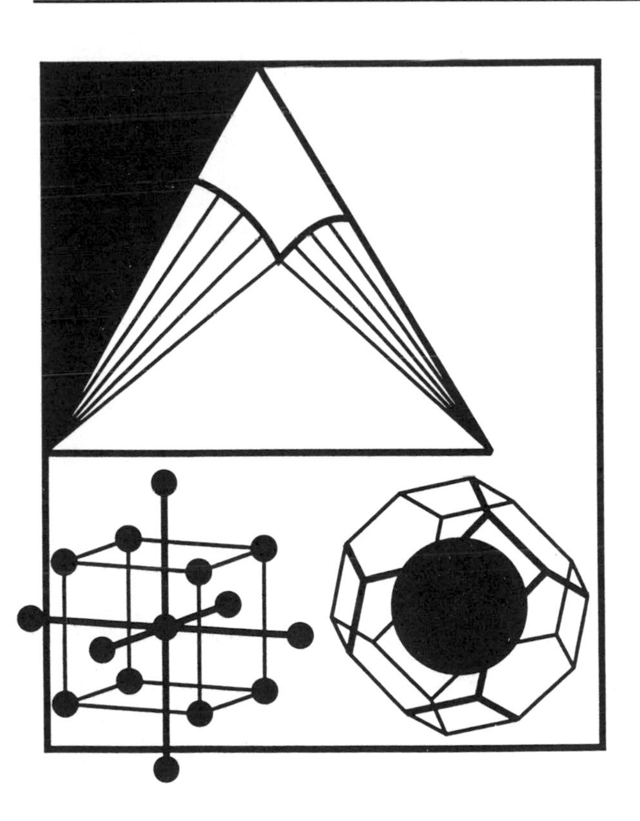

Chapter 12

Contents

Textbooks and Heaven only are Ideal;
Solidity is an imperfect state
Within the cracked and dislocated Real
Nonstoichiometric crystals dominate.
Stray Atoms sully and precipitate;
Strange holes, excitons wander loose; because
Of Dangling Bonds, a chemical Substrate
Corrodes and catalyzes—surface Flaws
Help Epitaxial Growth to fix adsorptive claws.

—John Updike *Midpoint*

Crystal Defects

We all know that perfect crystals do not exist, but this statement can be interpreted in many different ways. Do crystals contain defects, which may be defined as departures from perfect periodicity, because of accidents of growth and subsequent handling? The answer is yes. Characteristic defects, for example dislocations and grain boundaries, are formed this way but, at least in theory, these defects can be avoided by appropriate preparation and handling. Are there inherent crystalline defects in the sense that there is no way of eliminating them? Again, the answer is yes. Also there are dynamic disruptions of the crystal lattice (i.e., thermal vibrations) that do not completely disappear even at absolute zero because of the quantum mechanical zero-point energies. However, phonons do not really destroy the regular arrangement of the atomic positions but only smear them out.

Lattice defects play a role in the chemical behavior of solids that has no analog in liquid or gaseous phases. Specifically, they most frequently determine the mechanism and rate of solid state reactions. As a result, the study of crystal defects is the logical first step toward an understanding of solid state kinetics.

1. Definitions and Classification

Our concern in this chapter will be primarily with defects of atomic size, the two most important ones being vacancies (i.e., vacant lattice sites) and interstitials, atoms in nonlattice site positions. As explained in the next section, such defects are an equilibrium component of crystals at all finite temperatures and, therefore, are an

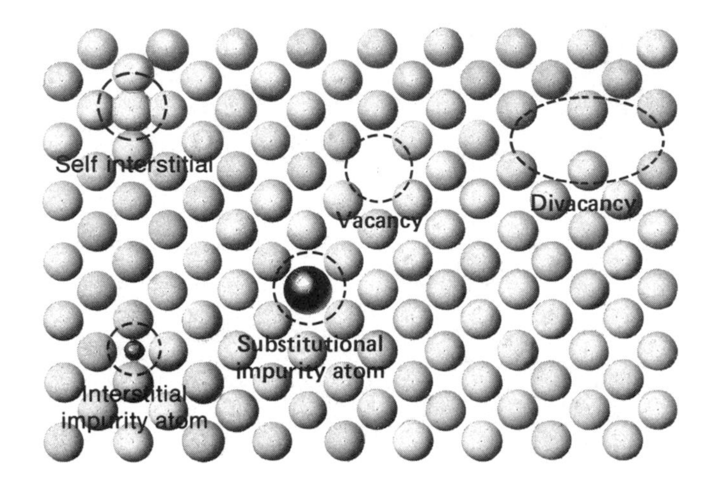

Figure 12.1.

Simple point defects in monoelemental substances, shown schematically. (After R. J. Borg and G. J. Dienes, *Solid State Diffusion*, Academic Press 1988.)

Table 12.1. Defects and physical properties and processes in crystals.

Defect	Dimensionality of Defect	Physical Property
Point defects	0	Color centers
vacancies		Diffusion
interstitials		Mechanical properties, Diffusion
Impurity atoms		Electrical properties
Line defects	1	Mechanical properties
dislocations		Crystal growth
Planar defects	2	Texture
grain boundaries		Fabrication, corrosion
stacking faults		Mechanical properties
Volume defects	3	Porosity
voids		Precipitation
second phase		Mechanical and magnetic properties
order–disorder		

innate component of the system. It is often convenient to include impurities in low concentrations as point defects, since they occupy either substitutional or interstitial positions and can interact with vacancies and interstitials. These simple point defects are illustrated in Fig. 12.1. In all these cases the periodic structure is not seriously altered, although defects do cause localized distortions. A convenient and useful classification of imperfections in crystals forming a hierarchy with respect to their dimensionalities is shown in Table 12.1. Along with the types of defects and examples are listed their important physical properties or processes that depend on their presence. Extended defects (dislocations) generally do not exert a profound effect upon chemical properties and thus lie outside the scope of this volume. They are treated in most solid state physics and materials science monographs, since they are of utmost importance with regard to mechanical properties

2. Thermodynamics of Point Defects in Monoatomic Solids

A. Vacant Lattice Sites

Let us begin by describing the generation of the simplest point defects, namely the creation of vacancies in a monoatomic crystalline solid. The process may be pictured as illustrated in Figs. 12.2 and 12.3, where we have taken an atom from the interior of the solid to the surface, leaving a vacancy behind. It is understood, of course, that these steps are repeated many times, so that new layers of the normal crystal are built up and the ensemble reaches the size to which we can apply standard thermodynamics of equilibrium vacancy generation. Note that in this reaction atoms are conserved but not lattice sites, a distinction useful in the clarity of presentation but unimportant in practice because of the very low concentrations encountered.

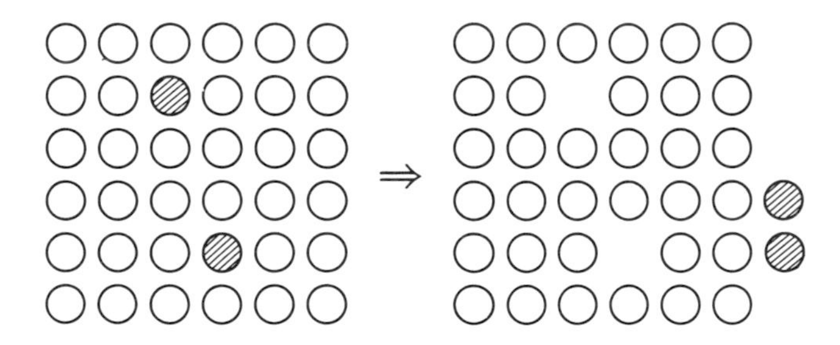

Figure 12.2.

Thermodynamic cycle for vacancy formation. Atoms are removed from the interior and replaced on the surface.

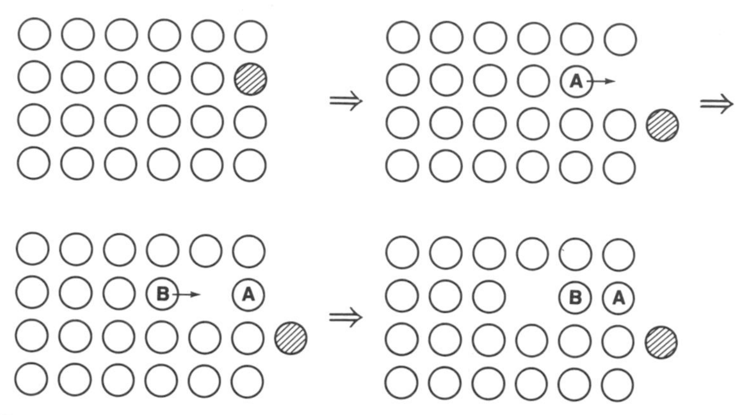

Figure 12.3.

Actual mechanism of vacancy formation. Atoms pop out onto the surface, usually an internal surface or dislocation, and the vacancy diffuses into the interior of the crystal.

It clearly takes energy to move an atom from the interior to the surface. This increase in the internal energy is balanced by the increase in configurational or mixing entropy. When we minimize the free energy with respect to the vacancy concentration, the equilibrium concentration as a function of temperature is obtained. Thus, it is the entropy of mixing that forces the creation of a certain number of vacant lattice positions at all temperatures above 0 K. Hence, vacancies are the natural result of thermodynamic equilibrium and not the result of any accident of growth or sample preparation. We can now proceed with the calculation of the equilibrium concentration of vacancies.

Let N_L be the total number of lattice sites and N_v the number of vacancies. The number of ways in which N_v vacancies can be arranged on N_L sites by elementary combinatory statistics is given by

$$\Omega = \frac{N_L!}{N_v!(N_L - N_v)!}. \tag{12.1}$$

The configuration or mixing entropy is defined by

$$\Delta S_{\mathrm{m}} = k \ln \Omega. \tag{12.2}$$

Application of Stirling's approximation for large numbers gives

$$\Delta S_{\mathrm{m}} = k[N_L \ln N_L - N_v \ln N_v - (N_L - N_v)\ln(N_L - N_v)], \tag{12.3}$$

and since $N_v \ll N_L$, we can write for the entropy of mixing,

$$\Delta S_m \cong [N_v \ln N_L - N_v \ln N_v]$$
$$= -kN_v \ln(N_v/N_L) = -RX_v \ln X_v. \tag{12.4}$$

As would be expected, this expression for noninteracting defects is the same as the entropy of mixing of an ideal solution (see Chapter 9), namely

$$\Delta S_m = -R[X_a \ln X_a + X_v \ln X_v] \cong -RX_v \ln X_v,$$

since the first term is negligible at low concentrations as $X_a \cong 1$. The complete expression for the Helmholtz free energy of the system can be written as

$$\Delta F_v = N_v[\Delta E_v - T\,\Delta S_v^{\mathrm{v}}] + RT\left[\frac{N_L - N_v}{N_L}\ln\frac{N_L - N_v}{N_L} + \frac{N_v}{N_L}\ln\frac{N_v}{N_L}\right], \tag{12.5}$$

where ΔS_v^{v} is the change in vibrational entropy arising from the change in the vibrational frequency spectrum around the vacant lattice site. Minimization of the free energy with respect to the vacancy concentration gives

$$\frac{\partial F_v}{\partial N_v} = 0 = \Delta E_v - T\,\Delta S_v^{\mathrm{v}} + RT\ln\frac{N_v}{N_L - N_v}. \tag{12.6}$$

Since this has been an equilibrium derivation all along, one can identify ΔE_v and ΔS_v as the standard internal energy and entropy of formation respectively, ΔE_v° and ΔS_v°. Thus, we finally obtain

$$\ln\frac{N_v}{N_L - N_v} = \ln X_v = -\frac{\Delta E_v^\circ - T\,\Delta S_v^\circ}{RT}, \tag{12.7}$$

or, since $N_v \ll N_L$, we can write for the atomic fraction of vacancies, X_v, the expression

$$X_v = \exp(-\Delta F_v^\circ/RT) \simeq \exp(-\Delta G_v^\circ/RT). \tag{12.8}$$

Equating ΔF_v° to ΔG_v° is an excellent approximation at low pressures (e.g., 1 atm). Equation (12.8) is often approximated by

$$X_v = e^{-\Delta E_v/RT}. \tag{12.9}$$

This is a very reasonable approximation since ΔS_v^{v} is small and since $\Delta H_v^\circ \cong \Delta E_v^\circ$. It is clear from this formula that X_v is zero at $T = 0$ and increases exponentially with increasing temperature. One must be careful, however, not to begin further thermodynamic derivations with (12.9) because the substitution of ΔE_v for ΔF_v° will lead to false results.

B. *Interstitial Atoms*

The process of creating an interstitial is just the reverse of generating a vacancy, namely the positioning of a surface atom into an interior interstitial position. It should be noted that the interstitial configuration is not always that of a simple inserted atom. In many metals, for example, the strain energy is minimized by the configuration of the split interstitial in which two atoms share in the displacement as illustrated in Fig. 12.4.

The calculation of the interstitial concentration is completely analogous to that for the vacancies. If there are N_I possible interstitial positions, and if ΔF_i° is the free energy required to move an atom from the surface to the interior, then the interstitial concentration at equilibrium is given by

$$N_i/N_I = X_i = \exp(-\Delta F_i^\circ/kT) \tag{12.10}$$

where N_i represents the number of interstitials. ΔF_i° is usually much larger than ΔF_v° since there is considerable crowding around an interstitial creating a substantial strain energy. Consequently, the equilibrium concentration of the self-interstitials is in general very small. The concentration of interstitial impurities, however, is often

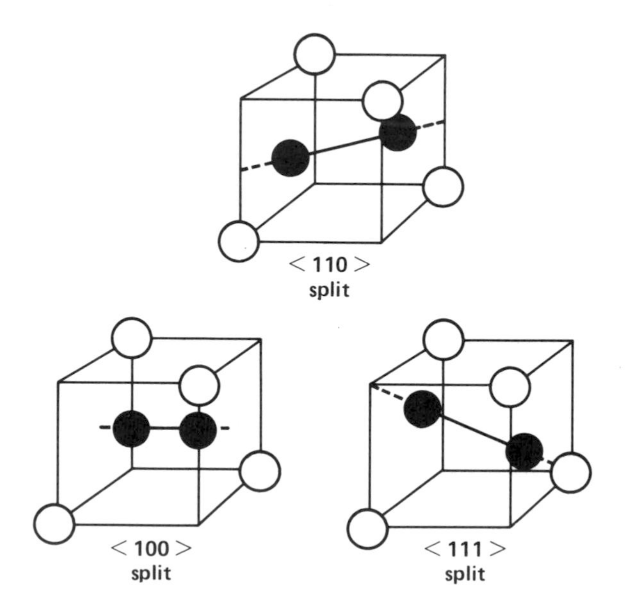

Figure 12.4.

Interstitial atoms in bcc structures; two atoms share the interstice with the "dumbell" axis along specific crystallographic directions.

significant and important, such as carbon in iron, which is steel, being a classical example. In this case, of course, the analogous laws of solubility come into play (see Chapter 9). The interstitial solutes are almost always much smaller than those of the host lattice.

C. Frenkel Pairs

The creation of an internal vacancy–interstitial pair is simply the promotion of an atom to an interstitial position leaving a vacancy behind. Widely separated vacancy–interstitial pairs are called Frenkel pairs. The phrase "widely separated" is important because close vacancy–interstitial pairs are likely to be unstable to annihilation by recombination. If the separation is less than a few atomic spacings, the vacancy and the interstitial automatically recombine under the influence of their interacting strain fields, which are of opposite sign. This separation beyond the range of mutual interaction allows us to apply independent statistics that assumes independent populations except for the restriction that the numbers of vacancies and interstitials are equal. This athermal recombination takes place without resorting to the normal thermally activated diffusion process within the *recombination volume.* This reaction is insignificant under equilibrium conditions but is an important annealing mechanism for radiation damage.

The derivation of the fundamental equation is only a little more complicated than for the case of isolated single defects. The total number of distinguishable ways of arranging N_v vacancies on N_L lattice sites and N_i interstitials on N_I interstitial sites is given by

$$\Omega = \frac{N_L!}{N_v!(N_L - N_v)!} \times \frac{N_I!}{N_i!(N_I - N_i)!}. \tag{12.11}$$

Remembering that the numbers of vacancies and interstitials are equal, we can write $N = N_v = N_i$. Substituting in (12.11) and applying Stirling's approximation, as before, gives

$$\Delta S_{\mathrm{m}} = RT[N_L \ln N_L - 2N \ln N + N_I \ln N_I - (N_L - N)\ln(N_L - N) - (N_I - N)\ln(N_I - N)] \tag{12.12}$$

for the configurational entropy. Minimizing the corresponding free energy with respect to the number of Frenkel pairs gives, after rearrangement,

$$N = [(N_L - N_v)(N_I - N_i)]^{1/2} \exp\left(\frac{\Delta S_{iv}^{\circ}}{2R}\right)\exp\left(\frac{-\Delta E_{iv}^{\circ}}{2RT}\right), \tag{12.13}$$

and as $N \ll N_L, N_I$ we can write, for the number of Frenkel pairs,

$$N_F = (N_L N_I)^{1/2} \exp(-\Delta F_{iv}^{\circ}/2RT). \tag{12.14}$$

The Frenkel pair concentration in most metals is very low at thermal equilibrium. However, their concentration becomes appreciable as a result of irradiation by energetic particles at low temperatures. As will be seen later, they can also affect the properties of ionic crystals.

One can also obtain (12.14) by applying the law of mass action. Writing the process as a thermodynamic cycle gives us

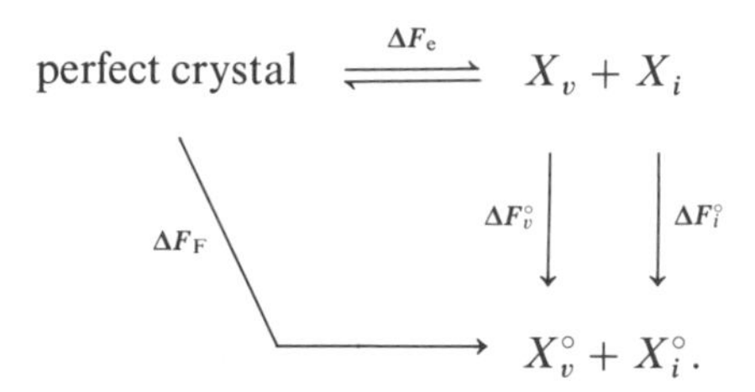

In this schematic $\Delta F_e = 0$ because the crystal is in equilibrium with its Frenkel defects, ΔF_v° and ΔF_i° are the standard free energies of formation of the vacancies and interstitials, respectively, and X_v° and X_i° are the unspecified standard states. ΔF_F is the free energy of the reaction that forms the Frenkel pairs. We now proceed to add all the terms in this closed cycle to obtain

$$\Delta F_F = \Delta F_v^{\circ} + \Delta F_i^{\circ} + \Delta F_e = \Delta F_v^{\circ} + \Delta F_i^{\circ},$$
$$= RT \ln(X_v/X_v^{\circ}) + RT \ln(X_i/X_i^{\circ}). \tag{12.15}$$

However, $X_v^{\circ} = X_i^{\circ} = 1$ by definition of the standard state. Hence, defining

$$\Delta F_v^{\circ} + \Delta F_i^{\circ} \equiv \Delta F_{iv}^{\circ},$$

we have

$$\Delta F_F = \Delta F_{iv}^{\circ} = RT \ln X_i X_v,$$

or, because $X_i = X_v = X_F$, the concentration of Frenkel pairs, we finally obtain

$$X_F = \exp(-\Delta F_{iv}^{\circ}/2RT), \tag{12.16}$$

because lattice sites are conserved, X_i, X_v, and X_F have the customary definition of mole fraction.

3. Defects in Ionic Crystals

Defects in ionic crystals present a somewhat more complicated picture because we have to consider two kinds of atoms, and we also have to maintain electrical neutrality. Thus, in simple salts of the MX type, for example the alkali halides, positive and negative defects are formed in equal numbers, as illustrated in Figs. 12.5 and 12.6. In compounds other than of one-to-one stoichiometry, the ratios of the positive and negative ion vacancies follow the stoichiometry in order to maintain electrical neutrality. The Schottky, or vacancy-type, defects dominate in the alkali halides, whereas the Frenkel-type defects, produced by the simultaneous creation of vacancies and interstitials, play an important role in silver salts.

The calculation of the concentration of Frenkel defects is exactly the same as before, since, in this case, electrical neutrality is automatically satisfied, and the positive and negative interstitials can be created completely independently. Thus, (12.14) and (12.16) are valid without modification.

The calculation for the Schottky defects also follows an almost identical development. Using the symbols cv and av to indicate cation and anion vacancies, respectively, we can write the free energy of formation for the simultaneous creation of

```
+  −  +  −  +  −  +  −
−  +  −  +  −  +  −  +
+  −  □  −  +  −  +  −
−  +  −  +  −  +  −  +
+  −  +  −  +  □  +  −
−  +  −  +  −  +  −  +
+  −  +  −  +  −  +  −
```

Figure 12.5.

Schottky defects, pairs of cation and anion vacancies, in an ionic crystal.

```
+  −  +  −  +  −  +  −  +  −
−  +  −  +  −  +  −  +  −  +
          ⊕
+  −  □  −  +  −  +  −  +  −
−  +  −  +  −  +  −  +  −  +
                   ⊖
+  −  +  −  +  □  +  −  +  −
−  +  −  +  −  +  −  +  −  +
+  −  +  −  +  −  +  −  +  −
```

Figure 12.6.

Frenkel defects, interstitial cations and anions with the corresponding vacancies in an ionic crystal.

both types of defects as

$$\Delta F = N_{cv}\,\Delta E_{cv} + N_{av}\,\Delta E_{av} - T(N_{cv}\,\Delta S^{v}_{cv} + N_{av}\,\Delta S^{v}_{av}) - RT\ln\left[\frac{N!}{N_{cv}!(N-N_{cv})!}\,\frac{N!}{N_{av}!(N-N_{av})!}\right], \tag{12.17}$$

where there are N cation and N anion sites, and ΔS^{v}_{cv} and ΔS^{v}_{av} reflect the changes in vibrational spectrum as a result of the introduction of the vacancies.

As stated already, $N_{cv} = N_{av} = N_s$, where N_s is the number of Schottky pairs. Making this substitution, applying Stirling's approximation, and minimizing the free energy with respect to N_s gives

$$\frac{N_s}{N - N_s} = \exp\left(-\frac{\Delta F^{\circ}_{s}}{2RT}\right), \tag{12.18}$$

where ΔF°_{s} is the free energy needed to remove an anion and a cation from two distant points in the interior to the surface where new crystal layers are formed. For $N_s \ll N$, we can simplify in the usual way to

$$X_s = \exp(-\Delta F^{\circ}_{s}/2RT). \tag{12.19}$$

An important effect arises in ionic crystals from the substitution, purposeful or inadvertent, of an ion of different valence for the host species. All compounds can dissolve foreign ions to some extent, and if the substitutional foreign ion differs in its charge state from the ion that it replaces, vacant lattice sites result. As an example, let a NaCl crystal dissolve a small amount of Ca^{2+}, as shown schematically in Fig. 12.7. For each substitution of a Na^{+} by a Ca^{+}, a second Na^{+} must be removed to preserve charge neutrality. Thus, lattice sites are conserved, but the ideal stoichoi-

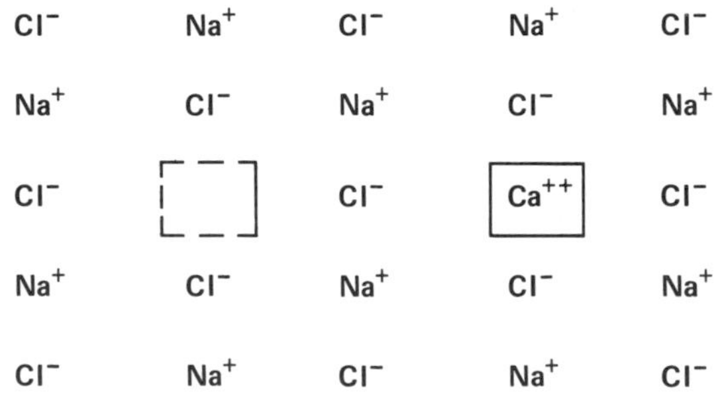

Cl^- Na^+ Cl^- Na^+ Cl^-

Na^+ Cl^- Na^+ Cl^- Na^+

Cl^- [ ] Cl^- Ca^{++} Cl^-

Na^+ Cl^- Na^+ Cl^- Na^+

Cl^- Na^+ Cl^- Na^+ Cl^-

Figure 12.7.

The substitution of Ca^{2+} for Na^{+} requires the formation of a Na^{+} vacancy in order to preserve electrical neutrality.

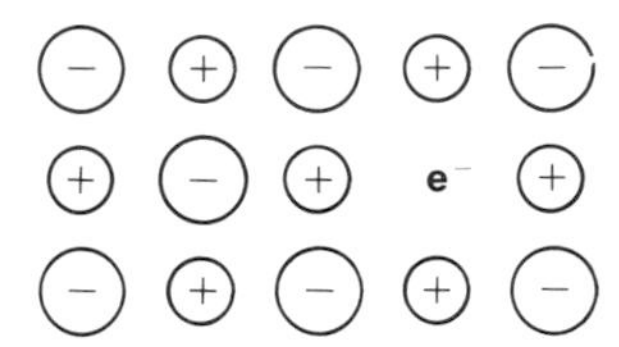

Figure 12.8

The *F*-center, an electron trapped at a negative ion vacancy. The electron is shared among the six nearest neighbors.

metric ratio, $(Na^+)/(Cl^-)$, is not. Defects thus introduced, in this case positive ion vacancies, are called *extrinsic* defects. They dominate at low temperatures, and the defect concentration is approximately independent of the temperature. At high temperatures the defects are primarily thermally generated, and this regime is called the intrinsic region. These thermally generated defects are the *intrinsic* ones, whose concentration depends exponentially on the temperature. Thus, by appropriate doping, one can control the defect concentration in the low-temperature regimes. Although solutes in the concentration range of parts per million, or even less, do not noticeably affect the thermodynamic properties, they can profoundly alter the defect concentrations. By so doing, miniscule quantities of impurities or dopants can completely change the chemical reactivity or electrical properties of solids.

The electric charge localized about defects in ionic crystals is also responsible for the formation of color centers. This process is illustrated schematically in Fig. 12.8 for the well-known *F*-center. A missing negative ion results in a positively charged defect and therefore will attract unattached or weakly bound electrons of which there can be many during, for example, x-ray irradiation. The trapped electron has quantized excited states which are responsible for the characteristic color. This is the simple and well verified model of the *F*-center in ionic crystals (*F* being derived from *Farbe*, the German word for color). Many other color centers are known that correspond to different defect-electronic configurations.

4. Defects in Semiconductors

Defect–electronic interactions, discussed briefly in the last section in connection with color centers, are particularly important in semiconductors. This is because, in addition to potential atomic defects, electrons, n, and p, positive holes in the valence band, are intrinsic defects in a perfect semiconductor. The process is analogous to the production of Frenkel pairs. Electrons are promoted to the conduction band by thermal agitation leaving holes in the valence band (see Chapter 7). Thus, the electrons are analogous to the interstitials, and the holes to the vacancies in the atomic case. However, in contrast to atomic defects, the electronic defects do not obey Boltzmann statistics, but have to be treated by the Fermi–Dirac distribution function (see Chapter 7).

In elemental semiconductors, such as silicon and germanium, atomic defects are formed in the manner discussed in detail in Section 12.2. However, there are enough electrons in the conduction band and holes in the valence band that we must take into account the interactions between atomic and electronic defects. The picture can get rather complicated as the atomic defects may acquire several different charge states. Let us discuss here a simple, but representative example, namely the negatively charged vacancy, v^-, in silicon. Here it is convenient to use the law of mass action, as presented earlier in this chapter (2C). The formation of such a charged defect involves the creation of electrons and holes, the creation of a neutral vacancy, and the trapping of an electron by the vacancy. Overall, then, the following equilibrium reactions are to be satisfied with the appropriate equilibrium constants and energies.

$$0 \rightleftarrows e^- + h^+ + E_g; \qquad (e^-)(h^+) = K_1,$$

where E_g is the band gap energy.

$$0 \rightleftarrows v^0 + \Delta H_v^0, \qquad v^0 = K_v, \tag{12.20}$$

where ΔH_v° is the enthalpy of neutral vacancy formation.

$$v^0 \rightleftarrows v^- + h^+ + E_A; \qquad (v^-)(h^+)/(v^0) = K_2',$$

where E_A is the energy required to extract a single charge from the valence band and to place it on a neutral vacancy, or using (12.20)

$$(v^-)(h^+) = K_2' K_v = K_2.$$

Electrical neutrality requires that the sum of the negative charges, i.e., those bound to vacancies and those that are free electrons in the conduction band, equal the positive holes in the valence band, viz.

$$e^- + v^- = h^+. \tag{12.21}$$

The general solutions, derived by straightforward algebra, are

$$\begin{aligned} (e^-) &= K_1/(K_1 + K_2)^{1/2}, \\ (h^+) &= (K_1 + K_2)^{1/2}, \\ (v^-) &= K_2/(K_1 + K_2)^{1/2}. \end{aligned} \tag{12.22}$$

The neutral vacancy v^0 has already been defined by (12.18). Interesting results are obtained from applying two simplifying approximations. If $K_1 \gg K_2$, the electron–hole equilibrium is hardly perturbed, and Eqs. (12.22) reduce to

$$(e^-) = (h^+) = K_1^{1/2} = Ae^{-E_g/2RT}, \tag{12.23}$$

$$(v^-) = \frac{K_2}{(K_1)^{1/2}} = \frac{K_2' K_v}{(K_1)^{1/2}} = \frac{K_2^{\circ\prime} K_v^\circ}{(K_1^\circ)^{1/2}} \exp \frac{1}{RT}\left[-E_A - \Delta H_v + \frac{E_g}{2}\right], \tag{12.24}$$

where the temperature dependence of the equilibrium constants has been made explicit. If, on the other hand, there is a high concentration of trapped electrons (that is, when $K_1 \ll K_2$), then Eqs. (12.22) reduce to

$$(e^-) = \frac{K_1}{(K_2' K_v)^{1/2}} = \frac{K_1^\circ}{(K_2^{\circ\prime} K_v^\circ)^{1/2}} \exp \frac{1}{RT}\left[\frac{E_A}{2} + \frac{\Delta H_v}{2} - E_g\right], \tag{12.25}$$

$$(v^-) = (h^+) = (K_s' K_v)^{1/2} = (K_2^{\circ\prime} K_v^\circ)^{1/2} \exp \frac{1}{RT}\left[-\frac{E_A}{2} - \frac{\Delta H_v}{2}\right]. \tag{12.26}$$

These relations are of practical importance because they can be used to interpret experiments under properly controlled conditions and to evaluate the appropriate energies.

5. Formation Energies of Vacancies and Interstitial Atoms

The energies or enthalpies of defect formation are difficult, if not impossible, to calculate *ab initio*. However, semiempirical calculations that rely heavily upon the application of electronic computers abound, and they provide a valuable insight into the physical phenomena that together comprise the overall value of the energy. We will examine the various contributions without delving into the highly specialized and intricate computational techniques required to produce them.

As discussed in Chapter 5, theoretical calculations of isolated bond energies assume the infinite separation of the constituent ions or atoms at 0 K as the standard state. Consequently, the removal of a single atom from a solid is given by

$$\varepsilon^\circ = (z/2)U(r), \tag{12.27}$$

where z is the number of nearest neighbors and $U(r)$ is the nearest-neighbor bond strength and is the sum of pairwise interactions. Obviously (12.27) could be expanded to include the atoms in more distant shells by including additional terms in $U(r)$. In this calculation we wish to conserve the number of atoms, so that the atom after its removal to effectively infinite separation from the solid is returned to the surface. The energy associated with the surface absorption is equivalent to the negative of the energy of sublimation, ε_s. Now the vacancy formation energy is given by

$$\varepsilon_v = \varepsilon^\circ - \varepsilon_s. \tag{12.28}$$

If one-half of the effective bonds are broken in removing the atom from the surface, then to a first approximation, $\varepsilon^0 = 2\varepsilon_s$. However, the atoms relax about the defect to relieve the strain energy, and this is the source of an additional contribution to ε_v. As expected, atoms in metals collapse inward about a vacancy while dilating outward when surrounding an interstitial. This leads to

$$\varepsilon_v = \varepsilon_s - E_r, \tag{12.29}$$

where E_r is the energy recovered by the relaxation of all the atoms about a vacancy that are significantly involved in the strain relief.

In an elemental solid the expected distortions around a vacancy and an interstitial are illustrated in Fig. 12.9. Shifting of the atomic equilibrium positions is not the only relaxtion, because the companion electrons must also rearrange, an effect much more difficult to calculate. Because of relaxation, ε_v in (12.29) is clearly less than the sublimation energy ε_s.

In comparison with metals, ionic crystals produce additional large Coulomb and polarization interactions. As already noted, these calculations are difficult, but much progress has been made in the last 40 years at various levels of sophistication. Some representative theoretical enthalpies of formation of vacancies are shown in Table 12.2. As expected, ΔH_v° is clearly less than the cohesive energy. A comparison of the values in this table shows very good agreement with the experimental ones of Section 12.7. It should be mentioned that the formation energy of self-interstitials in metals is expected to be large because of the severe dilational strain around such a defect, as illustrated in Fig. 12.9. For example, the formation energy of interstitial copper in copper is estimated to be about 3 eV, rendering the native interstitial concentration very small indeed.

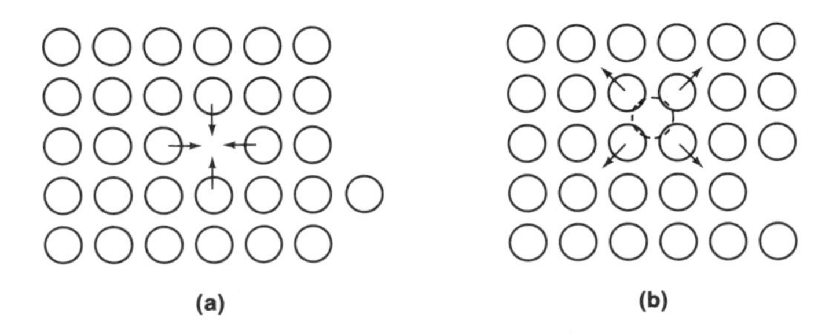

Figure 12.9.

Distortions in a metal around defects: (a) vacancy, (b) interstitial.

Table 12.2. Theoretical vacancy formation enthalpies (per mole) ΔH_v°, in eV.

Monoatomic Crystals	ΔH_v°	Ionic Crystals	ΔH_v°
Copper	1.0	Sodium chloride	2.1
Silver	0.9	Potassium chloride	2.2
Gold	0.8	Potassium iodide	1.9
Silicon	2.3	Rubidium chloride	2.2
Germanium	2.8		
Alkali metals	0.3–0.5		

6. Defect Complexes

Defects can combine with one another through the interactions of eleastic, electrostatic, and electronic forces. We have already mentioned that very close pairs of vacancies and interstitials will annihilate one another to relieve the localized strain. In this section we consider two types of defect reactions; the formation of divacancies, and the formation of vacancy–impurity complexes. We discuss these processes under equilibrium conditions and postpone the kinetics involved to Chapter 13.

A. *Divacancies*

The simplest complex is the divacancy, which consists of two vacancies in nearest—neighbor positions as illustrated in Fig. 12.1. This complex forms because its free energy of formation is less than that of two isolated vacancies. In metals this binding energy arises from a decrease in distortion and, hence, a decrease in strain energy when two vacancies are adjacent. There is also an electronic interaction favoring divacancy formation. In ionic crystals it is the pairing of positive and negative ion vacancies that is important, the electrostatic Coulomb interaction providing the major portion of the attractive force.

The equilibrium concentration of divacancies, v_2, may be calculated from statistical mechanics as in Section 12.2 or from the law of mass action; the latter is simpler. The reaction is

$$v + v \rightleftarrows v_2$$

or

$$(z/2)X_v^2 \rightleftarrows X_{v_2}. \tag{12.30}$$

The law of mass action gives for (12.30)

$$\Delta F_{\mathrm{B}}^{\circ} = -RT \ln K = -RT \ln \frac{X_{v_2}}{(z/2)X_v^2},$$

from which

$$X_{v_2} = (z/2)X_v^2 \exp(-\Delta F_{\mathrm{B}}^{\circ}/RT) \tag{12.31}$$

or

$$X_{v_2} = (z/2)e^{-2\Delta F_v^{\circ}/RT} e^{-\Delta F_{\mathrm{B}}^{\circ}/RT},$$

where the $z/2$ factor accounts for the fact that there are $zN/2$ pairs of potentially reactive lattice sites adjacent to each vacancy. $\Delta F_{\mathrm{B}}^{\circ}$ generally takes only negative values, since the binding energy is usually $\Delta E_{\mathrm{B}}^{\circ} = \Delta E_{v_2}^{\circ} - 2\,\Delta E_v^{\circ} < 0$, and, of course, $\Delta E_{v_2}^{\circ}$ and ΔE_v° are both positive, since work is required to form the defects. A somewhat more vague argument can be made for $2\,\Delta S_v^{\circ} > \Delta S_{v_2}$. These entropies stem from the alteration of the normal vibrational spectrum around lattice vacancies. To a first approximation, we can suppose the entropy to be proportional to the number of nearest-neighboring atoms to the defect since they are the vibrating entities. Because two nonadjacent vacancies have a greater number than when they combine to form a divacancy, it seems reasonable that $\Delta S_{v_2}^{\circ} - 2\,\Delta S_v^{\circ} = \Delta S_{\mathrm{B}}^{\circ} < 0$. These entropies are usually small, and (12.31) is often written as

$$X_{v_2} = (z/2)e^{-2\Delta E_v^{\circ}/RT} e^{+\Delta E_{\mathrm{B}}^{\circ}/RT} \tag{12.32}$$

Theoretically, values of $\Delta E_{\mathrm{B}}^{\circ}$ range from 0.3 to 0.6 eV in metals. We anticipate the discussions of Chapter 13 to point out here that the activation energies for divacancy motion in metals are in the same range and are often smaller than the activation energies for single vacancies, allowing them to play a significant role in diffusion processes. In alkali halides the binding energy is about 0.9 eV and is about equal to the migration energy.

B. Defect–Impurity Complexes

We shall illustrate by a simple example the relations describing vacancy–impurity association. The analog of (12.30) is the reaction

$$zI + v \rightleftarrows Iv, \tag{12.33}$$

where Iv is the vacancy–impurity complex.

The corresponding equilibrium constant is

$$K = \frac{(Iv)}{(zI)(v)}. \tag{12.34}$$

The inclusion of the multiplicative constant z in these expressions is necessary, just as in the case of the divacancies, because the actual reactants are the z nearest-neighbor positions to both the uncomplexed (I) and (v). Now using the familiar relationship between the standard free energy and the equilibrium constant for the chemical reaction as written above, we obtain

$$\Delta F_{Iv}^{\circ} = -RT \ln K.$$

Thus,

$$(Iv) = z(I)(v)e^{-\Delta F_{Iv}^{\circ}/RT}. \tag{12.35}$$

Substituting for v from (12.8) and recalling that

$$(I) = (I_0) - (Iv),$$

we obtain

$$(Iv) = \frac{z(I_0)e^{-(\Delta F_v^{\circ} + \Delta F_{Iv}^{\circ})/RT}}{1 + ze^{-(\Delta F_v^{\circ} + \Delta F_{Iv}^{\circ})/RT}}, \tag{12.36}$$

which by an argument identical to that used before for divacancies can be rearranged to the often-used approximate relation

$$(Iv) = \frac{z(I_0)}{e^{(\Delta F_v^{\circ} - \Delta E_{\mathrm{B}}^{\circ})/RT} + z}. \tag{12.37}$$

This expression can also be obtained by a straightforward application of Fermi–Dirac statistics, as discussed in Chapter 7 in connection with electron–hole creation. The method used here illustrates more clearly the consequences of not having $X_{Iv} \ll X_{I_0}$. In perhaps more familiar language, this illustrates the fact that when the number of available energy states does not greatly exceed the number occupied, Boltzmann statistics is invalid. If, on the other hand, $X_{I_0} \gg X_{Iv}$, then (12.37) becomes

$$X_{Iv} = zX_{I_0}X_v e^{+\Delta E_{\mathrm{B}}^{\circ}/RT}, \tag{12.38}$$

which is the same result as would be obtained from classical thermodynamics with the factor z effectively a correction to the concentration. If the impurity and vacancy concentrations are comparable,

$$(Iv) = \frac{z(I_0(v)e^{\Delta E_{\mathrm{B}}^{\circ}/RT}}{I + z(v)e^{\Delta E_{\mathrm{B}}^{\circ}/RT}}. \tag{12.39}$$

Again, in anticipation of Chapter 13, we may point out that vacancy–impurity complexes may facilitate diffusion, or they may act as trapping centers if the complexes are immobilized. The formulations of this section can be applied to more complex situations; for example, defect interactions with dislocations. Similar equations also emerge in describing segregation of impurities at grain boundaries as well as adsorption of gases on surfaces.

7. Experimental Determination of Crystal Defect Parameters

In this section we shall discuss representative experiments that illustrate the validity and use of the theoretical concepts discussed earlier. The most important parameters are clearly the formation energies. Representative values are collected in Table 12.3. We shall refer to the data of this table as we discuss the various experiments designed to measure these energies. Suffice it to say at this point that a comparison with the computational values of Table 12.2 indicates a rather gratifying agreement, perhaps better than we have any right to expect in view of the rather severe approximations assumed by the theoretical treatments. It should also be noted that the formation energy of the defects increases with the refractory character of the material, so that elements with high melting temperatures or enthalpies of vaporization have a high enthalpy of vacancy formation.

We shall examine two techniques for determining the equilibrium concentration of defects; specifically, changes in lattice expansion as a function of temperature, and quenching experiments. In recent years, positron annihilation techniques have been developed as an equilibrium method for measuring vacancy concentrations.

Table 12.3. Experimental vacancy formation enthalpies, in eV.

Crystal	ΔH_v°	Crystal	ΔH_v°
Cu	1.0	NaCl	2.5
Ag	1.0	KCl	2.5
Au	0.9	KBr	2.4
Al	0.7	KI	2.2
Pt	1.5	RbCl	2.0
Pb	0.5		
W	3.3		
Na	0.4		
Fe	1.5		
Ge	2.6		

A. Lattice Expansion

We can determine quite accurately the value of ΔH_v° from a comparison of high-temperature equilibrium measurements of the macroscopic volume, which is determined dilatometrically, and the atomic lattice spacings obtained from x-ray diffraction. These measurements give the true equilibrium values since they are carried out isothermally. For the fractional change in lattice parameter and length, we may write

$$\frac{3\,\Delta a}{a} = p(T) + r(T) + X(T),$$

$$\frac{3\,\Delta l}{L} = q(T) + s(T) + Y(T), \tag{12.40}$$

where the terms $p(T)$ and $q(T)$ represent the normal thermal expansion due to the increased vibration of the atoms, $r(T)$ and $s(T)$ represent the alteration in the frequency distribution because of interactions with the temperature-dependent defect concentration, and $X(T)$ and $Y(T)$ denote the thermal expansion arising from the varying defect concentration. For a crystal containing N atoms and n_v identical point defects, the term $X(T)$ can be written

$$X(T) = n_v(f\Omega)/N\Omega, \tag{12.41}$$

where $f < 1$ and is the numerical factor that accounts for the relaxation of the atoms about a vacancy that results in a volume less than the true atomic volume Ω. If there is not relaxation, f is equal to unity. The numerical factor f arising from the relaxation is the same for both the x-ray and volume measurements. This is because the local distortion is elastically propagated to the surface of the crystal, resulting in a commensurate change in the crystal volume. As shown by Eshelby[1] the x-ray measurement and the macroscopic dilatation are altered the same way by the defect strain fields. The term $Y(T)$ can, therefore, be written

$$y(T) = n_v(f + 1)\Omega/N\Omega \tag{12.42}$$

if one remembers that atoms are conserved in the measurement of the bulk thermal expansion. In terms of a model, one can visualize removing an atom from the interior, thus creating a vacancy of volume equal to $f\Omega$ and sticking the same atom on the surface of the crystal, thus creating, on the average, an additional volume Ω. Assuming

$$p(T) + r(T) = q(T) + s(T), \tag{12.43}$$

---

[1] J. D. Eshelby, *Solid State Physics* (F. Seitz and D. Turnbull, eds), vol. **3**, pp. 79–144, Academic Press, 1956.

which is, in fact, experimentally established, we can combine (12.40)–(12.43) to obtain

$$\frac{n_v}{N} = \pm 3\left(\frac{\Delta L}{L} - \frac{\Delta a}{a}\right), \tag{12.44}$$

which is the fundamental relationship used to calculate the relative vacancy concentration as a function of temperature.

A plus sign in (12.44) applies to the vacancy and a minus sign to the interstitial, since the processes are just the opposite of each other. The results of such measurements on gold are illustrated in Fig. 12.10. The temperature dependence of X_v is exponential in T^{-1} in full agreement with the expected Arrhenius relation. The slope, according to (12.9), gives the vacancy formation energy, ΔH_v°, which, according to this experiment, is 0.94 eV, in good agreement with the theoretical estimate listed in Table 12.2. Similar experiments for other elements also have shown good agreement.

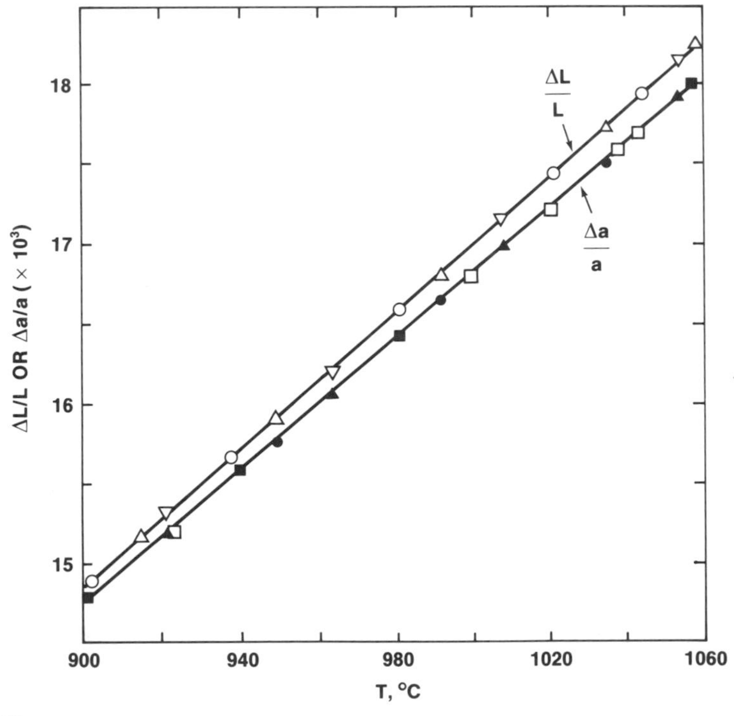

Figure 12.10.

Length expansion, $\Delta L/L$, and lattice parameter expansion, $\Delta a/a$, vs. temperature for gold in the 900 to 1060°C interval. $\Delta L/L > \Delta a/a$ corresponds to the thermal generation of vacancies. [From R. O. Simmons and R. W. Balluffi, *Phys. Rev.*, **125**, 862 (1962).]

B. Quenching

Rapid quenching from high temperatures is used to introduce excess vacancies by preserving the equilibrium concentration of the high-temperature state. This experiment is conceptually very simple, but is plagued by technical difficulties. The procedure is to quickly cool a wire from a succession of elevated temperatures and

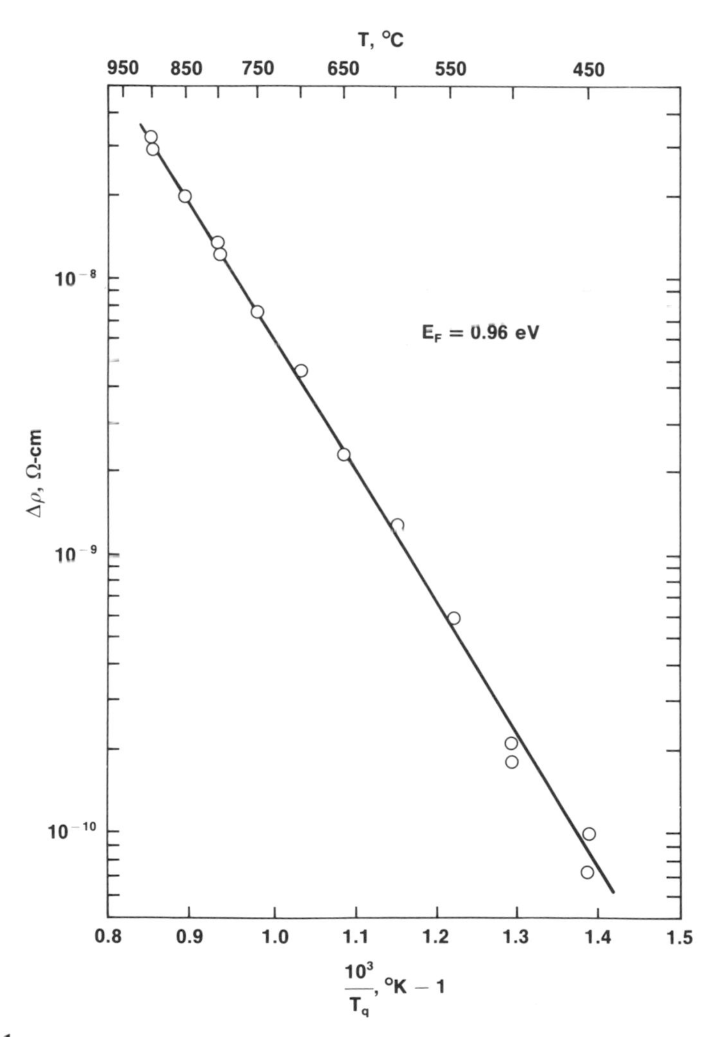

Figure 12.11.

Semilogarithmic plot of quenched-in resistivity vs. reciprocal of the absolute quench temperature. [From J. E. Bauerle and J. S. Koehler, *Phys. Rev.*, **107**, 1493 (1957).]

measure the successive changes in electrical resistance at liquid He temperature. At such a low temperature, $\sim$4.2 K, the normally large ohmic resistence due to phonon scattering of the electrons is suppressed and the residual resistence is proportional to the vacancy concentration, which in turn is a function of the initial temperature. High quenching rates are easy to obtain, but the slightest deformation of the wire, which can result from the thermal stresses generated by the differential cooling, is also recorded, but as a spurious change in electrical resistance. It is also necessary to make some correction for vacancy annihilation at the free surface of the wire, usually done by extrapolating measurements on wires of increasing size to infinite diameter. It might seem that very thick specimens would be best for minimizing this error. However, they also have maximal strain that can increase the resistivity,

Table 12.4. Monovacancy formation energies from quenching experiments.

Metal	E_F (eV)	Physical Property Measured
Gold	0.98 ± 0.03	Resistivity
	0.97	Resistivity
	0.98 ± 0.1	Resistivity
	0.95 ± 0.1	Resistivity
	0.79	Resistivity
	0.98	Length change
	0.97 ± 0.1	Stored energy
	0.98	Resistivity
Silver	1.10 ± 0.04	Resistivity
	1.10	Resistivity
	1.06	Resistivity
	1.04 ± 0.1	Resistivity
	1.0	Thermal electromotive force
Platinum	1.4 ± 0.1	Resistivity
	1.3 ± 0.1	Resistivity
	1.18	Resistivity
	1.4	Thermal electromotive force
Copper	1.0	Resistivity
Aluminum	0.76 ± 0.04	Resistivity
	0.79 ± 0.04	Resistivity
	0.76 ± 0.03	Resistivity
	0.76	Resistivity
Tin	0.51 ± 0.05	Resistivity
Magnesium	0.89 ± 0.06	Resistivity

Compilation from A. C. Damask and G. J. Dienes, *Point Defects in Metals*, Gordon and Breach, New York, 1963.

induced by the large radial thermal gradiants. Hence, the diameter of the wire is always a judicious compromise.

Let us again select gold as our example for obtaining data of high accuracy. It is an excellent substance for such experiments since it does not oxidize in air or react with water. Preliminary experiments indicated that the vacancies in gold migrate slowly at room temperature and water, an excellent quenching medium, could be used. High-purity gold wires of 0.016-in diameter were cooled at rates of about 4×10^{4}°C/sec by sudden immersion in water. The data thus obtained are the quenched in resistivity, $\Delta\rho$, as a function of the quench temperature T_q. The data are plotted in Fig. 12.11 as the logarithm of the quenched-in resistivity $\Delta\rho$ versus the reciprocal of the quench temperature, in order to test the expected Arrhenius relation (12.9). The result is clearly in agreement with the theory, since an excellent straight line is shown in Fig. 12.11 over a wide range of temperatures and $\Delta\rho$ values. The formation energy derived from the slope is 0.96 eV, a result in very good agreement with the value obtained in the lattice expansion experiment. The equilibrium vacancy concentration at the melting point is about 7×10^{-4} a value quite typical for many metals. The data so obtained for a variety of metals are compiled in Table 12.4.

Chapter 12 Exercises

1. Discuss the energy terms controlling vacancy formation. Why is the vacancy formation energy smaller than the cohesive energy?

2. Why is the equilibrium interstitial concentration, in general, much smaller than the equilibrium vacancy concentration? What factors might promote (or decrease) the concentration of equilibrium interstitials? At equal reduced temperatures (i.e., T/T_m), which would most likely have the largest and which the smallest concentration of Schottky defects, ionic or metallic compounds?

3. For a monatomic system with parameters $\Delta E_v^\circ = 20$ kcal/mole, $\Delta S_v^\circ =$ 1 kcal/mole, and $\Delta E_B^\circ = 4$ kcal/mole (the divacancy binding energy), calculate the monovacancy (v_1) and divacancy (v_2) concentrations at 1000 K and 500 K. Discuss the change in v_2/v_1 as a function of temperature.

4. From the data of problem 3, calculate the divacancy formation parameters $\Delta E_{v_2}^\circ$ and $\Delta S_{v_2}^\circ$.

5. Calculate the Schottky pair concentration in a simple salt of the NaCl type at 1000 K for a formation energy of 70 kcal/mole. What concentration of a positive

divalent ion substituting for the positive monovalent one would result in approximately the same vacancy concentration?

6. Show that, in principle, the enthalpy of vacancy formation can be obtained from plotting $\Delta C_p T^2$ versus T^{-1} where ΔC_p equals the change in C_p due solely to vacancy formation.

7. Let there be an impurity in Au whose binding energy to the vacancy is 0.25 eV (typical of silver as an impurity). Assume $\Delta S^\circ = 0$. Compare the results using equations (12.38) and (12.39) for $X_I^\circ = 10^{-6}$ at $T = 668$ K and $T = 1136$ K. Discuss the results.

8. The vacancy in Ge and Si is an acceptor (i.e., has a tendency to trap a negative charge). The reaction is

$$v^0 \rightleftarrows v^- + h^+,$$

where h^+ is the positive hole. If n is the concentration of free electrons when $[v^-]$ is the concentration of negatively charged vacancies, and n_i and $[v_i^-]$ are the same quantities in an intrinsic sample, show that

$$\frac{[v^-]}{[v_i^-]} = \frac{n}{n_i}.$$

(Hint: Make use of the law of mass action for electrons and holes: $np = n_i^2$.)

9. Let the concentration of singly ionized donor be N_D and that for acceptors N_A. Show that

$$\frac{n}{n_i} = \frac{N_D}{2n_i} + \left[1 + \frac{N_D^2}{4n_i^2}\right]^{1/2}$$

for donors, and

$$\frac{n}{n_i} = \frac{N_A}{2n_i} + \left[1 + \frac{N_A^2}{4n_i^2}\right]^{1/2}$$

for acceptors. (Hint: Make use of electrical neutrality.)

10. The formation of positive and negative ion vacancies in a monovalent ionic crystal is represented by the chemical equation

$$v_p + v_n \overset{K_1}{\rightleftarrows} \text{perfect crystal}.$$

If a divalent impurity is added to the crystal, the formation of complexes by the divalent ions and the positive ion vacancies is represented by

$$v_p + I \overset{K_2}{\rightleftarrows} C.$$

Using the notation

X_p = mole fraction of v_p
X_n = mole fraction of v_n
X_0 = mole fraction of v_p in a pure crystal
X_c = mole fraction of complexes
Y = mole fraction of added divalent impurity.

Show that

(1) $X_0 = 1/K_1^{1/2}$
(2) $Y = X_0[H(Z^2 - 1) + (Z - (1/Z)]$,

where H is defined by the two equilibrium constants as

$$H = K_2/K_1^{1/2},$$

and Z is the dimensionless quantity given by

$$Z = X_p/X_0.$$

(Hint: Use the condition of electrical neutrality.)

Additional Reading

A. C. Damask and G. J. Dienes, *Point Defects in Metals*, 2nd ed., Gordon and Breach (1971). A good introduction to the subject with discussion of experimental methods as well as theory.

J. H. Crawford, Jr., and L. M. Slifkin, eds., *Point Defects in Solids*, Vol. 1, *General and Ionic Crystals*, Plenum (1972); Vol. 2, *Semiconductors and Molecular Crystals*, Plenum (1975). Each chapter is written by an acknowledged expert. These volumes are a standard reference in the field.

F. A. Kroger, *The Chemistry of Imperfect Crystals*, Wiley (1964). Widely used both as a text and reference book. A large treatise that covers the fundamentals and defect-controlled phenomena mostly from a thermodynamic approach.

L. W. Barr and A. B. Lidiard, in *Physical Chemistry*, *An Advanced Treatise*, Vol. X, Chap. 3, Academic Press (1970), "Defects in Ionic Crystals" discusses energy calculations for various defects and also deals with migration (i.e., diffusion) energies.

R. A. Swalin, *Thermodynamics of Solids*, Wiley (1972) is a good and quite complete introductory treatment of the thermodynamics of defects; see Chaps. 13 and 14.

M. Lannoo and J. Bourgoin, "Point Defects in Semiconductors. I" in *Theoretical Aspects*, Springer-Verlag (1981) offers a quite complete coverage of the subject, including diffusion as well as an introduction to bonding, lattice dynamics, and quantum effects.

Chapter 13

Diffusion in Solids

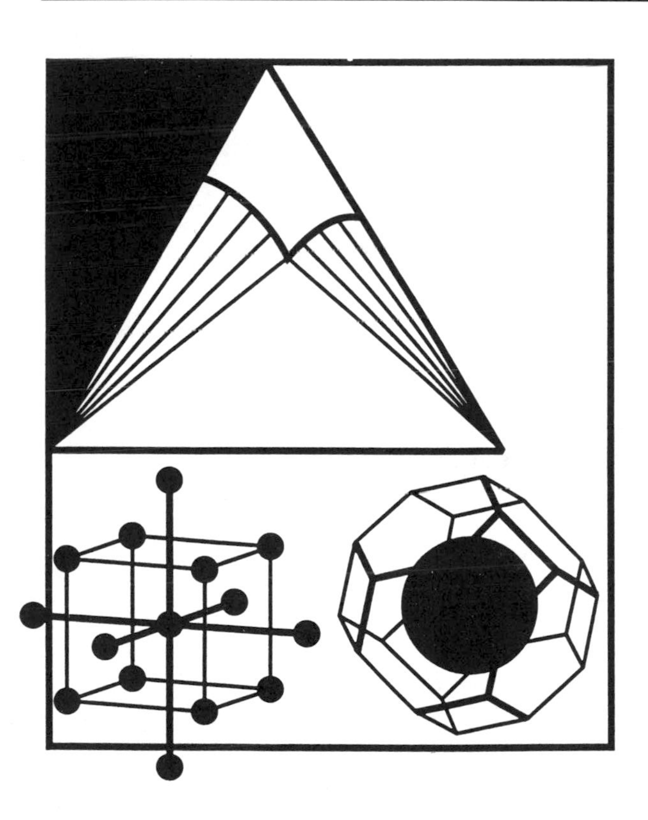

Chapter 13

Contents

A reasonable probability is the only certainty.

—E. W. Howe *Sinner Sermons*, quoted in J. R. Newman, *The World of Mathematics*

Diffusion in Solids

Solids, contrary to some popular ideas, are not inert systems, If they were, there would be no metallurgy, no ceramic engineering, and no semiconductor industry. On the other side of the coin, there would be no corrosion either. All of these important solid state processes depend on the exchange of atoms in the solid, that is, on solid state diffusion. As will be seen later, this chapter is a logical successor to Chapter 12, since diffusion in crystals proceeds by means of crystalline imperfections. The emphasis in this chapter is on mechanisms and diffusion-controlled processes with only a brief treatment of the phenomenological diffusion equations.

1. Basic Phenomenological Equations

The basic equations of diffusion, applicable to all states of matter, are purely phenomenological descriptions relating the rate of flow of matter to its concentration gradient. Let us illustrate the diffusion process in one dimension by Fig. 13.1. Three adjacent lattice planes are depicted in this figure, and the flow of material from plane x in either direction is indicated by the fluxes designated by J, which is the number of particles passing through a plane of unit area per unit time.

The number of tracer, or impurity, atoms per unit area at x is given by N_x, a quantity that is a function of time and position. The time rate of change of N_x is given by

$$\frac{\partial N_x}{\partial t} = \frac{N(t + \delta t, x) - N(t, x)}{\delta t}. \tag{13.1}$$

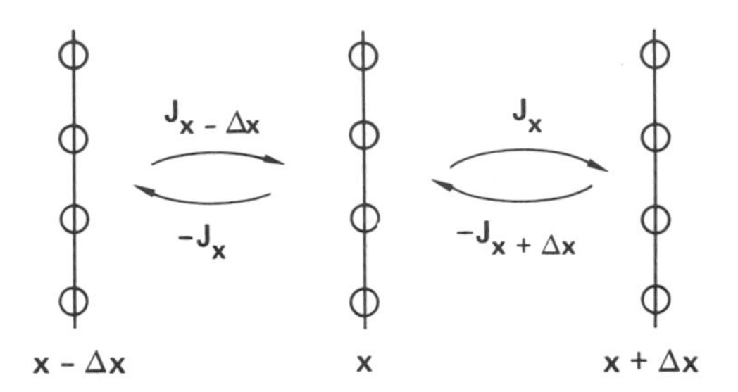

Figure 13.1.

Three adjacent lattice planes illustrating one-dimensional diffusion in which atoms exchange positions by jumping $\pm\Delta x$.

Now the total number of atoms arriving at a plane x is given by

$$N_+ = (\Gamma/2)[(N_{x-\Delta x})\delta t + (N_{x+\Delta x})\delta t], \tag{13.2}$$

and the number leaving is

$$N_- = \Gamma N_x \delta t. \tag{13.3}$$

Γ in these equations is the mean value of the elementary jump frequency. Subtracting (13.3) from (13.2) and multiplying below and above by Δx^2 gives

$$\frac{N_+ - N_-}{\delta t} = \frac{\delta N_x}{\delta t} = \frac{\Gamma}{2} \Delta x^2 \frac{N_{x-\Delta x} - 2N_x + N_{x+\Delta x}}{\Delta x^2}. \tag{13.4}$$

Upon converting to concentrations and passing to the limit (13.4) becomes

$$\frac{\partial C}{\partial t} = \frac{\Gamma}{2} \Delta x^2 \frac{\partial^2 C}{\partial x^2} = D \frac{\partial^2 C}{\partial x^2}, \tag{13.5}$$

the one-dimensional diffusion equation. This relation is, in effect, the result of taking the second finite difference in the concentration gradient and equating it to the time-dependent flow of atoms at a fixed point. We have identified the diffusion coefficient D with $(1/2)\Gamma\, \Delta x^2$ where Δx is, in fact, the distance of separation of nearest neighbors, a.

Generalization of the diffusion equation to three dimensions gives

$$\frac{\partial C}{\partial t} = D_{xx} \frac{\partial^2 C}{\partial x^2} + D_{yy} \frac{\partial^2 C}{\partial y^2} + D_{zz} \frac{\partial^2 C}{\partial z^2} = D_{ii} \nabla^2 C_i, \tag{13.6}$$

a relation usually simply referred to as the *diffusion equation*. For cubic crystals (13.6) simplifies to

$$D_{xx} = D_{yy} = D_{zz} = (1/6)a^2\Gamma, \tag{13.7}$$

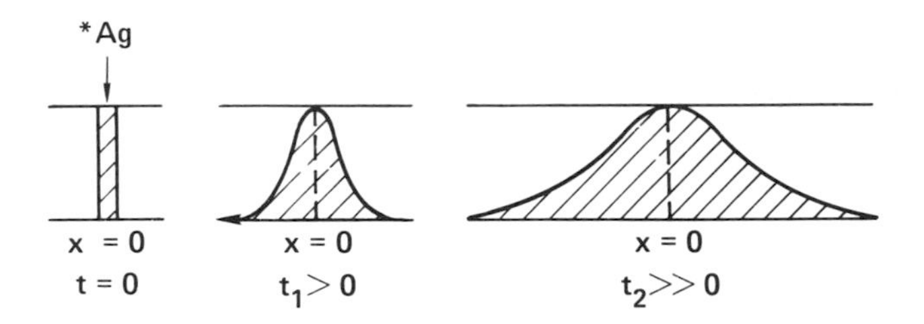

Figure 13.2.

The time sequence of diffusion profiles showing the tracer migration with its gaussian spatial distribution.

where Γ and a are taken along each of the principal crystallographic axes. The application of the general diffusion equations to specific problems requires the solution of a variety of boundary value problems, for which there are well-developed mathematical procedures. Basically, solutions of the partial differential equations have to be combined with specific boundary conditions. We shall illustrate the problems by some simple examples. For detailed solutions for a large variety of boundary conditions, the reader is referred to Crank.[1]

Let us consider a typical diffusion process, namely diffusion into a semi-infinite solid from a thin planar source, an often-used experimental arrangement. As a specific example, suppose a radioactive isotope, perhaps *Ag, is electroplated onto the flat surface of a silver bar and then another silver bar is welded to the first, sandwiching the tracer between the two. The appropriate solution of (13.5) for this case is given by

$$\frac{C(x, t)}{C_0} = \frac{1}{2\sqrt{\pi D t}} e^{-x^2/4Dt}, \tag{13.8}$$

where $C(x, t)$ is the concentration at x and t, and C_0 is the concentration at the origin ($x = 0, t = 0$); see Fig. (13.2). The function in (13.8) describes a gaussian distribution, and the spreading of the radioactive isotope is as illustrated in Fig. 13.2. The diffusion coefficient is usually calculated by plotting $\log(C/C_0)$ versus x^2 and obtaining $(Dt)^{-1}$ from the slope, t being the measured interval during which the diffusions occurred. For radioactive tracers, the logarithm of the counts per minute as a function of penetration, corrected for absorption, may be plotted directly.

[1] J. Crank, *The Mathematics of Diffusion*, Oxford, 1975.

Another often-used procedure is the diffusion of a gas into or out of a solid. The solution of the diffusion equation for diffusion out of a sphere of radius r is

$$\frac{\bar{C} - C_i}{C_f - C_i} = 1 - \frac{6}{\pi^2} \sum_{\nu=1}^{\infty} \frac{1}{\nu^2} \exp\left(\frac{-\nu^2\pi^2}{r^2} Dt \right), \tag{13.9}$$

where $\bar{C}$, C_i, and C_f are the average, initial, and final concentrations within the sphere. At sufficiently long times, all but the $\nu = 1$ term can be neglected, thus yielding the approximate solution

$$\frac{\bar{C} - C_i}{C_f - C_i} \cong 1 - \frac{6}{\pi^2} e^{-t/\tau}, \tag{13.10}$$

where $\tau = r^2/\pi^2 D$.

It is often not practical to measure the concentrations needed for the direct application of (13.9). One can use a modification of these equations when the total gas release is measured continuously as a function of time. The fractional release is then given by

$$\frac{M(t)}{M_\infty} = 1 - \frac{6}{\pi^2} \sum_{\nu=1}^{\infty} \frac{1}{\nu^2} \exp \frac{-\nu^2\pi^2 Dt}{r^2} \cong 1 - \frac{6}{\pi^2} e^{-t/\tau}, \tag{13.11}$$

where M_∞ is the total amount initially within the sample, and (13.11) can then be applied to the data to yield the value of D.

The diffusion process can also be treated on the basis of random walk, a description that explicitly involves the kinematics of atomic motion. The discussion outlined here, a rather standard derivation, follows the one given by Borg and Dienes.[2] Consider a particle executing a random walk in two dimensions as illustrated in Fig. 13.3, where the initial steps are of equal length $\mathbf{r}$ and where the vector $\bar{\mathbf{R}}$ connects the initial and final points. All jump directions are taken to have the same a priori probability and are assumed to be uncorrelated.

After n elementary jumps, $\bar{\mathbf{R}}$ is given by the vector equation

$$\mathbf{R}_n = \sum_{i=1}^{n} \mathbf{r}_i. \tag{13.12}$$

[2] R. J. Borg and G. J. Dienes, *Solid State Diffusion*, Academic Press, 1988.

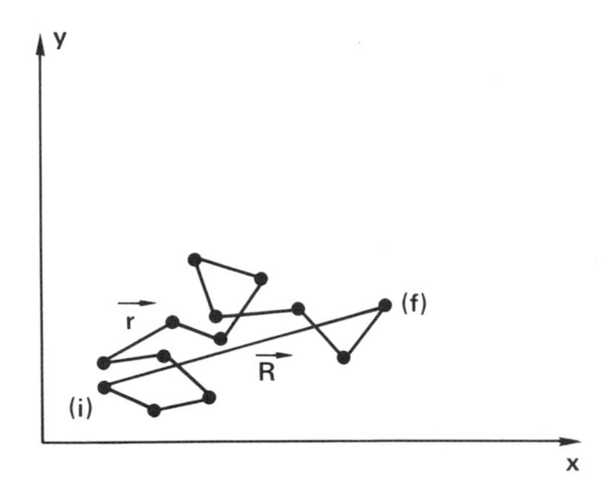

Figure 13.3.

Illustrating a particle executing a random walk of equal length jumps **r** from the start at (i) to the end at (f) with a total net displacement $\bar{\mathbf{R}}$.

Squaring both sides and introducing $\Theta_{i,i+j}$, the angle between the ith and $(i+j)$th jump leads to

$$\begin{aligned} R_n^2 &= \sum_{i=1}^{n} \mathbf{r}_i \cdot \mathbf{r}_i + 2\sum_{i=1}^{n-1} \mathbf{r}_i \cdot \mathbf{r}_{i+1} + 2\sum_{i=1}^{n-2} \mathbf{r}_i \cdot \mathbf{r}_{i+2} + 2\mathbf{r}_{n-1} \cdot \mathbf{r}_n \\ &= \sum_{i=1}^{n} r_i^2 + 2\sum_{j=1}^{n-1}\sum_{i=1}^{n-j} \mathbf{r}_i \cdot \mathbf{r}_{i+j} \\ &= nr^2\left(1 + \frac{2}{n}\sum_{j=1}^{n-1}\sum_{i=1}^{n-j} \cos\Theta_{i,i+j}\right), \end{aligned} \tag{13.13}$$

and since we are only interested in the average values of R_n^2, namely $\overline{R_n^2}$, we arrive at

$$\overline{R_n^2} = nr^2\left(1 + \frac{2}{n}\overline{\sum_{j=1}^{n-1}\sum_{i=1}^{n-j}} \cos\Theta_{i,i+j}\right). \tag{13.14}$$

The term containing the double sum vanishes because there are as many jumps with angular value $-\Theta_{i,i+j}$ as with $+\Theta_{i,i+j}$. This is the case for our postulated uncorrelated jumps. Correlation effects are discussed in Section 4 of this chapter. Thus, the most probable value of $\overline{R_n^2}$ is given by

$$\overline{R_n^2} = nr^2. \tag{13.15}$$

We shall now relate this result to the continuum equation for diffusion from an infinitely thin planar source into a semi-infinite solid, namely to (13.8) in the form

$$C(x) = (\alpha/\sqrt{t})e^{-x^2/4Dt}, \tag{13.16}$$

where $\alpha = m/2\sqrt{\pi D}$, m is the total amount of diffused material, and t is the time over which diffusion has occurred. The probability of $P(x)$ of finding an atom between x and $x + \Delta x$ is given by the fraction of atoms between these two points as shown in Fig. 13.4.

$$P(x)\,dx = \frac{C(x)\,dx}{\int_{-\infty}^{+\infty} C(x)\,dx} = \frac{\alpha}{t^{1/2}}\,e^{-x^2/4Dt}. \tag{13.17}$$

The integral in (13.17) is equal to unity, and the numerator is the distribution function $P(x)$. The average value of x^2 is calculated in the usual manner as

$$\overline{x^2} = \int_{-\infty}^{+\infty} x^2 P(x)\,dx = \frac{\alpha}{t^{1/2}} \int_{-\infty}^{+\infty} x^2 e^{-x^2/4Dt}\,dx. \tag{13.18}$$

Substituting $\xi^2 = x^2/4Dt$ yields

$$\int_{-\infty}^{+\infty} x^2 P(x)\,dx = \frac{\alpha}{t^{1/2}} \int_{-\infty}^{+\infty} (Dt)^{3/2}\xi^2 e^{-\xi}\,d\xi. \tag{13.19}$$

Substituting for α and setting $m = 1$ gives

$$\overline{x^2} = \frac{4Dt}{\pi^{1/2}} \int_{-\infty}^{+\infty} \xi^2 e^{-\xi^2}\,d\xi = \frac{4Dt}{\pi^{1/2}} \cdot \frac{\pi^{1/2}}{2} = 2Dt. \tag{13.20}$$

We can now identify $\overline{x^2}$ with $\overline{R_n^2}$ and write

$$\overline{R_n^2} = 2Dt = nr^2. \tag{13.21}$$

Now the time necessary for a particle to travel $\overline{R_n^2}$ in n jumps is given simply by

$$t = n\Gamma^{-1}, \tag{13.22}$$

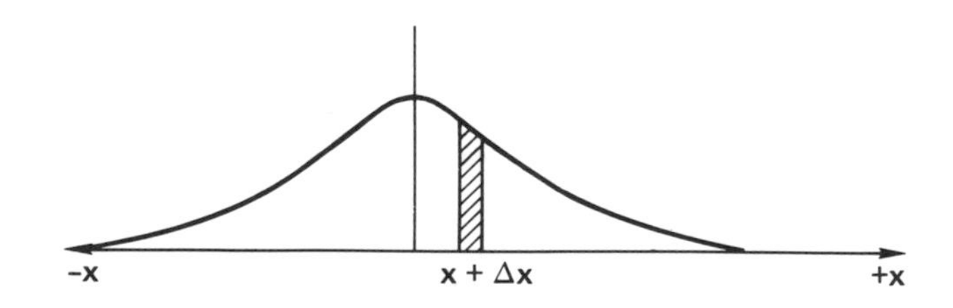

Figure 13.4.

where Γ, the mean jump frequency, is given by

$$\Gamma = 2D/r^2 \tag{13.23}$$

in one dimension and by

$$\Gamma = 6D/r^2$$

in three dimensions. These equations, which connect the continuum and discrete descriptions, give us a complete set of relations between the root-mean-square diffusion distance $(\overline{x^2})^{1/2}$, the total number of individual diffusive displacements n, the jump frequency Γ, and the diffusion coefficient.

2. Mechanisms

The simplest diffusion process is the motion of an interstitial atom from one interstitial site to another without involving any exchange with a matrix atom. The diffusion of interstitial impurity atoms is often of great practical importance, as, for example, the diffusion of carbon in iron which controls the formation of various carbon steels. Interstitial hydrogen diffusion occurs in many cases in many solids. Impurity diffusion is industrially of great importance also in the semiconductor industry.

As noted, simple interstitial jumping does not lead to any interchange of atoms. The simplest way to exchange two atoms is by direct interchange, as illustrated in Fig. 13.5. In this mechanism the two atoms, A and B, simply squeeze by each other. It turns out that for elemental substances, except in some unusual cases (self-diffusion in the basal plane of graphite and possibly in silicon), this process is energetically prohibitively expensive. Further, the A and B atoms clearly must diffuse at the same rate in this mechanism. As shown later in the experimental section, there are highly definitive experiments showing that the constituent atoms of alloys are characterized

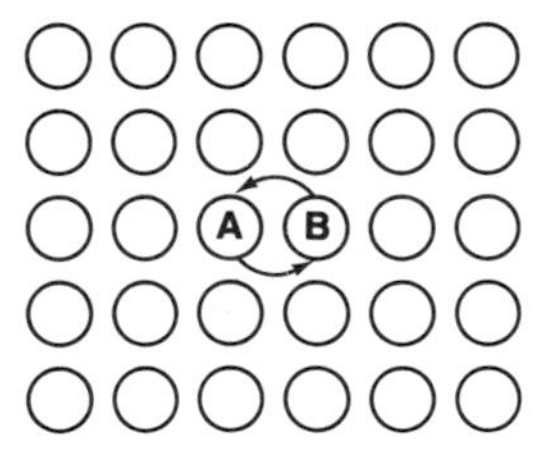

Figure 13.5.

The direct interchange of two atoms.

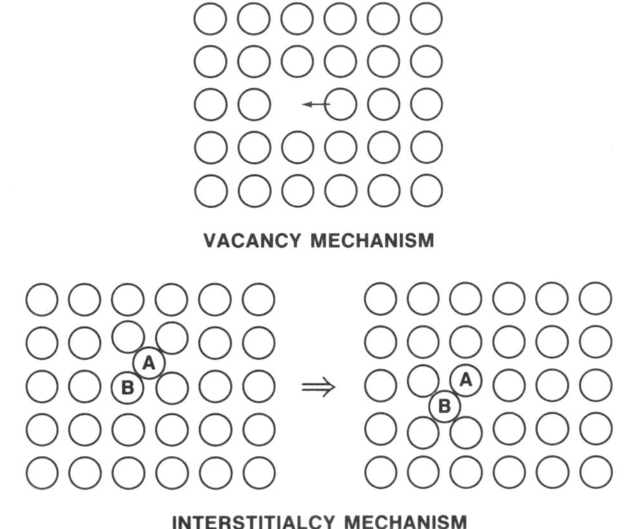

Figure 13.6.

The two most important mechanisms for substitutional diffusion.

by different diffusion coefficients. We shall, therefore, not discuss this mechanism further.

These considerations lead to the idea that substitutional diffusion depends on imperfections even though their concentration is generally very low (see Chapter 12). The two most important mechanisms are illustrated in Fig. 13.6. In both cases a defect acts as a catalyst for promoting atomic interchange in the sense that its concentration does not change during the diffusion process. It turns out that of the two the vacancy mechanism is by far the most important diffusion mechanism. Since the discussion of jump frequencies and their temperature dependence in the next section applies equally well to both mechanisms, we shall present the treatment in terms of vacancies with no loss of generality. It should be mentioned that there is ample experimental evidence for the dominant role of vacancies in diffusion, as discussed in Section 4.

3. Temperature Dependence and Activation Energies

The vacancy mechanism requires that the jump rate of an atom be given by the mean jump frequency times the probability that a vacancy is adjacent to the jumping atom. We already know from Chapter 12 how the concentration of vacancies varies as a function of temperature, and we must next determine what sort of a relation is followed by the jump frequency itself as a function of temperature. It is natural to discuss this process within the framework of absolute rate theory. An energy barrier must exist between the initial and final states of a jump as illustrated in Fig. 13.7. An equilibrium assembly of atoms or molecules contains a certain fraction of particles whose thermal energy exceeds that of the barrier and that can consequently cross the barrier. Using a physical chemical terminology, the fraction of atoms in the excited state, i.e., at the saddle point in Fig. 13.7, is given by the equilibrium constant for the reaction $X_0 \rightleftarrows X^{\ddagger}$ where ‡ designates activation. The activation energy then is related to the fraction of activated atoms by

$$X^{\ddagger} = X_0 \exp(-\Delta G^{\ddagger}/RT), \tag{13.24}$$

where X_0 is the fraction of atoms in the ground state, $X^{\ddagger}$ the fraction in the excited state, and $\Delta G^{\ddagger}$ the difference in the Gibbs free energy between these states. We now postulate that the fraction of atoms $X^{\ddagger}$ having free energy equal to or in excess of $\Delta G^{\ddagger}$ has a finite a priori probability of making a diffusive jump, which is designated by ω_0, and by analogy with standard chemical kinetics, we write for the specific rate constant

$$k_0 = \omega_0 \varepsilon^{-\Delta G^{\ddagger}/RT}. \tag{13.25}$$

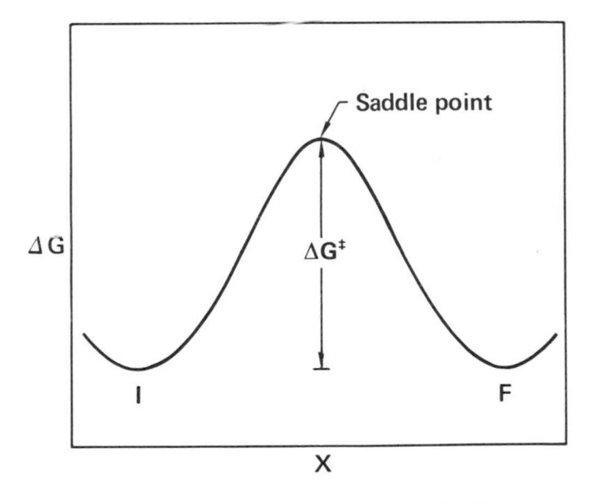

Figure 13.7.

A one-dimensional schematic picture of the energy surface connecting the initial state of the system, *I*, with the final state, *F*. For self-diffusion in simple crystal systems, these two states are identical, and the free energy of activation is the difference between the saddle point, located between *I* and *F*, and the ground state energies.

The jump frequency Γ is in effect the specific rate constant for the vacancy, for it is the probability of an atom making an elementary jump into a vacancy in unit time. Thus,

$$k_0 = \Gamma = \omega_0 \varepsilon^{-\Delta G^{\ddagger}/RT}. \tag{13.26}$$

The overall diffusion rate, as already noted, is the vacancy jump rate times the atomic fraction of vacancies, so multiplying by X_v and putting $D = \gamma\Gamma a^2$, where a is the lattice parameter and γ is a geometric constant gives

$$D = a^2\gamma\omega_0 e^{-\Delta G^{\ddagger}/RT} X_v = a^2\gamma\omega_0 e^{-(\Delta G^{\ddagger} + \Delta G_v^{\circ})/RT} \tag{13.27}$$

and, upon decomposing the free energy in the usual way, we obtain

$$D = \gamma a^2 \omega_0 e^{(\Delta S^{\ddagger} + \Delta S_v^{\circ})/R} e^{-(\Delta H^{\ddagger} + \Delta H_v^{\circ})/RT}. \tag{13.28}$$

This important result shows that D depends exponentially on the reciprocal of the absolute temperature with an overall activation energy that is the sum of the formation and motion energies.

In the foregoing derivation an inherently many body problem was reduced to a one-body model. As shown by Vineyard[3], the many-body aspects can be readily incorporated into the absolute rate theory. Consider the jump of an atom into a vacant lattice site at an average rate Γ. If there are $N/3$ atoms in the crystal, the number of degrees of freedom is N. In Fig. 13.8 the N-dimensional configurational space is illustrated schematically.

The solid contour lines represent hypersurfaces of constant potential energy. Point A represents a minimum in Φ corresponding to a lattice atom adjacent to a vacancy, with all other atoms in their equilibrium positions. If this atom has exchanged with the vacancy, and all other atoms have been relaxed back to their equilibrium positions, then an equivalent minimum exists in Φ at point B. In order to pass from point A to point B, the atom must surmount a potential barrier through saddle point P. The dotted line drawn through point P represents a unique $(N - 1)$-dimensional hypersurface S passing through point P and perpendicular to the contours of constant Φ everywhere else, and therefore separates region A from region B. A system in thermal equilibrium has a definite number of representative points in phase space Q_A, in the region to the "left" of S, and a definite number, I, crossing S from "left" to "right" per second. The transition rate Γ from A to B is given by

$$\Gamma = \frac{1}{\tau} = \frac{I}{Q_A}. \tag{13.29}$$

---

[3] G. H. Vineyard, *J. Phys. Chem. Sol.* **3**, 121 (1957).

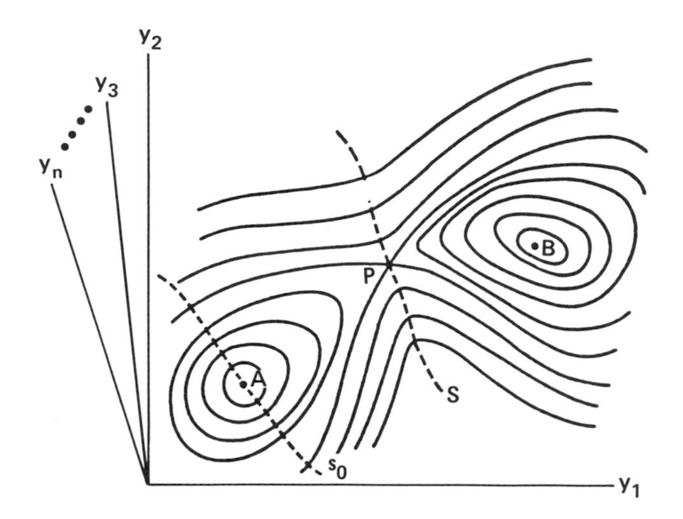

Figure 13.8.

N-dimensional configuration space showing, schematically, hypersurfaces of constant potential energy (solid lines) and imaginary constraining hypersurfaces (dotted lines). The saddle point is at P.

I and Q_A are evaluated by standard statistical mechanical methods[4,5]. Simplification is achieved by employing the theory of small vibrations. The final result is

$$\Gamma = \left(\prod_{j=1}^{N} v_j \Big/ \prod_{j=1}^{N-1} v'_j \right) e^{-[\Phi(P) - \Phi(A)]/kT}. \tag{13.30}$$

This relation is of a familiar form in which the transition rate is displayed as the product of an effective frequency v^* and an activation exponential $\Delta E^\ddagger$ given by

$$\Delta E^\ddagger = [\Phi(P) - \Phi(A)]. \tag{13.31}$$

In comparison with (13.26), one finds that

$$\Gamma = \omega_0 e^{-\Delta G^\ddagger/RT} = v^* \exp(-\Delta E^\ddagger/RT), \tag{13.32}$$

with

$$v^* = \left(\prod_{j=1}^{N} v_j \Big/ \prod_{j=1}^{N-1} v'_j \right). \tag{13.33}$$

---

[4] Vineyard, loc. cit.

[5] Borg and Dienes, *Solid State Diffusion*, Appendix C, Academic Press, 1988. loc. cit.

In this equation, the ν_j are the normal frequencies for vibrations about point A and the ν'_j for vibrations about the saddle point P. The effective frequency is the ratio of the product of the N normal frequencies of the system at the starting point of the transition to the $N-1$ normal frequencies of the system constrained in the saddle point configuration. Another way of viewing this result is to recall that the total number of degrees of freedom must be conserved. Thus, the translational motion of the diffusing atoms sacrifices a single vibrational degree of freedom.

One can also express the results in terms of the entropy of activation, $\Delta S^{\ddagger}$, which may be written as

$$\nu^* = \bar{\nu} e^{\Delta S^{\ddagger}/k},$$

and $\Delta S^{\ddagger}$ is given by

$$\Delta S^{\ddagger} = k \ln \left(\prod_{j=1}^{N-1} \nu_j^0 \Big/ \prod_{j=1}^{N-1} \nu'_j \right), \tag{13.34}$$

where ν_j^0 is the frequency of the ith normal mode for the system constrained to lie on the surface S_0 of Fig. 13.8.

This has been presented only in outline derivation in order to show that the exponential temperature dependence of diffusive motion is governed by a conventional activation energy, and this conclusion rests on a firm theoretical foundation. The actual calculation of migration activation energies is analogous to the calculation of the formation energies discussed in Chapter 12, but is even more difficult because it must deal with the more complex configuration at a saddle point. Some representative theoretical values are listed in Table 13.1 with the formation enthalpies repeated from Table 12.2. The corresponding overall diffusion enthalpies are easily obtained

Table 13.1. Theoretical vacancy formation and migration enthalpies, in eV

	ΔH_v^{f}	ΔH_v^{m}	$\Delta H_{\mathrm{a}}^{\mathrm{m}}$	$\Delta H_{\mathrm{c}}^{\mathrm{m}}$
Copper	1.0	1.0		
Alkali metals	0.3–0.5	Near zero		
Silicon	2.3	1.1		
Germanium	2.8	0.4		
Sodium chloride	2.1		0.7	0.4
Potassium Chloride	2.2		0.7	0.6
Rubidium chloride	2.2		0.7	0.6

ΔH_v^{f} = vacancy formation enthalpy.
ΔH_v^{m} = vacancy migration enthalpy.
$\Delta H_{\mathrm{a}}^{\mathrm{m}}$ = anion migration enthalpy.
$\Delta H_{\mathrm{c}}^{\mathrm{m}}$ = cation migration enthalpy.

as the sum of the formation and migration enthalpies [see (13.28)]. The symbol Q, which stands for the total activation energy, is frequently encountered in the older diffusion literature and is occasionally used in this chapter.

4. The Correlation Coefficient

The equations given so far are correct for interstitial jumping but are not quite right for the vacancy or interstitialcy mechanisms. The reason is that an atom that diffuses via vacancies does not execute a strictly random walk. Suppose we follow a particular distinguishable atom, such as an impurity or a tracer isotope, diffusing by means of a vacancy mechanism. Immediately following the exchange of the position of a vacancy with an atom, the two entities perforce remain nearest neighbors. At this instant the probability of the diffusing atom having a nearest-neighbor vacancy is unity rather than the much smaller random value given by zX_v. Consequently, the probability of this atom returning to its previous position is much greater than the random one.

Thus, in diffusing a given distance from its origin, the average moving atom "wastes" many of its jumps by simply going back and forth. Hence, more jumps are needed than would be if successive jumps were not highly correlated, as is the case for ideal interstitial diffusion. Consequently, (13.28) needs to be corrected. This is done by multiplying (13.28) by the correlation coefficient, which is a numerical correction factor, usually denoted by f. It is a purely geometrical factor that depends on the crystal structure and mechanism.

One may think that the correlation coefficient is nothing but a numerical nuisance factor, but this is not the case, since it does provide additional information about the diffusion process. To be more specific, it provides a means of investigating the individual jump frequencies of the host and tracer atoms. In essence, therefore, one takes advantage of the degree of failure of (13.28) to provide insights into the actual atomistics of the diffusion process. One can determine the correlation coefficient experimentally by using different isotopes and also investigate how the activation energy is partitioned between the host lattice and the jumping atom. Also, if the host and tracer are chemically different, one then can obtain information concerning the impurity–vacancy binding energy. It should be reemphasized that the diffusion of the vacancies themselves is uncorrelated in pure materials for after each jump there is again the same a priori probability for the succeeding jump that can be to any particular one of the z nearest neighbors, and hence random.

We return to (13.14) to begin our derivation of the correlation factor f.

$$\overline{R_n^2} = nr^2\left(1 + \frac{2}{n}\overline{\sum_{j=1}^{n-1}\sum_{i=1}^{n-j}\cos\theta_{i,i+j}}\right). \tag{13.14}$$

We recall that this equation was simplified for uncorrelated jumps by the vanishing of the double sum. This is no longer true for correlated jumps, since the values $\cos\theta_{i,i+j} = 0$ appears more often than random in the sum. Thus, a finite negative value for the double sum diminishes R_n at a fixed n by a certain amount. The correlation coefficient is defined by

$$f = \lim_{n\to\infty} \frac{\overline{R_n^2}\,(\text{tracer})}{\overline{R_n^2}\,(\text{vacancy})} = \lim_{n\to\infty}\left(1 + \frac{2}{n}\sum_{j=1}^{n-1}\cdots\sum_{i=1}^{n-j}\overline{\cos\theta_{i,i+j}}\right). \tag{13.35}$$

Thus, the calculation of f is, in reality, nothing more than the calculation of the average value of $\cos\theta_{i,i+j}$. This is a fairly elaborate procedure, which, after considerable algebra[6] leads to

$$\begin{aligned} f &= 1 + 2\,\overline{\cos\theta_1} + 2(\overline{\cos\theta_1})^2 + \cdots + 2(\overline{\cos\theta_1})^{(n-1)} \\ &= 1 + 2\sum_{j=1}^{n-1}(\overline{\cos\theta_1})^j, \end{aligned} \tag{13.36}$$

which, since $\cos\theta < 1$, converges to

$$f = 2\,\frac{1 - (\overline{\cos\theta_1})^{n-1}}{1 - \overline{\cos\theta_1}} - 1, \tag{13.37}$$

where

$$\lim_{n\to\infty}(\overline{\cos\theta_1})^{n-1} = 0,$$

and our final expression for the correlation coefficient is

$$f = \frac{1 + \overline{\cos\theta_1}}{1 - \overline{\cos\theta_1}} \tag{13.38}$$

The diffusion equation based strictly upon random walk must, therefore, be corrected by multiplying (13.28) by f, giving [with $\gamma = 1/2$ in (13.28)].

$$D = \frac{a^2}{2}\, f\omega_0 \exp\frac{\Delta S^\ddagger + \Delta S_v^\circ}{R}\exp\left[-\frac{\Delta H_v^\circ + \Delta H^\ddagger}{RT}\right]. \tag{13.39}$$

Correlation coefficients have been calculated for a variety of crystal structures and jump types. A compilation is given in Table 13.2.

---

[6] Borg and Dienes, chap. 3 loc. cit.

Table 13.2. Correlation factors for self-diffusion.[a]

Crystal Structure	Correlation Factor
Vacancy mechanism, two-dimensional lattices	
Honeycomb lattice	0.5
Square lattice	0.46694
Hexagonal lattice	0.56006
Vacancy mechanism, three-dimensional crystal structures	
Diamond	0.5
Simple cubic	0.65311
Body-centered cubic	0.72722
Face-centered cubic	0.78146
Hexagonal close-packed (with all jump frequencies equal)	0.78121 normal to c-axis 0.78146 parallel to c-axis
Interstitialcy mechanism (θ = angle between the displacement vectors of the atoms participating in the jump)	
NaCl, collinear jumps ($\theta = 0$)	0.666
NaCl, noncollinear jumps with $\cos\theta = 1/3$	0.9697
NaCl, noncollinear jumps with $\cos\theta = -1/3$	0.9643
Ca in CaF_2, collinear jumps ($\theta = 0$)	0.80
Ca in CaF_2, noncollinear jumps with $\theta = 90°$	1.0

[a] After J. R. Manning, *Diffusion Kinetics for Atoms in Crystals*, Chap. 3, Van Nostrand, 1968.

5. Representative Experimental Results

A. The Vacancy Mechanism

Various diffusion mechanisms were disussed in 13.2 and the importance of the vacancy mechanism emphasized. It is fitting to start our discussion of some representative experiments with the earliest direct experiments showing diffusion by means of vacancies, namely those by Smigelskas and Kirkendall.[7] In this experiment, as illustrated in Fig. 13.9, a brass specimen was coated with copper and fine inert Mo wires positioned at the interface. The specimen was heated to a temperature where diffusion is appreciable and, hence, measurable. The diffusion anneal was

[7] A. D. Smigelskas and E. O. Kirkendall, *Trans. AIME*, **171**, 130 (1947).

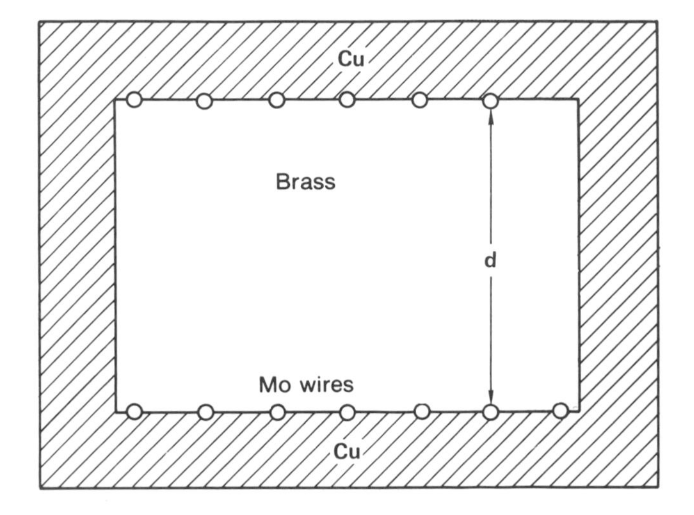

Figure 13.9.

The marker movement experiment of Smigelskas and Kirkendall. The distance between the opposing sets of Mo wires was found to decrease with the time of the diffusion anneal, thus demonstrating that Zn diffuses faster than Cu.

interrupted periodically in order to measure the separation d between markers and it was found to decrease with time. This shrinkage, usually called the Kirkendall effect, showed that zinc diffuses faster than copper, Zn diffusing outward from and Cu inward into the brass. As pointed out in Section 2, direct interchange of atoms requires that both species diffuse at the same rate. Thus, the Kirkendall effect eliminated this mechanism and gave strong experimental support to the vacancy mechanism. Clearly, in this experiment, new lattice sites were created exterior to the markers and vacancies interior to them. Thus, there is a net inward vacancy flow.

There exists by now an overwhelming number of independent methods, such as quenching and dilatation experiments, as discussed in Chapter 12, from which the validity of the vacancy mechanism can be, and has been, inferred. Thus, it is well established for many materials.

As already mentioned in Section 1, a widely used method for determining diffusion coefficients employs radioactive isotopes, giving results that are the easiest to interpret. The limitations of this technique arise from the lack of suitable isotopes, but it can be extended by selective activation analysis, or activation analysis combined with mass spectrometry. These techniques yield the concentration-independent tracer diffusion coefficient, since they are present at very low concentrations. When the radioactive tracer is of the same element as the matrix, the process is termed "self-diffusion", for example ^{22}Na in Na, ^{57}Fe in Fe, etc. Self-diffusion strictly occurs only in monoelemental substances, but the term is conventionally applied to the tracer diffusion of one of the components of an alloy or compound. In addition to tracer diffusion, there are several specialized techniques by which diffusion coefficients can be determined.[8]

Diffusion experiments carried out over a wide temperature range yield the diffusion coefficient as a function of temperature. These data usually obey the relation, often referred to as the Arrhenius relation,

$$D = D_0 \exp(-\Delta H^{\ddagger}/RT), \tag{13.39}$$

which is equivalent to (13.28) with all the preexponential factors combined in D_0.

B. Interstitial Diffusion

A classical example of a simple diffusion process over a very wide temperature range is the diffusion of carbon in α-iron. The diffusion coefficient for this process, a simple interstitial jumping of the carbon atom from site to site, has been measured by a variety of techniques over an unusually wide range. The data are shown in Fig. 13.10.

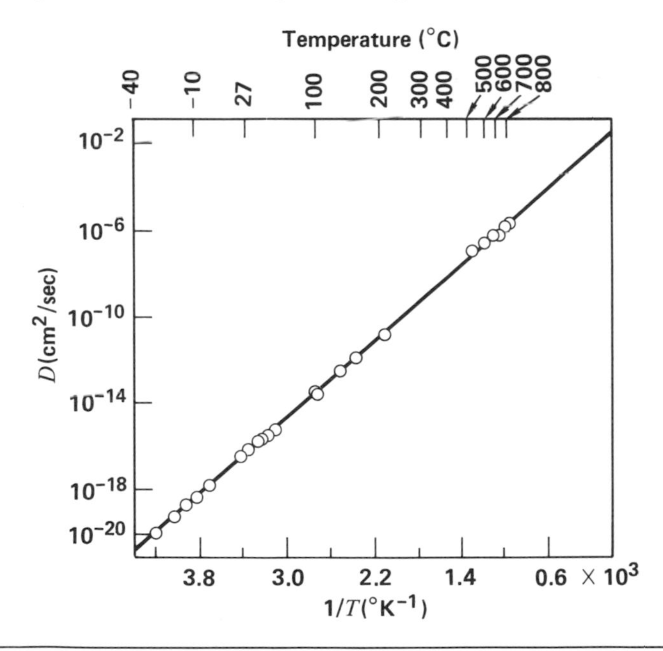

Figure 13.10.

Diffusion coefficient of carbon in bcc iron as a function of temperature: log D vs. $1/T$ plot.

[8] See, for example, Borg and Dienes, Chap. 11, loc. cit.

They are well represented by

$$D = 0.02e^{-20.1/RT}\ \text{cm}^2\ \text{sec}^{-1} \tag{13.40}$$

giving $D_0 = 0.02$ cm^2 sec^{-1} and $\Delta H^{\ddagger} = 20.1$ kcal. In contrast, iron self-diffusion is characterized by a value of 2 for D_0 and $\Delta H^{\ddagger} = 57.3$ kcal/mole, clearly much larger than that for carbon. One might compare the carbon to iron diffusion at 1000 K. The ratio is 1.3×10^6, a very large value. The validity of the theoretical exponential dependence of the diffusion coefficient on the temperature over a very wide range is to be noted.

C. Experimental Values of the Diffusion Parameters

Self-diffusion in metals is probably the next simplest atomic transport process. Almost all metals have face-centered cubic, body-centered cubic, and hexagonal close-packed crystal structures. Each atom bears the same geometric relation to its neighbors, and it is, therefore, sufficient to consider one type of lattice site. It is also true of metals

Table 13.3. Self-diffusion in elemental metals.

Element	Crystal Structure	T_m (K)	$\Delta H^{\ddagger}$ (kcal)	D_0 (cm^2/sec)	$D(T_m/2)$	$D(T_m)$	$D(300\ \text{K})$
K	bcc	337	9.36	0.16	1.1×10^{-13}	1.4×10^{-7}	2.4×10^{-8}
Na	bcc	371	11.5	0.72	2.0×10^{-14}	1.2×10^{-7}	3.0×10^{-9}
Cd[a]	hcp	594	18.2	0.12	4.8×10^{-15}	2.4×10^{-8}	6.5×10^{-15}
Cd[b]	hcp	594	19.1	0.18	1.6×10^{-15}	1.7×10^{-8}	2.2×10^{-15}
Al	fcc	933	34.0	1.71	2.0×10^{-16}	1.8×10^{-8}	2.8×10^{-25}
Cu	fcc	1110	50.4	0.1	1.4×10^{-21}	1.2×10^{-11}	1.8×10^{-38}
Ge	dc	1210	68.5	7.8	1.4×10^{-24}	3.3×10^{-12}	9.1×10^{-50}
Ag	fcc	1234	45.2	0.67	6.4×10^{-17}	6.5×10^{-9}	7.6×10^{-34}
Au	fcc	1336	42.1	0.09	1.5×10^{-15}	1.2×10^{-8}	1.8×10^{-32}
Si	dc	1683	115.8	1460	1.2×10^{-27}	1.3×10^{-12}	5.7×10^{-82}
Ni	fcc	1726	67.2	1.23	1.1×10^{-17}	3.8×10^{-9}	1.3×10^{-49}
Fe[c]	fcc		67.9	0.49	(1.9×10^{-17})	(3.0×10^{-9})	
Fe[d]	bcc	1809	57.4	1.39	1.8×10^{-14}	1.6×10^{-7}	2.0×10^{-42}
Pt	fcc	1825	68.1	0.33	1.6×10^{-17}	2.3×10^{-9}	7.6×10^{-51}
Mo	bcc	2883	104.5	0.13	1.8×10^{-17}	1.5×10^{-9}	8.7×10^{-78}

[a] $\parallel$ to hcp c-axis.
[b] $\perp$ to hcp c-axis.
[c] fcc Fe.
[d] bcc Fe.
From *Smithells Metal Reference Book*, 6th ed., 1983. Sect. 13.10, (E. A. Brandes, ed.), Butterworth.

Table 13.4. Experimental vacancy formation, migration, and diffusion enthalpies, in eV.

	ΔH_v^f	ΔH_v^m	ΔH_a^m	ΔH_c^m	ΔH_d
Cu	1.0	1.0			2.0
Ag	1.0	0.9			1.9
Au	0.9	0.9			1.8
Al	0.7	0.7			1.4
Pt	1.5	1.4			2.9
Pb	0.5	0.6			1.1
W	3.3	2.5			5.8
Na	0.4	Near zero			0.4
Fe	1.5	1.1			2.6
Ge	2.6	0.4			3.0
NaCl	2.5		0.9	0.6	1.9
KCl	2.5		0.9	0.7	2.0
KBr	2.4		0.9	0.6	1.8
KI	2.2		1.3	0.6	1.4
RbCl	2.0		1.5	0.5	1.3

Experimental migration energies for interstitials, in eV

	ΔH_i^m
He in KI	0.3
H in Ni	0.4
C in Fe	0.9

The entries are rounded off to the nearest 0.1 eV, the approximate estimated accuracy. Data from a variety of sources.

ΔH_v^f = vacancy formation enthalpy

ΔH_v^m = vacancy formation enthalpy

ΔH_a^m = anion migration enthalpy

ΔH_c^m = cation migration enthalpy

ΔH_d = diffusion enthalpy

ΔH_i^m = interstitial migration enthalpy

that impurities at low concentrations have a relatively small effect. Diffusion, because of its importance to the metallurgical industry, has been studied in metals more than in any other class of solids. Table 13.3 lists the values of $\Delta H^\ddagger$, D_0, and $D(T_m)$, the diffusion coefficient at the melting point, for a representative set of elemental metals.

We can now also separate $\Delta H^\ddagger$ into its two components, the vacancy formation and migration enthalpies, ΔH_v° and ΔH_v^m, since in many cases ΔH_v° is known from other experiments. Using the ΔH_v° data of Table 12.3, we can derive the migration enthalpies. A collection of experimental date is given in Table 13.4. These values are also in good agreement with the theoretically available values listed in

Table 13.1, giving additional support to the vacancy mechanism. It is also noteworthy that for most metals, the formation and migration enthalpies are of about the same magnitude, as is also implied by theory. Further, the interstitial migration energies are generally rather low compared with vacancy diffusion as would be expected.

D. Empirical Correlations

It is quite clear from Table 13.3 that D varies inversely as the melting temperature. We have already noted that D for vacancy diffusion is a function of the vacancy concentration and the vibrational frequencies. Both quantities reflect the bond strength, as does the melting temperature, although there is little similarity between melting and diffusion. Nevertheless, there is an empirical correlation between D and T_m, perhaps far better than should be expected, as shown in Fig. 13.11. This correlation is described by

$$Q(\text{kcal}) = 34T_m. \tag{13.41}$$

As might be expected, diamond cubic Ge, whose bonds are mostly covalent, does not fall on the same line as the metals. Clearly, the more refractory the material, the higher is its Q value. A slight improvement in the fit is obtained by plotting the fcc and bcc structures separately, but refinements of this interesting empirical correlation are no longer useful, since most elemental diffusion constants have been measured

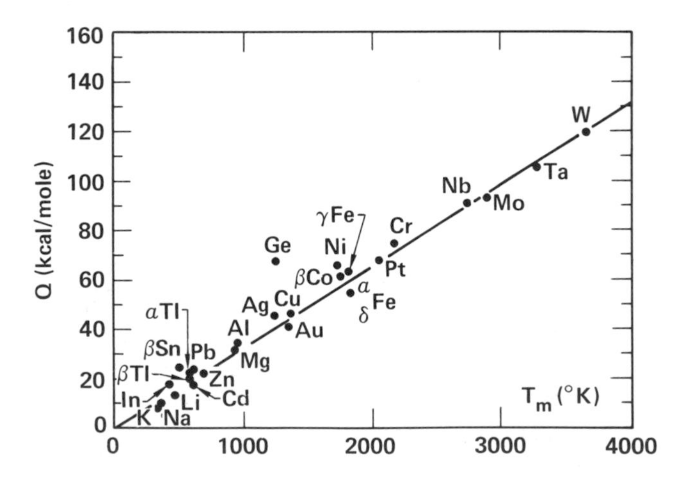

Figure 13.11.

Self-diffusion activation energy as a function of the melting temperature.

by now. It can also be seen from the data in Table 13.3 that values of D_0 are generally limited to a range of about 0.1 to 2 cm^2 sec^{-1} for vacancy diffusion in elemental metals, as is also true of their alloys. However, D_0 varies over a much wider range when other mechanisms and nonmetallic materials are included.

E. Intrinsic versus Extrinsic Defects

Ionic crystals cannot be successfully quenched because of their low thermal conductivities, and a different approach is needed to separate the total activation energy into its formation and migration components. In this regard, one takes advantage of doping experiments. Introduction of an aliovalent impurity creates an extrinsic region, wherein the vacancy concentration is controlled by the impurity as a result of the electrical neutrality requirement (see Fig. 12.7). Diffusion in the extrinsic region is independent of the defect formation energy, since the defects are at constant concentration and are independent of the temperature. The diffusion is therefore controlled solely by the migration energy. The total measured activation energy for diffusion, $\Delta H^{\ddagger}$ has the two components, the enthalpy of motion, $\Delta H_{\mathrm{m}}^{\ddagger}$ and the standard enthalpy of vacancy formation, ΔH_v°. Separating the vacancy concentration into its two components gives

$$D = D_0 e^{-\Delta H_{\mathrm{m}}^{\ddagger}/RT}(X_v^{\mathrm{t}} + X_v^{\mathrm{i}}), \tag{13.42}$$

where X_v^{t} stands for the concentration of the thermally generated vacancies, and X_v^{i} for the temperature-independent fraction controlled by the doping impurities, which must conform to the law of mass action. By lowering the temperature and/or by increasing the impurity concentration, we can render X_v^{t} negligible compared with X_v^{i}. Thus, in the extrinsic region, the temperture dependence of ln D yields $\Delta H_{\mathrm{m}}^{\ddagger}$, since X_v^{i} is independent of the temperature. Thus,

$$\frac{\partial \ln D}{\partial T^{-1}} = -\frac{\Delta H_{\mathrm{m}}^{\ddagger}}{R}. \tag{13.43}$$

The overall activation energy is obtained at high temperatures and relatively low impurity concentrations, since in the intrinsic regime the thermally generated vacancy concentration is a strong function of the temperature. Thus, the difference

$$\Delta H^{\ddagger} - \Delta H_{\mathrm{m}}^{\ddagger} = \Delta H_v^{\circ}$$

gives the vacancy formation enthalpy.

A classical experiment is shown in Fig. 13.12, introducing cadmium into NaCl as the dopant. The intrinsic and extrinsic regimes are very clearly exhibited by the data.

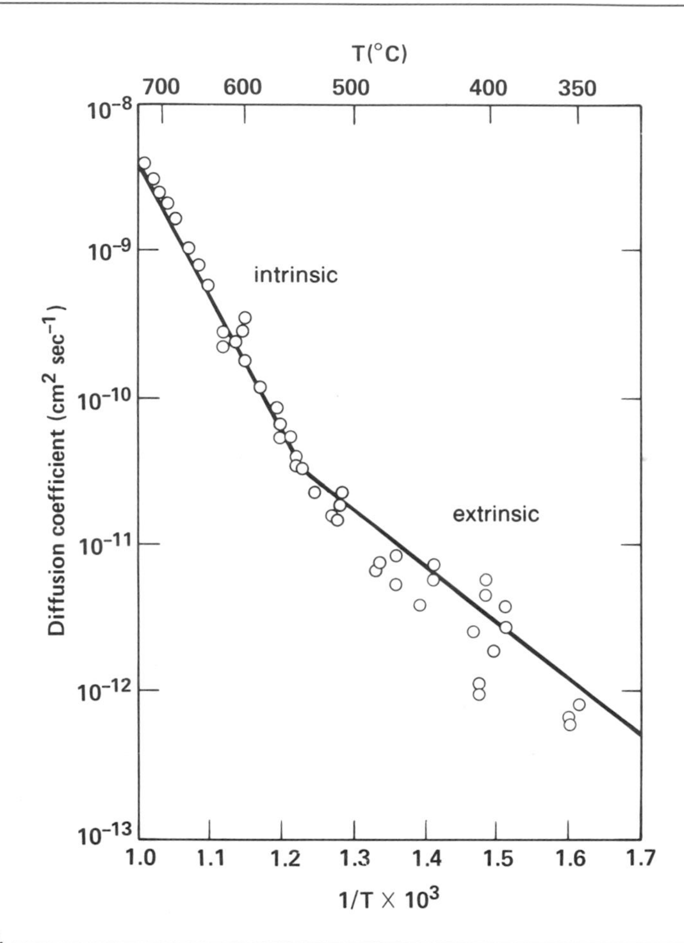

Figure 13.12.

Self-diffusion coefficient of sodium in crystals of NaCl doped with $CdCl_2$. [From D. E. Mapother, H. N. Crooks, and R. J. Maurer, *J. Chem. Phys.* **18**, 1231 (1950).]

From the slope of the extrinsic line we obtain a value of 0.8 eV for the cation migration energy, ΔH_c^m. From the overall Q value in the intrinsic region, $Q = 1.9$ eV, we derive a value of 2.2 eV for the Schottky formation enthalpy, ΔH_s, where $Q = (1/2)\Delta H_s + \Delta H_c^m$. [See (12.19) and (13.28).]

F. Oxygen Diffusion

We have discussed so far the very basic properties of diffusional processes with cations generally serving as examples. Let us look next very briefly at a rather practical

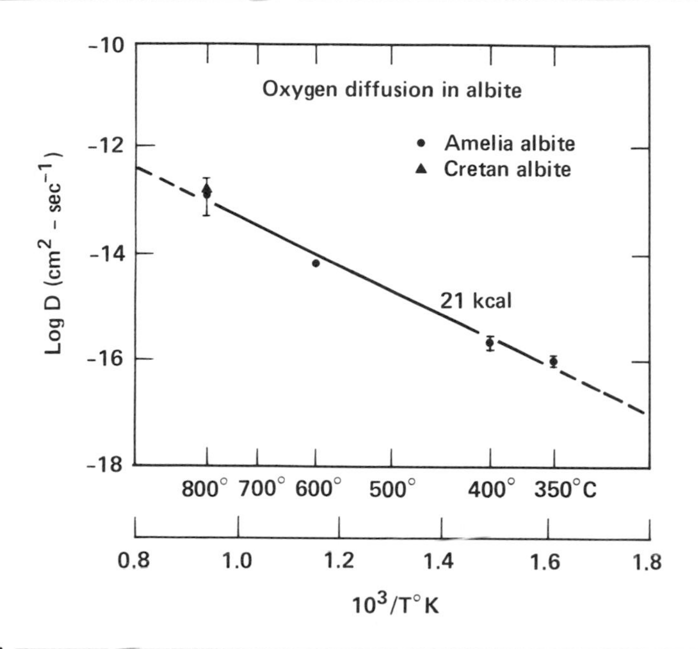

Figure 13.13.

An Arrhenius plot showing the temperature dependence of oxygen diffusion in albite. The Amelia and Cretan albites in this work were isotopically exchanged, and the $^{18}O/^{16}O$ gradient determined with an ion microprobe. Albite has the nominal formula $NaAlSi_3O_8$. [After Gilletti *et al.*, *Geochim. et Cosmochim. Acta*, **42**, 45 (1978) as cited in Ref. 2, p. 264.]

problem, namely the diffusion of oxygen. This is particularly important since oxygen is the most ubiquitous element in the earth's crust as well as in ceramic materials. Oxygen requires the use of special techniques since there is no suitable radioactive isotope. We shall illustrate here only the results using one method, the measurement of oxygen profiles by secondary ion spectroscopy, (SIMS), which uses energetic beams of heavy ions to sputter material from the specimen, which is then analyzed by mass spectrometry. The diffusion coefficient as a function of temperature in albite was measured by first diffusing ^{18}O into the mineral from a hydrothermal environment. The diffusion profile was then determined by SIMS. This technique can be used for total penetrations as small as 5×10^{-4} cm. The results for albite are shown in Fig. 13.13.

G. Diffusion as a Function of Atomic Order

The influence and importance of chemical bonding upon diffusion is very neatly illustrated by diffusion in ordering alloys, for example in CuZn. This alloy develops

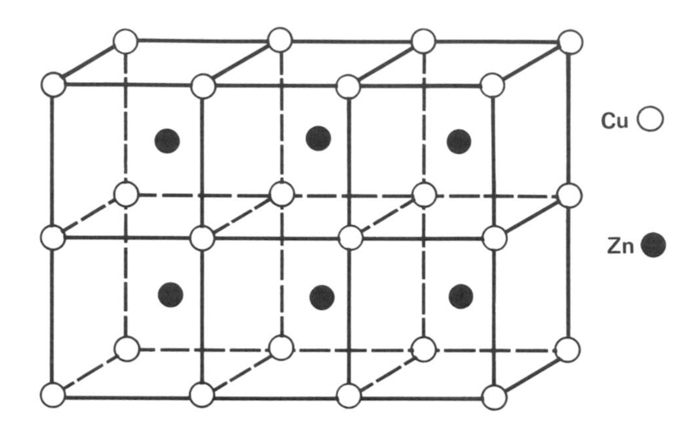

Figure 13.14.

Ordered structure of β-brass (CuZn).

an ordered structure, identified as β-brass, below a critical temperature T_c with the structure shown in Fig. 13.14. In the ordered CsCl configuration each copper is surrounded by eight zinc atoms, and vice versa. All near-neighbor bonds, therefore are Cu—Zn. Perforce, these bonds are stronger than the Cu—Cu and Zn—Zn bonds or we would not have the ordering reaction. Above T_c the atomic arrangement is random (with some residual short-range order near T_c). The foregoing bonding scheme suggests that it is energetically more expensive to form and move the vacancies in the ordered structure. Indeed, statistical theory yields an overall diffusion activation energy Q that increases with the second power of the order parameter s. This relation reads[9]

$$D = D_0 e^{-(1/RT)[Q(0) + \alpha s^2]}, \tag{13.44}$$

where the order parameter s varies between 0 and 1, being 1 at perfect order. The tracer diffusion coefficients for both Cu and Zn have been measured in this alloy both above and below T_c. These conclusions are well verified experimentally, as shown in Fig. 13.15, where the Cu and Zn self-diffusion coefficients are plotted as a function of $1/T$. Clearly, diffusion is slower at temperatures below T_c with a slope, and hence Q, larger than at temperatures above T_c. The transition is somewhat rounded because of the short-range order above T_c. The solid lines show the results of applying (13.44), where, of course, the order parameter s is itself a function of temperature.

[9] See, for example, L. A. Girifalco, *Statistical Physics of Materials*, Chap. 9, Wiley, 1973.

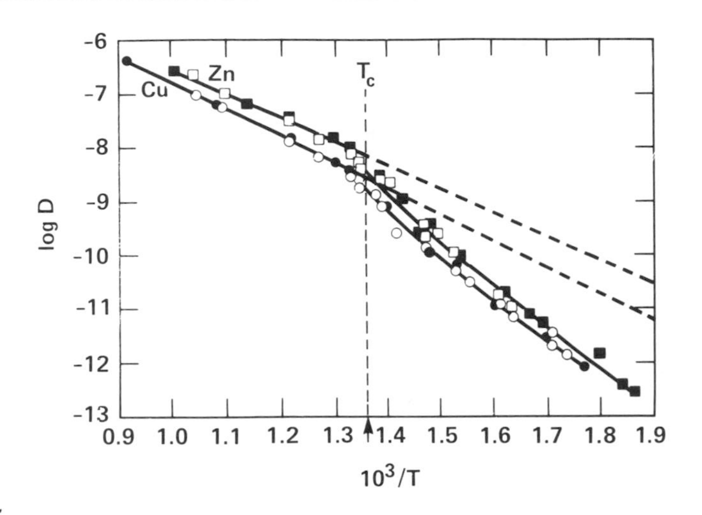

Figure 13.15.

Self-diffusion coefficients in β-brass. (O Zn, □ Cu data from A. B. Kuper, D. Lazarus, J. R. Manning, and C. T. Tomizuka, *Phys. Rev.*, **104**, 1536 (1956). ● Zn, ■ Cu data from P. Camagni, *Proc. 2nd Int. Conf. on Atomic Energy*, Vol. 20, Geneva, 1958, p. 11365.)

H. Surface Diffusion

Concluding this section on diffusion is a brief discussion of an unusual example of surface diffusion, namely diffusion in amorphous silicon. This topic is of current commerical interest but is included here primarily because it is an excellent pedagogical example of diffusion control by the chemistry of the internal surfaces. We examine and compare the diffusion of boron and hydrogen, both in amorphous silicon, A—Si, and in the crystalline state, C—Si. For the experimental procedures the reader is referred to the literature.[10] The diffusion data for boron in silicon are given in Table 13.5. The enormous difference in the values of D at 330°C for amorphous and crystalline silicon, which comprises some 15 orders of magnitude, implies a considerable difference in the diffusion mechanism in the two cases. This extreme difference has been attributed to the high density of subgrain boundaries in A—Si, which provide low-activation-energy, high-diffusivity pathways. This is the conventional, and often valid, view of diffusion in disordered regions, such as grain boundaries, where there are enhanced concentrations of vacancies. Diffusion in C—Si is presumed

[10] See, for example, Borg and Dienes, Chap. 9, loc. cit.

Table 13.5. Diffusion of boron in silicon.

Matrix	Temperature (°C)	D (cm^2 sec^{-1})
A—Si	200	6×10^{-18}
A—Si	330	2×10^{-16}
A—Si	400	6×10^{-14}
C—Si[a]	330	7×10^{-31}

[a] Extrapolated to 330°C from high-temperature data.

Table 13.6. The diffusion of hydrogen and deuterium in silicon.

System	D_0 (cm^2 sec^{-1})	Activation Energy (eV)
^{1}H in A—Si	1×10^{-3}	1.4
^{2}H in A—Si	1×10^{-3}	1.5
^{2}H in C—Si	9×10^{-3}	0.5

to be via the vacancy mechanism for which the activation energy must include the additional expenditure for vacancy formation.

The diffusion of hydrogen in silicon presents a completely different picture as shown by the data in Table 13.6. Clearly, the diffusion of hydrogen, with its low Q, is much faster in the silicon single crystal than in A—Si. It is known that A—Si contains an appreciable fraction of dangling, or unsaturated Si bonds, which serve as traps for the diffusing hydrogen. These dangling bonds are, of course, absent from a more or less perfect single crystal. Boron, to the contrary, does not appear to be attracted to such sites.

Thus, our two examples, boron and hydrogen, gave essentially opposite results for diffusion in amorphous and crystalline silicon and serve to emphasize the importance of the chemistry of the internal surfaces in an amorphous material.

6. Defect Reactions

The elementary point defects whose diffusion we discussed in this chapter, may undergo a variety of reactions, acting, for all practical purposes as chemical entities. The mobilities of the defects are, of course, involved in these processes, such as the diffusion of the defects to sinks where they are annihilated, or to traps where they become immobilized. They may also react with each other and with impurities to form various complexes. In a wider sense, these elementary reactions are often important in various solid state processes. In this section we shall discuss some representative examples.

A. First-Order Reactions

Let us consider the following situation. An excess of vacancies had been produced in a sample by, for example, quenching to low temperature. At some higher temperature the vacancies become mobile enough to reduce this concentration to the equilibrium value by migrating to external or internal sinks where they are annihilated. This process in its simplest form is characterized by a first-order reaction, ie., one in which the excess concentration decreases in proportion to its concentration. Thus,

$$\frac{dn}{dt} = -kn, \tag{13.45}$$

where k is a rate constant independent of n. The integral of this equation at constant temperature is

$$\frac{n}{n_0} = e^{-kt}. \tag{13.46}$$

These equations are approximate since in general there are transient exponentials at early times. If the initial defect distribution is uniformly random, the transients are less important than if it is highly localized, and the single exponential form is reached sooner. In practice, the simple single exponential form is often quite satisfactory, but its limitations must be kept in mind.

Another example of a first-order reaction is the annihilation of close pairs of vacancies and interstitials. These, among other defects, are produced by irradiation with the vacancy and the interstitial created in close proximity. The annihilation occurs by recombination with the more mobile species, say the interstitial, jumping into the vacancy. The energy barrier to recombination is smaller than that for diffusion, and hence most of the close pairs annihilate rather than dissociate. The highly correlated recombination of close pairs is governed by simple first-order kinetics, since the rate is a function only of the remaining concentration of the close pairs that are considered to be a single stable molecular species. This process is really "reaction"-controlled rather than diffusion-controlled, since long-range migration is obviously not involved.

An experimental example is shown in Fig. 13.16, where vacancies quenched into gold from elevated temperatures are allowed to migrate to sinks at lower temperatures. This isothermal annealing reaction was followed by measuring the decay in

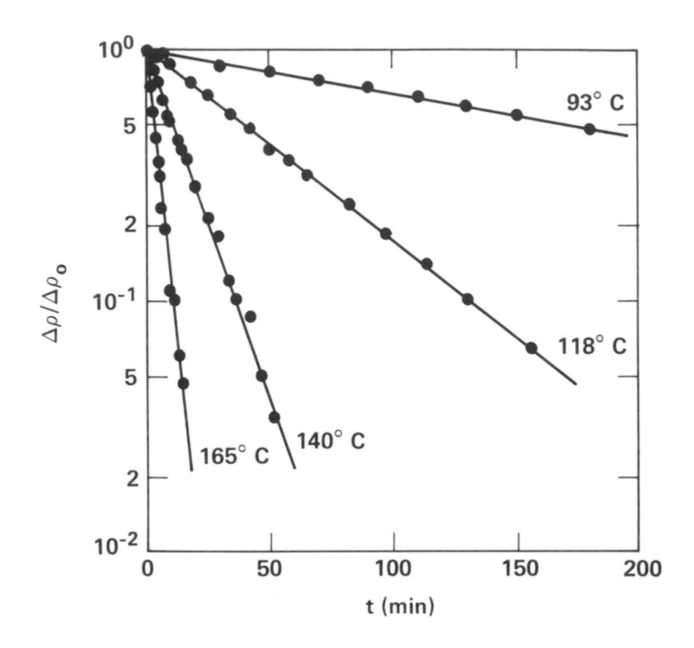

Figure 13.16.

Isothermal annealing curves obtained on 0.04 mm pure gold wires after quenching from 700°C. The ordinate is the residual resistivity. [After F. Cattaneo and E. Germagnoli, *Phys. Rev.*, **124**, 414 (1961).]

the quenched in residual resistivity, a quantity proportional to the vacancy concentration at low concentrations, as a function of time. The data in Fig. 13.16 clearly exhibit a first-order reaction that obeys (13.46) quite accurately.

From the slopes of these curves, the rate constant k of (13.46) is obtained as a function of temperature, and it, being proportional to the diffusion constant, is therefore expected to vary exponentially with $1/T$. This is how the activation energy for vacancy motion in gold was determined. The value thus obtained was 0.84 eV in very good agreement with theoretical estimates and other measurements of the monovacancy migration energy in gold (see Table 13.4). It should be noted that this is a completely independent measurement of the migration energy since it does not depend either on the formation or the overall diffusion activation energy.

B. Second-Order Reactions

The formation of divacancies is a typical second-order process. Because of its stability and expected high mobility, the divacancy can play an important role in defect reactions. Consider the following scheme for the annealing of excess vacancies involving divacancy formation, a likely event at high monovacancy concentrations.

$$
\begin{aligned}
v + v &\underset{k_{-1}}{\overset{k_1}{\rightleftharpoons}} v_2, \\
v &\xrightarrow{k_2} \text{sinks}, \\
v_2 &\xrightarrow{k_3} \text{sinks},
\end{aligned}
\tag{13.47}
$$

where v and v_2 are the concentrations (atomic fractions) of single vacancies and divacancies, and the k's are the respective rate constants.

The differential equations for these reactions are

$$
\begin{aligned}
\frac{dv}{dt} &= k_{-1}v_2 - k_1 v^2 - k_2 v, \\
\frac{dv_2}{dt} &= \frac{1}{2} k_1 v^2 - \frac{1}{2} \cdot k_{-1} v_2 - k_3 v_2.
\end{aligned}
\tag{13.48}
$$

The total vacancy concentration, n, is simply

$$n = v + 2v_2,$$

and, thus

$$\frac{dn}{dt} = \frac{dv}{dt} + 2\frac{dv_2}{dt} = -k_2 v - 2k_3 v_2. \tag{13.49}$$

This equation cannot be solved in closed form. However, local equilibrium is often quickly established between v and v_2, leading to the steady-state approximation given by

$$\frac{dv_2}{dt} = 0; \qquad v_2 = \frac{k_1 v^2}{k_{-1} + 2k_3}. \tag{13.50}$$

Substitution of v_2 into (13.49) gives

$$\frac{dn}{dt} = \frac{dv}{dt} = -k_2 v - \frac{2k_3 k_1 v^2}{k_{-1} + 2k_3}. \tag{13.51}$$

It has been established theoretically and experimentally that divacancies are far more mobile than monovacancies. Accordingly (13.51) can be simplified by letting $k_3 \gg k_2, k_1$. Under these conditions v_2 is very small since it anneals rapidly.

Consequently, $n = v$ approximately, and (13.51) becomes

$$\frac{dv}{dt} = -k_1 v^2, \tag{13.52}$$

a simple second-order decay reaction. The integral of (13.52) is, at a constant temperature,

$$1/v - 1/v_0 = k_1 t, \tag{13.53}$$

with the fractional decay v/v_0 given by

$$\frac{v}{v_0} = \frac{1}{1 + v_0 k_1 t}, \tag{13.54}$$

which is clearly a function of the initial concentration v_0, in contrast to the concentration-independent first-order decay of (13.46). The rate-determining constant

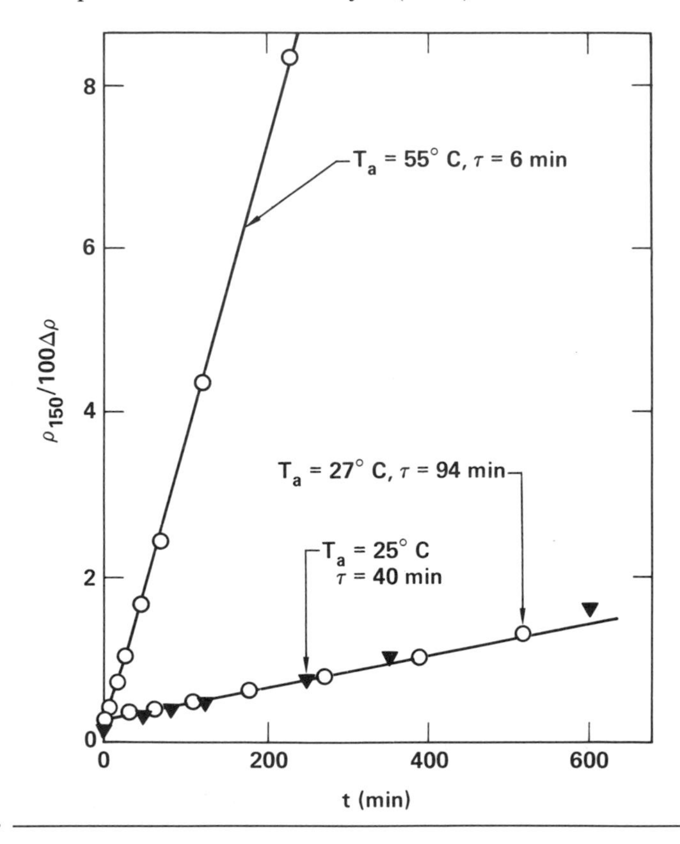

Figure 13.17.

The percentage change in resistivity during isothermal recovery is plotted reciprocally against the annealing time for gold quenched from 1000°C. Straight-line segments imply second-order kinetics. T_a is the annealing temperature, and τ is the half-time of the decay. [From W. Schule, A. Seeger, D. Schumacher, and K. King, *Phys. Stat. Solid*, **2**, 1005 (1962).]

in this reaction scheme is k_1, which represents the slow step. Thus, the slow step is the formation of the divacancy followed by its rapid migration to a sink.

Divacancy-facilitated annealing has been observed in gold when the quench was from a much higher temperature than in the experiment of Fig. 13.16, and hence resulted in a relatively high vacancy concentration. Such an experiment is illustrated in Fig. 13.17. The rapid quench from 1000°C left a sufficiently high concentration of vacancies, so that divacancy formation became important. The kinetics is clearly second-order since the data are plotted according to (13.53), and good agreement is observed with this equation. The controlling step is the divacancy formation, which, according to (13.53), is governed by rate constant k_1, namely the motion of monovacancies to form divacancies. The activation energy from Fig. 13.17 is consistent with this interpretation. Its value of 0.8 eV is in good agreement with other measurements of monovacancy migration energy in gold, as already cited.

The divacancy, because of its high mobility, often plays a role in self-diffusion itself. In metals it is believed to be an important mechanism at high temperatures, its role being manifested by high-temperature departure from the simple Arrhenius relation. We may illustrate this effect experimentally by the self-diffusion study of sodium as shown in Fig. 13.18. The extra diffusivity at high temperature—that is, over and above the straight Arrhenius line—may be attributed to the divacancy contribution to the diffusion.

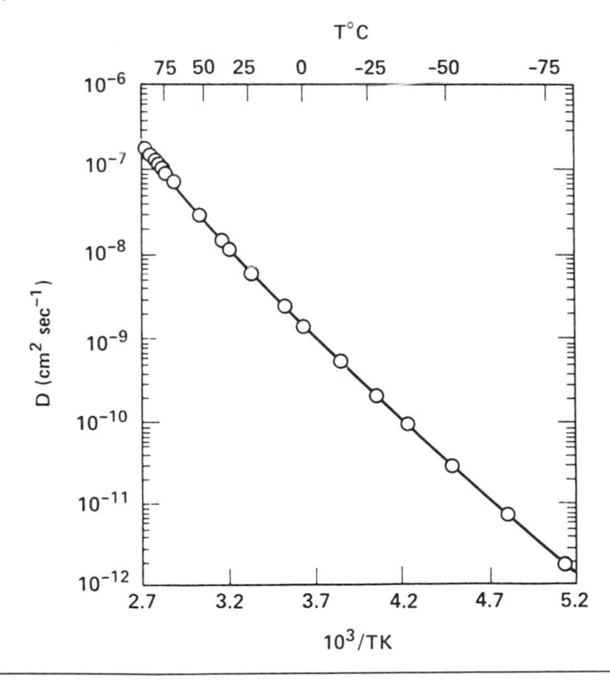

Figure 13.18.

The self-diffusion of Na in Na metal. [After J. N. Mundy, *Phys. Rev.* **B3**, 2431 (1971).]

C. Trapping Effects

A common interaction influencing defect reactions is that between defects and impurities, that is, trapping reactions when the resultant complexes are immobile. Theoretical treatments are available in the literature at various levels of sophistication. We shall treat here a simple trapping model,[11] often referred to as the "standard trapping model," which illustrates the essential features of the process. Let X and I represent any unbound defect and any uncomplexed impurity, respectively. The reactions to be considered are

$$\mathrm{X} + \mathrm{I} \underset{k_1}{\overset{k_{-1}}{\rightleftharpoons}} \mathrm{IX} \equiv \mathrm{C}, \tag{13.55}$$

$$\mathrm{X} \xrightarrow{k_2} 0,$$

where the first equation represents trapping with the formation of the complex, C, and the second describes diffusion out of the solid or disappearance at sinks.

The effective trap concentration is I, which includes the combinatory volume, usually the number of nearest-neighbor sites. All concentrations are given in mole or atomic fractions, and C, X, and I are all at very low concentrations. The complex C is assumed to be immobile. As in the divacancy case, one may assume quasi-equilibrium between C and X; that is, $k_1 \gg k_1, k_2$, which then implies

$$\frac{C}{X(I_0 - C)} = \frac{k_1}{k_{-1}} = K = K_0 e^{-\Delta H^\circ/T}, \tag{13.56}$$

where I_0 is the total trap concentration, ΔH° is the formation energy of the complex (in temperature units), and K_0 is taken as equal to 1 for convenience and because the vibrational entropy contribution is generally small. The only decay step is by means of rate constant, k_2, and therefore

$$\frac{dN}{dt} = -k_2 X, \tag{13.57}$$

where $N = C + X$.

Substituting for N and eliminating C via (13.56) yields

$$\left[\frac{1}{X} + \frac{KI_0}{X(1 + KX)^2}\right] dX = -k_2\, dt. \tag{13.58}$$

---

[11] A. C. Damask and G. J. Dienes, *Phys. Rev.*, **120**, 99 (1960).

This equation can be integrated, but it is more instructive to look at the limit where $C \ll I$, experimentally often the case. In this approximation

$$C/X = KI_0,$$

and substitution into (13.57) gives

$$(1 + KI_0)\frac{dX}{dt} = -k_2 X,$$

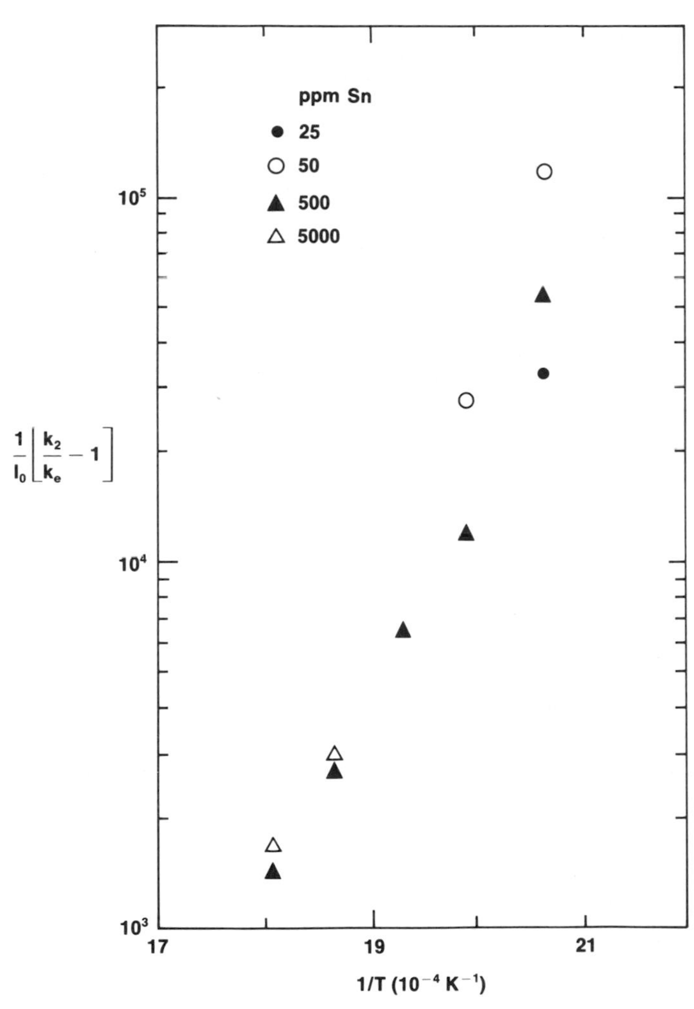

Figure 13.19.

Plot of the left-hand side of (13.61) vs. $1/T$ for the diffusion of Ag in Pb containing Sn as an impurity at the indicated concentrations illustrating simple trapping. [After C. K. Hu and H. B. Huntington, *Phys. Rev.* **B26**, 2783 (1982).]

and the effective rate constant for the decay is

$$k_e = k_2/(1 + KI_0). \tag{13.59}$$

In terms of the effective rate constant, one has then the decay of the total defect concentration as

$$N = X_0(1 + KI_0)e^{-k_e t} = N_0 e^{-k_e t}. \tag{13.60}$$

Rearrangement of (13.59) yields

$$K = \frac{1}{I_0}\left[\frac{k_2}{k_e} - 1\right] = e^{-\Delta H^\circ/T}. \tag{13.61}$$

The k_2/k_e ratio in (13.61) can be obtained from diffusion experiments, which have been performed on several metals by measuring diffusion in pure and doped systems. k_2 is obtained from the pure system ($I_0 = 0$), and k_e from one as a function of impurity concentration. A representative example is shown in Fig. 13.19 for the diffusion of silver in lead containing tin as an impurity, which acts as a trapping center. In this figure the logarithm of the left-hand side of (13.61) is plotted versus $1/T$. The linearity of the plot indicates that since (13.61) is obeyed, the simple "standard trapping model" is valid over a realistic range of impurity concentration. The scatter at the lower temperature may well arise from impurity clustering.

D. *Diffusion in Graphite*

A more complicated but well-documented trap-controlled diffusion occurs in graphite. Polycrystalline graphite, including pyrolytic graphite of near-theoretical density, is an example of a medium wherein bulk diffusion proceeds exclusively via a trapping mechanism for nearly all species of impurity diffusion.[12] The impurity particle is visualized as making a large number of normal jumps between traps. It is convenient and instructive to discuss this process by the random walk technique for one-dimensional diffusion. If $\bar{n}_t$ represents the mean number of jumps of the impurity out of the traps and $\bar{n}_r$ the number of rapid or free jumps between trapping events, then the impurity diffusion coefficient, the quantity being measured, is given by

$$D = \frac{l^2}{2t}(\bar{n}_t + \bar{n}_r), \tag{13.62}$$

[12] J. R. Wolfe, D. R. McKenzie, and R. J. Borg, *J. Appl. Phys.*, **36**, 1906 (1965).

where l is the unit length of elementary displacement in the graphite structure and t is the diffusion time. t can be expressed in terms of the mean jump frequencies, Γ, as

$$t = \bar{n}_t \Gamma_t^{-1} + \bar{n}_r \Gamma_r^{-1}, \qquad (13.63)$$

where we associate the longer residence time, and hence smaller frequency, Γ_t, with the trapped state and Γ_r represents the rapid or normal jumps. Substitution into (13.62) gives

$$D = \frac{l^2}{2} \frac{\bar{n}_t + \bar{n}_r}{\bar{n}_t \Gamma_t^{-1} + \bar{n}_r \Gamma_r^{-1}}. \qquad (13.64)$$

Let us denote the ratio $\bar{n}_r/\bar{n}_t$ by M and rearrange (13.63) to read

$$D = \frac{l^2}{2} \frac{1 + M}{\Gamma_t^{-1} + M\Gamma_r^{-1}}. \qquad (13.65)$$

If M is sufficiently large, so that $M\Gamma_r^{-1} > \Gamma_t^{-1}$, then we can approximate (13.65) as

$$D = \frac{l^2}{2}(1 + M)\Gamma_t \cong \frac{l^2}{2} M\Gamma_t = \frac{l^2}{2} M\nu \exp\left(\frac{-\Delta G_t^\ddagger}{RT}\right), \qquad (13.66)$$

where $\Delta G_t^\ddagger$ is the activation free energy for detrapping.

We have assumed in this approximation that the temperature dependence of Γ_r, if any, is small compared with that of the detrapping step. According to this approximation, free diffusion is so fast that the rate-determining step is the detrapping of the diffusing particle with its characteristic energy and attempt frequency ν. The pre-exponential value is greatly enhanced by the factor M. One can also show[13] that in many cases (13.66) is expected to be valid even if there is a distribution of detrapping energies, and an experimental observation of a single activation energy over a limited range of temperature should not be taken as proof of its uniqueness.

The diffusion of uranium in various types of graphites, a representative example, is illustrated in Fig. 13.20. Arrhenius behavior is observed as expected from (13.66). The enthalpy and D_0 values are compiled in Table 13.7.

The unusually large values of D_0, the total preexponential factor, are consistent with the detrapping picture and are a direct consequence of the fast diffusion between traps, which naturally leads to long diffusion lengths between trapping events. According to the data of Table 13.7, D_0 is very large in polycrystalline graphite, or for diffusion parallel to the basal plane in pyrolytic graphite, namely in the 10^4 to

[13] J. R. Wolfe *et al.* loc. cit.

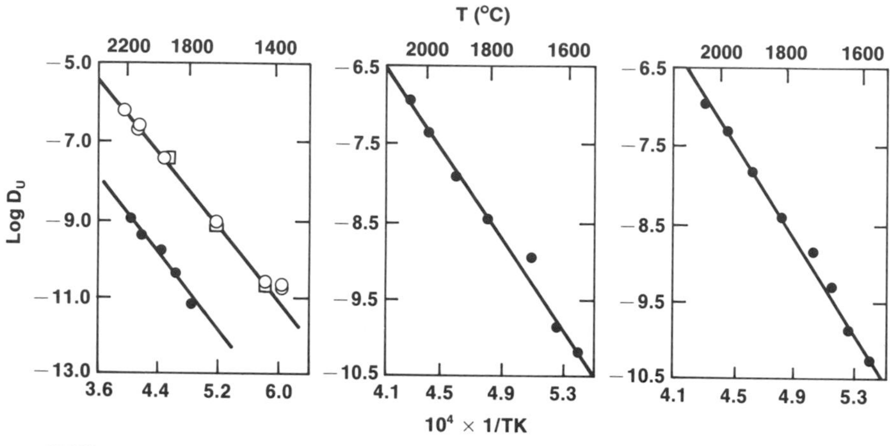

Figure 13.20.

The diffusion coefficients of uranium in (left) pyrographite {○ ⟨*a*⟩ direction ● ⟨*c*⟩ direction}, in (center) Z.T.A., in (right) S-700 graphite as functions of temperature. Pyrographite has nearly theoretical density and is essentially a single crystal, Z.T.A. is polycrystalline high-density material, whereas S-700 is highly porous. Nevertheless, the Arrhenius equation is obeyed by all three graphites. (After Wolfe *et al.*, loc. cit.)

Table 13.7. Activation energies and values of D_0 for uranium diffusion in various types of graphite.

Graphite	Q (kcal)	D_0 (cm^2 sec^{-1})
PG (a)	115.0 ± 2	6.76×10^3
PG (c)	129.5 ± 5	3.85×10^2
ZTA	135.1 ± 3	6.22×10^5
S-700	136.6 ± 3	9.29×10^5

10^6 range. As a matter of fact, these very large values may be taken as strong evidence for a trapping mechanism.

Chapter 13 Exercises

1. Show that Eq. (13.8) is a solution of (13.5), the one-dimensional diffusion equation.
2. Discuss the determination of vacancy migration energies in ionic crystals by doping experiments.

3. Discuss hydrogen diffusion in crystalline and amorphous silicon.

4. Under what conditions does the excess vacancy concentration anneal out proportional to the square of its concentration?

5. Discuss the effect of trapping as exemplified by Fig. 13.19.

6. Why is diffusion enhanced by disorder as exemplified by Fig. 13.15?

7. Suppose there are two pathways for the diffusion of the same species in a given medium. If the value of D_0 is the same for both but the values of $\Delta H^{\ddagger}$ differ by 2, 5, and 10 kcal, respectively, what percentage of the diffusive jumps will occur via each path?

8. The diffusion coefficient for gas dissolved in a particular solid is given by

$$D = 0.15\ (\mathrm{cm^2\ sec^{-1}})\ \exp\left[-\frac{20\ \mathrm{kcal}}{RT}\right].$$

How long will it take for one-half of the gas to diffuse from a sphere of 1 cm diameter at 200°C? at 400°C?

9. The unit jump distance of diffusing atoms increases with increasing temperature beause thermal expansion increases the atomic spacing. Consequently, a determination of $\Delta H^{\ddagger}_{\mathrm{m}}$, which requires values of D at least at two temperatures, would appear to require a temperature dependence for D_0. Show rigorously that this is not so, and that one need not correct for thermal expansion.

10. Compare quantitatively at $T = 800, 1000, 1200$°C the flux of interstitial diffusing atoms with those diffusing substitutionally in an elemental bcc lattice. $\Delta H^{\ddagger}_{\mathrm{m,i}}$ for interstitial diffusion is 0.2 eV, and $\Delta G^{\circ}_{\mathrm{i}} = 2.5$ eV. $\Delta G^{\circ}_{v} = 1.0$ eV and $\Delta H^{\ddagger}_{\mathrm{m},v} =$ 1.0 eV. Assume $D^{i}_0 = 0.1$ and $D^{v}_0 = 1.0$

11. It is estimated that the fuel cladding of a boiling water reactor sustains about 0.5 dpa (displacements per atom) during one year's operation. How does this compare with the number of thermally activated displacements if the average value of D for the cladding is 10^{-10} cm^2 sec^{-1} Assume the interatomic spacing of the cladding is 4 Å. Can you explain why radiation-produced defects can degrade the mechanical integrity of a solid, whereas a much greater number of equilibrium defects do not?

12. Prove that

$$C(x, t) = \frac{\alpha}{t^{1/2}} \int_{-\infty}^{-\infty} f(x') \exp\left[\frac{-(x - x')^2}{4Dt}\right] dx'$$

is a solution of the one-dimensional diffusion equation

$$\frac{\partial C}{\partial t} = D \frac{\partial^2 C}{\partial x^2},$$

where $f(x')$ is the initial distribution of the diffusing species at $t = 0$.

13. Given the face-centered square planar structure shown in the figure, calculate the correlation coefficient, assuming

   a. the vacancy, □, and the impurity tracer atom, ⊗, are always nearest neighbors and never dissociate (tight binding approximation).
   b. no additional vacancies are ever bound to the tracer.
   c. the vacancy and tracer can be either nearest or next-nearest neighbors.
   d. the vacancy and tracer exchanged positions on the previous jump leading to the configuration as shown.

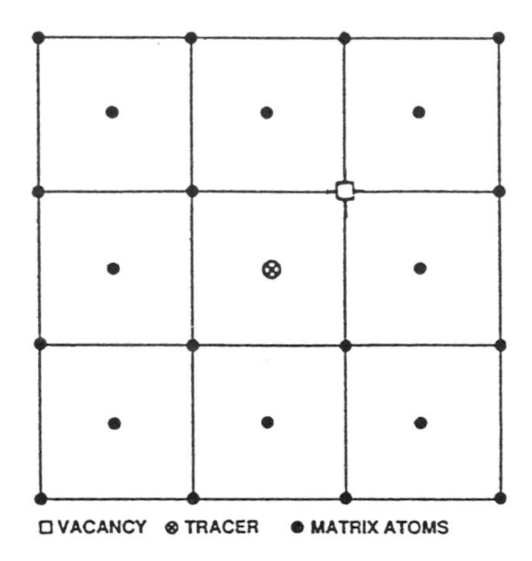

[Hint: Sum the probabilities of the vacancy being taken to any of the four nearest-neighbor positions in 1, 2, 3, . . . , n jumps ($n = 4$ is sufficiently large). Multiply each by the appropriate $\cos \theta$ and use (13.38) to calculate f.]

Additional Reading

R. J. Borg and G. J. Dienes, *Solid State Diffusion*, Academic Press (1988). A modern text on diffusion in solids.

P. G. Shewmon, *Diffusion in Solids*, McGraw-Hill, New York (1963). Chapter 1 provides a clear and concise introduction to the diffusion equations, including derivations.

J. Crank, *Mathematics of Diffusion*, Clarendon Press, Oxford (1975). As the title states this volume deals exclusively with the solution of differential equations appropriate to various boundary conditions, i.e., the mathematics of diffusion. It does not deal with the physics or chemistry of the diffusion, but is an invaluable reference for the practitioner.

J. R. Manning, *Diffusion Kinetics for Atoms in Crystals*, Van Nostrand (1968). Chapters 1 and 2 give derivations of the fundamental equations, including the relation between diffusion and random walks. The mathematics are nicely related to atomic motion in a clear, understandable way.

L. A. Girifalco, *Atomic Migration in Crystals*, Blaisdell (1964). Provides a lucid discussion of the basic physics with an absolute minimum of mathematics.

A. C. Damask and G. J. Dienes, *Point Defects in Metals*, Gordon and Breach (1963). A clearly written introduction to the subject of defect diffusion in elemental metals.

C. P. Flynn, *Point Defects and Diffusion*, Clarendon Press (1972). A detailed mathematical discussion.

A. S. Nowick and J. J. Burton, eds., *Diffusion in Solids*, Academic Press (1975). A multiauthor advanced treatise.

G. E. Murch and A. S. Nowick, eds., *Diffusion in Crystalline Solids*, Academic Press (1984). A multiauthor advanced treatise. Diffusion in Si and Ge is thoroughly explored by Frank, Gösele, Mehrer, and Seeger in Chap. 2.

J. H. Crawford, Jr., and L. M. Slifkin, eds., *Point Defects in Solids*, Vol. 2, "Semiconductors and Molecular Crystals," Plenum (1975). Contains three chapters dealing in depth with defects and diffusion in semiconductors. Chapter 1 by J. W. Corbett and J. C. Bourgoin discusses defect creation in semiconductors emphasizing radiation damage. Chapter 2 by H. C. Casey, Jr., and G. L. Pearson treats diffusion, and Chapter 3 by O. L. Curtis, Jr., examines the effect of point defects on electrical and optical properties.

Chapter 14

Phase Transitions

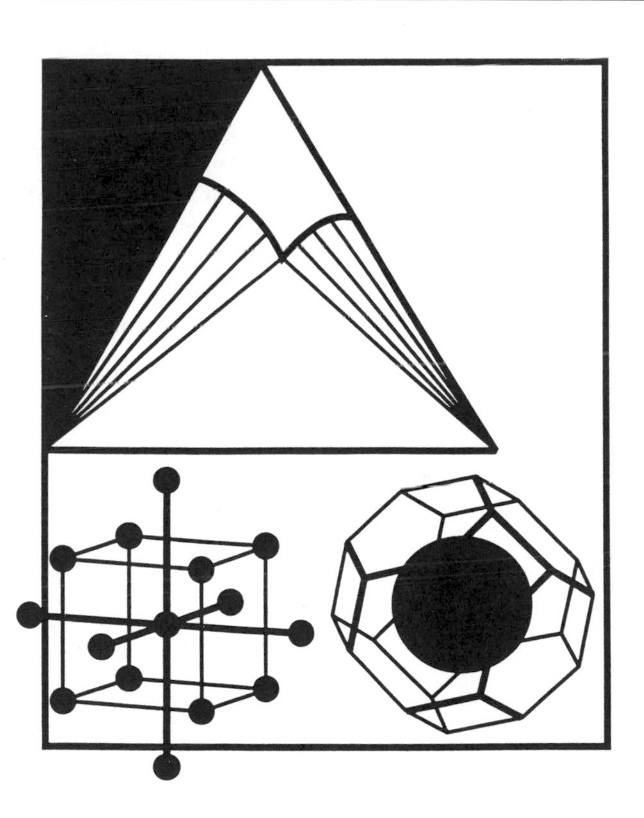

Chapter 14

Contents

I'm trying to learn 'em my experience.

—Casey Stengel, former manager, New York Mets

Phase Transitions

In this chapter we shall be concerned with structural and phase transformations at equilibrium and, hence, describable by thermodynamics, although approach to equilibrium in solids is often a slow process and we need to be aware of the kinetics of transformation. There is a wide variety of important structural and phase transformations, such as melting, magnetic transitions, and order–disorder transformations, and they can be classified phenomenologically according to the order of the transition. The order, in turn, is defined by the discontinuity in the free-energy derivatives at the transition temperature. A discontinuity in any of the first derivatives of the extensive state functions defines a first-order transition (e.g., melting). If the first derivatives are continuous, but any one of the second derivatives shows a discontinuity, then we are dealing with a second-order transition (e.g., ferromagnetic to paramagnetic). A first-order transition, of necessity, is accompanied by a finite isothermal heat. This term frequently causes confusion when it is simply stated that a second-order transition does not possess a latent heat. A second-order transition is also accompanied by an enthalpy change derived strictly from the transition but unlike a latent heat, it is evolved or absorbed over a finite temperature range, whereas the former occurs isothermally at the transition temperature, T_{tr}. These are compared graphically in Fig. 14.1. These are the two important classes although it is perfectly possible, in principle, to have higher-order transitions.

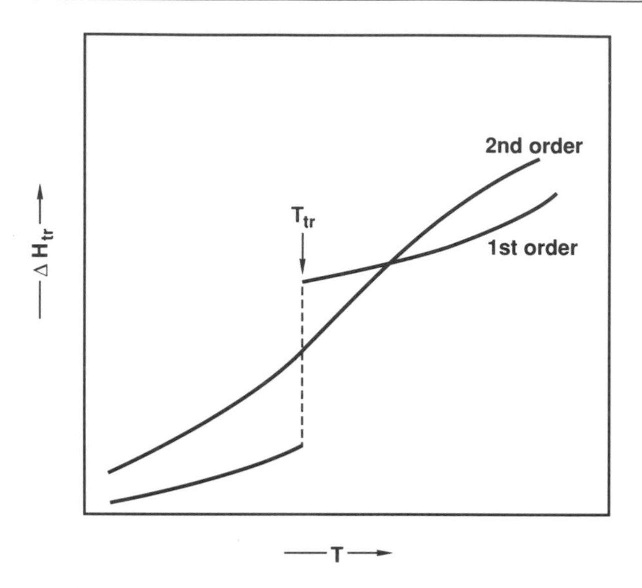

Figure 14.1.

The enthalpies of substances with first- and second-order phase transitions. T_{tr} also corresponds to the inflection point of the second order transition. Different experimental methods will often give slightly different values of T_{tr}.

1. Melting

A. Thermodynamics

The everyday phenomenon of melting is the prototype of a first-order transformation. For any phase transition, the free energy is continuous at the transition temperature, since, at equilibrium, we must have equality of the free energies between phases; hence, at the solid–liquid transition $G_s = G_l$. As already noted, if the first derivatives of G are discontinuous, then the transition is defined as of first order. The standard thermodynamic relations

$$\left(\frac{\partial G}{\partial T}\right)_p = -S \tag{14.1}$$

$$\left(\frac{\partial G}{\partial P}\right)_T = V, \tag{14.2}$$

$$\frac{\partial(G/T)}{\partial(1/T)} = H \tag{14.3}$$

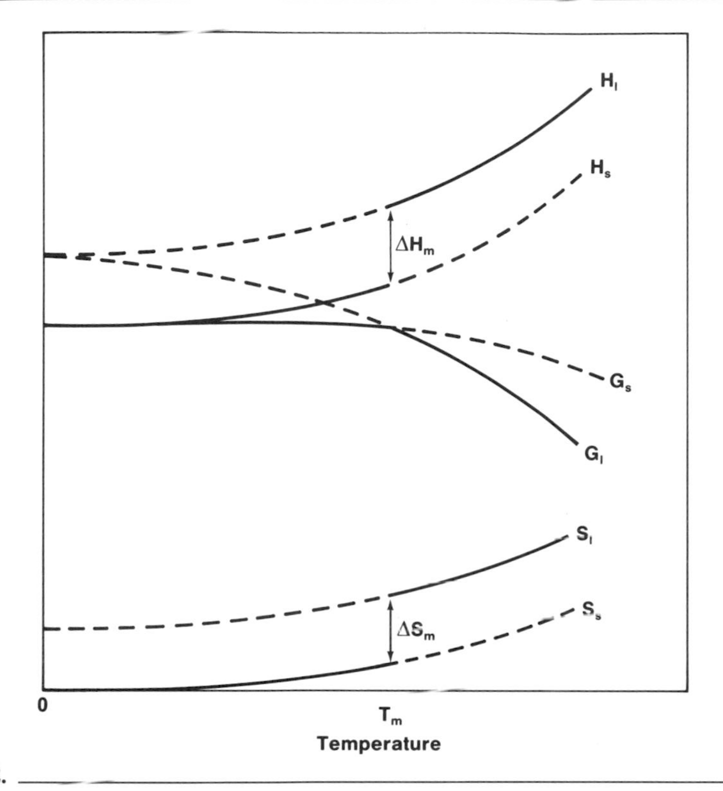

Figure 14.2.

Schematic diagram of changes in free energy, heat content, and entropy of melting.

tell us that we can expect discontinuities in the entropy, volume, and internal energy. The energetics of the melting process at constant pressure may be illustrated as in Fig. 14.2. The solid is stable below T_m, the melting temperature; i.e., $G_s < G_l$, with the opposite inequality above T_m. There is a discontinuous change in H, with ΔH being the latent heat of fusion with a corresponding discontinuity in the entropy.

If the appearance of the alternative phase can be prevented, then, in principle, the free energy of either phase can be investigated even in temperature and pressure regions where it is thermodynamically unstable. Undercooling (sometimes referred to as supercooling, which appears to be a self-contradictory word)—that is, cooling below the equilibrium solidification temperature—is a well-known and well-studied phenomenon. It is, however, very difficult, or impossible, to superheat a solid above T_m, except under very uncommon conditions.[1] The most likely explanation of this

[1] Such as ice heated by a pressure pulse. Note, however, that nucleation is not instantaneous in the interior. The ice experiment showed an incubation time.

asymmetry has to do with the kinetics of the transformation. Solidification from the melt proceeds by means of nucleation and growth of the solid particles, and the nucleation process can be very slow in a pure liquid. If a tiny particle of the solid (i.e., seed crystal) is added to the undercooled liquid, immediate solidification generally results. Impurities in the melt often play a similar role and prevent undercooling. In contrast, nucleation of the liquid phase is apparently instantaneous, most likely because the surface of the solid is already much strained, or even disordered, and hence liquidlike even below the melting temperature.

From the equality of the free energy at the transition temperature, we may write

$$dG_s = dG_l, \tag{14.4}$$

and, therefore,

$$V_s\,dP - S_s\,dT = V_l\,dP - S_l\,dT, \tag{14.5}$$

and, upon rearranging, we obtain

$$\frac{dP}{dT} = \frac{S_l - S_s}{V_l - V_s} = \frac{\Delta S_m}{\Delta V}. \tag{14.6}$$

This relation can also be expressed in terms of ΔH_m since, during a reversible equilibrium change,

$$\Delta S_m = \Delta H_m / T_m, \tag{14.7}$$

where ΔH_m is the molar heat of fusion. Thus,

$$\frac{dP}{dT} = \frac{\Delta H_m}{T\,\Delta V}. \tag{14.8}$$

These relations are known as the Clausius–Clapeyron equations and apply to any equilibrium between two phases (see also Chapter 9, Section 7B). In the case of evaporation rather than fusion, one simply changes V_s to V_l and V_l to V_v in (14.6), and ΔH then becomes the latent heat of evaporation.

For most substances, $V_l > V_s$ and $S_l > S_s$, so that dP/dT is positive and an increase in pressure results in an increase in the melting temperature. The opposite response in water is well known, since at 0°C the density of ice is less than that of water, and increasing the pressure lowers the melting point of ice.

A combination of the Clausius–Clapeyron equation and Raoult's law may be used to derive the expression for the change in the freezing point by a dissolved solute. The pertinent relation is

$$\frac{d \ln X_1}{dT} = \frac{\Delta H_f}{RT^2}, \tag{14.9}$$

where X_1 is the mole fraction of the solvent ($X_1 = 1$ for pure solvent) and ΔH_f is the molar heat of fusion. Integration gives, with T_0 the freezing point of the pure solvent,

$$\ln X_1 = \frac{-\Delta H_f}{R}\left(\frac{1}{T} - \frac{1}{T_0}\right) \cong -\frac{\Delta H_f\,\Delta T}{RT_0^2}, \tag{14.10}$$

where $\Delta T = T_0 - T$. For dilute solutions,

$$\Delta T = RT_0^2 X_2/\Delta H_f, \tag{14.11}$$

where $X_2 = 1 - X_1$ is the mole fraction of solute.

B. *Melting at High Pressures*

For large changes in pressure, ΔH and ΔV become functions of the pressure. Considerable research has been done at very high pressures on the melting of solids of simple crystal structures.[2] The results are often expressed by the semiempirical Simon equation, usually given in the form

$$(P - P_0)/a = (T/T_0)^c - 1, \tag{14.12}$$

where P_0 and T_0 are the coordinates of the triple point for the solid while a and c are empirical parameters. For many solids, when $P \gg P_0$, a convenient logarithmic form, viz.

$$\ln (P + a) = c \ln T + \ln a - c \ln t_0 = c \ln T + b, \tag{14.13}$$

may be used. At the origin, the slope of the Simon equation must reduce to the Clausius–Clapeyron relation. Thus,

$$\frac{dP}{dT} = \frac{c(a + P)}{T} = \frac{\Delta S}{\Delta V} = \frac{\Delta H}{T\,\Delta V}. \tag{14.14}$$

Although some crystals of simple structure obey these equations to high accuracy, others do not.

C. *Statistical Mechanics of Melting*[3]

We have already described, in Fig. 14.2 and its discussion, the large increase in entropy upon melting. A simplified statistical mechanical model is illuminating in

---

[2] See, for example, A. R., Ubbelohde, *Melting and Crystal Structure*, Clarendon Press, Oxford, 1965.
[3] See, for example, J. C. Slater, *Introduction to Chemical Physics*, Chap. XVI, McGraw-Hill, New York, 1939.

justifying this entropy increase. There is a multiplicity of distinguishable states in the liquid arising from the nonlocalization of the atomic or molecular units. The solid is quite reasonably characterized, for our present purposes, by a single level. For N molecules and w complexions the numbers of states is w^N. The partition function may then be written

$$Z = \exp(-NE_s/kT) + w^N \exp(-NE_l/kT), \tag{14.15}$$

where E_s and E_l are the internal energies of the solid and liquid, respectively. The free energy (Gibbs equals Helmholtz at zero pressure) may then be written

$$G = -kT \ln Z. \tag{14.16}$$

If there is no liquid present, the partition function would consist of the first term only, giving

$$G_s = NE_s. \tag{14.17}$$

Conversely, for the liquid we have

$$G_l = NE_l - NkT \ln w. \tag{14.18}$$

Since we expect the internal energy of the liquid to be larger than that of the solid, it is clearly the entropy term in (14.18) that stabilizes the liquid above the melting point. In this simple model, according to (14.17) and (14.18), the G_s line of Fig. 14.2 would be the same as the horizontal H_s line, and G_l would be replaced by a straight line of negative slope, giving a very sharp intersection at T_m. The plot of the total free energy according to the more accurate expressions, (14.15) and (14.16), represents a continuous curve that bends sharply, but not discontinuously, through a small temperature range with discontinuous derivatives as shown in Fig. 14.2.

D. Theories and Mechanisms of Melting

A single accepted atomic theory and mechanism of melting does not exist. It is surprising that we do not understand the details of such a ubiquitous phenomenon at the atomic or molecular level. The problem really resides in our poor knowledge of the structure of liquids. Although we have a very good quantitative description of solids, we do not have one for its partner in the phase transition, viz. the liquid. What seems intuitively clear is that, because of the first-order nature of the transition accompanied by a large entropy change, we are dealing with a catastrophic disruption of the solid. This is the process addressed by the one-phase models, which attempt to establish, at least semiempirically, a solid state criterion for melting while ignoring the melt phase, although a thermodynamic description of melting must take account of both. However, in the absence of an appropriate atomic description of the liquid, this has been the usual approach.

One well-known theory is the *critical vibration theory* proposed by Lindemann in 1910. The vibrational instability in the Lindemann model arises from the increasing amplitude of vibrations with increasing temperature. At some amplitude, at a critical fraction of the interatomic distance, the vibrations are assumed to interact to such an extent that the crystal becomes mechanically unstable. The mechanism of this final step is not specified but may perhaps be related to the vanishing of the shear modulus. (There are, of course, other mechanical instabilities that do not result in melting.) In any case, this is the critical vibration model, which may be quantified as follows.

The mean vibrational energy for a force proportional to the displacement x is given by (see also Chapter 3, Section 3)

$$U = \int_0^{r_c} Cx\,dx = \frac{1}{2} Cr_c^2, \tag{14.19}$$

where C is the elastic stiffness constant and r_c is the critical amplitude as a fraction of the interatomic distance. The restoring force is related to the frequencey ν and the mass of the atom m by

$$2\pi\nu = (C/m)^{1/2}. \tag{14.20}$$

The total energy, potential plus kinetic, is

$$U_f = Cr_c^2, \tag{14.21}$$

which, since melting usually occurs at a high enough temperature to be in the classical region, we may equate to $3kT_m$, and combining with (14.20) and (14.21) gives

$$T_m = 4\pi^2\nu^2 r_c^2\, m/3k. \tag{14.22}$$

The frequency is frequently assumed to be that given by the Einstein model.[4] r_c may be replaced by a fraction of the molar volume of the solid, and (14.22) may thus be reduced to

$$T_m = \kappa\nu^2 V_s^{2/3}\, m, \tag{14.23}$$

where the general constant κ may be expected to be the same for crystals of similar structure, an expectation only approximately fulfilled. Overall, however, Lindemann's

[4] This simplified treatment assumes harmonicity. In real systems there are large anharmonic effects as $T \rightarrow T_m$.

formula provides a convenient means of correlating various physical properties of a crystal to T_m.

In Lindemann's theory, the mechanisms of the disruption of the solid remained unspecified. The known disordered nature of liquids led to the idea of a catastrophic increase in disorder as the possible mechanism of melting. We shall mention briefly two approaches. In the first one, the concentration of elementary point defects, vacancies, and/or interstitial atoms, may reach a value at which the material is highly disordered. Clearly, we require a discontinuous change in the defect concentration from a very low value in the crystal to a large value in the melt. This cannot occur in the simple "linear" treatment given in Chapter 12 where the defect formation energy was assumed to be independent of the defect concentration. In order to obtain a first-order transition the formation energy needs to be at least quadratic in the defect concentration.[5] Similarly, as in the Lennard-Jones–Devonshire model,[2] the energy of defect formation may be assumed to depend strongly on the volume of the solid, which increases with temperature, causing a corresponding decrease in the energy of defect formation. One can again obtain a first-order transition this way. Experimental evidence is meager, but some of these ideas appear to be applicable to sublattice melting in ionic crystals, at least in a phenomenological sense.[5] In turn, sublattice melting is known to be important for "superionic" conduction.

The second scheme, which has had some success, considers a catastrophic generation of dislocations leading to collapse of the crystal.[6] Again, one must have a cooperative interaction to lower the dislocation energy as its concentration increases. The dislocation mechanism appears to be applicable to two-dimensional melting, has some experimental support, and has been considerably elaborated upon. The role of the dislocation mechanism in ordinary three-dimensional melting is far less developed and remains rather speculative.

E. Some Empirical Correlations

The melting points of solids vary by more than a factor of 20 degrees kelvin, a very wide range. Many attempts have been made over the years to find correlations with other physical properties in order to reduce this variation and possibly to obtain some insights into the relevant relations.[7] We discussed in Chapter 13 the relation of atomic

---

[5] See, for example, D. O. Welch and G. J. Dienes, *J. Phys. Chem. Solids*, **38**, 311 (1977).

[6] See, for example, the review by R. M. J. Cotterill, *Phys. Bull.* **32**, 285–7 (1981), in which transitions to the glassy state are also discussed.

[7] See, e.g., ref. 2, chap. 3, loc. cit.

mobility to cohesive energy and melting. Since melting involves disruption of the lattice, it is natural to look for correlation with the cohesive or sublimation energy.

Table 14.1 contains melting data for a wide variety of solids, including the melting

Table 14.1. Melting data.

Solid	ΔH_c (kcal/mole)	T_m (K)	$\frac{\Delta H_c}{T_m}$	$\frac{\Delta H_m}{\Delta H_c}$	Solid	ΔH_c (kcal/mole)	T_m (K)	$\frac{\Delta H_c}{T_m}$	$\frac{\Delta H_m}{\Delta H_c}$
				Monovalent metals					
Li	39	459	0.085		Cu	81	1357	0.060	0.038
Na	26	371	0.070	0.024	Ag	68	1234	0.055	0.039
K	20	336	0.060	0.026	Au	92	1336	0.069	0.033
Rb	19	312	0.061	0.026					
Cs	19	302	0.063	0.027					
				Divalent metals					
Mg	36	923	0.039	0.034	Ba	49	1123	0.044	
Ca	48	1083	0.044		Cd	27	594	0.046	0.054
Sr	47	1073	0.044		Zn	27	692	0.039	0.051
					Hg	15	234	0.064	0.037
				Trivalent metals					
Al	55	932	0.059	0.038					
Ga	52	303	0.172	0.026					
Tl	40	576	0.069	0.018					
	Tetravalent metals					Pentavalent metals			
Si	108	1693	0.064		Bi	48	544	0.088	0.053
Ge	85	1232	0.069		Sb	54	903	0.060	0.088
Pb	47	601	0.078	0.026					
Sn	78	505	0.154	0.025					
				Ionic crystals					
NaF	194	1265	0.153	0.037	CaF_2	402	1633	0.246	
KF	186	1133	0.164	0.033	$CaCl_2$	296	1047	0.283	
NaCl	153	1073	0.143	0.039	BaF_2	400	1553	0.258	
KCl	153	1043	0.147	0.039	$BaCl_2$	312	1232	0.253	
AgCl	127	729	0.174		$CdCl_2$	178	841	0.212	
				Molecular crystals					
H_2	0.22	14	0.016	0.13	CO	1.9	68	0.028	0.11
H_2O	11	273	0.040	0.13	CO_2	6.4	217	0.030	0.31
NH_3	7.1	198	0.036	0.26	CH_4	2.3	90	0.026	0.10
A	1.9	83	0.023	0.15	N_2	1.7	63	0.027	0.13

ΔH_c (kcal/mole) = energy of vaporization or cohesive energy.
T_m(K) = melting point.
$\Delta H_c/T_m$ = ratio of cohesive energy to melting point (kcal/mole^{-1} deg^{-1}).
$\Delta H_m/\Delta H_c$ = ratio of latent heat of melting to the cohesive energy.
Data from various sources.

point T_m, the heat of vaporization or cohesive energy ΔH_c, the ratio $\Delta H_c/T_m$, and the ratio of the latent heat of melting ΔH_m to the cohesive energy ΔH_c. The ratio $\Delta H_c/T_m$ varies far less than do the melting points, ranging for most metals from 0.04 to 0.06, with a few notable exceptions. The ratios are high for Hg, Sn, and Ga, undoubtedly related to their complex crystal structure, which apparently facilitates melting. The corresponding ratios are higher for ionic crystals, ranging from 0.14 to 0.28, reflecting the close packing and tight binding in these materials. The more loosely bound molecular crystals are characterized by relatively low ratios, and within any given group of crystals the ratios are very nearly constant, as is clearly shown by the data in the table, and hence are characteristic of the type of bonding. The $\Delta H_m/\Delta H_c$ ratios are far less discriminating, running about 0.03 to 0.04 for both metals and ionics. They are high, however, in the 0.10 to 0.30 range for molecular crystals. In any case, the point is that fusion requires only a small part of the cohesive energy, and strong bonding is retained in the liquid phase.

2. Order–Disorder Transitions

There is a large variety of transitions described as of the order–disorder type. In the present context, order implies a periodicity or symmetry of arrangement that is a departure from randomness as characterized by disorder. The unifying treatment is in terms of an order parameter that varies from unity, describing perfect order, to zero signifying randomness. The ordered structures are stable at low temperature and become disordered at elevated temperatures. The transitions may be of first or higher order. We discussed in Chapter 4, Section 1F an order–disorder process, namely, electric dipole alignment as a function of temperature.

All types of order fall into one of two general categories called short-range and long-range order, respectively. Substances with long-range order (LRO) possess short-range order (SRO) as well. Several kinds of parameters have been devised to describe the degree of order, and we will limit our discussion to perhaps the simplest examples.

Beginning with short-range order, we can define a parameter σ that describes the fraction of nearest-neighboring atoms that are of a specific kind, or have their magnetic moments aligned in specified directions, or, if they be neighboring molecules, are oriented in a specific way with respect to one another. Consider, for example, an equiatomic binary alloy AB and assume that perfect order, for a one-dimensional model, alternates A and B as follows:

—A—B—A—B—A—B—A—B—,

so that each A atom has two B atoms as nearest neighbors, and vice versa in the state of perfect order. For the random state, half the neighbors would be A and the other half perforce, B. The SRO parameter is now defined by

$$\sigma = \frac{X_{AB} - X_{AB}^{r}}{X_{AB}^{po} - X_{AB}^{r}}, \tag{14.24}$$

where X_{AB} is the actual fraction of AB pairs in a particular solid solution, and X_{AB}^{r} and X_{AB}^{po} are the respective values for the random state and that having perfect order. As a result of this definition, $0 \leq \sigma \leq 1$, with $\sigma = 0$ corresponding to the random state and $\sigma = 1$ for perfect order. Naturally, more complicated arrays can be formulated with various values of X_{AB}^{po}, e.g.,

—A—A—A—B—B—B—A—A—A—B—B—B—,

or for $X_{\uparrow\uparrow}$ and $X_{\uparrow\downarrow}$,

$-\uparrow-\uparrow-\uparrow-\uparrow-\uparrow-\uparrow-\uparrow-\uparrow-,$

representing the aligned magnetic moments of a perfectly ordered ferromagnet and

$-\uparrow-\downarrow-\uparrow-\downarrow-\uparrow-\downarrow-\uparrow-\downarrow-\uparrow-,$

showing a simple linear antiferromagnetic configuration. This scheme is easily generalized to complicated three-dimensional structures since one need only account for the nature of the nearest neighbors.

Clearly, σ correlates the states of the nearest neighbors and is in fact the simplest of all pair correlation parameters. The next logical step is to extend this concept to include atoms or molecules more distant from the central fiducial one. Such parameters are derived from more complicated pair correlation functions than (14.24), but the underlying concept remains the same.

The long-range order parameter s is defined by the fraction of atoms, molecules, or magnetic moment directions that are on the *right* crystallographic sublattice leading to perfect order. The easiest way to visualize this is to again consider an equiatomic alloy that has the bcc crystal structure. The body-centered cubic lattice can be subdivided into two interpenetrating primitive cubic sublattices, the cube centers of one being the cube corners of the other. Let us label one sublattice α and the other, β. Assuming that perfect order places all A atoms on the α lattice and hence all B atoms on the β, we define the LRO parameter in terms of fractional occupation by

$$s = \frac{X_{A\alpha} - X^{r}_{A\alpha}}{X^{po}_{A\alpha} - X^{r}_{A\alpha}}, \tag{14.25}$$

where the superscripts have the same meaning as in the case of short-range order. Clearly using B atoms on β sites to define s gives the same result. As before, $0 \leq s \leq 1$.

The difference between long- and short-range order is not a trivial one, and both can have profound and differing effects upon the chemical and physical properties of solids.

A. Orientational Order–Disorder Transition

The rotation of NH_4 in NH_4Cl may be considered a prototypic example of orientational order–disorder. The NH_4 ion is close to spherical since it is tetrahedrally coordinated and rotates rather freely at elevated temperatures as a result of thermal excitation. However, the lowest energy state of the ensemble below the transition temperature requires all the NH_4^+ molecules to assume the same orientation, either 1 or 2, as shown in Fig. 14.3. As is characteristic of all second-order transitions, this one has associated with it a specific heat anomaly, i.e., a lambda point. The degree of order is a measure of the relative concentrations of NH_4^+ in the 1 and 2 positions; perfect order corresponds to all having the same orientation, and complete randomness to one half in the 1 and the other half in the 2 orientation.

Each NH_4 has six NH_4 first neighbors. Let N_{11} be the number of nearest-neighbor pairs with orientation 1. At complete order, when all pairs are in orientation 1, we

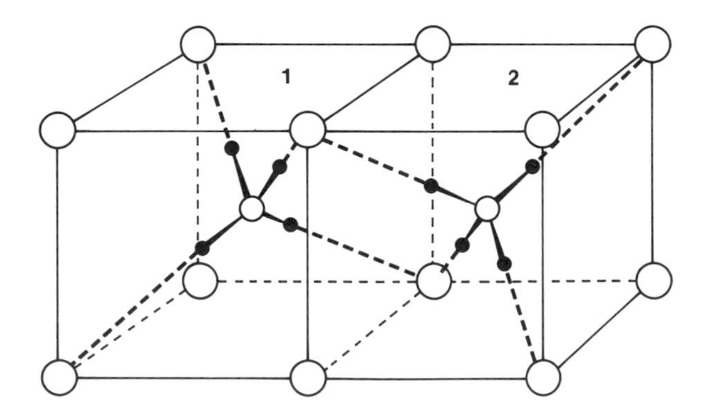

Figure 14.3.

NH_4Cl at high temperature showing the two possible NH_4^+ ion orientations. (After G. Burns, *Solid State Physics*, Chap. 11, Academic Press, 1985.)

have $N_{11} = 6N/2 = 3N$ and $N_{22} = 0$, where N is the number of molecules (we divided by 2 to avoid counting the pairs twice). At other than perfect order ($s \neq 1$), N_{11} is given by

$$N_{11} = \frac{N}{2}(1+s)\frac{3}{2}(1+s) = \left(\frac{3N}{4}\right)(1+s)^2, \tag{14.26}$$

where s is the order parameter. We arrived at this formula by the following reasoning; to find the number of nearest neighbor pairs N_{11} when both have the 1 orientation, we multiplied the number of NH_4^+ ions with the 1 orientation, $(N/2)(1+s)$, by one-half of the probability that one of six neighbors is also in orientation 1, namely $(3/2)(1+s)$. Note that at $s = 1$ we have $N_{11} = 3N$, and at $s = 0$ it becomes $N_{11} = 3N/4$, which is the random probability when the 1 and 2 orientations are equally populated. We find the number of 11, 22, and 12 pairs by identical arguments and obtain the relations

$$N_{11} = \frac{3N(1+s)^2}{4},$$
$$N_{12} = \frac{3N(1-s^2)}{4}, \tag{14.27}$$
$$N_{22} = \frac{3N(1-s)^2}{4},$$

with obviously correct limits for the order parameter $s = \pm 1$ and 0 ($+1$ for N_{11}, -1 for N_{22}). We now assume, as an approximation, that the internal energy depends only on the relative orientation of near-neighbor pairs, omitting the longer-range interactions. Thus,

$$E = N_{11}\varepsilon_{11} + N_{22}\varepsilon_{22} + N_{12}\varepsilon_{12}, \tag{14.28}$$

where ε_{11} is the energy of a 11 pair, etc. (i.e., the energy required to break the designated bonds). In this model the 11 and 22 orientations are, of course, equivalent in the near-neighbor approximation, and $\varepsilon_{11} = \varepsilon_{22}$. Substitution of (14.27) into (14.28) gives

$$E = (3N/4)(E_0 + s^2\bar{E}), \tag{14.29}$$

where

$$E_0 = \varepsilon_{11} + \varepsilon_{12} + \varepsilon_{22},$$
$$\bar{E} = \varepsilon_{11} + \varepsilon_{22} - \varepsilon_{12}.$$

By standard combinatorial analysis, we can write the number of ways of arranging $(N/2)(1 + s)$ ions with the 1 orientation and $(N/2)(1 - s)$ ions with the 2 orientation on N sites as

$$\Omega_c = \frac{N!}{[(N/2)(1 + s)]![(n/2)(1 - s)]!}. \tag{14.30}$$

The configurational entropy is given by

$$S = k \ln \Omega_c,$$

which, after using Stirling's approximation, may be written from (14.30) as

$$S = kN \ln 2 - (kN/2)[(1 + s)\ln(1 + s) + (1 - s)\ln(1 - s)]. \tag{14.31}$$

Minimizing now the free energy, $F = E - TS$, with respect to the order parameter s, gives, after rearranging,

$$\ln \frac{1 + s}{1 - s} = -\frac{3\bar{E}s}{kT}. \tag{14.32}$$

From (14.32) the temperature can be calculated for any value of s. For small values of s we may expand the left-hand side and obtain

$$s(s + s^3/3 + \cdots) = -3\bar{E}s/kT. \tag{14.33}$$

The linear terms define the critical temperature T_c as

$$T_c = -3\bar{E}/2k, \tag{14.34}$$

which shows immediately that $\bar{E}$ must be negative in order to have a positive T_c and, hence, a real phase transition.

Taking the derivative of s with respect to T in (14.33) while preserving only the first two terms gives

$$\frac{ds}{dT} = \frac{9\bar{E}}{4skT^2} \to -\infty \qquad \text{as } s \to 0, \tag{14.35}$$

where $\bar{E}$ is negative. This shows that s starts from 0 at $T = T_c$ with an infinite negative slope. The s versus reduced temperature T/T_c is shown schematically in Fig. 14.4, a typical second-order behavior with no discontinuity. The specific heat, using (14.29), is given by

$$C_v = \left(\frac{\partial E}{\partial T}\right)_v = \left(\frac{dE}{ds}\frac{ds}{dT}\right)_v = \frac{3N\bar{E}s}{2}\frac{ds}{dT}. \tag{14.36}$$

The result for this model is sketched in Fig. 14.5 and shows the specific heat anomaly characteristic of a second-order phase transition. The actual behavior of ammonium

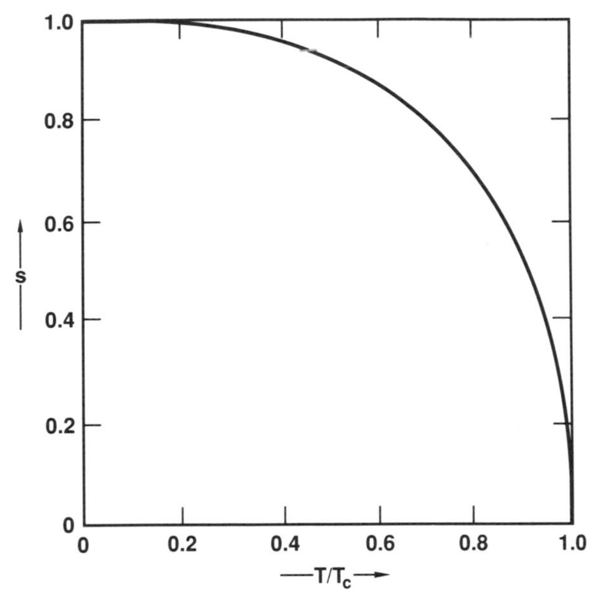

Figure 14.4.

Sketch of the variation of the order parameter s with the reduced temperature T/T_c according to (14.32).

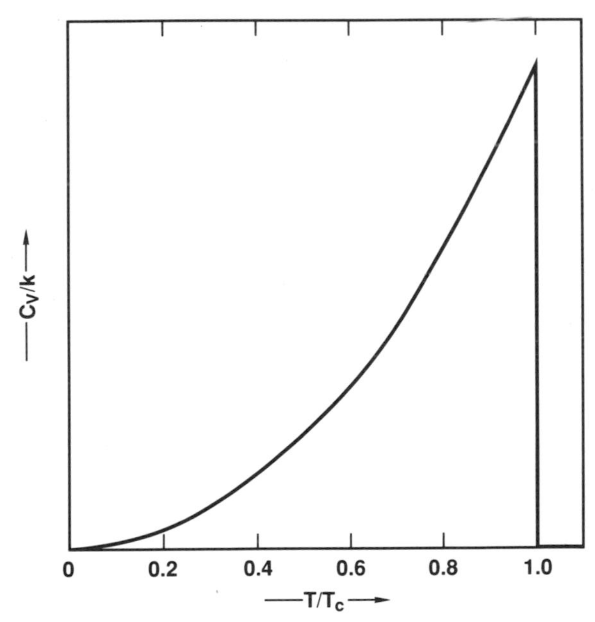

Figure 14.5.

Sketch of the configurational heat capacity vs. the reduced temperature.

chloride is somewhat more complicated than that of this idealized model since its specific heat anomaly is not ideally sharp because of the persistence of short-range order at temperatures just above T_c. This persistence of SRO above T_c is characteristic of second-order transitions

B. Long-Range Order–Disorder Transitions in Alloys

Many alloy systems that are random solid solutions at high temperatures become ordered at low temperatures, with the individual constituents seeking specific lattice sites. A typical equimolar alloy system is copper–zinc, which forms an ordered phase with the CsCl structure, as illustrated in Fig. 14.6. In this structure at complete order there are two interpenetrating sublattices labeled α and β. At complete order, all A atoms (say Cu) are on sublattice α and all B atoms (say Zn) are on the β sublattice. At the other extreme, complete disorder, there are equal numbers of A and B on α and β sites, which is then a random bcc solid solution.

This system, and any equimolar AB system, can be treated by the methods of the previous section with rather minor modifications. It is of interest, however, to approach the properties of such a system from a more chemical point of view, namely by the application of the law of mass action.[8] This method is quite general and is applicable to other than equimolar binary systems. Further, this scheme can also

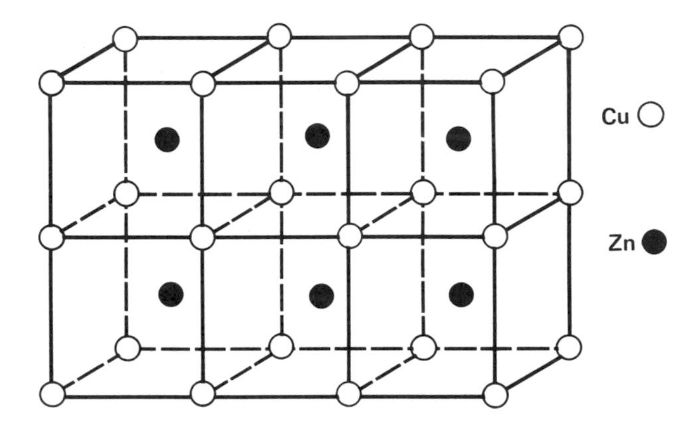

Figure 14.6.

Ordered structure of β-brass (CuZn).

[8] G. J. Dienes, *Acta. Met*, **3**, 549 (1955).

serve as the starting point for an analysis of the rate of ordering. Let there be N_B B atoms on N_B β sites and N_A A atoms on N_A α sites and designate the number of wrong atoms by $N_B^\alpha = N_A^\beta$. The order–disorder transformation is describable by the "chemical" equation

$$A^\alpha + B^\beta \underset{k_2}{\overset{k_1}{\rightleftarrows}} B^\alpha + A^\beta. \tag{14.37}$$

The left-hand side of (14.37) represents the ordered state, and disordering proceeds from left to right. The corresponding fractional concentrations are

$$\frac{N_A - x}{N}, \frac{N_B - x}{N} \cdot \frac{x}{N}, \frac{x}{N}$$

where

$$x = N_B^\alpha = N_A^\beta \qquad \text{and} \qquad N = N_A + N_B.$$

At equilibrium,

$$K = \frac{k_1}{k_2} = \frac{(x/N)^2}{(N_A/N - x/N)(N_B/N - x/N)}. \tag{14.38}$$

Substitution of $N_A/N = \gamma$, then $N_B/N = 1 - \gamma$, leads to

$$K = \frac{x^2/N^2}{\gamma(1 - \gamma) - x/N + (x/N)^2}. \tag{14.39}$$

The Bragg–Williams[9] order parameter s is given by

$$x/N = (1 - s)\gamma(1 - \gamma), \tag{14.40}$$

with obvious limits at $s = 1$ and $s = 0$, with the random probability of wrong pairs given by $\gamma(1 - \gamma)$.

In terms of s the equilibrium condition becomes

$$K = \frac{\gamma(1 - \gamma)(1 - s)^2}{s + \gamma(1 - \gamma)(1 - s)^2}. \tag{14.41}$$

k_1 and k_2 may be expressed as follows. Let the potential barrier in going from disorder to order be denoted by $E^\ddagger$ where the usual approximation, viz. $F^\ddagger = E^\ddagger$ is made, and the energy difference between the ordered and disordered states by E. Then

$$k_1 = v_1 e^{-(E+E^\ddagger)/RT}, \qquad k_2 = v_2 e^{-E^\ddagger RT}, \tag{14.42}$$

---

[9] See, for example, the classical review paper by F. C. Nix and W. Shockley, *Rev. Mod. Phys*, **10**, 1. (1938).

where ν_1 and ν_2 have the dimensions of frequency and the equilibrium constant K is given by

$$K = \frac{k_1}{k_2} = \frac{\nu_1}{\nu_2} \exp\left(\frac{-E}{RT}\right). \tag{14.43}$$

Following Bragg and Williams, we let $\nu_1 = \nu_2$; i.e., we neglect any changes in vibrational entropy. Further, we shall assume that the order-disorder energy E varies linearly with the LRO parameter s. Thus

$$E = E_0 s, \tag{14.44}$$

where E_0 is the energy difference between the fully ordered and completely disordered states.

For an AB alloy, $\gamma = 1/2$, and the equilibrium relation (14.38) becomes

$$\ln \frac{1+s}{1-s} = \frac{E_0 s}{2RT}. \tag{14.45}$$

This relation is clearly of the same form as (14.32), and a plot of s versus T would look the same as Fig. 14.4. By going through the statistical mechanics[3] as in deriving (14.32), we would have arrived at (14.45) with $E_0 = 8(E_{11} + E_{22} - 2E_{12})$. As before $ds/dT \to \infty$ as $s \to 0$, and T_c is given by

$$T_c = E_0/4RT, \tag{14.46}$$

with E_0 positive for a real transition from the ordered to the random state.

Formula (14.41), derived on the basis of the law of mass action, is more general and has been used for γ values not equal to 1/2. As a matter of fact, $\gamma = 1/2$ is a rather special case, since only for this value does the s versus T curve start up with infinite slope from $s = 0$ and is the transition, therefore, of second order. In general, ds/dT is positive at $s = 0$, as can be shown by expanding for small s and taking the derivatives, and the transition is a first-order one. Let us illustrate such cases by discussing briefly another classical order–disorder system, $AuCu_3$, whose ordered and disordered structures are shown in Fig. 14.7.

For an AB_3 alloy, $\gamma = 1/4$ and the equilibrium condition, with the same assumptions as before, becomes

$$\ln\left(\frac{16s}{3(1-s)^2} + 1\right) = \frac{E_0 s}{RT}. \tag{14.47}$$

The results of an actual calculation using (14.47) with $E_0/R = 1000$ K are shown in Fig. 14.8. Over a part of the temperature range s versus T is double-valued and stable solutions must be decided based on free-energy arguments. In Fig. 14.8

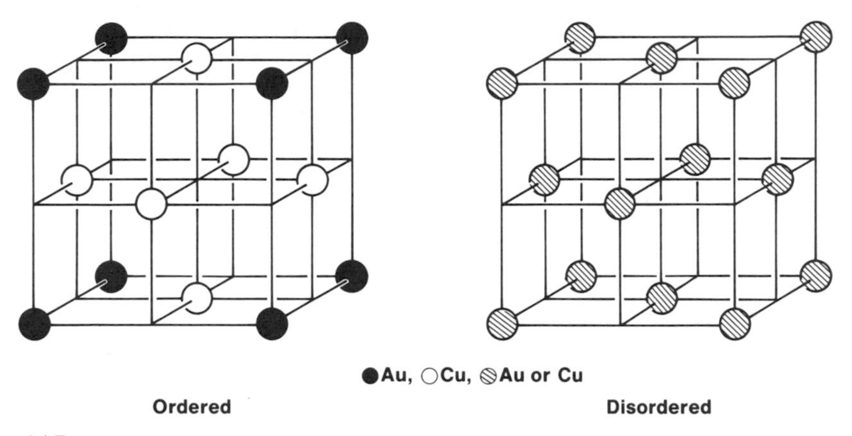

Figure 14.7.

The structure of ordered and disordered $AuCu_3$.

the vertical line at 205 K, which is the critical temperature, was obtained by equating the free-energy minima (analogous to Fig. 9.6), which occurred at $s = 0.467$ and at $s = 0$. Above T_c the thermodynamically stable state is at $s = 0$, but up to the tip of the curve (at 207.2 K in the figure) there is still a free-energy barrier in going from a finite s to $s = 0$ (in addition to any barrier to atomic migration). Thus, it should be possible to superheat the material up to this temperature. The tip of the curve at

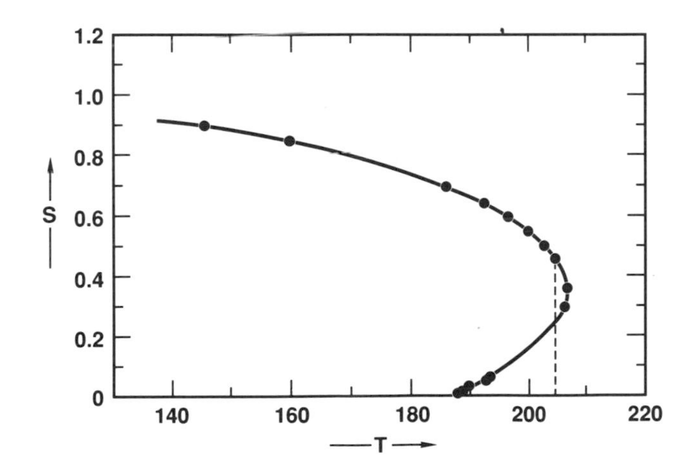

Figure 14.8.

Equilibrium s vs. T curve for an AB_3 alloy of the $AuCu_3$ structure calculated by means of (14.47) with $E_0 = 1000$ K.

$s = 0.35$ represents the limit of possible superheating. These characteristics are clearly those of a first-order phase transition with a discontinuity and a latent heat.

C. Kinetics of Ordering Transitions

The formalism used for the order–disorder equilibrium can be adapted to describe the ordering kinetics. Using (14.37) and the fractional concentrations, we can write for the overall reaction rate

$$\frac{d(x/N)}{dt} = k_1\left(\frac{N_A}{N} - \frac{x}{N}\right)\left(\frac{N_B}{N} - \frac{x}{N}\right) - k_2\left(\frac{x}{N}\right)^2, \tag{14.48}$$

which, upon substituting from (14.40) and (14.41), becomes (letting $\nu_1 = \nu_2 = \nu$)

$$\frac{ds}{dt} = \nu e^{-E^{\ddagger}/RT}[\gamma(1-\gamma)(1-s)^2 - e^{-E_{0}s/RT}[s + \gamma(1-\gamma)(1-s)^2]]. \tag{14.49}$$

This basic kinetic equation differs from an ordinary second-order chemical rate equation by the dependence of the second exponential term on s. There is no analytical solution to this equation, and it has to be treated numerically. Equation

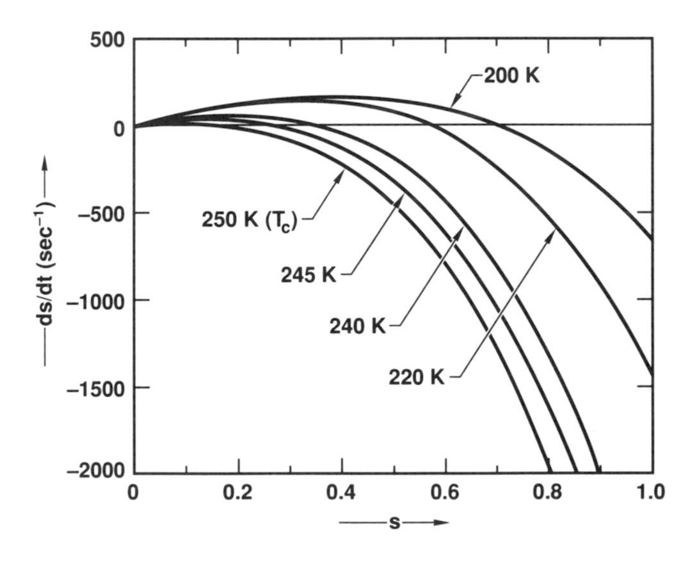

Figure 14.9.

Schematic diagram for the rate of change of long-range order vs. long-range order for an AB alloy for several temperatures with the vacancy interchange mechanism. (After Vineyard, loc. cit.)

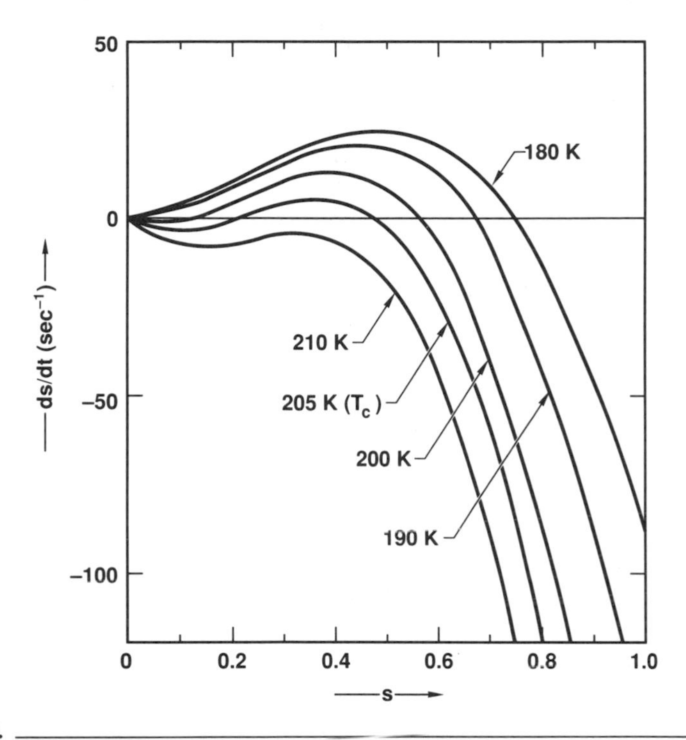

Figure 14.10.

Schematic diagram for the rate of change of long-range order vs. long-range order for an AB_3 alloy for several temperatures with the vacancy interchange mechanism. (After Vineyard, loc. cit.)

(14.49) as it stands is somewhat unrealistic in not including the actual interchange mechanism, which is known to be the vacancy. When this feature is included[10] the treatment becomes more complex but the numerical solutions in both cases show the same essential features[10,11]. These are illustrated schematically in Fig. 14.9 for an AB alloy and in Fig. 14.10 for an AB_3 alloy. Let us discuss the AB case first, where the critical temperature was taken to be 250 K. The rate ds/dt at $s = 0$ is always zero. The other zero values of ds/dt correspond to the equilibrium values at the different temperatures. Because of the zero rate at $s = 0$, ordering must start by fluctuations. However, since ds/dt as a function of s has a positive slope at all temperatures below T_c, large fluctuations should not be needed to initiate ordering in the AB case.

[10] G. H. Vineyard, *Phys. Rev.*, **102**, 981 (1956).

[11] Dienes, loc. cit.

The situation is quite different in the AB_3 case, illustrated in Fig. 14.10 for a T_c value of 205 K shown by the equilibrium curve of Fig. 14.8. At sufficiently low temperatures, below T_c, the curves are similar to the equiatomic case. In the vicinity of T_c, however, there are virtual disordering rates, negative ds/dt values, for an alloy starting at $s = 0$ for small values of s. This means that s has to change discontinuously from $s = 0$ to a finite s value before ordering can take place. Thus, long induction periods are indicated, since large fluctuations are required to start the ordering reaction. Thus the ordering process in this case reminds one of nucleation and growth, but it should be emphasized that the need for fluctuations is a simple consequence of this "chemical" rate theory for this alloy. This behavior is, of course, correlated with the multiple roots in the equilibrium equation (14.47) for this alloy, and hence, with the first-order transition, as illustrated in Fig. 14.8

The equilibrium and kinetic treatments show clearly, by the difference of the behavior of the AB and AB_3 alloys, the profound influence of stoichiometry on the order–disorder process.

3. Magnetic Order

As mentioned in the introduction to Chapter 14, Section 2, the statistical treatment of magnetism follows the same mathematical principles as all other order–disorder phenomena. Given that the atoms or ions possess magnetic moments associated with uncompensated electron spins, there are three principal configurations that can occur spontaneously in the absence of an applied external field. A completely random spin arrangement gives rise to paramagnetism. Paramagnetic substances exposed to a suitably strong external field will partially order magnetically. The degree of magnetic order as a function of field strength is given by the Langevin function, which was derived for electric dipoles in Chapter 4, Section 1F. For the purpose of the present discussion, ferro- and antiferromagnetism are more to the point. Ferromagnetism is the result of the spontaneous ordering of the atomic moments, so that they are all collinear with their north and south poles all in the same direction relative to the crystal coordinates. Antiferromagnetism can be any one of an essentially infinite number of ordered static configurations that result in a zero net macroscopic moment because of internally cancelling spin directions. These three elementary arrangements are shown for a single primitive cubic cell in Fig. 14.11. (It should be remarked that simple monoelemental primitive cubic structures actually cannot support long-range magnetic order.) Also, it should be noted that there are many more complicated spin geometries than the elementary ones portrayed in Fig. 14.11 (e.g., spiral spin lattices, ferromagnetic arrays, spin glasses, etc.). However, ferromagnetism will suffice to illustrate the basic principles.

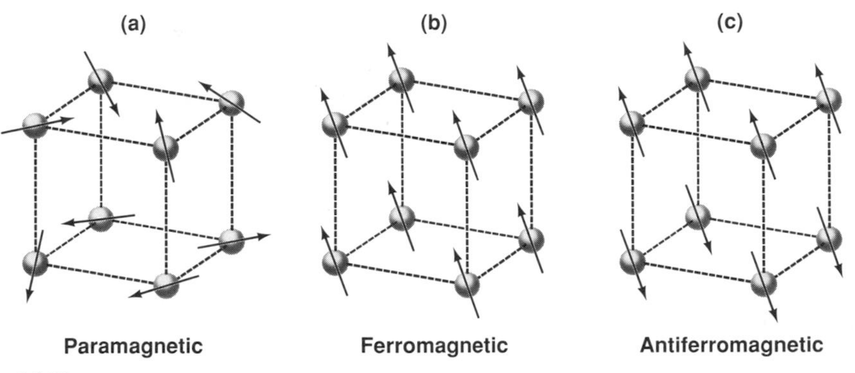

Figure 14.11.

Elementary configurations of atomic moments that give rise to (a) paramagnetic, (b) ferromagnetic, and (c) antiferromagnetic behavior.

As in all second-order phase transitions, magnetic ordering is accompanied by an anomalous specific heat (i.e., a so-called *lambda point*). This is shown for Ni, which is ferromagnetic, in Fig. 14.12, along with all the various contributing components.

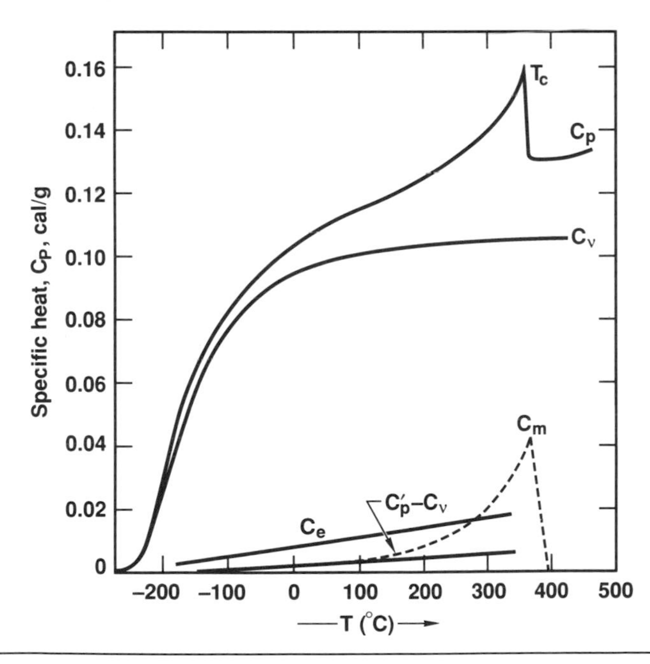

Figure 14.12.

Heat capacity curves for Ni as a function of temperature: C_p, the total heat capacity at constant pressure, C_v the vibrational heat capacity, C_m the magnetic heat capacity, C_e the electronic heat capacity, and $C'_p - C_v$, in which $C'_p = C_p - C_m$ and C_v is the heat capacity at constant volume. (R. M. Bozorth, Van Nostrand, p. 736, 1951.)

The $C'_p - C_v$ term accounts for the energy used for the thermal expansion, and all others are self-explanatory. Below T_C, the Curie temperature, long-range spin order develops spontaneously, and it increases with decreasing temperature. Just above T_C short-range order persists, as evidenced by the less than vertical decrease in C_p with T.

A. Paramagnetism—The Two-Level Model

The two-level model of paramagnetism[12] reproduces the results obtained for the NH_4^+ ion in Section 2A. Consider a paramagnetic substance in which the atoms have a moment consisting of a single electron spin, and, hence, a moment of one Bohr magneton μ_β, viz., $\mu_\beta = e\hbar/2mc = 0.927 \times 10^{-20}$ erg/gauss. We assume that the atomic moment can take but two orientations: one opposing the applied field **H**, and the other with its poles aligned in the same sense as **H**. Clearly, the lower energy state places the north pole of the atomic dipole in opposition to the direction of **H**. The energy levels in a magnetic field are

$$\varepsilon_J = m_J g \mu_\beta \mathbf{H}, \tag{14.50}$$

where m_j is the azimuthal quantum number and g is the so-called Landé g-factor or spectroscopic splitting factor. m_j takes the values $-J, -(J-1), \ldots, (J-1), J$, where m_j is the total angular momentum for a single spin with no orbital angular momentum, $m_j = \pm 1/2$, and g is taken to be equal to 2, though actually 2.0023. Consequently, the allowable energy levels are

$$E = \pm \mu_\beta \mathbf{H}. \tag{14.51}$$

Schematically, this is shown in Fig. 14.13. The equilibrium concentrations of the spin-up, N_-, and spin-down, N_+, populations are now

$$\frac{N_+}{N} = \frac{e^{\mu_\beta \mathbf{H}/kT}}{e^{\mu_\beta \mathbf{H}/kT} + e^{-\mu_\beta \mathbf{H}/kT}},$$

$$\frac{N_-}{N} = \frac{e^{-\mu_\beta \mathbf{H}/kT}}{e^{\mu_\beta \mathbf{H}/kT} + e^{-\mu_\beta \mathbf{H}/kT}}. \tag{14.52}$$

[12] The subsequent development follows C. Kittel, *Introduction to Solid State Physics*, 3rd ed., pp. 434–435, Wiley, New York, 1967.

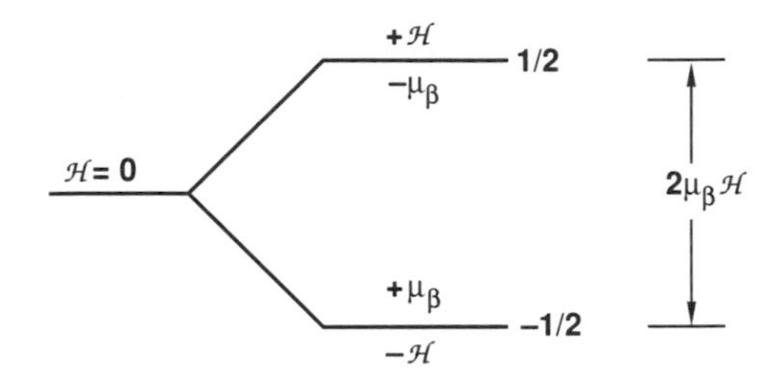

Figure 14.13.

The basic energy diagram for the two-level model.

Defining $x \equiv \mu_\beta \mathbf{H}/kT$, we can calculate the total macroscopic magnetism M of the ensemble by

$$M = (N_+ - N_-)\mu_\beta = N\mu_\beta \frac{e^x - e^{-x}}{e^x + e^{-x}} = N\mu_\beta \tanh x. \tag{14.53}$$

The relative magnetization referenced to the saturation magnetization is given by

$$M/M_{\text{sat}} = \tanh x. \tag{14.54}$$

If M_{sat} corresponds to perfect spin order, then M/M_{sat} can be considered to be a long-range order parameter with the usual limits $0 \leq M/M_{\text{sat}} \leq 1$.

B. The Brillouin Function

A more general and realistic treatment of paramagnetism than the two-level model is provided by the so-called Brillouin function. A magnetic field **H** will cause the magnetic moments of an ensemble of atoms to precess about the pole axis of the field. The allowable quantized magnetic energy levels are dictated by the permissible values of J, which derive from the vector addition of the spin and orbital angular momenta.[13] The magnetic potential energy is then given in increments of $g\mu_\beta \mathbf{H}$, where the m_J values correspond to allowable values of J. At all finite temperatures, the thermal energy produces randomness in the spin directions in opposition to the magnetic coupling, which strives to align the moments. Consequently, if we assume

[13] The vector model of the atom is given in all atomic physics texts and most modern texts on physical chemistry and so will not be repeated here.

a classical Maxwellian distribution, we can write for the magnetization of N atomic moments[14].

$$M = N \sum_{-J}^{J} m_J g \mu_\beta e^{m_J g \mu_\beta \mathbf{H}/kT} \Big/ \sum_{-J}^{J} e^{m_J g \mu_\beta \mathbf{H}/kT}. \tag{14.55}$$

Substituting $x = g\mu_\beta \mathbf{H}/kT$ allows (14.55) to be rewritten as

$$M = Ng\mu_\beta \frac{d}{dx}\left(\ln \sum_{-J}^{J} e^{m_J x}\right). \tag{14.56}$$

Since we assume $m_J g \mu_\beta \mathbf{H}/kT > 1$, we cannot expand the exponential terms in the usual manner, but writing the summation as a geometric progression

$$e^{-Jx}(1 + e^x + e^{2x} + \cdots + e^{2Jx}),$$

and using the formula for its sum leads to

$$M = Ng\mu_\beta \frac{d}{dx}\left[\ln e^{-Jx}\left(\frac{1 - e^{(2J+1)x}}{1 - e^x}\right)\right]. \tag{14.57}$$

Remembering that $\sinh x = (e^x - e^{-x})/2$, substituting into (14.57), and performing the differentiation, we arrive at

$$\begin{aligned} M &= Ng\mu_\beta J\left[\frac{2J+1}{2J}\coth\left(\frac{2J+1}{2J}\right)y - \frac{1}{2J}\coth\frac{y}{2J}\right], \\ &= Ng\mu_\beta J\mathbf{B}_J(y), \end{aligned} \tag{14.58}$$

where B_J is the Brillouin function and

$$y = Jg\mu_\beta \mathbf{H}/kT.$$

As $J \to \infty$ and, hence, all orientations of the moments become possible,

$$\mathbf{B}_J(y) = \coth y - \frac{1}{y} = L(y) \qquad (J \to \infty), \tag{14.59}$$

where $L(y)$ is the classical Langevin function (4.34).

Equation (14.58) may also be written as

$$M(T)/M(0) = \mathbf{B}_J(y), \tag{14.60}$$

---

[14] The derivation follows A. H. Morrish, *The Physical Principles of Magnetism*, p. 71, Wiley, New York, 1965.

where $M(0)$ is the saturation magnetization at absolute zero. For $J = 1/2$ and $g = 2$,

$$M(T)/M(0) = \mathbf{B}_J(y) = \tanh y. \tag{14.61}$$

This is the same result derived for the two-level model (14.53), as well as for β-brass, by an obvious change in (14.45), which also is a two-level system.

In some cases, such as those exemplified in Fig. 14.14, the Brillouin function accurately predicts the field and temperature dependence of the magnetization. The orbital angular momentum of all three ions, viz. Gd^{3+}, Fe^{3+}, and Cr^{3+}, is quenched so that $J = S$, where S is the vector sum of the spins of the outer electrons.

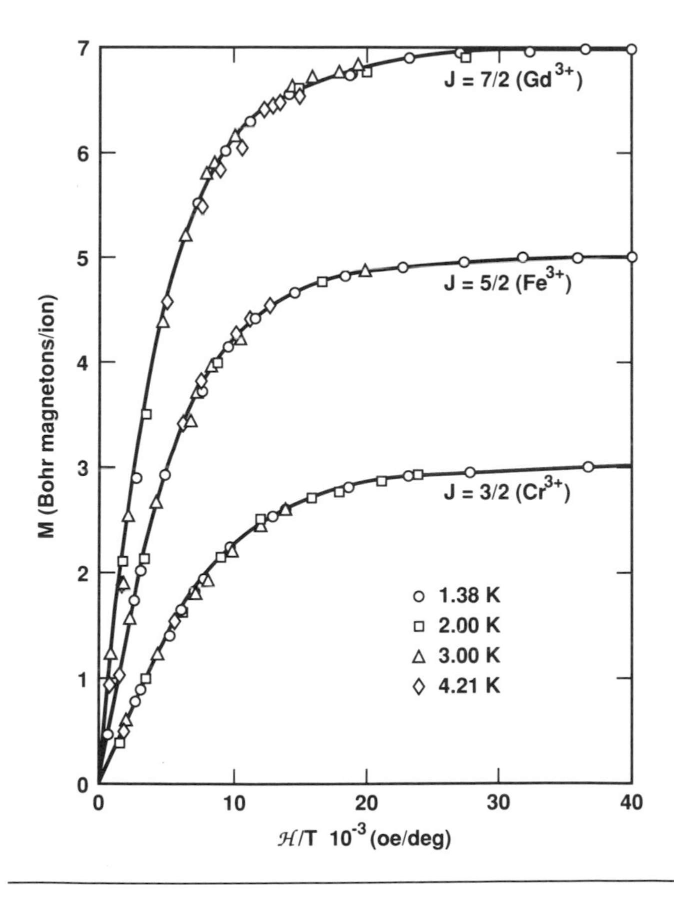

Figure 14.14.

The agreement between the experimental values and the respective Brillouin functions (solid lines). Note the approach to magnetic saturation in each case as $\mathbf{H} \to \infty$. [After W. E. Henry, *Phys. Rev.*, **88**, 559 (1952)].

Consequently, the saturàtion magnetization is equal to the number of unpaired electrons per ion expressed in units of Bohr magnetons. The concept of saturation is important to our discussion of order–disorder because fractional saturation in ferro- and paramagnets is equated to the degree of order.

C. Ferromagnetism

The formalism that led to the Brillouin function can be applied to ferromagnetism as well as paramagnetism. To do this, the external applied field is replaced by a fictitious internal field called the *molecular field* $\mathbf{H}_M$, which is assumed to be directly proportional to the macroscopic magnetization.[15,16] Thus,

$$\mathbf{H}_M = \lambda M(T), \tag{14.62}$$

where λ is the molecular field constant. Then y of (14.58) becomes

$$y = Jg\mu_\beta \lambda M(T)/kT. \tag{14.63}$$

Referring to (14.58), we can reason that the saturation magnetization is given by

$$M(0) = Ng\mu_\beta J, \tag{14.64}$$

since $B_J(y) \to 1$ as $y \to \infty$ when $T \to 0$. This gives the maximum .value of the magnetization. The simultaneous solution for (14.63) and (14.64) is

$$\frac{M(T)}{M(0)} = \frac{kT}{N\lambda g^2 \mu_\beta^2 J^2}\, y. \tag{14.65}$$

The value of $M(T)/M(0)$ must simultaneously satisfy (14.58) and (14.65). This solution can be obtained graphically by plotting $M(T)/M(0)$ from both equations versus y. The value of T that satisfies both corresponds to T_c, the Curie temperature below which ferromagnetic long-range spin order commences. This is illustrated in Fig. 14.15.

Another common way of plotting $M(T)/M(0)$ is shown in Fig. 14.16.

---

[15] All models, such as the molecular field model, strictly apply only within a single magnetic domain.

[16] The atomic magnetic moments of the transition elements are caused by the electrostatic repulsion between electrons on the same atom. The repulsion energy is lowered when the electrons are more widely separated and, of course, are in accordance with the Pauli exclusion principle. This energy lowering is called the exchange energy and, in free atoms, gives rise to Hund's rule, which says that the configuration that maximizes the spin is the ground state. However, maximizing the spin of atoms in molecules can lead to a loss of bond energy, which stems from the pairing of electrons with opposite spins. The situation is entirely analogous to that between high- and low-spin states as discussed in Chapter 6, Section 4C.

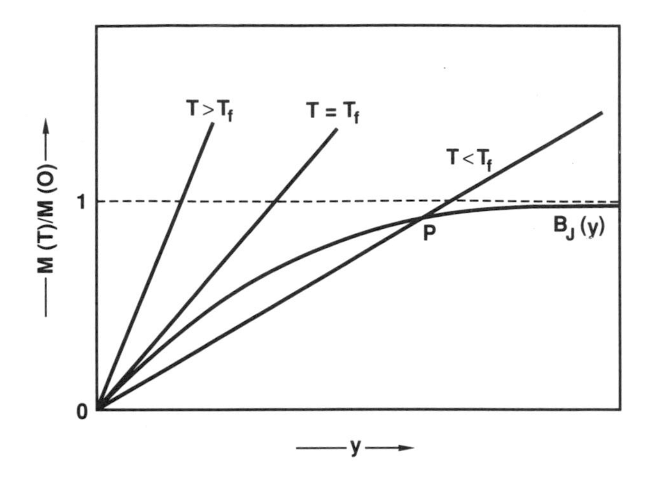

Figure 14.15.

The graphical determination of T_C. The temperature at which the straight line given by (14.65) becomes tangential to $B_J(y)$ as $y \to 0$ defines T_C. (Adapted from A. H. Morrish, p. 263, loc. cit.)

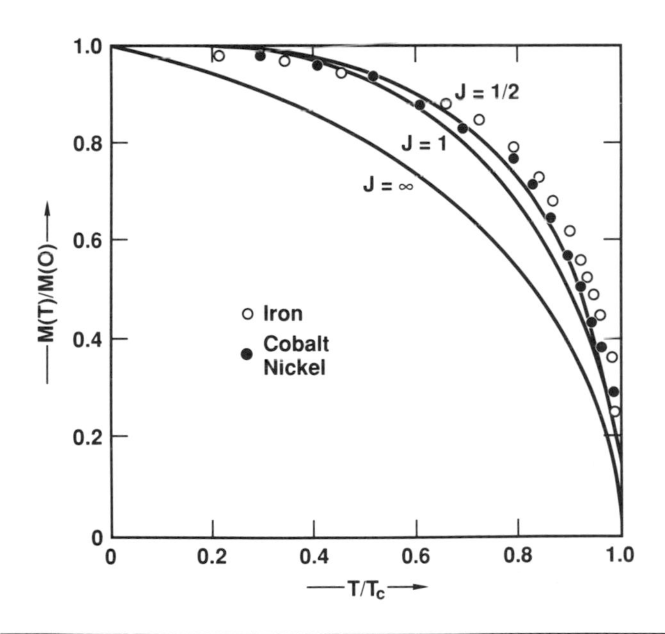

Figure 14.16.

Reduced values of the magnetization as a function of reduced temperature. The solid lines are obtained from molecular field theory. [F. Tyler, *Phil. Mag.*, **II**, 596 (1931), as presented by A. H. Morrish, p. 264, loc. cit.]

The value of $M(T)/M(0)$ can be regarded as a long-range spin-order parameter. In most cases of second-order phase transitions, short-range order generally persists to temperatures well above the commencement of long-range order. In the case of Fe, short-range spin order has been detected by neutron diffraction up to $\sim 100°C$ above T_c. Bulk magnetization measurements clearly cannot measure SRO because of the influence of the applied field upon the magnetic order in the paramagnetic region.

4. Spinodal Decomposition

We have discussed phase diagrams with a miscibility gap in Chapter 9 (see Fig. 9.9). We here use this system to illustrate spinodal decomposition. At temperatures within the two-phase region, the free energy versus composition has two minima (see also Fig. 9.6), which, for clarity, are repeated here in the pertinent phase and free-energy diagrams of Fig. 14.17. The phase diagram shows clearly the two-phase region (α and

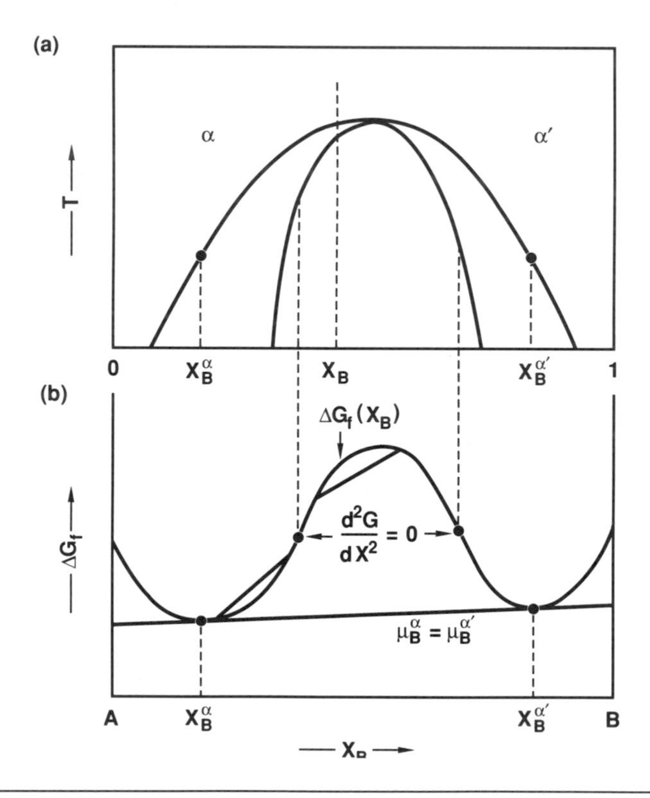

Figure 14.17.

Phase diagram (a) and free energy diagram (b) for a solid solution with a miscibility gap. Diagram (b) at $T = T'$. (See Fig. 9.9.).

α'), and the free-energy diagram is an enlarged region of Fig. 9.6 corresponding to the temperature T' as denoted on the phase diagram.

The singular points identified in Fig. 14.17 are the inflection points with $d^2G/dx^2 = 0$ and are referred to as the spinodes. The spinodes play an important role in the decomposition of the solid solution. The corresponding region between them in the phase diagram is delineated by the vertical spinodal lines. Suppose we quench a solution with starting composition X_B inside the spinodal. Initially the composition will be the same everywhere, and its free energy will be given by ΔG_f on the free-energy diagram. However, this composition is unstable with respect to small fluctuations in composition, giving B-rich and B-poor regions since such fluctuations decrease the total free energy. This is because the slope of ΔG_f versus X_B decreases past the inflection point with increasing X_B. The process will continue until the equilibrium compositions X_B^α and $X_B^{\alpha'}$ are reached. These solutions are in equilibrium with one another, since component B has the same chemical potential in both. In this process no nucleation is required; since the system is spontaneously

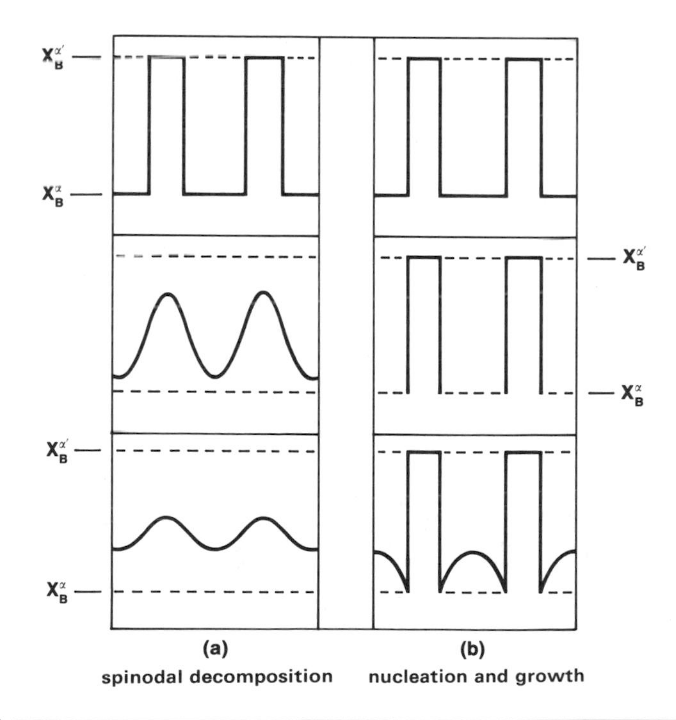

Figure 14.18.

Schematics of the composition modulation for inside (a) and outside (b) the spinodal as a function of time. Discrete discontinuities in composition between α and a' give rise to a surface free-energy term in (b) but not for (a), where the composition varies continuously.

unstable and there is no barrier to the fluctuation. The transition rate is controlled by interdiffusion and results finally in precipitates of X_B^{α} are $X_B^{\alpha'}$.

Consider now a solid solution quenched outside the spinodal region. This produces a metastable alloy, since the slope of ΔG_f versus X_B increases with a small fluctuation in composition. In this case we can decrease the free energy only by forming nuclei of a composition very different from that of the matrix, presumably near compositions X_B^{α} and $X_B^{\alpha'}$, a process that requires overcoming a surface energy barrier. The nuclei then grow to finally form the corresponding precipitates.

The microstructures produced by the two processes are rather different in their appearance and develop as illustrated schematically by Fig. 14.18, where we sketch the composition modulation as a function of time after quenching inside and outside the spinodal regions. The compositional modulation resulting from spinodal decomposition is diffuse with no sharp interfaces between B-rich and B-poor regions. In contrast, nuclei are characterized by very sharp interfaces, since their composition is far from that of the matrix. Limits on the wavelength λ of the modulation in spinodal decomposition can be estimated theoretically, since it depends primarily on the composition gradient and the minimum occurs at the spinodal point. (The maximum rate is expected to occur at this point, since the time scale for the

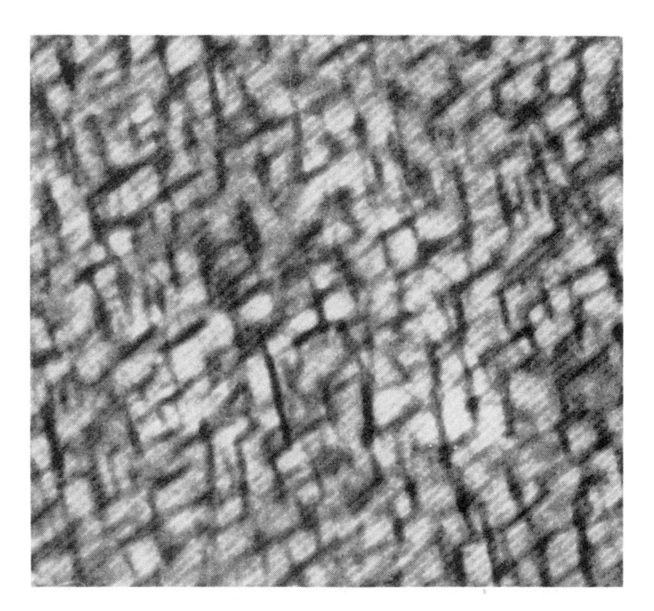

Figure 14.19.

A coarsened spinodal microstructure in Al–22.5 at .% Zn–01% Mg solution treated 2 h at 400°C and aged 20 h at 100°C. Thin-foil electron micrograph (×314.000). (After K. B. Rundman, *Metals Handbook*, 8th ed., Vol. 8, American Society for Metals, p. 184, 1973.)

decomposition reaction is proportional to λ.[17]) Also, the minimum wavelength generally decreases with increasing undercooling. A representative spinodal microstructure is illustrated in Fig. 14.19.

In the nucleation and growth process, the spacing of the precipitate particles depends primarily on the nucleation rate. High nucleation rate coupled with relatively slow growth results in highly dispersed precipitates.

5. Precipitation by Nucleation and Growth

As explained in Section 4, cooling a binary solid solution outside the spinodal (see Fig. 14.17) leads to a decrease of the free-energy only by forming nuclei of a composition very different from the matrix. Thus, separation into, and precipitation of, phases X_B^{α} and $X_B^{\alpha'}$ begins with the nucleation and the subsequent growth of the precipitate particles.

Classical nucleation theory assumes the composition of the nucleus to be constant and only the size to vary. Decomposition into X_B^{α} and $X_B^{\alpha'}$ lowers the free-energy of the system, but the initial nuclei of these compositions are quite unstable because their large surface-to-volume ratios introduce a correspondingly large surface free-energy, which is the positive component of the free-energy of nucleation. The free-energy of formation of a spherical embryo of radius r is given by

$$\Delta G_{\mathrm{f}} = (4/3)\pi r^3\,\Delta G_v + 4\pi r^2\,\Delta G_s, \tag{14.66}$$

where ΔG_v is the difference in the bulk free-energy of the undercooled metastable matrix and the emerging equilibrium phase, and ΔG_s is the interfacial free energy. The surface term is always positive, but ΔG_v becomes increasingly negative as the temperature becomes lower.

At a fixed temperature, with ΔG_v and ΔG_s constant, the r^3 volume term increases faster than the r^2 surface term, and there is a critical radius, r_{c}, determined by the maximum in the free-energy as

$$\frac{\partial\,\Delta G_{\mathrm{f}}}{\partial r} = 0 = 4\pi r_{\mathrm{c}}^2\,\Delta G_v + 8\pi r_{\mathrm{c}}\,\Delta G_s. \tag{14.67}$$

Rearranging (14.67) yields

$$r_{\mathrm{c}} = -2\,\Delta G_s/\Delta G_v, \tag{14.68}$$

---

[17] J. Burke, *The Kinetics of Phase Transformations in Metals*, pp. 118–124, Pergamon Press, 1965.

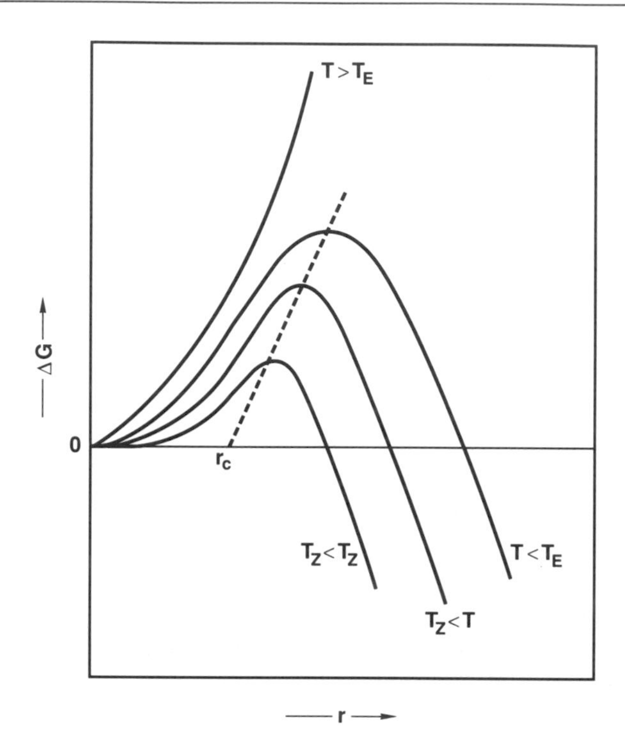

Figure 14.20.

The free energy of formation of spherical embryos as a function of the radius for a series of temperatures.

and, hence,

$$\Delta G_{\mathrm{f}} = 16\pi \, \Delta G_s^3/3 \, \Delta G_v^2 \, . \tag{14.69}$$

Figure 14.20 traces the free-energy path accompanying particle growth as a function of temperature.

As noted ΔG_v becomes increasingly more negative with decreasing temperature since the stable phase at low temperature is the precipitate. If T_{E} is the equilibrium transformation temperature, then there are no stable nuclei above T_{E} and ΔG_f increases without limit with increasing r. The surface free-energy is known to vary rather slowly with temperature. Thus, there is no homogeneous nucleation at or above this temperature. The equilibrium[18] number of nuclei of critical size per unit

[18] Virtual equilibrium is assumed here based on quasi-equilibrium kinetics.

volume is given by

$$N_c = N \exp(-\Delta G_f/RT), \tag{14.70}$$

where N is the number of atoms per unit volume.

The next step in the precipitation process is the growth of nuclei of size $r > r_c$. In the simplest model, the rate of condensation of the atoms on the surface of spherical particles is assumed to be essentially instantaneous so that precipitation is controlled by diffusion to the interface. In Fig. 14.21, C_I is the initial concentration lying somewhere between X_B^α and $X_B^{\alpha'}$ of Fig. 14.17. The equilibrium concentration of the matrix directly at the interface is C_E, and X_B^α is the equilibrium concentration of component B in the α phase. As a first approximation C_E is taken to the independent of r which is only valid for particles that have exceeded their initial submicroscopic dimensions. Consider now a single isolated particle of instantaneous radius R in a semi-infinite supersaturated solid solution. As the interface advances a small distance dR, the number of additional atoms precipitated is $C_B\, dR$, where C_B is the concentration of B in the α phase. Of these $C_E\, dR$ were supplied by the interface and the remainder arrived by diffusion. Thus, one can write for the flux at the interface

$$(C_B - C_E)\frac{dR}{dt} = D\left(\frac{\partial C}{\partial r}\right)_{r=R}, \tag{14.71}$$

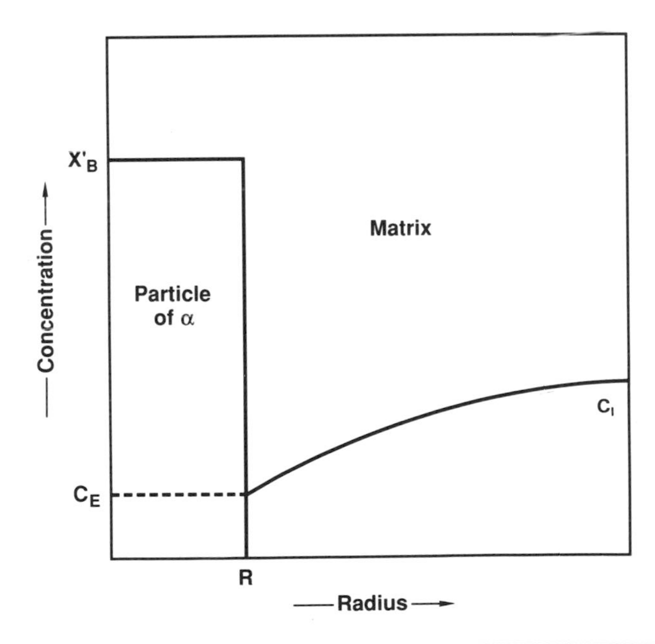

Figure 14.21.

The concentration of solute in and around a growing precipitate at constant temperature.

where D is the diffusion coefficient of the solute atoms. For small degrees of supersaturation we can employ the steady-state approximation. The steady-state solution of the equation for diffusion through a spherical shell of inner radius R and outer radius R' is

$$\frac{\partial C}{\partial r} = \frac{C' - C_E}{1/R - 1/R'} \cdot \frac{1}{r^2}, \tag{14.72}$$

where C' is the concentration at R'. Setting $R' = \infty$ and $C' = C(t)$ gives, for the gradient at the interface, by substitution into (14.72),

$$\left(\frac{\partial C}{\partial r}\right)_{r=R} = \frac{C(t) - C_E}{R}. \tag{14.73}$$

Combining (14.71) with (14.73) yields

$$R\frac{dR}{dt} = D\left[\frac{C(t) - C_E}{C_B - C_E}\right]. \tag{14.74}$$

If W is the fraction already precipitated, then

$$\frac{C(t) - C_E}{C_I - C_E} = 1 - W, \tag{14.75}$$

and the rate of growth for one of a set of competing particles is given by

$$R\frac{dR}{dt} = D\left(\frac{C_I - C_E}{C_B - C_E}(1 - W)\right). \tag{14.76}$$

Initially, t is small and $C(t) = C_I$, $(1 - W) = 1$, and then, by integration, R^2 is proportional to the time. The particle volume then is expected to grow proportionally to $t^{3/2}$.

More complicated expressions are needed at longer times when interparticle interference becomes important. The calculations must also be modified for the growth of nonspherical particles. Rather more elaborate theories have been proposed, but fortunately they, as well as many experiments, are empirically quite well described by the rate equation

$$y = 1 - e^{-(kt)^n}, \tag{14.77}$$

where y is the fraction transformed and k and n are empirical parameters. The value of n characterizes the model processes as shown in Table 14.2. Equation (14.77) is variously known as the Johnson–Mehl and Avrami relation.[19] It has been applied

[19] See, for example, Burke, particularly Sections 2.5, 2.6, and 7.5, loc. cit.

Table 14.2. Values of n in (14.77) for different models.[a]

Model	n
Diffusion-controlled growth of a fixed number of particles	3/2
Growth of a fixed number of particles limited by the interface process	3
Diffusion-controlled growth of cylinders in axial direction only	1
Diffusion-controlled growth of discs of constant thickness	2
Growth on dislocations	2/3
Nucleation at a constant rate and diffusion-controlled growth	5/2
Growth of a fixed number of eutectoid cells	3
Nucleation at a constant rate and growth of a eutectoid	4

[a] After Burke, loc. cit.

in describing and analyzing a wide variety of solid state experimental kinetic data. The constant k in (14.77) is not a conventional rate constant since the corresponding dy/dt involves both y and t. Thus, its temperature dependence may or may not be simple.

A closely related differential equation is often used in the form

$$\frac{dy}{dt} = k_m(1 - y)y^{(m-1)/m}, \tag{14.78}$$

which defines another rate constant, k_m, and an exponent m different from n in (14.77). Equation (14.78) is a true rate equation, and k_m is a true rate constant. The temperature dependence of k_m may then be taken to be governed by the usual Arrhenius relation; i.e.,

$$k_m = Ae^{-E_A/RT}, \tag{14.79}$$

where E_A is the activation energy.

6. Martensitic Transformations

There is an important class of phase transformations involving diffusionless processes without thermally activated atomic motion or change in composition. The major representatives in this category are called the martensitic transformations. The

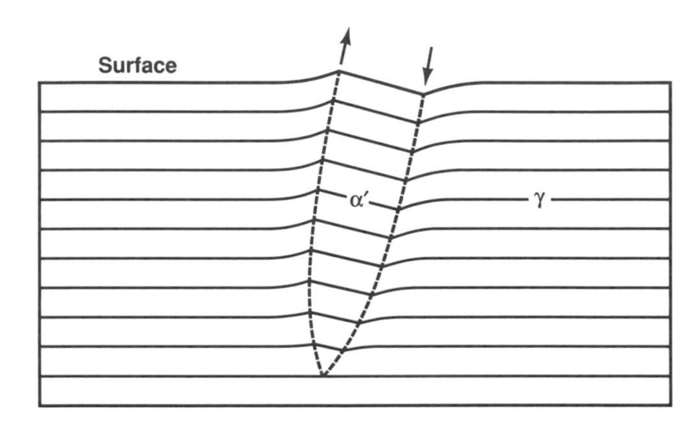

Figure 14.22.

Schematic illustration of surface tilting by a martensite particle. (After D. A. Porter and K. E. Easterling, *Phase Transformation in Metals and Alloys*, Van Nostrand Reinhold, Chap. 6, 1987.)

mechanism is displacive, and such phase changes consist of small cooperative displacements of atoms from their equilibrium positions, thus changing the crystal symmetry. The rate of growth of martensite achieves velocities comparable with that of sound. Shape and volume are both altered, and the deformation is a combination of simple shear and contraction or expansion.

The nucleation sites are believed to be localities of high internal strain derived from dislocation tangles. A premartensitic structure is recognizable from electron microscopy and has been correlated with a change in the vibrational spectrum (for Ni–Al) by neutron diffraction. The martensite crystals usually form as lenslike small platelets and tend to tilt out of the surface to accommodate a shape change, as illustrated in Fig 14.22. These surface relief patterns are characteristic of the martensitic transformation, and a typical sequence of such a transformation is illustrated in Fig. 14.23.

A. The Fe–Ni System

We shall illustrate the principles basic to this type of transformation with the iron–nickel system, emphasizing primarily the essential thermodynamics rather than the intricate geometry, crystallography, and particle nucleation. The stable γ phase at high temperatures is fcc. The low-temperature stable phase is either α, which is bcc, or a mixture of α and γ, depending on the bulk composition; the martensite is tetragonal.

Let us first of all look at the free-energy diagram as a function of temperature for such an alloy at a specific composition X_{Ni}. If a diffusionless transformation occurs,

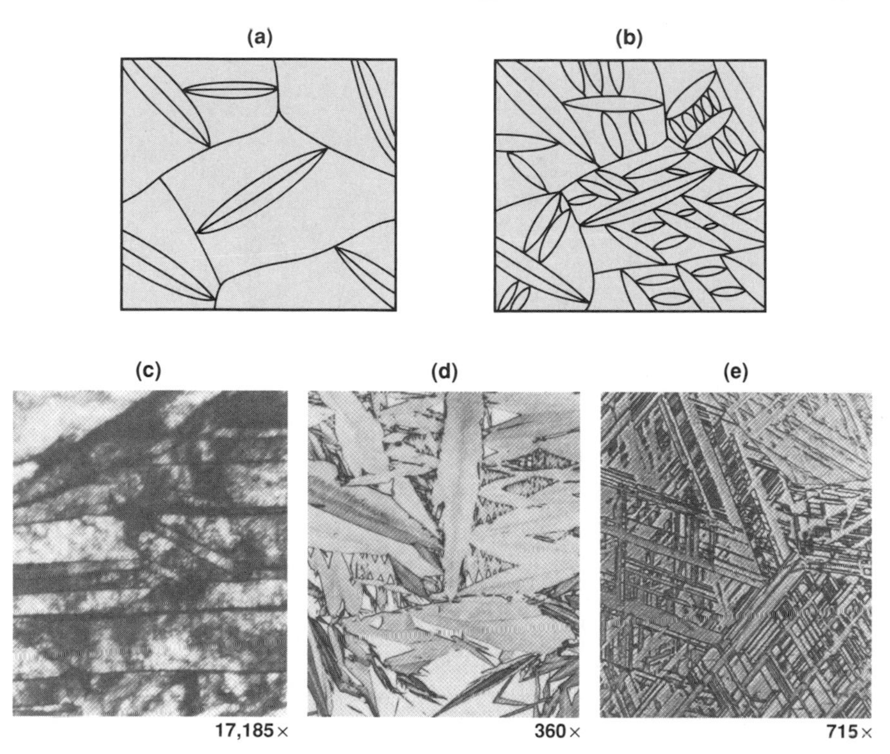

Figure 14.23.

(a), (b) Growth of martensite with increasing cooling below M_s. (c)–(e) Different martensite morphologies in iron alloys: (c) lath, (d) plate, (e) sheet, (Figures a, b after Porter and Easterling, loc. cit.; figures c, d, e from *Metals Handbook*, Vol. 8, American Society for Metals, pp. 198, 199.)

the composition of both phases is fixed and the free energy is a function only of the temperature, as shown in Fig. 14.24. At temperature T_0, the free energies of γ and α' are equal. The notation α' means that this phase has a distorted bcc structure (tetragonal) but not the equilibrium α composition that would be attained by long-range diffusion. Thus, α', the martensite, is metastable with respect to α and $G_{\alpha'} > G_\alpha$. A portion of the corresponding phase diagram is sketched in Fig. 14.25, where the solid circle shows the α' point corresponding to T_0. Undercooling is *required* for the spontaneous transformation to α'. This is the temperature difference $M_s - T_0$, as illustrated in Fig. 14.24, where M_s is called the martensite start temperature. At this temperature the lowering of the free-energy by transformation is large enough to overcome the forces opposing the phase change (strain energy, surface energy). Upon further cooling, temperature T_s of Fig. 14.25 is reached, which

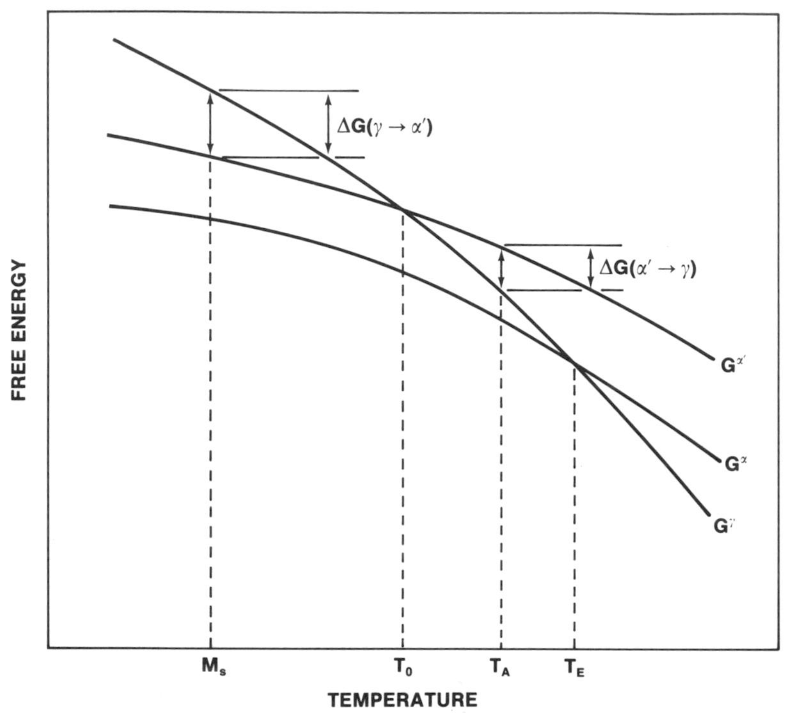

Figure 14.24.

Schematic plot of G_γ, $G_{\alpha'}$, and G_α for an Fe–Ni alloy versus temperature.

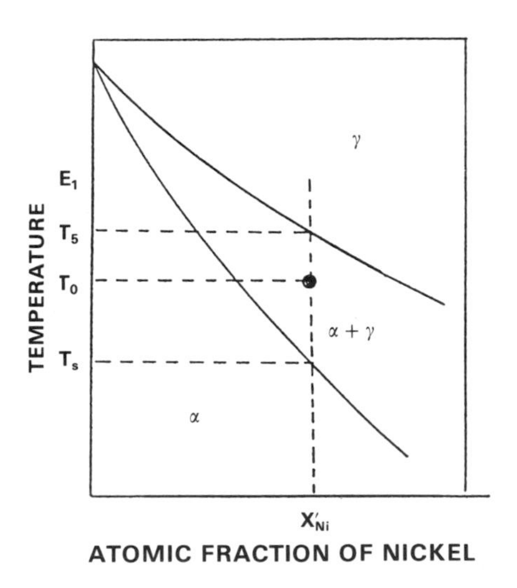

Figure 14.25.

Sketch of part of the Fe–Ni phase diagram.

is the equilibrium temperature below which alpha is the stable phase. The high-temperature point T_A in Fig. 14.24 represents the spontaneous transformation of α' to γ so that $T_A - T_0$ is the corresponding superheating interval. T_E in this figure is the equilibrium temperature for the transformation of the α to the γ phase.

Schematic free-energy–composition diagrams are shown in Fig. 14.26, which illustrate the cooling of an alloy of composition X'_{Ni} from T_4 to T_1 and below. At

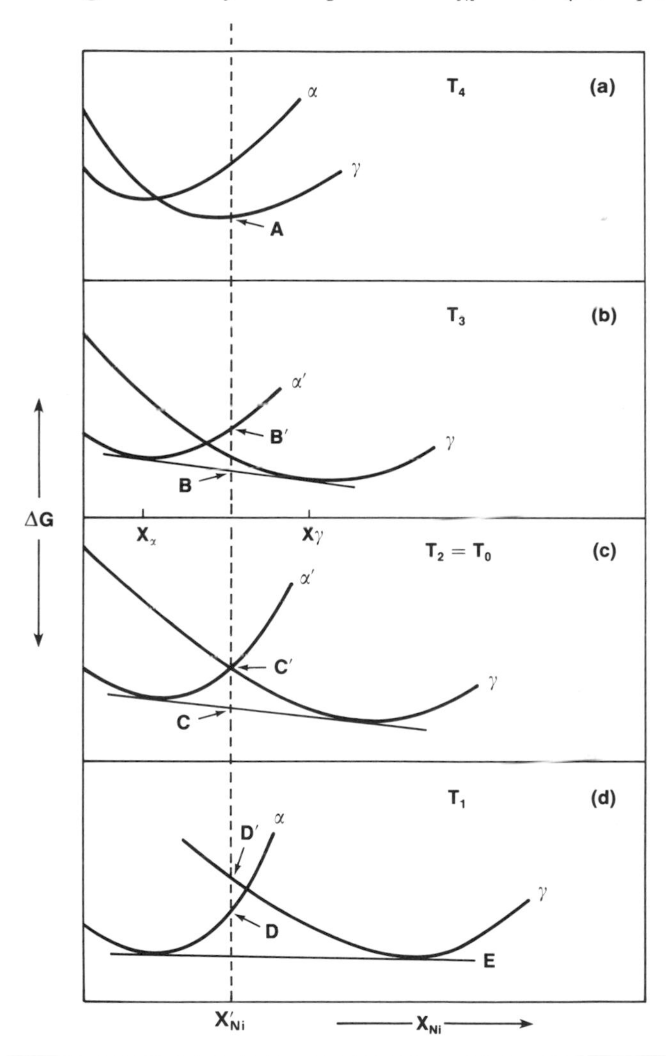

Figure 14.26.

Schematic free-energy–composition diagrams for the Fe–Ni system at four temperatures decreasing from top to bottom. (Reprinted with permission from L. Kaufman and M. Cohen, *Thermodynamics and Kinetics of Martensitic Transformations*, Copyright 1958, Pergamon Press.).

the high temperature T_4, the stable phase is γ with a free-energy value A. As the system is cooled to T_3, it wants to decompose into X_α and X_γ, but if diffusion is too slow for this to occur, the system remains quenched as γ with the free energy at B. α' of composition X'_{Ni} cannot form at this temperature because its free energy B' is higher than B. The α'–γ free-energy curves cross at T_2, and, therefore, T_2 is the same as T_0 of Fig. 14.24, where, in the absence of diffusion, the free energy is C'. At temperatures below T_0, again with diffusion not occurring, the α' phase is more stable than γ since its free energy at D is less than at D'. It is, of course, metastable relative to α or to the decomposition into X_α and X_γ. At T_1 and below, the alloy may transform martensitically to α' if ΔG of the transformation is larger than the forces opposing the phase change; i.e., if $T < M_s$. At this and lower temperatures, the diffusion rate is so small that there is rapid production of α' by martensitic precipitation.

B. The Fe–C System

Because it is a necessary component of steel, Fe–C martensite has been by far the most studied of all such systems. Nevertheless, there is no universally accepted detailed mechanism for this particular transformation. Because of its extreme importance a brief, qualitative description follows.

Pure bcc Fe transforms above 910°C to fcc, γ-Fe. The latter can dissolve, interstitially, more carbon than the low-temperature bcc phase. This is because the octahedrally coordinated fcc site (see Figs. 1.13 and 1.14) is larger than its tetrahedral counterpart in the bcc structure (problem 2, Chapter 1). Consequently, given a sufficiency of dissolved C, an fcc solid solution quenched to lower temperatures in the bcc regime becomes supersaturated with respect to carbon. If the final temperature is below a critical temperature that depends upon composition, body-centered tetragonal martensite forms. The amount is variable, depending upon composition, final temperature, and quenching rate. Because of the local strain associated with the martensitic phase, it becomes the essential hardening agent for all steels. Alloying with other transition elements will also affect the mechanical properties of steel, but martensite is the universally essential ingredient.

7. Glasses

Glasses have been known and used since antiquity. We are all fully familiar with a piece of glass, but a definition of a glass is difficult; we like the one given by Morey[20]

[20] G. W. Morey, *The Properties of Glass*, Reinhold, New York, 1938.

> A glass is an inorganic substance in a condition which is continuous with, and analogous to, the liquid state of the substance, but which, as the result of having been cooled from a fused condition, has attained so high a degree of viscosity as to be for all practical purposes rigid.

This definition is suitable here since it is a physical–chemical definition, and we shall be concerned primarily with thermodynamics and structure. The restriction to inorganic materials is somewhat arbitary, but again, this book is limited to such materials.

Let us start by looking at a representative volume versus temperature response of a glassy material as illustrated in Fig. 14.27. Cooling down from the liquid state, we come to the temperature T_m, which corresponds to the melting point of the thermodynamically stable crystalline state. Crystallization is, however, prevented by the high viscosity of the melt and, hence, the low mobility of the constituent atoms. The volume

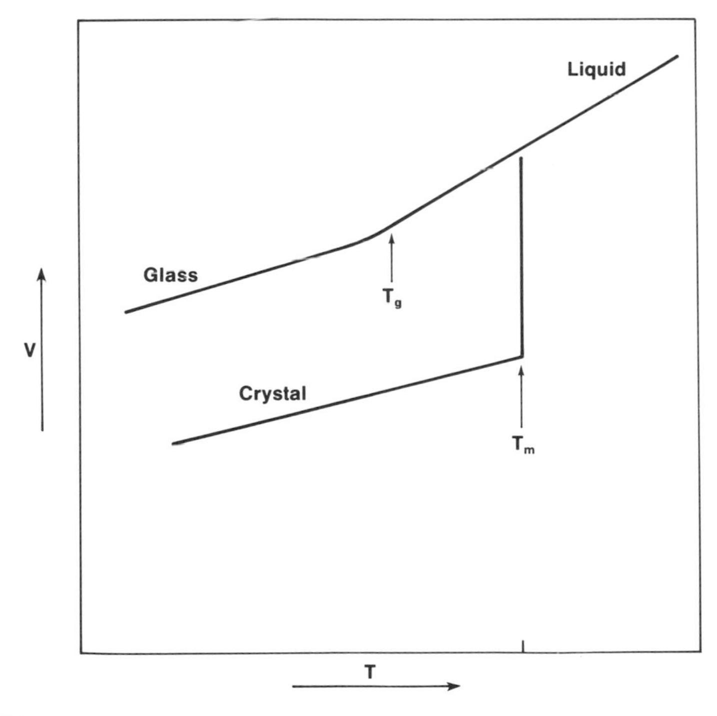

Figure 14.27.

Schematic diagram of the volume as a function of temperature for liquid, glass, and crystal. T_m is the melting point of the crystal, and T_g is the glass temperature.

continues to shrink in this undercooled liquid, with the slope becoming more gradual as the viscosity increases, and a glassy solid finally results. The region of rapid slope change is usually associated with T_g, the "glass temperature."

Thermodynamically, the foregoing process is a second-order transition to a metastable state, and there is no discontinuity in the volume as a function of temperature in contrast to the first-order melting of a crystal at T_m. There are, however, discontinuities at T_g in the derivatives, such as the coefficient of thermal expansion and the specific heat. The classical example for the rapid change in the specific heat is illustrated by the C_p versus T curve for vitreous silica in Fig. 14.28.

As already alluded to, the viscosity of the material plays a crucial role in glass formation. Near the glass temperature T_g, it is about 10^{13} poise while at T_m it is about 10^2 poise. In the T_m region, one would expect crystalline nuclei to form, but in these glass-forming materials this is a difficult and slow process. The driving force for nucleation is the difference between the surface and bulk-free energies (14.66). The measured difference in the enthalpies of formation of the glassy and crystalline states is very small; for cristobalite and glassy silica they are, in fact, the same. Because of their similar structures, the entropic difference is also expected to be small. As a

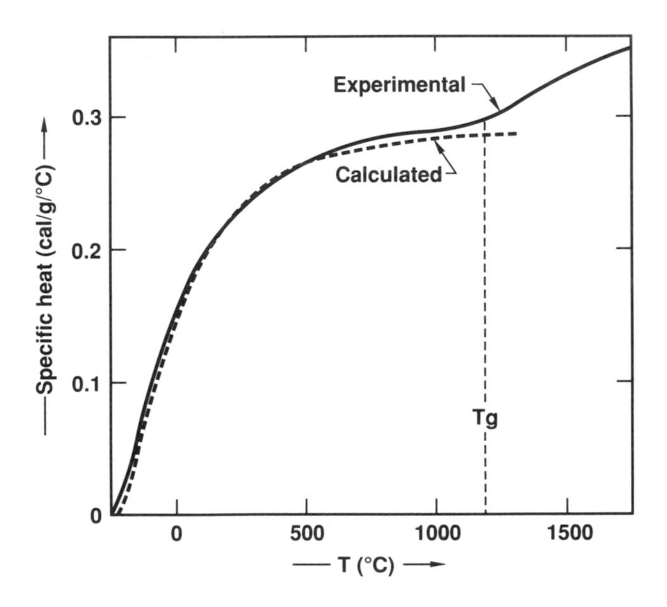

Figure 14.28.

Experimental and calculated values of the specific heat of vitreous silica. [After H. T. Smyth, H. S. Skogen, and W. B. Harsell, *J. Amer. Ceram. Soc*, **36**, 327 (1953).]

consequence, the overall driving force is inferred to be small. To this must be added that the surface free-energy term is also small, otherwise quick cooling would not be necessary for glass formation. This brings us back to viscosity and slow molecular mobility. The critical region for crystallization is between T_m and T_g. As we approach the low mobility region at T_g, any crystallization, now potentially driven by easier nucleation, is prevented by the high viscosity. The maximum crystallization rate is between viscosity values of 10^6 to 10^8 poise. Glass is usually worked at viscosity values of 10^3 to 10^5 poise and then cooled rapidly enough to prevent crystallization. Below T_g most glasses, although thermodynamically metastable, normally will persist for very long times in the vitreous state.

Pertaining to the structure of glasses, we shall limit our discussion to inorganic oxide glasses and examine their structure within the framework of the Zachariasen[21] theory. In this theory the short-range order arrangement of the atoms or ions in the glass and the crystal is about the same, with the major difference between the two states residing in the long-range order, which is absent in the vitreous network state. These two states are illustrated schematically in Fig. 14.29 for polyhedra in two dimensions. The condition for the existence of a network is that the corresponding crystal structure be relatively open and that, by only slight distortions of the bond angles, the molecules may be joined in the irregular network structure.

Simple rules formulated by Zachariasen as the criteria for network formation in inorganic oxides, X_nO_m, may be stated as follows:

1. An oxygen atom may be linked to no more than two X atoms.
2. The number of oxygen atoms surrounding each X atom must be small (3 or 4).
3. The groups of oxygen atoms (i.e., polyhedra) share corners only, not edges or faces. Further, if the network is to be three-dimensional, at least three corners of the polyhedra must be shared with other polyhedra.

Violation of conditions 1 and 2 leads to compact and rigid structures that would randomize with difficulty. If condition 3 is violated, then the polyhedra would be so strongly fixed relative to each other that the required flexibility for network formation would be lost.

Detailed x-ray studies by Warren and co-workers confirmed experimentally many aspects of the network theory.[22] Overall, the Zachariasen–Warren picture is quite a satisfactory phenomenological description of glasses. The most familiar glass-forming

[21] W. H. Zachariasen, *JACS*, **54**, 3841 (1932).

[22] B. E. Warren, *Z. Kristallogr.*, **86**, 349 (1933); B. E. Warren, H. Krutter, and O. Morningstar, *J. Amer Ceram. Soc*, **19**, 202 (1936).

(a)

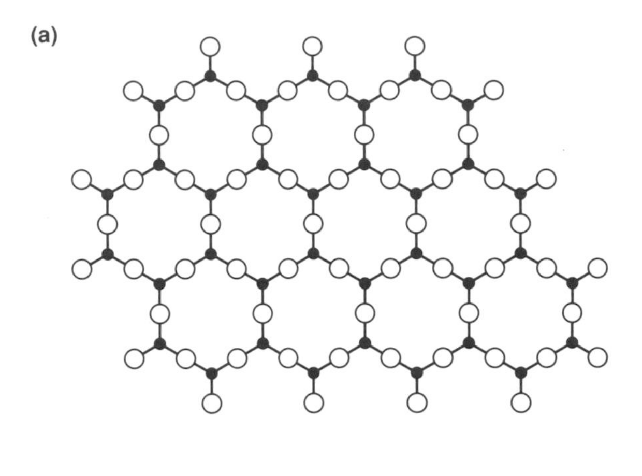

(b)

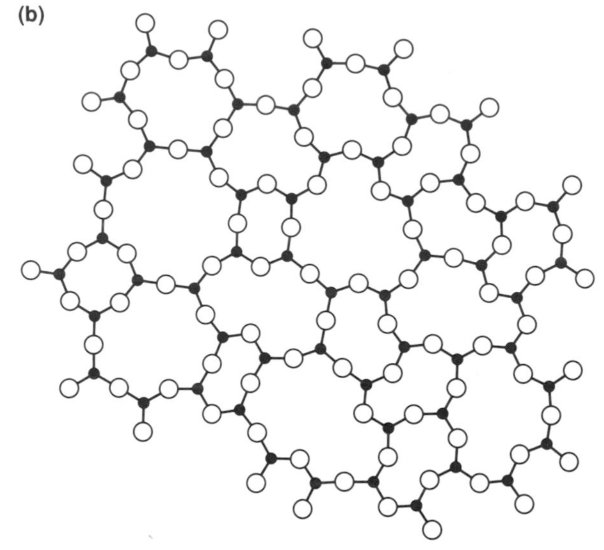

Figure 14.29.

Schematic representation of a network of polyhedra: (a) crystalline state, (b) as a glass.

oxides, which do obey Zachariasen's rules, are B_2O_3, SiO_2, GeO_2, P_2O_5, Al_2O_3, and Sb_2O_3.

There are many other materials that form glasses,[23] including such improbable candidates as pure carbon and alloys.

---

[23] See, for example, the review by J. M. Stevels, "The Structure and Physical Properties of Glass," *Handbuch der Physik*, Vol. XIII, pp. 510–645, 1962.

Table 14.3. Several glass compositions (percent by weight).

	SiO_2	Na_2O	MgO	CaO	Al_2O_3	B_2O_3
Window glass	72	12.6	2.6	11	1	
A bottle glass	74	17	3.5	5.0		
Pyrex brand	80	4.0			2.2	13

The pure binary oxide glasses are usually modified by fusing them with alkaline oxides. These additions are important for a variety of reasons, such as lowering the glass working temperature, imparting leaching resistance, preparation of colored glass, etc. In Table 14.3, we show some typical glass compositions. Where should one place these extra atoms, the network modifiers? Research by Warren and co-workers showed that the arrangements are as indicated schematically in Fig. 14.30. Some oxygens are now bonded to only one silicon, and the average ring size has been increased. The charge on the nonbridging oxygen is compensated by the added cations within the rings.

Another very physicochemical approach to network formation is that by Winter[24] who suggested that the average number of p electrons per atom is of importance. The group VI elements with four p electrons per atom are the only ones known to form monatomic glasses. They also form the well-known binary glasses with elements of groups III to V, provided the ratio of the total number of p electrons, n_p, to the total number of atoms n_a, is larger than 2. The same criterion applies to glasses formed by the elements of group VII with those of groups II to V. This point is illustrated for sodium silicates in Table 14.4.

This theory is of interest in relating network formation directly to chemical bonding. The p electron forms directed covalent bonds with ease, which, in turn, tend to lead to open structures and high viscosities, the obvious ingredients of glass formation. The theory is an approximate one with numerous exceptions, but it is a useful guide to the chemistry of glass formation.

We should mention briefly metallic glasses, usually prepared by very rapid quenching, a process known for over 30 years.[25] Just as in the case of inorganic

---

[24] A. Winter, *C. R. Acad. Sci. Paris*, **240**, 73 (1955).

[25] See, e.g., Rodney Cotterill, *The Cambridge Guide to the Material World*, Chap. 11, Cambridge University Press, New York, 1985.

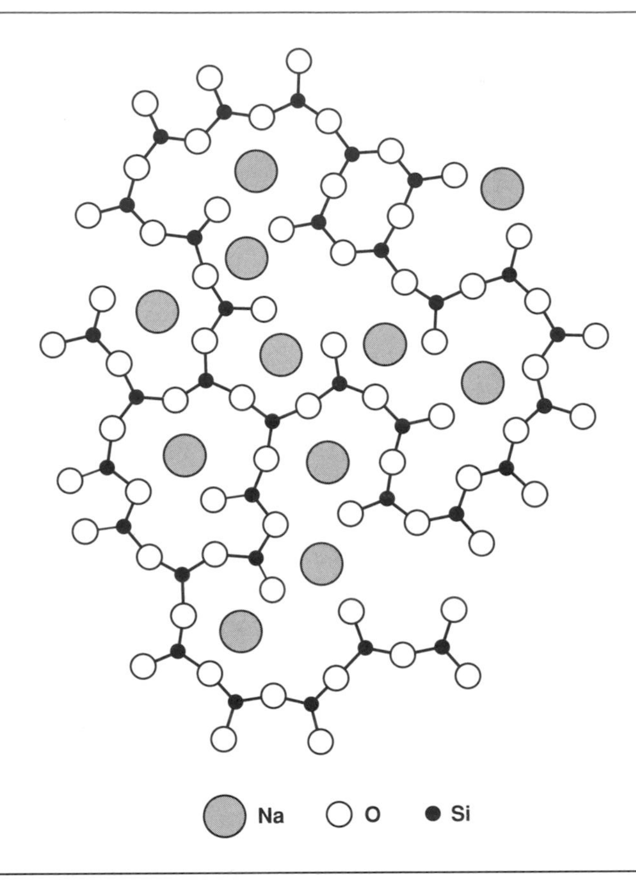

Figure 14.30.

Schematic representation in two dimensions of the structure of soda–silica glass. [After B. E. Warren and J. Biscoe, *J. Amer. Ceram. Soc.*, **21**, 259 (1938).]

Table 14.4. The ratio n_p/n_a for a number of sodium silicates in its relation to their glass-forming abilities.[a]

Composition	SiO_2	Na_2O_1 $2SiO_2$	Na_2O_1 $2CaO_1$ $3SiO_2$	Na_2O_1 SiO_2	$3Na_2O_1$ $2SiO_2$	$2Na_2O_1$ SiO_2
Number of atoms (n_a)	3	9	16	6	15	9
Number of p electrons (n_p)	10	24	42	14	32	18
Ratio n_p/n_a	3.3	2.66	2.63	2.33	2.1	1
	Easy to obtain in vitreous state			Easy devitrification		No glass formation

[a] After Stevels, loc. cit.

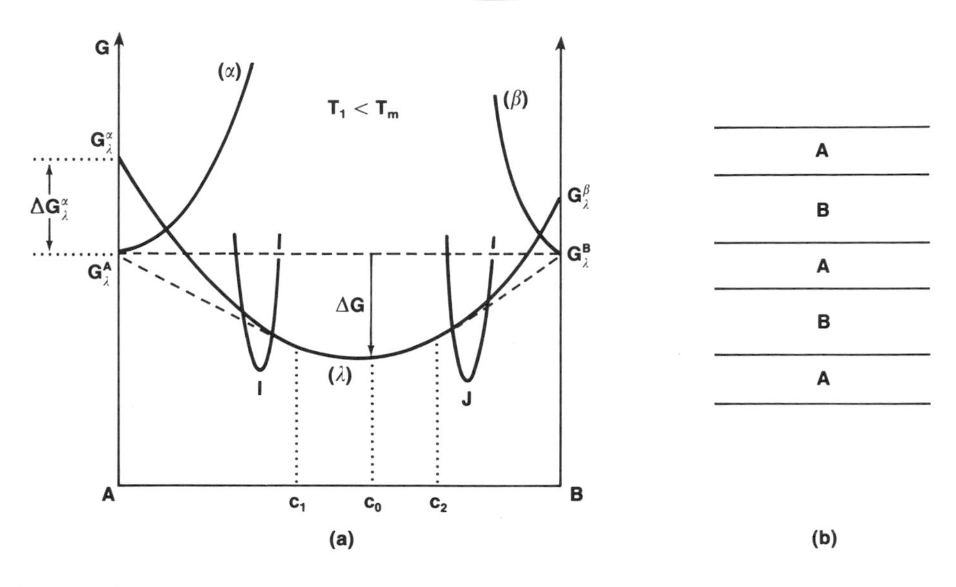

Figure 14.31.

Upon amorphization of a crystalline multilayer of elements A and B(b), the Gibbs free energy of the system decreases by ΔG. I and J denote intermetallic compounds. (From Gerl and Guilmin, loc. cit.)

glasses, one has to quench fast enough to cross the temperature interval ΔT between T_m the melting temperature of the alloy, and the glass temperature T_g. The composition of these glasses is normally near deep eutectics, where ΔT is small enough to be traversed by reasonable quench rates. Because metal atoms diffuse much more rapidly than molecules, the quench rates necessary for the formation of metallic glasses must be extremely rapid. Rates of the order of 10^6 °C/sec are frequently attained.

More recently, techniques have been developed for preparing amorphous alloys by interdiffusion.[26] It is sometimes possible to obtain an amorphous alloy by diffusion from multilayers of crystalline solids prepared by vapor deposition. The thermodynamics is illustrated in Fig. 14.31.[27] Driven by the increase in configurational entropy, amorphization can, sometimes, lower the free energy. If the nucleation and growth of the intermetallic compounds I and J can be avoided by keeping the

[26] See, e.g., W. L. Johnson, *Progr. Mater. Sci.*, Pergamon Press, **30**, 81 (1966).

[27] M. Gerl and P. Guilmin, *Solid State Phenomena*, Vols. 3, 4, Masson, pp. 215–1124, 1988.

temperature low enough and yet high enough for significant interdiffusion, an amorphous alloy may be produced by an isothermal solid state reaction. This type of amorphization has been observed in a number of systems.

8. The Photographic Process

The photographic process, as used in practice[28], is intricate and complex. This complexity is well illustrated by Fig. 14.32, where a scanning electron micrograph of a color film is shown.[29] In this section we discuss briefly the fundamental steps as they lead to the nucleation and growth of specks of silver and represent, in a sense, a phase transition induced by exposure to light. The essential ingredient of a photographic emulsion is silver bromide in the form of microcrystals that have usually been sensitized, for example, by sulfide compounds to form "sensitivity centers" at which the latent image will form upon exposure to light. A photographic developer then reduces the silver ions to silver in the microcrystals that contain a latent image, and thereby produce the visible picture. The latent image speck acts as a catalyst in this process by accepting electrons from the developer. The enhancement of this submicroscopic latent image speck produced by only a few photons to a much larger grain of silver is the heart of the photographic process. The amplification of the optical signal by this catalytic action is about 10^8 to 10^9.

Our understanding of the reactions leading to the latent image is based on the Gurney–Mott theory proposed in 1938.[30,31] According to this model, exposure to light creates a photoelectron and a positive hole. The photoelectron migrates to and is captured by a deep electron trap, the sensitivity center. The positive hole is most likely trapped at a Br^- and thus produces Br^0. A randomly migrating interstitial silver ion finds the trapped electron and becomes reduced to a silver atom, Ag^0, by this reaction. This single silver atom center is neither stable nor developable, but repetition of the foregoing steps at the same place leads to the formation of the latent

[28] See, for example, J. F. Hamilton, *Adv. Phys.*, **37**, 359–441 (1988).

[29] After T. Tani, *Physics Today*, **42**, 37 (1989).

[30] R. W. Gurney and N. F. Mott, *Proc. Roy. Soc. London, Ser. A*, **164**, 151 (1938).

[31] N. F. Mott and R. W. Gurney, *Electronic Processes in Ionic Crystals*, Chap. 7, Oxford, 1940.

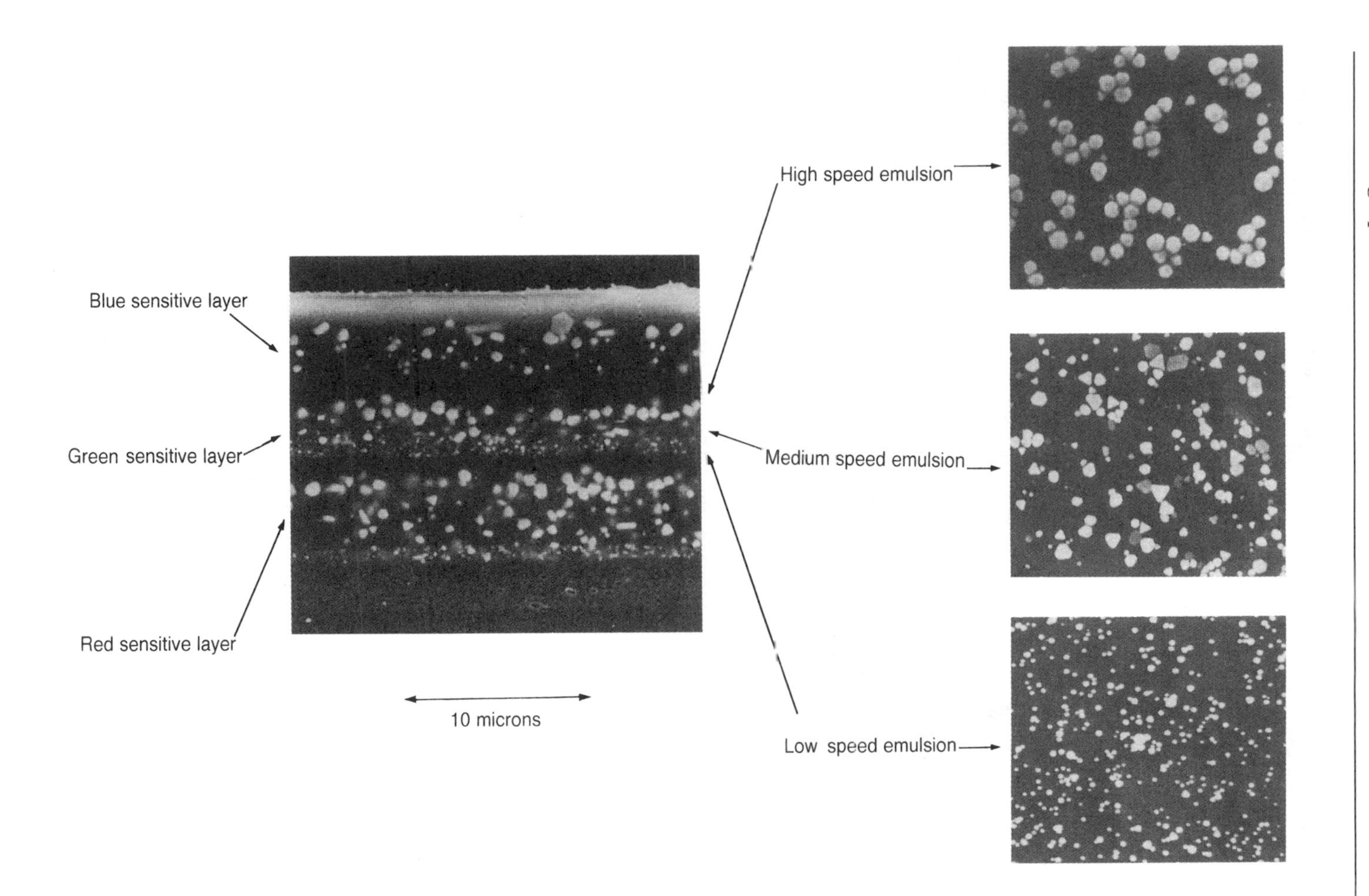

Figure 14.32.

Scanning electron micrograph of the cross section of a color negative film. Its 19-μm thickness consists of 14 layers with different functions, containing a total of 1.2×10^7 well-designed silver halide microcrystals and more than 100 kinds of organic functional compounds. (After Tani, loc. cit.) Each color sensitive layer is itself a multilayer of varying grain size. The larger grains of AgBr will develop latent images in response to lower levels of illumination as the probability of producing Ag^0 increases with grain size.

image by this reaction. These steps may be described by the following chemical equations with T representing the sensitivity center electron trap.

$$\text{photon} \rightarrow e^- + h^+, \tag{a}$$

$$e^- + \mathrm{T} \rightarrow e_\mathrm{T}^-, \tag{b}$$

$$e_\mathrm{T}^- + \mathrm{Ag}_i^+ \rightarrow \mathrm{Ag}^0 + \mathrm{T}. \tag{c}$$

Repetition of (a) to (c) n times at the same center yields the latent image speck $(\mathrm{Ag}^0)_n$. Several recent studies indicate that the minimum number of silver atoms to form a stable and developable latent image is 4.[32] The dimer center, $n = 2$, is found to be stable but not developable.

There are several unusual properties of silver halides that are crucial to the preceding process. First of all, the band gap of most ionic crystals is much too wide for electron–hole generation by visible light, but in silver halides it is relatively small. In AgBr the absorption rises rapidly in the blue region of the visible light spectrum. the transition being an "indirect" one from the inverted valence band, which resulted from the interaction of the silver d states with the halide p states to the conduction band. Next, we need to invoke the intrinsic defect structure of the silver halides, which is mainly Frenkel pairs (i.e., interstitial silver ions and vacant lattice sites). The interstitial silver ions are known to migrate with a very low activation energy (in the 0.02 and 0.04 eV range), and this high mobility is needed for an efficient Ag^+–e_T^- reaction. However, the fraction of Frenkel pairs in bulk silver halides at room temperature is very low. What saves the situation is that the defect concentration is higher by a factor of 100 or more[33] in microcrystals. This amplification arises from the surface charge and the associated subsurface electric field in microcrystals. These are then the unusual solid state properties of the silver halides essential to the operation of the photographic process.

Chapter 14 Exercises

1. A fictitious element has a density of 7.8 g cm^{-3} in the solid state at its melting temperature T_m of 1500 K; the density of the liquid is 7.5 g cm^{-3}. If the atomic weight is 50 and the heat of melting is 4.1 kcal/mole, calculate the change in T_m caused by (a) reducing the ambient pressure from 1 to 0 atm, (b) increasing the pressure from 1 to 10^4 atm, (c) increasing the pressure from 1 to 3×10^5

---

[32] Hamilton, loc. cit.

[33] Hamilton, loc. cit.

atm. (d) What additional data are needed to make these calculations more accurate?

2. Derive the expression

$$\left(\frac{\partial E}{\partial T}\right)_P = C_p - P\left(\frac{\partial V}{\partial T}\right)_P.$$

3. Derive a general thermodynamic expression for the equilibrium vapor pressure of a condensed phase as a function of applied pressure

4. The element Hg boils at 630 K. Calculate the equilibrium vapor pressure at 300 K.

5. Derive an expression for the dependence of solubility upon pressure for a binary (a) ideal solution and (b) regular solution.

6. A substance α has an allotropic transition to β at high pressure and 273 K. The same transition occurs at low pressure at a much lower temperature. Which has the larger molar volume, V_α or V_β?

7. Calculate the magnitude of the configurational entropy ΔS_c when 20% of the Cu and Zn atoms in β-brass are on the "wrong" sublattice sites.

8. Derive an expression for the free energy of nucleation of a solid from a liquid, assuming the liquid to be ideal, whereas the solid is a regular solution.

9. Consider a solid in which all atoms possess the same finite magnetic moments and have z nearest neighbors (e.g., Fe, Co, or Ni). Suppose the moments are confined to but two collinear orientations with respect to their nearest neighbors, viz. parallel, $\uparrow\uparrow$, or antiparallel, $\uparrow\downarrow$. The magnetic energy is then given by $E_m = -J/2(n_1 - n_2)$, where n_1 is the number of $\uparrow\uparrow$ pairs, n_2 is the number of $\uparrow\downarrow$ pairs, and J is a constant. The LRO parameter is assumed given by the fractional bulk magnetization, $s = M(T)/M_0$, where M_0 is the magnetization for perfect ferromagnetic alignment. This model, the zeroth approximation, is the so-called Ising model and is analogous to the Bragg–Williams model for binary alloys. Derive the expression

$$\ln\frac{1+s}{1-s} = \frac{2Jz}{kT}\,s.$$

10. By considering the free energy required to create disorder, show than an infinite one-dimensional chain of magnetic atoms cannot be ferromagnetic.

11. A very simple nucleation consists of the formation of trimers from monomers via dimers, that is

$$X_1 + X_1 \xrightarrow{k_1} X_2$$

and

$$X_2 + X_1 \xrightarrow{k_2} X_3.$$

If X_2 is in steady state, then show that

a. $X_2 = \frac{1}{2}\frac{k_1}{k_2} X_1$

b. $\frac{1}{X_1} - \frac{1}{X_1^0} = \left(\frac{3}{2}\right)k_1 t.$

12. Derive equation (14.9) by combining the Clausius–Clapeyron relation with Raoult's law. (Hint: Assume that the vapor pressure obeys the ideal gas law and that its molar volume is large compared with that of the liquid.)

13. Suggest a mechanism that explains why Ag halides become more photochemically sensitive as the grain size increases. (see Fig. 14.32).

Additional Reading

B. A. Bilby and J. W. Christian, (1956). *The Mechanism of Phase Transformation in Metals and Alloys*, Inst. of Metals, London.

J. E. Burke, (ed.), *Progress in Ceramic Science*, Vol. 1, Pergamon Press, Oxford, (1960).

J. E. Burke, (ed.) (1962) *Progress in Ceramic Science*, Vol. 11, Pergamon Press (1962).

J. W. Christian, (1965) *The Theory of Phase Transformations in Metals and Alloys*, Pergamon Press, Oxford.

J. W. Christian, (1983), *The Theory of Phase Transformation in Metals and Alloys*, Wiley-Interscience, New York.

M. Cohen, (1958). "The Nucleation of Solid-State Transformation," *Trans. A.I.M.E.* **171**.

V. D. Frechette, (ed.) (1960). *Non-crystalline Solids* Wiley, New York.

G. Jones, (1956). *Glass*, Menthuen & Co., Ltd., London.

A. G. Khachaturian, (1983). *Theory of Structural Transformation in Solids*, Wiley-Interscience, New York.

W. D. Kingery, (ed.) (1959). *Kinetics of High Temperature Processes*, John Wiley & Sons, New York; Chapman & Hall, London.

Paul G. Shewmon, (1963) *Transformation in Metals* McGraw-Hill, New York.

L. E. Tanner and W. A. Soffa, (eds.) (1988). Proc. ASM/MSD Symp. on "Pretransformation Behavior Related to Displacive Transformations," *Metall. Trans.* **18A**. Many recent advances are covered in this collection.

Appendix A

van der Waals Forces

Van der Waals forces have been alluded to in several places in the text, commencing with the crystalline structures of Se, graphite, and the rare gases described in Chapter 1. We have justified the uncertainty principle and the existence of a zero-point vibrational energy in other appendices. This places us in a position to understand the source of van der Waals forces, and we follow a derivation first attributed to Born.[1]

As explained qualitatively in Chapter 4, Section 1F, the oscillating electric dipole of an atom or molecule induces an electric field in proximate atoms or molecules. The sign of the induced dipole is oppositely directed to the sign of the first one; hence, the force between the two is always attractive. Following Born, we consider the simple model of two linear electrical oscillators a distance R apart, as illustrated in Figure A.1. The positive charges $(+e)$, corresponding to the nuclei, are kept fixed while the negative charges $(-e)$, corresponding to the electron clouds, are allowed to vibrate along the x-axis with displacements x_1 and x_2.

We consider the positive charges to be fixed while the electron cloud vibrates in accordance with the uncertainty principle. The excursions of the oscillating electrons are labeled x_1 and x_2, where the subscripts refer to atoms 1 and 2. The potential energy of the Coulomb interactions of the four charges is given by

$$V = \frac{e^2}{R} + \frac{e^2}{R + x_1 + x_2} - \frac{e^2}{R + x_1} - \frac{e^2}{R + x_2}, \qquad \text{(A.1)}$$

[1] M. Born, *Atomic Physics*, Blackie and Son, Ltd. (1947).

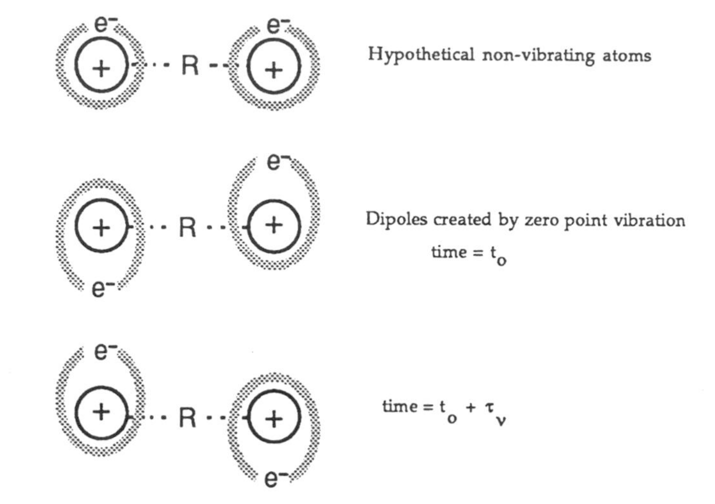

Figure A.1.

Two coupled linear electrical oscillators.

where the first two terms account for the mutual repulsion between the two nuclei $(+e)$ and between their respective electron clouds $(-e)$. The last two terms in (A.1) describe the attractive potentials between the nuclei and electrons of the opposite atom. This equation can be cast in a form that can be expanded as follows: let

$$y = \frac{x_1 + x_2}{R}, \frac{x_1}{R}, \frac{x_2}{R}.$$

Then write

$$V = \frac{e^2}{R}\left[1 + \frac{1}{1 + (x_1 + x_2)/R} - \frac{1}{1 + x_1/R} - \frac{1}{1 + x_2/R}\right]. \tag{A.2}$$

Each term in the brackets is of the form

$$1/(1 + y) = 1 - y + y^2 - \cdots,$$

and expanding each of them and collecting all terms up to and including the quadratic leads to

$$V = 2e^2 x_1 x_2 / R^3. \tag{A.3}$$

In addition to the interatomic potentials summarized by (A.1), one must include the restoring forces due to the intra-atomic potentials exerted by the nuclei upon their own electrons. They are given by $e^2 x_1/2k$ and $e^2 x_2/2k$, where $(2k)^{-1}$, as defined in Appendix G, equals $\overline{\delta x^2}$, the root-mean-square displacement derived from the uncertainty principle. Adding these terms plus the kinetic energy to (A.3) brings us to

$$E = \frac{1}{2m}(p_1^2 + p_2^2) + \frac{e^2}{2k}(x_1^2 + x_2^2) + \frac{2e^2 x_1 x_2}{R^3}. \tag{A.4}$$

A more general form of (A.4) is obtained if we reference the displacements to the principal axes by the transformation

$$x_s = \frac{1}{\sqrt{2}}(x_1 + x_2), \qquad x_a = \frac{1}{\sqrt{2}}(x_1 - x_2). \tag{A.5}$$

This also necessitates transforming the axes of the momenta, viz.

$$p_s = \frac{1}{\sqrt{2}}(p_1 + p_2), \qquad p_a = \frac{1}{\sqrt{2}}(p_1 - p_2). \tag{A.6}$$

Now substituting (A.5) and (A.6) into (A.4) leads to

$$\begin{aligned} E &= \frac{1}{2m}(p_s^2 + p_a^2) + \frac{e^2}{2k}(x_s^2 + x_a^2) + \frac{e^2}{R^3}(x_s^2 - x_a^2) \\ &= \left[\frac{1}{2m}p_s^2 + \left(\frac{e^2}{2a} + \frac{e^2}{R^3}\right)x_a^2\right] + \left[\frac{1}{2m}p_a + \left(\frac{e^2}{2a} - \frac{e^2}{R^3}\right)x_a^2\right]. \end{aligned} \tag{A.7}$$

The expressions in the brackets correspond to the energies of two uncoupled oscillators vibrating with their respective frequencies

$$\nu_s = \frac{1}{2\pi}\sqrt{\frac{e^2}{m}\left(\frac{1}{k} + \frac{2}{R^3}\right)}, \qquad \nu_a = \frac{1}{2\pi}\sqrt{\frac{e^2}{m}\left(\frac{1}{k} - \frac{2}{R^3}\right)}. \tag{A.8}$$

In the usual quantized form the respective energies are given by

$$E_{n_s} = h\nu_s(n_s + 1/2), \qquad E_{n_a} = h\nu_a(n_a + 1/2). \tag{A.9}$$

The total energy of the ground state is the sum of E_{n_s} and E_{n_a} for $n_s = n_a = 0$, which is

$$E_{00} = (1/2)h(\nu_s + \nu_a). \tag{A.10}$$

Substituting for ν_s and ν_a from (A.8) and expanding yields

$$\begin{aligned} E_{00} &= \frac{h}{4\pi}\left[\sqrt{\frac{e^2}{m}\left(\frac{1}{k} + \frac{2}{R^3}\right)} + \sqrt{\frac{e^2}{m}\left(\frac{1}{k} - \frac{2}{R^3}\right)}\right] \\ &= \frac{h}{2\pi}\sqrt{\frac{e^2}{mk}}\left[1 - \frac{k^2}{2R^6} + \cdots\right]. \end{aligned} \tag{A.11}$$

Thus, the additional potential energy that arises from the oscillating charge varies as the inverse sixth power of the internuclear separation and is negative, which is to say attractive.

The number of terms contained by equations having the form of (A.1) depends upon the number of interacting charges and, hence, varies depending upon the nature of the participating atoms or molecules. The important point is that the R^{-6} dependence of E_{00} remains the same in all cases. (See, for example, Pauling and Wilson, *Introduction to Quantum Mechanics*, McGraw-Hill, pp. 387–388, 1935, for additional information and references.)

Appendix B

The Grüneisen Constants

The basic definition of the Grüneisen constant comes from the equation of state

$$\gamma = (P - P_0)V/3RT, \tag{3.11}$$

where

$$\gamma = -\frac{1}{3N}\sum_{j}^{3N}\frac{d\ln \nu_j}{d\ln V} \tag{3.8}$$

in the high-temperature limit.

Over the years, various approximations have been developed for γ. We have already cited, in Eq. (3.16), the thermodynamic gamma, γ_{th}, as

$$\gamma_{th} = \alpha V/\chi C_V, \tag{3.16}$$

where α is the thermal expansion and χ is the compressibility. The great advantage of this formula is that it is defined in terms of macroscopic variables and is obtainable, in principle, without any knowledge of the interatomic forces. γ_{th} may also be written, by means of the thermodynamic definitions of α and χ, as

$$\gamma_{th} = \frac{-V}{C_V}\left(\frac{\partial P}{\partial V}\right)_T\left(\frac{\partial V}{\partial T}\right)_P = \left(\frac{V}{C_v}\right)\left(\frac{\partial P}{\partial T}\right)_V. \tag{B.1}$$

We may now relate γ_{th} to the γ of the Mie–Grüneisen equation of state as given by (3.11). Differentiating (3.11) at constant volume gives

$$\left(\frac{\partial P}{\partial T}\right)_V = \frac{3R}{V}\left[\gamma + T\left(\frac{\partial \gamma}{\partial T}\right)_V\right], \tag{B.2}$$

and, therefore,

$$\gamma_{\text{th}} = \frac{3R}{C_V}\left[\gamma + \left(\frac{\partial \gamma}{\partial T}\right)_V\right]. \tag{B.3}$$

These two gammas are equal only in the classical limit, when $C_V = 3R$ and when gamma is independent of the temperature. However, as discussed later in this appendix, these two gammas usually do not differ greatly at elevated temperatures.

Other approximations to (3.8) have been suggested over the years. Slater[1] related γ to the maximum frequency $\nu_{\max}$ of the Debye approximation to the frequency spectrum and expressed $\nu_{\max}$ approximately as

$$\nu_{\max} = 1/V^{1/3}(\chi\rho)^{1/2}, \tag{B.4}$$

where ρ is the density. Putting $\rho \propto 1/V$ and substituting into

$$\gamma = \frac{-d \ln \nu_{\max}}{d \ln V} \tag{B.5}$$

gives

$$\gamma_{\text{SL}} = -\frac{1}{6} + \frac{1}{2}\frac{d \ln \chi}{d \ln V}, \tag{B.6}$$

where we identified the gamma of this approximation as γ_{SL}, the Slater gamma. From standard thermodynamics, γ_{SL} is finally given by

$$\gamma_{\text{SL}} = -\frac{2}{3} - \frac{1}{2} V \frac{(\partial^2 P/\partial V^2)_T}{(\partial P/\partial V)_T}. \tag{B.7}$$

For completeness, we also give here a similar approximation due to Dugdale–MacDonald:[2]

$$\gamma_{\text{DM}} = -\frac{1}{3} - \frac{1}{2} V \frac{(\partial^2(PV^{2/3})/\partial V^2)_T}{(\partial(PV^{2/3})/\partial V)_T}. \tag{B.8}$$

Finally, Leibfried and Ludwig[3] suggested that the average over the normal modes of the logarithmic derivative be approximated by one-half of the logarithmic

---

[1] J. C. Slater, *Introduction to Chemical Physics*, Chap. 13, McGraw-Hill, New York, 1939.

[2] J. S. Dugdale and D. K. C. MacDonald, *Phys. Rev*, **89**, 832 (1953).

[3] G. Leibfried and W. Ludwig, *Solid State Physics*, (F. Seitz and D. Turnbull), eds., vol. 12, p. 276. Academic Press, New York, 1961.

derivative of the averaged squared frequency. This gamma, γ_{LL}, is then given by

$$\gamma_{LL} = -\frac{1}{2}\frac{d}{d \ln V}\left\{\ln\left[\frac{N}{3}\sum_{j} v_j^2\right]\right\}, \tag{B.9}$$

an expression that can be readily evaluated from the second derivatives of an interatomic potential, if such a potential is known.

One can arrive at some judgment about the various gammas from molecular dynamic simulations of the equation of state for various potentials.[4] Such calculations indicate that γ_{LL} describes best the simulations obtained for Lennard–Jones and Morse potentials. γ_{SL} and γ_{DL} are far less satisfactory. These simulations also indicated the γ_{th} and γ generally differ by only a few percent at elevated temperatures. Primarily, this is because as the temperature increases the $(\partial\gamma/\partial T)_V$ term decreases, resulting in considerable compensation.

Anderson[5] suggested another useful dimensionless parameter, δ_T, defined as

$$\delta_T = -\frac{1}{\alpha B_T}\left(\frac{\partial B_T}{\partial T}\right)_P, \tag{B.10}$$

where α is the volume thermal expansivity and B_T is the isothermal bulk modulus. The usefulness of δ_T is that it is independent of temperature at high temperatures and varies little from material to material. One can show after some thermodynamic manipulations that δ_T is related to the thermodynamic γ_{th} by

$$\delta_T = \frac{\partial \ln \gamma_{th}}{\partial \ln V} + \beta_T - 1, \tag{B.11}$$

where

$$\beta_T = \left(\frac{\partial B_T}{\partial P}\right)_T. \tag{B.12}$$

δ_T is also approximately independent of pressure. Assuming a power law for γ as a function of volume,

$$\gamma = \gamma_0(V/V_0)^q, \tag{B.13}$$

and we obtain

$$\delta_T = \beta_T - (1 - q), \tag{B.14}$$

where β_T and q may be taken as independent of pressure.

---

[4] D. O. Welch, G. J. Dienes, and A. Paskin, *J. Phys. Chem. Solids*, **39**, 589 (1974).
[5] O. L. Anderson, *J. Geophys. Res.*, **72**, 3661 (1967).

It is also of interest to derive the equivalent expressions for the adiabatic bulk modulus B_S, using the relation $B_S = B_T(1 + \alpha\gamma T)$. The result is

$$\delta_T = \delta_S + \gamma. \tag{B.15}$$

Another thermodynamic formulation of the Grüneisen constant that is frequently used for hydrodynamic problems is

$$\gamma \equiv V\left(\frac{\partial P}{\partial E}\right)_V. \tag{B.16}$$

Appendix C

Zero-Point Energy of the Harmonic Oscillator

The derivation begins with Hooke's law:

$$F = m\frac{d^2x}{dt^2} = -Cx, \tag{C.1}$$

which states, that in one dimension, the restoring force in tension upon an object of mass m is proportional to the displacement x (Chapter 3, Section 3). A general solution to (C.1) is

$$x = A\sin(\omega t + \delta), \tag{C.2}$$

where $\omega = \sqrt{C/m}$, where C is the elastic stiffness constant and m is the mass of the oscillator; A and δ can be considered simply as constants. A standing wave is produced when the object vibrates with a frequency of $\omega/2\pi$ equal to the reciprocal of the characteristic period of oscillation. Rewriting (C.1) in terms of the potential energy [see (3.25)] leads to

$$E_{\text{pot}} = \frac{Cx^2}{2}.$$

Now substituting into (6.9), the one-dimensional Schrödinger equation, leads to

$$\frac{d^2\psi}{dx^2} + \frac{2m}{\hbar^2}\left(E - \frac{1}{2}m\omega^2x^2\right)\psi = 0. \tag{C.3}$$

To solve (C.3), we adopt the trial function

$$\psi_0 = e^{-\alpha x^2}, \tag{C.4}$$

where $\alpha = m\omega/2\hbar$. Then

$$\frac{d\psi_0}{dx} = -2\alpha x e^{-\alpha x^2} \tag{C.5}$$

and

$$\frac{d^2\psi_0}{dx^2} = -2\alpha x e^{-\alpha x^2} + 4\alpha^2 x^2 e^{-\alpha x^2} = \frac{-2m}{\hbar^2}\left(\frac{1}{2}\hbar\omega - \frac{1}{2}m\omega^2 x^2\right)\psi_0. \tag{C.6}$$

It can be seen that (C.6) is the same equation as (C.3) if $E = (1/2)\hbar\omega$. It has not been shown that this is the ground state. However, ψ_0 has no nodes; therefore, ω is the minimum frequency corresponding to the minimum energy. Thus, we conclude that the ground state is

$$E_0 = \frac{\hbar\omega_0}{2}. \tag{C.7}$$

We can pursue the same line of reasoning to obtain the first excited state. In this case the trial wave function is

$$\psi_1 = xe^{-\alpha x^2}, \tag{C.8}$$

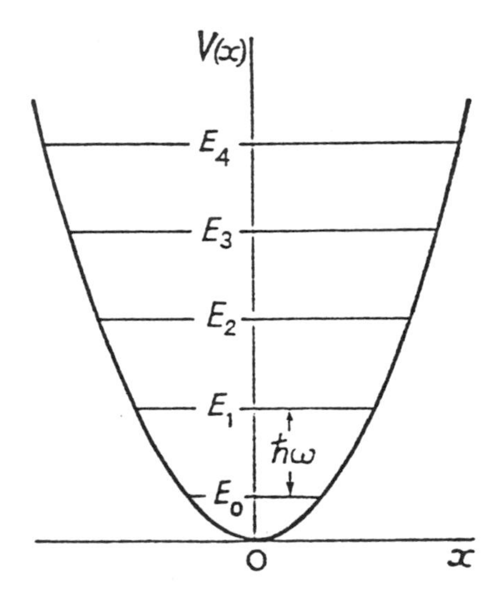

Figure C.1.

The energy levels for a harmonic oscillator, i.e., a particle in a one-dimensional parabolic potential energy well, are given by $E_n = (n + 1/2)\hbar\omega = (n + 1/2)\hbar\nu_0$.

which has a node at $x = 0$. Twice differentiating (C.8) leads to

$$\frac{d^2\psi_1}{dx^2} = -6\alpha x e^{-\alpha x^2} + 4\alpha^2 x^3 e^{-\alpha x^2} = \frac{-2m}{\hbar^2}\left(\frac{3}{2}\hbar\omega - \frac{1}{2}m\omega^2 x^2\right)\psi_1. \qquad \text{(C.9)}$$

Once again, by comparing (C.9) with (C.3), it is clear that $E_1 = (3/2)\hbar\omega$. The successively higher vibrational energy states increasing in magnitude by a single quantum $\hbar\omega$ are represented in Fig. C.1.

Appendix D

Partial Differential Equations and Boundary Value Problems: The Vibrating String

Linear partial differential equations of the second order play a very important role in theoretical physics and chemistry. To illustrate how such equations are handled mathematically,[1] we shall discuss here a representative, simple example, namely the equations describing the transverse vibrations of a string displaced transversely and held constant at both ends. The partial differential equation describing the system is

$$\frac{\partial^2 y}{\partial t^2} = c^2 \frac{\partial^2 y}{\partial x^2}, \tag{D.1}$$

where y is the transverse displacement and $c = 1/v$, where v is the velocity of propagation. The description of the problem is incomplete unless we also define the boundary conditions. The string of length L is stretched between the points $(0, 0)$ and $(L, 0)$, is displaced into position $f(x) = y$, and is released from rest with no external forces acting upon it. The boundary conditions are

$$y(0, t) = 0; \quad y(L, t) = 0, \qquad t \geq 0, \tag{D.2}$$

$$y(x, 0) = f(x), \qquad 0 \leq x \leq L, \tag{D.3}$$

$$\frac{\partial y(x, 0)}{\partial t} = 0, \qquad 0 \leq x \leq L. \tag{D.4}$$

[1] Partial differential equations and boundary value problems are discussed in many standard advanced calculus texts. For a detailed treatment, see, e.g., R. V. Churchill, *Fourier Series and Boundary Value Problems*, McGraw-Hill, New York, 1941.

The standard procedure is to find particular solutions that satisfy (D.1), (D.2), and (D.4) and then determine a linear combination of these equations to satisfy condition (D.3). This last step relies on an important theorem of linear systems, namely that any linear combination of solutions is also a solution for the system.

We proceed now to find particular solutions by the method of separation of variables. Let

$$y = X(x)T(t).$$

Substitution into (D.1) gives

$$\frac{\partial^2 T}{T\partial t^2} = c^2 \frac{\partial^2 X}{X\partial x^2}, \tag{D.5}$$

which can be valid if, and only if, each side is equal to a constant, say k. The result is a set of *ordinary* differential equations given by

$$\frac{d^2X}{dx^2} - kX = 0, \tag{D.6}$$

$$\frac{d^2T}{dt^2} - kc^2T = 0. \tag{D.7}$$

In order to satisfy (D.2) and (D.4), we must have

$$X(0) = 0; \qquad X(L) = 0 \tag{D.8}$$

and

$$\frac{dT(0)}{dt} = 0. \tag{D.9}$$

The general solution of (D.6) is,

$$X = c_1 e^{xk^{1/2}} + c_2 e^{-xk^{1/2}}, \tag{D.10}$$

where c_1 and c_2 are arbitrary constants. A positive k cannot satisfy both conditions of (D.8), and k, therefore, must be negative. Let k be given by

$$k = -b^2,$$

which leads to

$$X = c_1 e^{ibx} + c_2 e^{-ibx}, \tag{D.11}$$

where $i = (-1)^{1/2}$.

We use now the relations between complex exponential and trigonometric functions, derived as follows. By standard series expansion

$$e^{ix} = 1 - \frac{x^2}{2!} + \frac{x^4}{4!} + \cdots + i\left(x - \frac{x_3}{3!} + \frac{x^5}{5!} \cdots\right),$$

which is immediately recognizable as the sum of the *cosine* and *sine expansions*. Thus,

$$e^{ix} = \cos x + i \sin x,$$

and, by an identical argument,

$$e^{-ix} = \cos x - i \sin x.$$

Equation (D.11) may, therefore, be rewritten as

$$X = A \sin bx + B \cos bx, \tag{D.12}$$

where A and B are again arbitrary constants. Since $X(0) = 0$, we must have $B = 0$.
To satisfy $X(L) = 0$ with A finite, we must have

$$\sin bL = 0, \tag{D.13}$$

which can be satisfied only with values of b given by

$$b = n\pi/L \qquad (n = 1, 2, 3, \ldots). \tag{D.14}$$

We have arrived at a very important result, namely that the boundary conditions introduced *discreteness* into the problem. The solutions for X are then

$$X = A \sin(n\pi x/L). \tag{D.15}$$

(Note that $n = -1, -2, \ldots$ lead to no new solutions.)
Equation (D.7) now reads, since k became $-n^2\pi^2/L^2$,

$$\frac{d^2T}{dt^2} + \frac{c^2n^2\pi^2}{L^2} = 0,$$

which, together with $dT(0)/dt = 0$, has the solution

$$T = C \cos(n\pi ct/L)$$

Thus, all the functions

$$A_n \sin(n\pi x/L)\cos(n\pi ct/L) \qquad (n = 1, 2, 3, \ldots) \tag{D.16}$$

are solutions of the partial differential equation (D.1) and the boundary conditions (D.2) and (D.4).

By the linearity theorem noted earlier, any finite linear combination of these solutions will also satisfy (D.1), (D.2), and (D.4). At $t = 0$ the solution will reduce to a finite linear combination of $\sin(n\pi x/L)$. We can satisfy boundary condition (D.3) if such a series can be equated to the function $f(x)$. The infinite series of functions

$$y = \sum_{1}^{\infty} A_n \sin\left(\frac{n\pi x}{L}\right)\cos\left(\frac{n\pi ct}{L}\right) \tag{D.17}$$

also satisfies equations (D.1) and conditions (D.2) and (D.4), provided it is convergent, represents a continuous function, and is termwise differentiable with respect to either x or t. It will also satisfy condition (D.3), provided the numbers A_n can be determined such that $f(x)$ of (D.3) is given by

$$f(x) = \sum_{1}^{\infty} A_n \sin\left(\frac{n\pi x}{L}\right). \tag{D.18}$$

It is the result of Fourier's theorem that if such an expansion of $f(x)$ is possible, then the A_n are given by

$$A_n = \frac{2}{L}\int_0^L f(x)\sin\left(\frac{n\pi x}{L}\right)dx. \tag{D.19}$$

The series in (D.18) with the A_n defined by (D.19) is known as the *Fourier sine series* of the function $f(x)$. This series converges to $f(x)$ in the $0 \leq x \leq L$ interval with only moderate restrictions on the function $f(x)$, and is of almost unrestricted used in practical problems in physics and chemistry.

Appendix E

Polarization Catastrophe and the Shell Model

In this appendix we shall show first that if the polarization interactions are included in the electrostatic terms, as is standard in ionic crystal theory, and if the repulsive interaction is kept a function of r only, then there are some serious difficulties with the ionic model. In essence, the cohesive energy may become infinite, and, hence, the crystal may collapse for certain ranges of the interionic distance and the polarizabilities. We briefly go through the arguments in order to indicate why models such as the shell model were invented to avoid the catastrophe.

Let us rewrite the energy of interaction of a pair of ions in the general form

$$E(r) = -\frac{Z^2e^2}{r} - \frac{Z(\mu_+ + \mu_-)e}{r^2} - \frac{2\mu_+\mu_-}{r^3} + \frac{\mu_+^2}{2\alpha_+} + \frac{\mu_-^2}{2\alpha_-} + E_r(r), \tag{E.1}$$

where the polarization interactions have been included [see (4.58) and (4.62)] and where $E_r(r)$ is the repulsive interaction energy, which we kept, for the present, as a function of r only. In this equation, as discussed in Chapter 4, Section 2E, r is the interionic separation, the μ's are the dipole moments, and the α's are the free-ion polarizabilities.

μ_+ and μ_- can be determined by applying the equilibrium conditions

$$\frac{\partial E}{\partial \mu_+} = 0, \qquad \frac{\partial E}{\partial \mu_-} = 0.$$

The results are

$$\mu_+ = \frac{Ze\alpha_+}{r^2}\frac{[1 + 2\alpha_-/r^3]}{[1 - 4\alpha_+\alpha_-/r^6]}, \qquad \mu_- = \frac{Ze\alpha_-}{r^2}\frac{[1 + 2\alpha_+/r^3]}{[1 - 4\alpha_+\alpha_-/r^6]}. \tag{E.2}$$

Substitution into (E.1) gives, after considerable algebra,

$$E(r) = E_r(r) - \frac{Z^2e^2}{r} - \frac{Z^2e^2}{2r}\left[\frac{\alpha_+ + \alpha_-}{r^3} + \frac{4\alpha_+\alpha_-}{r^6}\right]\left[1 - \frac{4\alpha_+\alpha_-}{r^6}\right]^{-1}. \quad \text{(E.3)}$$

Clearly, as r approaches $(4\alpha_+\alpha_-)^{1/6}$, the energy diverges: i.e., $E(r) - E_r(r) \to \infty$. Thus, the inclusion of the polarization interactions in the pairwise potential can result in a polarization catastrophe.

If the repulsive interaction is described by the simple Born–Mayer exponential function, then (E.3) has no solution. With the inverse power law for repulsion, there is always a solution, but at a very small r and, hence, with an enormous polarization energy. Such polarization catastrophes can be avoided by the use of the simple shell

Table E.1. Interaction parameters for the alkali halides.[a]

Molecule	A_{2+} (eV)	A_{+-} (eV)	A_{--} (eV)	b $Å^{-1}$	α_+ $Å^3$	α_- $Å^3$	Q_+	Q_-	r^+ Å	r^- Å
LiF	98.99	229.24	420.88	3.3445	0.00	1.10	—	1.35	0.82	1.18
LiCl	49.84	324.71	1677.50	2.9240	0.00	3.42	—	2.65		1.58
LiBr	42.95	378.16	2639.73	2.8329	0.00	4.67	—	2.55		1.72
LiI	18.77	163.21	1125.31	2.3256	0.00	7.15	—	2.50		1.91
NaF	316.65	260.40	200.64	3.0303	0.18	1.10	5	1.35	1.17	
NaCl	423.51	1254.62	3484.40	3.1546	0.18	3.42	5	2.65		
NaBr	257.04	1025.12	3828.27	2.9412	0.18	4.67	5	2.55		
NaI	113.19	611.38	3092.95	2.5907	0.18	7.15	5	2.50		
KF	1515.77	523.38	169.42	2.9586	0.83	1.10	10	1.35	1.46	
KCl	1555.21	1786.91	1924.80	2.9674	0.83	3.42	10	2.65		
KBr	1637.96	2791.28	4450.75	2.9851	0.83	4.67	10	2.55		
KI	1001.38	2796.70	7329.87	2.8169	0.83	7.15	10	2.50		
RbF	4203.64	969.90	209.57	3.0488	1.40	1.10	5	1.35	1.57	
RbCl	2398.79	4529.25	3376.53	3.1447	1.40	3.42	5	2.65		
RbBr	3434.08	4035.45	4450.75	2.9851	1.40	4.67	5	2.55		
RbI	3246.31	6699.07	13011.10	2.9674	1.40	7.15	5	2.50		
CsF	52336.97	6155.13	677.06	3.5461	2.44	1.10	5	1.35	1.72	
CsCl	27187.18	13849.04	6592.15	3.3557	2.44	3.42	5	2.65		
CsBr	25173.40	19866.21	14706.59	3.3333	2.44	4.67	5	2.55		
CsI	16195.97	23611.48	32222.17	3.2051	2.44	7.15	5	2.50		

[a] After D. O. Welch, O. W. Lazareth, G. J. Dienes, and R. D. Hatcher, *J. Chem. Phys.*, **64**, 835 (1976). With A_i and r_i taken from M. P. Tosi and F. G. Fumi, *J. Phys. Chem. Solids*, **25**, 45 (1964). See also M. P. Tosi, *Solid State Physics* F. Seitz and D. Turnbull, eds. vol. 16, Table XX, pp. 1–113, Academic Press, New York, 1964.

model proposed by Dick and Overhauser.[1] We have already discussed in Chapter 4, Section 3C the physical arguments leading to this model. Mathematically, if we use an effective interionic separation in the expression for the repulsive energy and make it proportional to the dipole moment, as in (4.61),

$$\mathbf{r}^{\text{eff}} = \mathbf{r}_{ij} + \frac{\mathbf{u}_i}{Q_i} - \frac{\mathbf{u}_j}{Q_j}, \tag{4.61}$$

then the catastrophes are eliminated. The net effect is that the repulsive interaction increases with increasing polarization. Thus, E_r in (E.1) becomes a function of the dipole moments, $E_r(r, \mu)$, and the final equation is

$$E(r) = -\frac{Z^2e^2}{r} - \frac{Z(\mu_+ + \mu_-)e}{r^2} - \frac{2\mu_+\mu_-}{r^3} + \frac{\mu_+^2}{2\alpha_+} + \frac{\mu_-^2}{2\alpha_-} + A \exp\left[-b\left(r + \frac{\mu_+}{Q_+} - \frac{\mu_-}{Q_-}\right)\right], \tag{4.62}$$

as in the text.

The interaction parameters in (4.62) can be obtained by using equilibrium conditions and by fitting a very large number of crystal and molecular experimental data, as discussed in Chapter 4, Section 3C. The interaction parameters for the alkali halides are listed in Table E.1.

---

[1] B. G. Dick, Jr., and A. W. Overhauser, *Phys. Rev.*, **112**, 90 (1958).

Appendix F

Operators

Any mathematical operation can be represented by a characteristic symbol, which is then called an operator. The differential operator is the one of most interest to us here. Let D stand for the operation d/dx. As an example, in the use of D, consider a linear differential equation with constant coefficients conveniently written as

$$(D^n + a_1 D^{n-1} + \cdots + a_{n-1} D + a_n) y = f(x). \tag{F.1}$$

The expression in the parentheses is a linear differential operator of order n. It is not an algebraic expression multiplying y, but is a symbol signifying that certain operations of differentiation are to be performed on the function y. For example,

$$\begin{aligned} D(D-2)\ln x &= (D^2 - 2D)\ln x = D^2 \ln x - 2D \ln x \\ &= \frac{d^2 \ln x}{dx^2} - \frac{2d \ln x}{dx} = \frac{-1}{x^2} - \frac{2}{x}. \end{aligned} \tag{F.2}$$

The advantage of operator notation is that linear differential operators with constant coefficients formally obey the laws valid for polynomials, and the operator D can, therefore, be manipulated algebraically. One has to be careful, of course, when multiplication with a variable is involved. For example,

$$Dxf(x) = f(x) + x\frac{df}{dx},$$

Table F.1

p	P
q_k	q_k.
p_k	$\frac{h}{2\pi i}\frac{\partial}{\partial q_k}$
m_k	$\frac{h}{2\pi i}\left(q_i\frac{\partial}{\partial q_m} - q_m\frac{\partial}{\partial q_i}\right)$
$H(p, q)$	$H\left(\frac{h}{2\pi i}\frac{\partial}{\partial q}, q\right)$

but

$$xDf(x) = x\frac{df}{dx}. \tag{F.3}$$

Operators play a very important role in quantum mechanics.[1] One of the postulates of quantum mechanics is that for any physically observable quantity, p, there corresponds an operator P. The choice of correct operator must rely on trial. The classical to quantum mechanical correspondence may be illustrated by the assignments in Table F.1. q_k here is any Cartesian coordinate, p_k is the associated momentum, and $H(p, q)$ is the energy in Hamiltonian form. The operator corresponding to m_k, the angular momentum, was constructed by expressing it in terms of spatial coordinates and linear momenta as

$$m_z = xp_y - yp_x, \tag{F.4}$$

and then replacing each classical variable by the corresponding quantum mechanical operator; similarly for H.

The fundamental postulate of quantum mechanics expressed in terms of an operator equation is

$$P\psi_\lambda = p_\lambda\psi_\lambda. \tag{F.5}$$

In words: The only possible values that a measurement of the observable p can yield are the eigenvalues p_λ of Eq. (F.5). A prominent example is provided by the

[1] We follow here, in part, the discussion by Lindsay and Marganau in *Foundations of Physics*, Chap. 9, Wiley, New York, 1936.

Hamiltonian or energy function, where, if the energy of a system in state ψ is E, then

$$H\psi = E\psi, \tag{F.6}$$

the expectation value of H must be E, since the energy is constant. This result follows, however, from an equation of the same form as (6.45); thus

$$\langle H \rangle = \int \psi^* H \psi \, d\tau, \tag{F.7}$$

which is equal to E because of equation (F.6), assuming the normalization of ψ. The H in the integrand is, of course, a differential operator. It is not difficult to see that a similar result must hold for any quantity that is a constant of motion.

This suggests that the expectation value of any dynamical variable should be given by an equation like (6.45), whether the operator that appears in the integrand is a function of position only or a differential operator. We will not attempt to prove this, but will take it as a further fundamental postulate of quantum mechanics to be justified by results.

We may illustrate the use of this operator equation by the simple example of the vibrating string, which we have already treated by standard methods in Appendix D. The differential equation of a vibrating string is

$$\frac{\partial^2 y}{\partial x^2} - \frac{1}{v^2}\frac{\partial^2 y}{\partial t^2} = 0. \tag{F.8}$$

If we are concerned with harmonic solutions, we put $y = f(x)e^{2\pi i \nu t}$ and obtain the ordinary differential equation

$$\frac{d^2 f}{dx^2} + \frac{4\pi^2 \nu^2}{v^2} f = 0, \tag{F.9}$$

an operator equation of the form of (F.5). Here, the operator P is d^2/dx^2, and the eigenvalue p is the combination of constants $4\pi^2\nu^2/v^2$. There is a solution for every p of the familiar type

$$f = A \sin(\sqrt{p}x + \gamma), \tag{F.10}$$

where A and γ are arbitrary constants. We also need to satisfy the boundary conditions

$$f(0) = 0; \qquad f(L) = 0. \tag{F.11}$$

The first condition is satisfied by $\gamma = 0$, the second by

$$p^{1/2}L = n\pi. \tag{F.12}$$

Thus, the only permissible values of p are given by

$$p_n = n^2\pi^2/L^2. \tag{F.12}$$

As already noted in Appendix D, the boundary conditions introduced discreteness into the problem. The p_n of (F.11) are the eigenvalues of the operator $P = d^2/dx^2$ for the foregoing boundary conditions. Thus, solutions consistent with the boundary conditions select values of p_λ, the eigenvalues. The functions ψ_λ in (F.5) are the eigenfunctions.

Appendix G

The Uncertainty Principle

The general form of Heisenberg's uncertainty principle is stated by either

$$\Delta p\, \Delta x \geq h/4\pi \qquad \text{or} \qquad \Delta E\, \Delta t \geq h/4\pi,$$

where p is the linear momentum and the remaining symbols have their usual meanings. An elementary derivation of the *exact form*, viz. $\Delta p\, \Delta x = h/4\pi$, attributed to Max Born[1] begins by assuming the necessity of the zero-point energy of a harmonic oscillator, as was established in Appendix C. The uncertainty in the mean square of the position coordinate of a vibrating object, $\overline{\delta x^2}$, can be described by the gaussian error function, so that

$$\psi_0 = ke^{-kx^2/2}, \tag{G.1}$$

where $k = 4\pi^2 m\nu_0/h$ and the subscript zeros designate the ground state of the system. Hence, we can write for the mean square deviation, remembering that $\overline{\delta x} = 0$,

$$\overline{\delta x^2} = \overline{x^2} = \frac{\int_{-\infty}^{\infty} x^2\psi_0^2\, dx}{\int_{-\infty}^{\infty} \psi_0^2\, dx} = \frac{1}{2k} = \frac{h}{8\pi^2 m\nu_0}. \tag{G.2}$$

However, the total energy of the oscillator is given by

$$E = p^2/2m + m(2\pi\nu_0)^2 x^2/2. \tag{G.3}$$

[1] M. Born, *Atomic Physics*, App. XXXII, p. 357, Blackie & Son, (1947).

Now let us assume that the energy is determined exactly. Then

$$\overline{\delta p^2} = m^2(2\pi v_0)^2\overline{\delta x^2} = \tfrac{1}{2}hv_0 m, \tag{G.4}$$

which rederives the finite value of the zero-point energy. Proceeding with the derivation of the uncertainty principle, we write, from (C.4) and (C.2),

$$\delta\overline{p^2}\delta\overline{x^2} = h^2/16\pi, \tag{G.5}$$

thus finally giving $\delta p \delta x = h/4\pi$, which is the desired result. It is not much more difficult to derive the general form of the uncertainty principle, but it is rather lengthy and the reader is referred to almost any text on elementary wave mechanics.

The uncertainty principle has its origins in the delocalizing effect of assigning a wavelike behavior to such objects as electrons.

Appendix H

Euler's Equation and Thermodynamics

Euler's equation has to do with the characteristics of homogeneous functions. These functions play an important role in thermodynamics, since *extensive* variables are always homogeneous functions of one another in the first degree. A homogeneous function is defined by the relation

$$F(ax_1, ax_2, \ldots, ax_n) = a^n F(x_1, x_2, \ldots, x_n), \tag{H.1}$$

where a and n are constants and F is a homogeneous function of degree n in the variables $x_1, \ldots, x_n$. Euler's theorem states that

$$x_1 \frac{\partial F}{\partial x_1} + x_2 \frac{\partial F}{\partial x_2} + \cdots + x_n \frac{\partial F}{\partial x_n} = nF(x_1, x_2, \ldots, x_n). \tag{H.2}$$

The proof is as follows. For convenience of notation, substitute $x_1' = ax_1$, $x_2' = ax_2$, etc., in (H.1) and differentiate with respect to a to obtain

$$x_1 \frac{\partial F}{\partial x_1'} + x_2 \frac{\partial F}{\partial x_2'} + \cdots + x_n \frac{\partial F}{\partial x_n'} = na^{n-1} F(x_1, x_2, \ldots, x_n). \tag{H.3}$$

Since a is an arbitrary constant, we can set it equal to 1 so that $x_1' = x_1$, $x_2' = x_2$, etc., and (H.2), Euler's equation, follows immediately.

It is worthwhile to arrive at the same result in a somewhat different way. Differentiate F with respect to x_i to yield

$$a \frac{\partial F}{\partial x_i'} = a^n \frac{\partial F}{\partial x_i}, \tag{H.4}$$

and, therefore,

$$\frac{\partial F}{\partial x_i'} = a^{n-1}\frac{\partial F}{\partial x_i}.$$

Substitution into (H.3) permits the cancellation of a^{n-1} and leads directly to (H.2).

Let us apply these results now to any extensive property X, with temperature and pressure kept constant. X is a homogeneous function of degree 1 [$n = 1$ in (H.2)]. A simple example is the volume V of a liquid as a function of its mass M, a direct proportionality since $V(aM) = aV(M)$. Consider X now as a function of the concentration of components $N_1, N_2, \ldots, N_n$ at constant T and P. By Euler's equation we have, since $n = 1$,

$$N_1\frac{\partial X}{\partial N_1} + N_2\frac{\partial X}{\partial N_2} + \cdots + N_n\frac{\partial X}{\partial N_n} = X(N_1, N_2, \ldots, N_n), \tag{H.5}$$

and by defining the partial quantities as

$$\bar{X}_1 = \frac{\partial X}{\partial N_1}, \qquad \bar{X}_2 = \frac{\partial X}{\partial N_2}, \ldots, \bar{X}_n = \frac{\partial X}{\partial N_n},$$

we write

$$N_1\bar{X}_1 + N_2\bar{X}_2 + \cdots + N_n\bar{X}_n = X(N_1, N_2, \ldots, N_n). \tag{H.6}$$

By (H.4) the $\bar{X}_i$ are homogeneous functions of degree zero. Thus,

$$\begin{gathered} N_1\frac{\partial \bar{X}_1}{\partial N_1} + N_2\frac{\partial \bar{X}_1}{\partial N_2} + \cdots + N_n\frac{\partial \bar{X}_1}{\partial N_h} = 0, \\ \vdots \\ N_1\frac{\partial \bar{X}_n}{\partial N_1} + N_2\frac{\partial \bar{X}_n}{\partial N_2} + \cdots + N_n\frac{\partial \bar{X}_n}{\partial N_n} = 0. \end{gathered}$$

We now use the fact that the order of differentiation is irrelevant, and hence,

$$\frac{\partial}{\partial N_i}\left(\frac{\partial X}{\partial N_j}\right) = \frac{\partial^2 X}{\partial N_i\,\partial N_j} = \frac{\partial}{\partial N_j}\left(\frac{\partial X}{\partial N_i}\right), \tag{H.7}$$

and, therefore,

$$\frac{\partial \bar{X}_i}{\partial N_j} = \frac{\partial \bar{X}_j}{\partial N_i}. \tag{H.8}$$

Substitution into (H.7) gives

$$N_1 \frac{\partial \bar{X}_1}{\partial N_1} + N_2 \frac{\partial \bar{X}_2}{\partial N_1} + \cdots + N_n \frac{\partial \bar{X}_n}{\partial N_1} = 0,$$

$$\vdots$$

$$N_1 \frac{\partial \bar{X}_1}{\partial N_n} + N_2 \frac{\partial \bar{X}_2}{\partial N_n} + \cdots + N_n \frac{\partial \bar{X}_n}{\partial N_n} = 0,$$

and, therefore, we can replace the ∂N_i to get

$$N_1 \, d\bar{X}_1 + N_2 \, d\bar{X}_2 + \cdots + N_n \, d\bar{X}_n = 0,$$

or, in a more usual notation,

$$\sum_i N_i \, d\bar{X}_i = 0, \tag{H.9}$$

a relation valid for any extensive property. In the special case when X is the Gibbs free energy G, we obtain

$$\sum_i N_i \, d\mu_i = 0, \tag{H.10}$$

where μ_i is the chemical potential. Equation (H.10) is the Gibbs–Duhem–Margules equation.

It is worthwhile to show how to arrive at (H.9) by a more intuitive derivation often used in solution theory when discussing partial molar quantities. Since X is a function of $N_1, N_2, \ldots, N_n$, its differential, at constant temperature and pressure, may be written

$$dX = \bar{X}_1 \, dN_1 + \bar{X}_2 \, dN_2 + \cdots + X_n \, dX_n = \sum_i \bar{X}_i \, dN_i. \tag{H.11}$$

We can build up a system, at constant T and P, by the simultaneous addition of dN_1, dN_2, etc., in the constant proportions required by the system. During this process, because of the constancy of the composition, the $\bar{X}_i$ remain constant. We may integrate (H.11) to obtain

$$X = \bar{X}_1 N_1 + \bar{X}_2 N_2 + \cdots + \bar{X}_n N_n = \sum_i \bar{X}_i N_i. \tag{H.12}$$

We can differentiate (H.12) now without imposing the artificial constraint used in its derivation. Thus,

$$dX = \sum_i \bar{X}_i \, dN_i + \sum_i N_i \, d\bar{X}_i. \tag{H.13}$$

Comparison with (H.11) immediately gives (H.9).

Author Index

Subject Index